Polymer Materials for Drug Delivery and Tissue Engineering

Polymer Materials for Drug Delivery and Tissue Engineering

Editors

Ariana Hudita
Bianca Gălăţeanu

MDPI

Basel • Beijing • Wuhan • Barcelona • Belgrade • Novi Sad • Cluj • Manchester

Editors
Ariana Hudita
Department of Biochemistry
and Molecular Biology
University of Bucharest
Bucharest
Romania

Bianca Gălăţeanu
Department of Biochemistry
and Molecular Biology
University of Bucharest
Bucharest
Romania

Editorial Office
MDPI
St. Alban-Anlage 66
4052 Basel, Switzerland

This is a reprint of articles from the Special Issue published online in the open access journal *Polymers* (ISSN 2073-4360) (available at: www.mdpi.com/journal/polymers/special_issues/polymer_mater_drug_delivery_tissue_engineering).

For citation purposes, cite each article independently as indicated on the article page online and as indicated below:

Lastname, A.A.; Lastname, B.B. Article Title. *Journal Name* **Year**, *Volume Number*, Page Range.

ISBN 978-3-0365-8949-7 (Hbk)
ISBN 978-3-0365-8948-0 (PDF)
doi.org/10.3390/books978-3-0365-8948-0

Contents

About the Editors vii

Preface ix

Ariana Hudiță and Bianca Gălățeanu
Polymer Materials for Drug Delivery and Tissue Engineering
Reprinted from: *Polymers* **2023**, *15*, 3103, doi:10.3390/polym15143103 1

Bianca Galateanu, Ariana Hudita, Elena Iuliana Biru, Horia Iovu, Catalin Zaharia and Eliza Simsensohn et al.
Applications of Polymers for Organ-on-Chip Technology in Urology
Reprinted from: *Polymers* **2022**, *14*, 1668, doi:10.3390/polym14091668 5

Kokila Thirupathi, Thi Tuong Vy Phan, Madhappan Santhamoorthy, Vanaraj Ramkumar and Seong-Cheol Kim
pH and Thermoresponsive PNIPAm-co-Polyacrylamide Hydrogel for Dual Stimuli-Responsive Controlled Drug Delivery
Reprinted from: *Polymers* **2022**, *15*, 167, doi:10.3390/polym15010167 32

Cristina Busuioc, Andrada-Elena Alecu, Claudiu-Constantin Costea, Mihaela Beregoi, Mihaela Bacalum and Mina Raileanu et al.
Composite Fibers Based on Polycaprolactone and Calcium Magnesium Silicate Powders for Tissue Engineering Applications
Reprinted from: *Polymers* **2022**, *14*, 4611, doi:10.3390/polym14214611 47

Arunnee Sanpakitwattana, Waraporn Suvannapruk, Sorayouth Chumnanvej, Ruedee Hemstapat and Jintamai Suwanprateeb
Cefazolin Loaded Oxidized Regenerated Cellulose/Polycaprolactone Bilayered Composite for Use as Potential Antibacterial Dural Substitute
Reprinted from: *Polymers* **2022**, *14*, 4449, doi:10.3390/polym14204449 59

Nils Wilharm, Tony Fischer, Alexander Hayn and Stefan G. Mayr
Structural Breakdown of Collagen Type I Elastin Blend Polymerization
Reprinted from: *Polymers* **2022**, *14*, 4434, doi:10.3390/polym14204434 76

Shafi Ullah, Abul Kalam Azad, Asif Nawaz, Kifayat Ullah Shah, Muhammad Iqbal and Ghadeer M. Albadrani et al.
5-Fluorouracil-Loaded Folic-Acid-Fabricated Chitosan Nanoparticles for Site-Targeted Drug Delivery Cargo
Reprinted from: *Polymers* **2022**, *14*, 2010, doi:10.3390/polym14102010 91

Wafa Shamsan Al-Arjan, Muhammad Umar Aslam Khan, Hayfa Habes Almutairi, Shadia Mohammed Alharbi and Saiful Izwan Abd Razak
pH-Responsive PVA/BC-*f*-GO Dressing Materials for Burn and Chronic Wound Healing with Curcumin Release Kinetics
Reprinted from: *Polymers* **2022**, *14*, 1949, doi:10.3390/polym14101949 105

Waleed Y. Rizg, Khaled M. Hosny, Bayan A. Eshmawi, Abdulmohsin J. Alamoudi, Awaji Y. Safhi and Samar S. A. Murshid et al.
Tailoring of Geranium Oil-Based Nanoemulsion Loaded with Pravastatin as a Nanoplatform for Wound Healing
Reprinted from: *Polymers* **2022**, *14*, 1912, doi:10.3390/polym14091912 121

Ilya L. Tsiklin, Evgeniy I. Pugachev, Alexandr V. Kolsanov, Elena V. Timchenko, Violetta V. Boltovskaya and Pavel E. Timchenko et al.
Biopolymer Material from Human Spongiosa for Regenerative Medicine Application
Reprinted from: *Polymers* **2022**, *14*, 941, doi:10.3390/polym14050941 . . . . **137**

Viera Khunová, Mária Kováčová, Petra Olejniková, František Ondreáš, Zdenko Špitalský and Kajal Ghosal et al.
Antibacterial Electrospun Polycaprolactone Nanofibers Reinforced by Halloysite Nanotubes for Tissue Engineering
Reprinted from: *Polymers* **2022**, *14*, 746, doi:10.3390/polym14040746 . . . . **150**

Abeer Aljubailah, Saad M. S. Alqahtani, Tahani Saad Al-Garni, Waseem Sharaf Saeed, Abdelhabib Semlali and Taieb Aouak
Naproxen-Loaded Poly(2-hydroxyalkyl methacrylates): Preparation and Drug Release Dynamics
Reprinted from: *Polymers* **2022**, *14*, 450, doi:10.3390/polym14030450 . . . . **162**

Nikolay Zahariev, Maria Marudova, Sophia Milenkova, Yordanka Uzunova and Bissera Pilicheva
Casein Micelles as Nanocarriers for Benzydamine Delivery
Reprinted from: *Polymers* **2021**, *13*, 4357, doi:10.3390/polym13244357 . . . . **193**

Chavee Laomeephol, Apichai Vasuratna, Juthamas Ratanavaraporn, Sorada Kanokpanont, Jittima Amie Luckanagul and Martin Humenik et al.
Impacts of Blended *Bombyx mori* Silk Fibroin and Recombinant Spider Silk Fibroin Hydrogels on Cell Growth
Reprinted from: *Polymers* **2021**, *13*, 4182, doi:10.3390/polym13234182 . . . . **206**

Muhammad Shahid Latif, Abul Kalam Azad, Asif Nawaz, Sheikh Abdur Rashid, Md. Habibur Rahman and Suliman Y. Al Omar et al.
Ethyl Cellulose and Hydroxypropyl Methyl Cellulose Blended Methotrexate-Loaded Transdermal Patches: In Vitro and Ex Vivo
Reprinted from: *Polymers* **2021**, *13*, 3455, doi:10.3390/polym13203455 . . . . **220**

Sevakumaran Vigneswari, Tana Poorani Gurusamy, Wan M. Khairul, Abdul Khalil H.P.S., Seeram Ramakrishna and Al-Ashraf Abdullah Amirul
Surface Characterization and Physiochemical Evaluation of P(3HB-*co*-4HB)-Collagen Peptide Scaffolds with Silver Sulfadiazine as Antimicrobial Agent for Potential Infection-Resistance Biomaterial
Reprinted from: *Polymers* **2021**, *13*, 2454, doi:10.3390/polym13152454 . . . . **237**

Ionuț-Cristian Radu, Cătălin Zaharia, Ariana Hudiță, Eugenia Tanasă, Octav Ginghină and Minodora Marin et al.
In Vitro Interaction of Doxorubicin-Loaded Silk Sericin Nanocarriers with MCF-7 Breast Cancer Cells Leads to DNA Damage
Reprinted from: *Polymers* **2021**, *13*, 2047, doi:10.3390/polym13132047 . . . . **255**

About the Editors

Ariana Hudita

Ariana Hudita Ph.D. is a scientific researcher in the Department of Biochemistry and Molecular Biology of the University of Bucharest, as well as in the Department of Drug Control of the University of Medicine and Pharmacy "Carol Davila". She is a dedicated researcher with a strong passion for cancer research, her primary focus being the development of innovative approaches to improve the current landscape of cancer-patient management. Her research interests encompass nanomedicine, in vitro 3D cancer modeling, tumor microenvironment characterization and manipulation, and the development of liquid biopsy protocols. These research directions are sustained by active collaborations with specialists and young researchers from diverse biomedical fields, multidisciplinary being the foundation of the research team of which Ariana Hudita Ph.D. is part. Moreover, recognizing the paramount importance that a multidisciplinary team has on cancer research, Miss Ariana Hudita co-founded the Romanian National OncoHub conference, and is currently the President of the Organizing Committee of this event that aims to bridge the gap between science and medicine, forging a path toward a brighter future in the fight against cancer.

Bianca Gălăţeanu

Bianca Galateanu Ph.D. is an Associate Professor at the University of Bucharest and also a Scientific Researcher in the Department of Biochemistry and Molecular Biology at the same university. She is an experienced researcher in the biomedical field of research, with over 15 years of experience in (i) adipose tissue engineering for regenerative purposes, (ii) the development of liquid biopsy approaches for cancer patients, (iii) the study of nano delivery systems for the targeted delivery of chemotherapy, (iv) colorectal cancer disease in vitro modeling, and (v) tumor microenvironment manipulation. Lastly, she is co-founder of OncoHub and President of the Romanian National OncoHub Conference, where famous scientists gather every year to approach multidisciplinary topics in the field of oncology and cancer research.

Preface

In recent years, the biomedical engineering field has seen remarkable advancements, focusing mainly on developing novel solutions for enhancing tissue regeneration or improving therapeutic outcomes. One of the key components of the latest developments is polymers (natural, synthetic, or blended formulations) due to their indisputable wide range of properties and functionalities that transform them into ideal materials for designing drug delivery systems or scaffolds for tissue engineering. This reprint, devoted to the topic of "Polymer Materials for Drug Delivery and Tissue Engineering", gathers a multidisciplinary collection of articles that merge the knowledge and expertise of worldwide researchers that present their latest research in the field of polymer-based materials. This reprint also brings together articles that report original research on the design and synthesis of novel drug delivery systems and scaffolds or that provide insights into polymer interaction. Polymers are valuable tools for the development of drug delivery systems and biomimetic scaffolds due to their tailorable design, versatility, attractive physiochemical properties, and excellent biocompatibility. With ongoing advancements in polymer science and personalized medicine, polymers hold great promise in revolutionizing the current landscape of the biomedical field, providing a real opportunity for designing personalized biomedical products that can be easily adjusted to meet patients' needs.

This reprint, entitled "Polymer Materials for Drug Delivery and Tissue Engineering", is a tribute to all the dedicated scientists engaged in the field of polymer materials for drug delivery and tissue engineering applications. The editors express their gratitude to all the contributing authors of this Special Issue as their insightful research and commitment to advancing the field have not only elevated the quality of our Special Issue but also contributed significantly to the broader scientific community. Also, we acknowledge all the members of the *Polymers* Editorial Office for their indispensable support and invaluable work in the development of this Special Issue.

Ariana Hudita and Bianca Gălăţeanu
Editors

polymers MDPI

Editorial

Polymer Materials for Drug Delivery and Tissue Engineering

Ariana Hudiță * and Bianca Gălățeanu

Department of Biochemistry and Molecular Biology, University of Bucharest, 91-95 Splaiul Independentei Street, 050095 Bucharest, Romania; bianca.galateanu@bio.unibuc.ro
* Correspondence: ariana.hudita@bio.unibuc.ro

In recent years, the biomedical engineering field has seen remarkable advancements, focusing mainly on developing novel solutions for enhancing tissue regeneration or improving therapeutic outcomes. One of the key components of the latest developments is polymers (natural, synthetic, or blended formulations), due to their indisputable wide range of properties and functionalities that transform them into ideal materials for designing drug delivery systems or scaffolds for tissue engineering.

The Special Issue "Polymer Materials for Drug Delivery and Tissue Engineering" gathers a multidisciplinary collection of articles that merge the knowledge and expertise of worldwide researchers that present their latest research in the field of polymer-based materials. This Special Issue brings together articles that report original research on the design and synthesis of novel drug delivery systems and scaffolds or highlight insights into polymer interaction [1–5]. To improve the current knowledge and open up new perspectives in the use of biopolymer blends, Wilharm et al. [6] characterized the interaction of collagen and elastin fibers during polymerization and revealed that elastin is incorporated homogeneously into the collagen fibers. The results contribute significantly to designing elastin-based biomaterials with or without actuatoric applications. In another study [7], a bioactive scaffold obtained based on demineralized human spongiosa Lyoplast was tested to reveal its microstructure and biochemical properties. The reported findings validate the biopolymer for its further use in developing scaffolds for articular hyaline cartilage tissue engineering, as the material preserves the typical hierarchical porous structure and the presence of collagen and other extracellular matrix proteins, with no cytotoxic-mediated effect in vitro on chondroblasts. Moreover, Laomeephol et al. [8] revealed the advantages of blending *Bombyx mori* silk fibroin (SF) and recombinant spider silk protein eADF4(C16) into a hydrogel that showed superior mechanical and biological performance when compared with the characteristics of individual materials.

Furthermore, this Special Issue reveals other biomedical applications of polymers such as the development of organ-on-a-chip platforms. With a focus on the urology field, Galateanu et al. [9] presented an overview of the use of polymers in developing microfluidic devices for creating physiological organ biomimetic systems for the kidney, bladder, and prostate using lithography, or bioprinting, representing valuable tools for understanding disease mechanisms or developing more effective drugs.

For tissue engineering purposes, polymers are widely used for the design and fabrication of biomimetic scaffolds that can substitute the defective tissue while maintaining a suitable environment to promote healing [10]. Depending on the intended application, a polymer-based scaffold can be designed to mimic the natural extracellular matrix (ECM) and therefore provide mechanical support for cellular adhesion, proliferation, and differentiation in a 3D manner that resembles the complex architecture of the targeted defective tissue. Scaffolds can be easily tuned depending on the requirements imposed by the intended applications in terms of mechanical properties, degradation rates, and surface characteristics, this versatility being the main reason why the same polymer can be used for the development of a variety of scaffolds that meet different tissue engineering

check for updates

Citation: Hudiță, A.; Gălățeanu, B. Polymer Materials for Drug Delivery and Tissue Engineering. *Polymers* **2023**, *15*, 3103. https://doi.org/10.3390/polym15143103

Received: 7 July 2023
Accepted: 10 July 2023
Published: 21 July 2023

applications. Moreover, the scaffold can be tailored to encapsulate and act as a reservoir for soluble bioactive molecules that can be released in a controlled manner or as a response to cues provided by the biological microenvironment. For example, polycaprolactone (PCL) was employed as a starting material for the development of multiple scaffolds that respond to different potential tissue restorative applications [11–13]. Busuioc et al. [11] manufactured bioactive and bioresorbable composite materials based on PCL and calcium magnesium silicate powders through an electrospinning technique to aid and sustain bone tissue regeneration. PCL-based scaffolds were also developed to feature antibacterial properties, as infections are frequently associated with tissue injuries. An oxidized regenerated cellulose/PCL bilayered composite (ORC/PCL) [12] incorporating different concentrations of cefazolin was screened to select the best formulation that can act as a synthetic dural substitute, with proper antibacterial activity. Despite the observations that drug incorporation triggers mechanical and physical changes in comparison with pristine scaffolds, the ORC/PCL composite loaded with 2.5 g cefazolin was selected as a promising formulation resembling the microstructure and physical and mechanical properties of the drug-unloaded composite while featuring antibacterial activity sustained for up to 4 days. Khunova et al. [13] also proposed the use of PCL for designing a bioactive scaffold that can act as a potential substrate to prevent bacterial infections associated with wound healing. Using electrospinning, PCL nanofibers reinforced by halloysite nanotubes (HNTs) loaded with erythromycin were manufactured, as a strategy to overcome the low solubility of the drug in aqueous solutions. The addition of HNTs in the PCL nanofibers improved the mechanical properties of the final scaffold, which presented a superior Young modulus and tensile strength and a strong antibacterial effect on both Gram-negative (*Escherichia coli*) and Gram-positive (*Staphylococcus aureus*) bacteria.

For wound healing applications, Al. Arjan and collaborators [14] proposed a pH-sensitive dressing prepared by blending bacterial cellulose (BC) with polyvinyl alcohol (PVA) and graphene oxide (GO), which was studied as a potential drug delivery platform after curcumin loading. The GO content impacts the physical–mechanical performance of the composite hydrogel in terms of the biodegradation rate, and mechanical and hydrophilicity properties. In vitro studies on Gram-positive and Gram-negative bacteria, as well as on cancer cells, revealed a dual role of the pH-responsive PVA/BC-*f*-GO dressing material that exhibits both antimicrobial and anticancer activities.

As in tissue engineering applications, the biocompatibility and biodegradability of polymers are essential features that promote their use for drug delivery system development, such as nanoparticles, micelles, and hydrogels [15]. Drug delivery systems improve drug solubility, protect the drug load from degradation, decrease the adverse effects associated with free-drug administration, and can be easily tuned to feature a controlled release of the drug suitable for the desired application. A strategy for modulating the release of the loaded drugs is the design of stimuli-responsive drug delivery systems. Such an example is the temperature- and pH-responsive poly(N-isopropyl acrylamide)-co-poly(acrylamide) (PNIPAM-co-PAAm) drug delivery system fabricated by Thirupathi and collaborators [16]. To validate the drug delivery behavior, curcumin was employed as a model drug, with the results showing a nearly complete release of the cargo in the presence of combined pH and temperature.

Moreover, especially for drug delivery systems developed for cancer treatment, functionalization strategies can be employed to direct nanoshuttles to tumor cells, thus increasing the therapeutic efficacy of the drug cargo while reducing toxicity [15]. For colorectal cancer management, Ullah et al. [17] designed chitosan nanoparticles loaded with 5-fluorouracil (5-FU) and decorated with folic acid (FA) to improve folate receptor affinity. Their results revealed the enhanced cytotoxicity of the 5-FU-FA chitosan nanoparticles compared to free 5-FU and non-functionalized 5-FU chitosan nanoparticles in Caco2 tumor cell cultures. However, functionalization strategies are not mandatory for developing effective nanoshuttles for anticancer drug delivery, since nanosized drug delivery systems accumulate preferentially in tumor cells, due to the abnormalities of the tumor vasculature.

A nanosystem based on *Bombyx mori* silk sericin was successfully developed for doxorubicin delivery and induced cytotoxicity and genotoxicity in MCF-7 tumor cells selected as an in vitro model for breast cancer [18].

In conclusion, polymers are valuable tools for the development of drug delivery systems and biomimetic scaffolds due to their tailorable design, versatility, attractive physiochemical properties, and excellent biocompatibility. With ongoing advancements in polymer science and personalized medicine, polymers hold great promise in revolutionizing the current landscape of the biomedical field, providing a real opportunity for designing personalized biomedical products that can be easily adjusted to meet patients' needs.

Author Contributions: Conceptualization, A.H. and B.G.; Writing–original draft preparation, A.H. and B.G. Writing–review and editing, A.H. and B.G., supervision A.H. All authors have read and agreed to the published version of the manuscript.

Funding: This research received no external funding.

Acknowledgments: The guest editors want to acknowledge the authors and reviewers for their invaluable contribution and scientific excellence, which sustained the success of this Special Issue. Also, we acknowledge all the members of the *Polymers* Editorial Office for their indispensable support and the invaluable work in the development of this article collection by the in house Editor.

Conflicts of Interest: The authors declare no conflict of interest.

References

1. Rizg, W.Y.; Hosny, K.M.; Eshmawi, B.A.; Alamoudi, A.J.; Safhi, A.Y.; Murshid, S.S.; Sabei, F.Y.; Al Fatease, A. Tailoring of Geranium Oil-Based Nanoemulsion Loaded with Pravastatin as a Nanoplatform for Wound Healing. *Polymers* **2022**, *14*, 1912. [CrossRef] [PubMed]
2. Aljubailah, A.; Alqahtani, S.M.; Al-Garni, T.S.; Saeed, W.S.; Semlali, A.; Aouak, T. Naproxen-Loaded Poly (2-Hydroxyalkyl Methacrylates): Preparation and Drug Release Dynamics. *Polymers* **2022**, *14*, 450. [CrossRef] [PubMed]
3. Zahariev, N.; Marudova, M.; Milenkova, S.; Uzunova, Y.; Pilicheva, B. Casein Micelles as Nanocarriers for Benzydamine Delivery. *Polymers* **2021**, *13*, 4357. [CrossRef] [PubMed]
4. Latif, M.S.; Azad, A.K.; Nawaz, A.; Rashid, S.A.; Rahman, M.H.; Al Omar, S.Y.; Bungau, S.G.; Aleya, L.; Abdel-Daim, M.M. Ethyl Cellulose and Hydroxypropyl Methyl Cellulose Blended Methotrexate-Loaded Transdermal Patches: In Vitro and Ex Vivo. *Polymers* **2021**, *13*, 3455. [CrossRef] [PubMed]
5. Vigneswari, S.; Gurusamy, T.P.; Khairul, W.M.; HPS, A.K.; Ramakrishna, S.; Amirul, A.-A.A. Surface Characterization and Physiochemical Evaluation of P(3HB- Co-4HB)-Collagen Peptide Scaffolds with Silver Sulfadiazine as Antimicrobial Agent for Potential Infection-Resistance Biomaterial. *Polymers* **2021**, *13*, 2454. [CrossRef] [PubMed]
6. Wilharm, N.; Fischer, T.; Hayn, A.; Mayr, S.G. Structural Breakdown of Collagen Type I Elastin Blend Polymerization. *Polymers* **2022**, *14*, 4434. [CrossRef] [PubMed]
7. Tsiklin, I.L.; Pugachev, E.I.; Kolsanov, A.V.; Timchenko, E.V.; Boltovskaya, V.V.; Timchenko, P.E.; Volova, L.T. Biopolymer Material from Human Spongiosa for Regenerative Medicine Application. *Polymers* **2022**, *14*, 941. [CrossRef]
8. Laomeephol, C.; Vasuratna, A.; Ratanavaraporn, J.; Kanokpanont, S.; Luckanagul, J.A.; Humenik, M.; Scheibel, T.; Damrongsakkul, S. Impacts of Blended Bombyx Mori Silk Fibroin and Recombinant Spider Silk Fibroin Hydrogels on Cell Growth. *Polymers* **2021**, *13*, 4182. [CrossRef]
9. Galateanu, B.; Hudita, A.; Biru, E.I.; Iovu, H.; Zaharia, C.; Simsensohn, E.; Costache, M.; Petca, R.-C.; Jinga, V. Applications of Polymers for Organ-on-Chip Technology in Urology. *Polymers* **2022**, *14*, 1668. [CrossRef]
10. Socci, M.C.; Rodríguez, G.; Oliva, E.; Fushimi, S.; Takabatake, K.; Nagatsuka, H.; Felice, C.J.; Rodríguez, A.P. Polymeric Materials, Advances and Applications in Tissue Engineering: A Review. *Bioengineering* **2023**, *10*, 218. [CrossRef] [PubMed]
11. Busuioc, C.; Alecu, A.-E.; Costea, C.-C.; Beregoi, M.; Bacalum, M.; Raileanu, M.; Jinga, S.-I.; Deleanu, I.-M. Composite Fibers Based on Polycaprolactone and Calcium Magnesium Silicate Powders for Tissue Engineering Applications. *Polymers* **2022**, *14*, 4611. [CrossRef] [PubMed]
12. Sanpakitwattana, A.; Suvannapruk, W.; Chumnanvej, S.; Hemstapat, R.; Suwanprateeb, J. Cefazolin Loaded Oxidized Regenerated Cellulose/Polycaprolactone Bilayered Composite for Use as Potential Antibacterial Dural Substitute. *Polymers* **2022**, *14*, 4449. [CrossRef] [PubMed]
13. Khunová, V.; Kováčová, M.; Olejniková, P.; Ondreáš, F.; Špitalský, Z.; Ghosal, K.; Berkeš, D. Antibacterial Electrospun Polycaprolactone Nanofibers Reinforced by Halloysite Nanotubes for Tissue Engineering. *Polymers* **2022**, *14*, 746. [CrossRef] [PubMed]
14. Al-Arjan, W.S.; Khan, M.U.A.; Almutairi, H.H.; Alharbi, S.M.; Razak, S.I.A. PH-Responsive PVA/BC-f-GO Dressing Materials for Burn and Chronic Wound Healing with Curcumin Release Kinetics. *Polymers* **2022**, *14*, 1949. [CrossRef] [PubMed]

15. Ginghină, O.; Hudiță, A.; Zaharia, C.; Tsatsakis, A.; Mezhuev, Y.; Costache, M.; Gălățeanu, B. Current Landscape in Organic Nanosized Materials Advances for Improved Management of Colorectal Cancer Patients. *Materials* **2021**, *14*, 2440. [CrossRef] [PubMed]
16. Thirupathi, K.; Phan, T.T.V.; Santhamoorthy, M.; Ramkumar, V.; Kim, S.-C. PH and Thermoresponsive PNIPAm-Co-Polyacrylamide Hydrogel for Dual Stimuli-Responsive Controlled Drug Delivery. *Polymers* **2022**, *15*, 167. [CrossRef] [PubMed]
17. Ullah, S.; Azad, A.K.; Nawaz, A.; Shah, K.U.; Iqbal, M.; Albadrani, G.M.; Al-Joufi, F.A.; Sayed, A.A.; Abdel-Daim, M.M. 5-Fluorouracil-Loaded Folic-Acid-Fabricated Chitosan Nanoparticles for Site-Targeted Drug Delivery Cargo. *Polymers* **2022**, *14*, 2010. [CrossRef] [PubMed]
18. Radu, I.-C.; Zaharia, C.; Hudiță, A.; Tanasă, E.; Ginghină, O.; Marin, M.; Gălățeanu, B.; Costache, M. In Vitro Interaction of Doxorubicin-Loaded Silk Sericin Nanocarriers with Mcf-7 Breast Cancer Cells Leads to DNA Damage. *Polymers* **2021**, *13*, 2047. [CrossRef] [PubMed]

polymers

Review

Applications of Polymers for Organ-on-Chip Technology in Urology

Bianca Galateanu [1], Ariana Hudita [1,*], Elena Iuliana Biru [2,*], Horia Iovu [2,3], Catalin Zaharia [2], Eliza Simsensohn [4], Marieta Costache [1], Razvan-Cosmin Petca [4] and Viorel Jinga [4]

1 Department of Biochemistry and Molecular Biology, University of Bucharest, 91-95 Splaiul Independentei Street, 050095 Bucharest, Romania; bianca.galateanu@bio.unibuc.ro (B.G.); marieta.costache@bio.unibuc.ro (M.C.)
2 Advanced Polymer Materials Group, Department of Bioresources and Polymer Science, University Politehnica of Bucharest, 1-7 Gh. Polizu Street, 011061 Bucharest, Romania; horia.iovu@upb.ro (H.I.); catalin.zaharia@upb.ro (C.Z.)
3 Academy of Romanian Scientists, Ilfov Street, 50044 Bucharest, Romania
4 "Carol Davila" University of Medicine and Pharmacy Bucharest, 050474 Bucharest, Romania; eliza.simsensohn@gmail.com (E.S.); razvan.petca@umfcd.ro (R.-C.P.); viorel.jinga@umfcd.ro (V.J.)
* Correspondence: ariana.hudita@bio.unibuc.ro (A.H.); iuliana.biru@upb.ro (E.I.B.)

Abstract: Organ-on-chips (OOCs) are microfluidic devices used for creating physiological organ biomimetic systems. OOC technology brings numerous advantages in the current landscape of preclinical models, capable of recapitulating the multicellular assemblage, tissue–tissue interaction, and replicating numerous human pathologies. Moreover, in cancer research, OOCs emulate the 3D hierarchical complexity of in vivo tumors and mimic the tumor microenvironment, being a practical cost-efficient solution for tumor-growth investigation and anticancer drug screening. OOCs are compact and easy-to-use microphysiological functional units that recapitulate the native function and the mechanical strain that the cells experience in the human bodies, allowing the development of a wide range of applications such as disease modeling or even the development of diagnostic devices. In this context, the current work aims to review the scientific literature in the field of microfluidic devices designed for urology applications in terms of OOC fabrication (principles of manufacture and materials used), development of kidney-on-chip models for drug-toxicity screening and kidney tumors modeling, bladder-on-chip models for urinary tract infections and bladder cancer modeling and prostate-on-chip models for prostate cancer modeling.

Keywords: organ-on-chip; tumor-on-chip; polymeric microfluidic devices; kidney-on-chip; bladder-on-chip; prostate-on-chip

Citation: Galateanu, B.; Hudita, A.; Biru, E.I.; Iovu, H.; Zaharia, C.; Simsensohn, E.; Costache, M.; Petca, R.-C.; Jinga, V. Applications of Polymers for Organ-on-Chip Technology in Urology. *Polymers* **2022**, *14*, 1668. https://doi.org/10.3390/polym14091668

Academic Editor: Ángel Serrano-Aroca

Received: 26 February 2022
Accepted: 18 April 2022
Published: 20 April 2022

1. Introduction

Regardless of the therapeutic area of new emerging therapies and novel agents entering the market, nephrotoxicity is a major challenge, as the kidney is the second target of drugs and chemicals after the liver. More than 25% of the adverse effects in today's pharmacotherapy are caused by nephrotoxic effects [1]. Of these, 20% are reported during postmarket surveillance [2], as early stages of drug development fail to deliver relevant output with this respect [3]. Consequently, the poor correlation between the preclinical and clinical outcomes has led to the failure of most drugs before reaching the patient [4]. Preclinically approved drugs have been withdrawn a few times due to the side effects observed in the clinical trials [5,6]. Therefore, the pharmaceutical industry is under high pressure to speed up the drug-development process and to design new cures that are very effective in humans with reasonable costs [7].

The traditional in vivo tests on animal models are costly and often fail to accurately predict the efficiency and toxicity in humans due to the species' different metabolic responses to specific agents and the variations in some genes' expressions, such as cytochrome

P450 genes [8]. Consequently, the poor similarity between the physiological environment in animals and human bodies, which may alter the results of drug efficiency in various diseases, is a major barrier for future use of in vivo tests on animals [9]. With respect to cancer research, animal models in particular lack predictability [10] since they do not recreate the exact human tumor microenvironment (TME) and may exhibit different cell biology and cancer behavior when tumorous cells interact locally with stromal cells. In addition, the ethics concerns of sacrificing animals are a significant barrier in testing many discovered drugs on animals [11,12]. In 2021, the European Parliament agreed with a large majority to ban experiments on animals, which have killed about 12 million animals in 2017, revealing the importance of finding alternatives for biological assays developed with other procedures than sacrificing animals [13].

Despite tremendous research efforts and advanced medical treatments, cancer continues to be one of the most frequent causes of death in the world. In 2020, 18.1 million new cases and 9.5 million cancer-associated deaths were reported worldwide [14]. According to American Institute for Cancer Research, more than 50% of cancer cases were reported in men. Prostate, bladder, and kidney cancers account for an estimated 23% of all cancers diagnosed in men in the last two years. Furthermore, recent World Health Organization (WHO) statistics are estimating a ~55% increase of overall cancer cases from 2020 to 2040 [15]. Alarmingly, the mortality rate is expected to increase by ~65% by 2040, highlighting the urge to find and acquire more efficient anticancer remedies.

The main difficulties in cancer research are forming an effective in vitro TME able to accurately recapitulate the local tissue in which the tumor is forming [16]. Conventional preclinical in vitro models for anticancer drug screening are generally classified in 2D cell cultures and 3D cell architectures and have been extensively exploited as simple and cost-efficient methods to simulate cancer propagation and drug response [17]. The 3D cancer models deliver a helpful substitute to animals, but they still do not consider the dynamic environment of the human tissues or organs. However, they do not reproduce the complex assemblage of the human 3D cells from living organs to properly elucidate the cancer cell migration and invasion, also taking into consideration the mechanical forces (such as hydrostatic pressure, fluid shear stress, breathing motions in lung) naturally occurring in human bodies. Nonetheless, neither of these systems is transporting a blood or nutrient-rich medium through an endothelium-lined vasculature, limiting the real prediction of tissue–tissue interactions and circulating immune cells during therapeutic drug dosage [18,19].

Standard cell-culture techniques fail to provide insights into complex multifaced interactions that take place in a multiorgan system. The need to transport fluids containing pharmaceutical compounds through models of different pathological conditions while accurately simulating physiological processes is challenging the scientific media to engineer new technology able to replace animal testing.

Microfluidics is the science and technology of systems that allow the processing and manipulation of microscale fluids (10^{-9} to 10^{-18} L) using channels with sizes of tens to hundreds of micrometers [20]. By combining microelectronics with structural analysis and molecular biology, microfluidics leads to a deeper understanding of the mechanism by which the cellular, biochemical, and physiochemical environment indicate tumor sensitivity and resistance to therapy [21].

Recently, new devices known as organ-on-chips (OOCs), which are able to recapitulate the multicellular assemblage, tissue–tissue interactions, and to replicate human pathologies and the appropriate physical TME, have emerged as a practical cost-efficient solution for tumor-growth investigation and anticancer-drug screening by combining the microfluidic technology with 3D cell-culture procedure to simulate the entanglement of the cells as in their native environment [19,22,23]. OOCs are compact and easy-to-use microphysiological functional units that recapitulate the native function and the mechanical strain that the cells experience in the human bodies, allowing the development of a wide range of applications such as disease modeling or even the development of diagnostic devices. However, important features of the membranes involved in the fabrication of OOC compartments to allow cells' structural support and nutrient transportation are often poorly investigated. Nowadays, both synthetic and natural polymers are explored for the manufacturing process of advanced OOC microdevices, being able to replicate various organ bionic pathophysiological models. Poly(dimethylsiloxane) is one of the most employed synthetic polymers used for lung, liver, heart, and multi-organ-on-chip (MOOC) membranes in microfluidic devices due to its extraordinary high transparency and flexibility. However, it is not a degradable material able to contribute to the formation of the natural extracellular matrix (ECM). Alternative biopolymers with higher biocompatibility, such as collagen-based materials containing cell-growth factors and hormones, have been used for OOC fabrication to simulate the physiological behavior of living organs. Despite significant advances, many polymeric materials still do not meet the mechanical properties of the in vivo organs and do not exhibit optimal cytocompatibility suitable for accurate pharmaceutical screening or dynamic simulation of cancer cell behavior.

This review brings to attention the specifications and fabrication methods for OOCs and the importance of polymeric porous materials used in OOCs in relation to cell behavior. In this context, the current work aims to review the scientific literature in the field of microfluidic devices and materials designed for urology applications in terms of OOC fabrication (principles of manufacture and materials used), development of kidney-on-chip models for drugs toxicity screening, and kidney tumors modeling, bladder-on-chip models for urinary tract infections and bladder cancer modeling, and prostate-on-chip models for prostate cancer modeling.

2. Fabrication of Organ-on-Chip

2.1. Principle and Manufacture of Organ-on-Chips

Microfluidic technologies have rapidly developed in the past years in terms of fabrication methodology, materials involved, and complexity of the systems to faithfully respond to the medical requirements. OOCs are microfluidic cell-culture micromachines that can recapitulate an organ-level response to medical treatment and reconstruct physiological dynamics observed in native human tissues, for instance, physiological flow, biomechanical motions, nutrient transportation, and drug delivery [22], in a convenient manner, delivering a platform with a new opportunity for oncology research. As biology-inspired engineered microdevices [24], the OOCs for tumor investigation must enable a series of possibilities: (i) the introduction of pharmaceutical compounds or reagents as fluids with the similar dynamic flow as for biological fluids; (ii) the ability to perfuse these fluids around on the chip, and to combine and mix them; (iii) the introduction of other sensors or devices for monitoring the results, such as detectors for bioanalysis.

By incorporating tumor organoids in microfluidic devices, "tumor-on-chip" (TOCs) models that allow the reconstruction of the TME are created. These chips enable a deeper understanding of the tumor mechanism in vivo, which runs to enhanced preclinical evaluation of drug efficiency. Human solid tumors are highly heterogeneous [16], owning a complex microenvironment with a dense ECM, abnormal vessels, various stromal cells, or different immune-type cells [22]. Additionally, the nearby stromal tissues of the tumor act as an active source (and reservoir) of different cytokines and growth factors that affect the tumor development and pharmacological feedback. Several studies have shown that the complex variety of the cellular microenvironment may impact in some respect the tumor behavior, including tumorigenesis, angiogenesis, tumor invasion, metastasis, and endurance to therapeutic products. Many compounds generated by tumorous or stromal cells determine the propagation dynamics of solid tumors. Moreover, the 3D nature and the size of the tumor proved to be an utmost concern in the proper understanding of tumor dynamics showing a direct connection between the size of a tumor and its aggressiveness and the ability of a drug to be delivered to it [25–27]. Consequently, cancer is a true suite of complex pathologies that share some common elements and presents few differences that seriously complicate the choice of satisfactory treatment. Unlike in vivo tumor-growth microenvironments, in vitro cancer models are typically investigated under atmospheric conditions not specific to the living organs. In an in vitro metastasis cell culture, migrated tumor cells have to be subjected to varying microenvironments and oxygen gradients to mimic the activity of in vivo intravasation [28]. Tumor chips combine micromachining and cell biology to manipulate the external conditions and precisely mimic physiological environments, such as dynamic mechanical stress, fluid shear, oxygen, and drug concentration gradients and cell patterning to reflect the full picture of tumor formation and growth mechanism [29,30].

As TOCs resulted from the need to investigate appropriately cancerous cells in their specific TME, the fabrication technology is the same as for conventional OOCs, whereas the healthy organ cells were replaced by diseased cells. The microdevices are named chips since they were originally engineered using micromanufacturing techniques used in computer-microchip production [20]. Microfluidic systems can be employed to form tumor chips with a single-line channel by engaging cells from a single source, or more complex organ chips that associate two or more tissue categories that can be interposed right through a porous ECM-coated membrane or an ECM gel that fills one or more micronetworks [31]. The viability of the cells can be preserved over prolonged time intervals (up to several months) by perfusing the culture medium either across the endothelium-lined vascular microchannels, parenchymal microchannels, or both. More importantly, the culture medium can be substituted by blood for several hours through endothelialized vascular channels [32].

There are two steps required to be taken into consideration for the proper design of a tumor chip: (1) to comprehend the fundamentals essential for the physiological function of the aimed organ, and afterward to establish the key factors such as various cell types, structures, and the organ's particular physiological microenvironment; (2) to construct a cell-culture device relying on the identified features. Different procedures have been embraced to build tumor-chip kits, among which the most extensively engaged are photolithography and soft lithography [33,34], replica molding [35], microcontact printing, and bioprinting techniques [36–38].

2.1.1. Lithography

Lithography represents an etching process applied in microfabrication to project parts on a thin layer or the form of a substrate (also named a wafer) and is commonly divided into three categories: photolithography (or UV lithography), soft lithography, and replica molding. Photolithography employs light to transfer a geometric shape from a photomask to a photosensitive chemical photoresist on the wafer. In the first step, masks are required that correspond to the targeted constructions [22,34]. The manufacturing process continues with the deposition of a spin-coated photoresist layer on a wafer that can be corroded by chemical reagents, for instance, silicon, glass, or quartz, and the photoresist is subjected to UV photopolymerization. Afterward, the pattern is moved to the wafer and is etched to achieve a microfluidic chip with microflow channels. Although extensively employed, the lithography-manufacturing processes are costly due to the cleanroom requirements, the necessity of multiple masks, and time-consuming multiple processing steps. Recently, Kasi and coworkers proposed a rapid-prototyping organ-chip device using maskless photolithography [39]. The authors reported a simplified method that describes a rapid and cleanroom-free microfabrication compatible with soft lithography for fast-prototyping organ-chip devices in a maximum of 8 h. Soft lithography used for tumor-chip microfabrication involves in the first stage the preparation of a microchannel mold on a silicon substrate by photolithography [40], followed by the use of a liquid polymer (commonly polydimethylsiloxane, PDMS) to discharge the mold to acquire an optically transparent elastomeric stamp with microstructures. In the end, different complex 3D microchannels are achieved on various polymer wafers by transferring the pattern from the stamp. Ferreira et al. have recently developed a fast alternative to soft lithography to manufacture OOCs based on PDMS with integrated microactuators. The novel protocol decreases the complexity and number of steps, and is more time and cost-efficient compared to complex multilayered microfluidic devices [41].

Replica molding represents the technology in which a patterned silicon mold is employed, followed by the pouring step of a liquid polymer (usually PDMS) onto the mold for thermal crosslinking [42]. Later, the PDMS instrument is removed from the substrate and fixed on a clean, smooth wafer (for instance glass) to achieve a microfluidic chip with microfluidic networks.

The microcontact-printing technology is very similar to the replica-molding technique [36]. It is differentiating only by the supplementary steps further used to manipulate the pattern of cultured cells by printing the PDMS stamp on the wafer with biomolecules (such as proteins) in a designed pattern so that the cells on the membrane can be modeled as well by adjusting the pattern of the printed proteins [43,44]. However, even though microfluidic devices have been successfully fabricated by lithography or related techniques, the procedures are still costly and time-consuming. Additionally, these procedures are able only to produce the microchip itself, while other components such as microtissues, mechanical stimuli, or result detectors need to be separately produced.

Recent advances in lithography-based fabrication techniques have emerged in high-throughput technologies with a standardized format, being more user-friendly and affordable for large preclinical research studies. The OrganoPlate system (Mimetas, Leiden, The Netherlands) is a commercial compact microfluidic device that can process 96 independent cultures using a standard microtiter plate. The microdevice replaces the inner membrane with a gel-media boundary for transepithelial transportation showing good results for the investigation of fluidic diffusion, tumor-cell invasion, and aggregation and toxicity assays [2]. The main technologies used for the microfabrication of OOCs are shown in Figure 1.

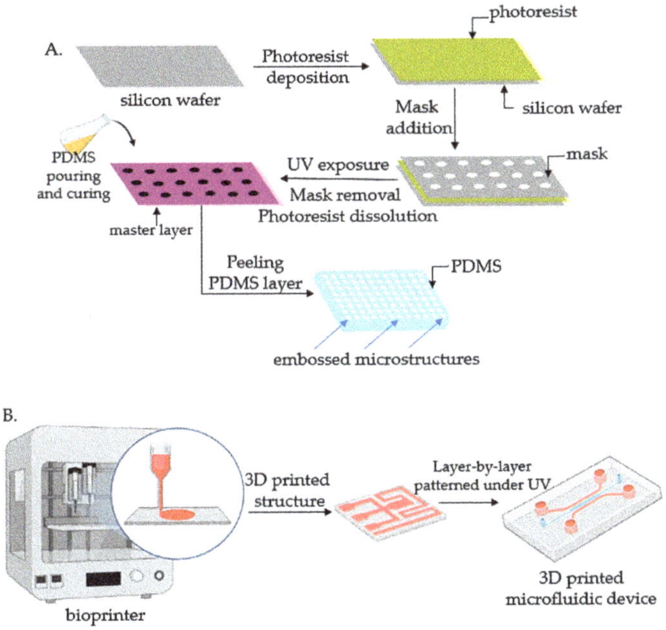

Figure 1. Fabrication technologies for OOCs: (**A**) The fabrication of micropatterned slabs of PDMS through photolithography; (**B**) Schematic 3D-printing process for the fabrication of microfluidic devices.

Injection molding allows fast replication of polymeric (micro)structures with great surface quality. Generally, the injection-molding process requires a few basic steps: (a) the material is melted and the molds are compressed together, and (b) the material is injected into the mold and cooled down prior to removal [34]. Although it has the advantage of large-scale production of OOCs, the fabrication process requires the careful adjustment of multiple parameters such as injection pressure, injection speed, melting temperature, etc. [45], to ensure high-quality production. Thermoplastic materials such as polystyrol [46] were used for the fabrication of liver-on-a-chip microdevices. One of the main disadvantages of this technique refers to the limited area of materials that can be used, as most of the polymers show thermal shrinkage during the fabrication [45]. However, the injection-molding process is reducing the time of fabrication and the final costs and therefore is predominantly used for the production of the commercially available elements of OOCs.

Laser-ablation methods, such as micromachining or computer numerical-control (CNC) micromachining, are surpassing the limitations of manual control during the fabrication of OOC microdevices. Laser micromachines use a laser to engrave the OOC device, and the process is applicable to a wide range of materials such as metals, glass or polymers. Shaegh et al. [47] designed a rapid prototyping method to produce microfluidic chips from thermoplastics with patterned microvalves combining laser ablation and thermal-fusing bonding. In this study, a CO_2-assisted laser micromachine was used to pattern and cut PMMA layers covered with polyurethane film in order to generate a gas-actuated microvalve for microfluidic lab-on-a-chip applications. CNC micromachining is a fully automated manufacturing process in which the machines are operated via numerical control, wherein a computer software dictates the shape of the desired object. The laser-processing techniques have the advantage of rapidly creating precise and complex geometries of microdevices. Moreover, the laser-ablation techniques may be combined with the previously discussed methods for rapid prototyping or master models in order to achieve the desired microfluidic system, but require high technical knowledge for operating and expensive machinery to perform OOC fabrication [34].

2.1.2. Bioprinting

Bioprinting technology has emerged in the last years as a versatile tool based on layer-by-layer addition that can be easily used for tumor-chip manufacturing [48–50]. A significant advantage is related to a wider range of materials and cells that can be printed simultaneously onto a substrate of cell-compatible biomaterials to construct 3D composite blocks with good spatial resolution and reproducibility [51,52]. Bioprinting also comprises a large group of procedures that can be divided into fused-deposition modeling (FDM), stereolithography, inkjet printing, and laser-assisted bioprinting [53]. Bioprinting procedures are appealing due to numerous advantages that accommodate the tumor-chip fabrication. Bioprinting technology can reconstruct the diverse TME and the 3D nature of the tumor by using bioinks that form cell conglomerations that can include various cell types, such as tumor-related fibroblasts, immune cells, and endothelial cells, to create vascular systems [54]. In addition, layer-by-layer bioprinting enables the formation of the biomimetic microenvironment for a heterogenous supply of biologically important proteins and growth factors that are essential to managing cancer-cell signaling, growth, and invasion [55]. Furthermore, the bioprinting technique allows for the printing of cells quickly and precisely in microfluidic chips, pattern vasculature, and model biological barriers. Vascularization channels play a significant role as they are vital to preserve tissue activities or to differentiate tissue parts. The tumor vasculature is much distinct from the blood vessels of healthy tissues, showing alterations in heterogeneity, multidirectional blood flow, permeability, and unordered distribution all over the diseased cells [56]. Proper vascularization of specific cancer-type cells has constantly been questioned in the production of functional in vitro tumor-cell cultures. Bioprinting technology manages to overcome these difficulties, showing the ability to replicate the abnormalities encountered in tumor tissue by building miniaturized pathophysiological models and offering control of features at the same size scale of living cells [57–59]. The main polymers used as bioinks for OOCs by bioprinting are summarized in Table 1.

Table 1. The main polymers used as bioinks for OOCs by bioprinting.

Bioink Composition	Bioprinting Method	OOC Model	References
Collagen	extrusion	Lung, gut	[60,61]
Gelatin	extrusion	kidney	[62,63]
Methacrylate gelatin (GelMa)	extrusion	Vessel, liver	[64,65]
Alginate	extrusion	Heart	[66,67]
Cellulose	extrusion	tumor, liver	[68,69]
Polyethyleneglycol (PEG) Poly ε caprolactone (PCL)	inkjet, extrusion	Colon tumor	[59,65]

2.2. Material Requirements for the Fabrication of Organ-on-Chips

Many biomaterials and fabrication methods have been recently developed to meet the OOCs' functions. While designing the OOC microdevice, the organ or tissue characteristics and behavior must be considered to accurately simulate the in vivo motions, cell proliferation, or drug responses. Considering the organ characteristics aimed to be mimicked, different types of OOCs can be created.

However, although each type of OOCs may function differently, the main components remain the same. The OOC system is generally equipped with two compartments: first, the blood vessel compartment, which contains the endothelial cells; and secondly, the organ compartment, in which the cells of the investigated organ are introduced. These compartments are usually separated by a porous polymeric membrane able to provide cell communication between the two compartments of the device. Thus, the polymeric membranes play an essential role in the successful functioning of the OOC system, acting as interfaces with selective permeability for cell adhesion and cell separation.

The culture medium included in OOCs system may be artificially produced or naturally derived from living organs. The natural medium mainly contains biological fluids such as blood, plasma, etc., and organ fragments. The artificial environment is fabricated by involving nutrients that mimic the physiological conditions, such as salts, oxygen and CO_2 gasses, and blood substitutes. Thus, the cell-culture medium will provide a continuous supply of nutrients and the porous polymeric membrane will deliver the cells of interest that are cultured on the other side of the membrane. In addition to the nutrient-rich fluid perfusion and oxygen ventilation from the blood-vessel compartment, in the organ compartment, mechanical forces can be applied to the polymeric membrane to simulate the peristalsis respiratory motions from the in vivo organs.

Despite the significant progress in material science, only a few materials can recapitulate the physiological conditions of living organ testing. First and foremost, the materials involved in the construction of the OOCs microdevices should ensure the necessary biocompatibility to provide a biologically safe and nontoxic environment that allows cell migration without generating any inflammatory response or exacerbated immunogenicity. Polymeric materials are desirable because of their great structural similarity to the ECM.

Besides biocompatibility, an essential requirement for the polymeric membrane that separates the OOCs compartments is the similarity with the ECM in order to support the cell attachment and diffusion through its pores from one side of the interface to the other compartment. Proteins such as collagen, fibronectin, and vitronectin are ideal candidates as they are involved in biological processes and play an important role in the cell-adhesion mechanism. Moreover, it was shown that the surface roughness of the membrane could affect cell behavior. In the case of synthetic polymers such as PMMA layers, it was observed that the cell adhesion and spreading of vascular cells improved with higher surface roughness, but proliferation was not affected, and it was indicated that the cell adhesion is dependent on the protein adsorption [70]. In the case of PDMS, surface roughness and surface energy play a key role in the cell-attachment process. It was observed that higher surface energy promotes the formation of stronger cell–ligand bonding, leading to improved cell growth and proliferation. However, the cell-membrane interaction drastically decreases above critical surface energy and critical roughness ratio. Thus, the study on the cellular behavior of HeLa and MDA MB 231 cells on a rough PDMS surface revealed that the optimum conditions for cell adhesion, growth, and proliferation are obtained at moderate surface energy and intermediate roughness ratios [71].

Additionally, the hydrophilicity of the polymeric membrane represents an important factor in determining the protein adsorption and, therefore, cell adhesion. It was observed that the hydrophobicity of the polymeric membrane could cause the absorption of small nonspecific molecules such as drugs or biomolecules from the cellular media, misleading the experimental results [72]. For this reason, most synthetic polymers have been submitted to chemical or physicochemical surface modifications to increase the hydrophilicity of the materials by enhancing their wettability properties. PMMA structures have been modified with different hydrophilic functionalities such as aminated polyethyleneglycol [73] or submitted to oxygen plasma treatment for the activation of the PMMA surface [74] to control the cell adhesion. The cellular adhesion is strongly determined by the surface wettability and roughness of the polymer, while the cell attachment is influenced by the cell type [75]. Premnath and colleagues [76] developed a simple approach to laser-modify the surface of a silicon chip to adjust cell adhesion and proliferation, and showed that the cervical cancer cells' behavior is modulated to migrate onto untreated sites. Transparent polyethersulfone (PES) membranes have proven to be with optimized morphology, and have showed improved adhesion and cell viability compared to the commercial hydrophilic polytetrafluoroethylene (PTFE) [77]. Moreover, the increase in hydrophilicity of the polymeric layers enhanced the cell adhesion as well as the cell-adhesive proteins. It is worth mentioning that hydrophilicity/hydrophobicity balance can also control the essential nutrient diffusion through the polymeric membrane from the OOCs compartments and generally, water contact-angle measurements are performed to determine the wettability properties of the polymers (Table 2).

Table 2. The main polymers employed in the fabrication of membranes in OOCs microdevices.

Polymer	Chemical Structure	Contact Angle with Water	Young's Modulus	Application
Polydimethylsiloxane (PDMS)	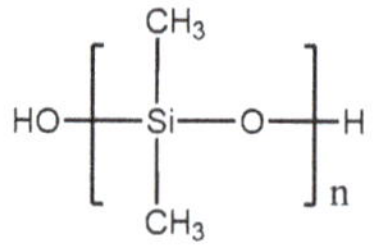	107° [78]	Variable from kPa to MPa	Cardiovascular [79] Kidney [80] Liver [81] Lung [82]
Poly (bisphenol-A-carbonate) (PC)	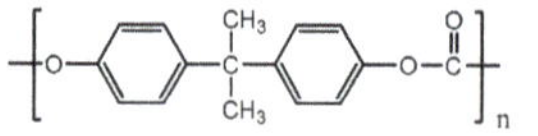	~85° [83]	~1.2 GPa [84]	Tumor vasculature [85] Colon and breast cancer [86]
Poly (ethylene terephthalate) (PET)	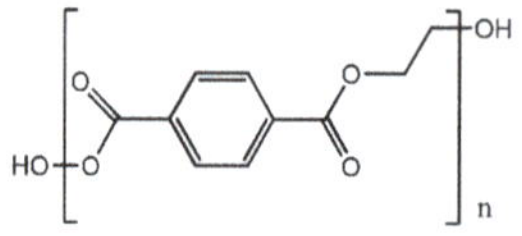	83° [87]	4.7 GPa [88]	Gut [89] Kidney [90] Liver [91]
Polylactic acid (PLA)	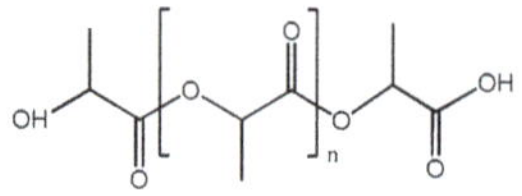	~75° [92]	3.1 GPa [93]	Endothelial barrier [94]

Table 2. *Cont.*

Polymer	Chemical Structure	Contact Angle with Water	Young's Modulus	Application
Poly (ε-caprolactone) (PCL)	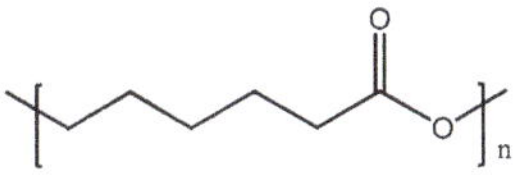	120° [95]	~400 MPa [96]	Blood-brain barrier [97]
Polytetrafluoroethylene (PTFE)	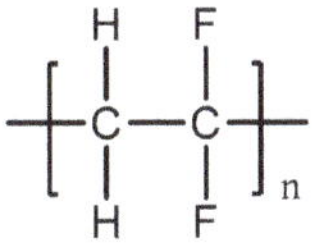	108° [98]	392 MPa [99]	Liver [100]

As the surface properties such as surface roughness and wettability influence the first stages of cell adhesion and migration, the mechanical properties of the polymeric membrane are modulated by the material's stiffness and influence the later stages of cell growth. Investigations such as tensile strength, compressive stress, and wettability are generally achieved to determine if the polymer properties are suitable for the application. Thus, the mechanical performances of the interface membrane from the OOCs should be chosen accordingly to the native organ characteristics and to mimic the occurring physiological stimuli present in the replicated environment. Moreover, the porosity of the membrane layer represents an essential feature as it provides cell communication between the chip compartments and allows nutrient transportation to the cells. Pore size and shape also determine the space available for the cells to migrate, and small pores mainly reduce the cell adhesion, but increase the barrier function of cells.

2.3. Polymers for Organ-on-Chips

Materials used to manufacture OOCs play a crucial role as they may directly interact with the biological fluids and cellular system, affecting the experiment results. Although several types of polymeric structures have been employed, only a few polymers exhibit the required specifications needed to fabricate the OOCs, and both synthetic polymers and biopolymers have been investigated.

In the last few years, synthetic and natural polymers have been used for OOC applications as their performances depend on the polymer structure, porosity, transparency, flexibility, stiffness, etc. Synthetic polymers such as polydimethylsiloxane (PDMS) [101], polycarbonate (PC) [102], polyethyleneterephtalate (PET) [103], aliphatic polyesters, polyurethanes [104], etc., have been employed for the preparation of porous-compartment-separation membranes due to their easily adjusting porosity, surface roughness, and mechanical properties. Exhibiting improved biocompatibility, natural polymers such as collagen, gelatin, or polysaccharides have attracted significant attention lately due to their faithful replication of the native tissues with channel interconnections that allow the perfusion of oxygen and nutrients to simulate the natural behaviors of cell differentiation, spreading, and adhesion along the separating membrane [105]. However, natural polymers show batch-to-batch composition variability and lower mechanical properties, thus affecting the reproducibility of the experiments. However, each employed biomaterial still exhibits weaknesses to achieve the optimal OOC performances.

Polydimethylsiloxane (PDMS) is an elastomeric polymer and one of the most employed synthetic materials for organ-chip production, as it exhibits optical transparency, biocompatibility, gas-permeation properties, flexibility, and allows permanent microscopic observation of 3D cell structures for real-time assessment of tumor behavior and response to therapy [101,106,107]. Microfluidic systems made from PDMS offer a much closer physiological microenvironment to that of the living TME from a biomechanical point, exhibiting porosity and significantly lower stiffness. However, PDMS exhibits an important limitation in chemical screening due to its hydrophobicity, which leads to nonspecific absorption of small molecules such as drugs or pharmaceutical compounds [108,109]. Nevertheless, the limitations of PDMS materials have been overcome through surface modification treatments with another polymer that enhances the hydrophilicity of the PDMS-based materials to facilitate cellular adhesion and migration in microfluidic compartments. One of the most commonly employed surface-modification approaches consists of oxygen-plasma treatment by converting some hydrophobic methyl groups into hydroxyl functionalities via oxidation. It was reported that the hydrophilicity of the oxidized PDMS after plasma treatment can increase by almost 30° [110].

Moreover, the wettability behavior of PDMS materials can be improved by coating the polymer layers with other hydrophilic polymers. The proteins from ECM composition are generally used to achieve the natural moiety of PDMS to ensure cell attachment and proliferation. ECM proteins, such as collagen and fibronectin, create covalent bonding onto the PDMS surface and thus facilitate cell adhesion by modifying the surface roughness of the synthetic polymer [111]. Although the hydrophilicity and biocompatibility are significantly increased this way, dissociation processes during protein coating are associated with this approach [112]. Other biopolymers, such as gelatin [113], fibronectin [114], or hydrophilic synthetic polymers such as polydopamine [115,116] and polyethylene glycol [117] have been successfully employed for modifying the surface roughness of PDMS and managed to enhance the cellular adhesion for short-term cellular culture.

Polycarbonate (PC) structures are commonly used polymeric materials for the OOCs fabrication and as porous membranes for Transwell® inserts. The PC structures are hydrophobic, exhibit high transparency, and are inert and stable under biological processes [118]. Similar to PDMS, the surface of PC is commonly modified by gas plasma treatment or protein coating [119]. However, PC structures exhibit higher rigidity than PDMS. They are not suitable for stretchable membranes or soft substrates.

Other thermoplastics are tested, such as polyolefins [103], styrene-ethylene butylene styrene (SEBS) [120], polyurethanes [121], and copolymers [122], but there is still a demand for novel materials in this domain.

Polyolefins such as cyclic olefin polymers (COP) and cyclic olefin copolymers (COC) are thermoplastic materials that have also been employed for the manufacturing of microfluidic devices, being less gas-permeable than PDMS. COP and COC are lipophobic materials and do not exhibit unspecific adsorption of small molecules, allowing their use in drug-development and diffusion experiments [122]. Moreover, they exhibit optical transparency, good chemical resistance to polar solvents, thermal resistance, and reproductible mass production.

Polymethylmethacrylate (PMMA) is another frequently used material as a substrate [123,124]. Kang and coworkers [125] demonstrated that PMMA chemically attached to porous orbitally etched polyethylene terephthalate (PTFE) membranes is impermeable to small hydrophobic compounds and more consistent results concerning the anticancer vincristine drug cytotoxicity on human lung adenocarcinoma cells cultured in PDMS-based chip. Thus, PMMA shows promising features for tumor chip microdevices fabrication as the small molecules cannot infiltrate the material.

Aliphatic polyesters such as polylactic acid (PLA) and poly (ε-caprolactone) (PCL) have been used to mimic the in vitro blood–brain barrier, functioning as membranes in OOCs microdevices. Both polymers are biodegradable and hydrophobic. Although their surface can be chemically modified to enhance the wettability and cell proliferation, these polymers have been sparsely employed as they tend to degrade under biological processes releasing acidic degradation products that could affect the cellular medium [123].

To fulfill the requirements of mechanical actuatable OOCs, the employed materials need to be optically transparent, flexible, easily moldable, and nonabsorptive to drugs and cell nutrients. Moreover, it is necessary to find sustainable alternative materials. At present, the manufacturing process of OOCs and experimental validation remain high-priced so that less-expensive and reusable materials should be employed.

Gelatin is an animal protein obtained by collagen hydrolysis, and is very much employed in drug-delivery and tissue-engineering applications [126] due to its ability in drug transportation and cell-growth promotion [127]. Gelatin exhibits biocompatibility and flexibility, and forms complexes with proteins, growth-factor nucleotides, and biopolymers, and can be shaped in colorless gels [128]. Additionally, photo-crosslinked gelatin methacrylate (GelMa) is highly used in tissue engineering and shows interest in OOC formation [129,130].

Collagen-based ECM gel (known as Matrigel) is a commercial product that contains ECM hydrogel made of tumor-derived basement membrane proteins employed for cell culture [131]. Matrigel gives structural and signal-transduction functions. Tumor cells exhibit aggressive behavior in Matrigel medium, and it is frequently used to assess the malignancy of cancer cells and observe the mechanism of tumor growth [132].

Bacterial cellulose paper has also attracted interest for tumor-chip applications as it has good biocompatibility, and it is a naturally derived polymer that is densely vascularized through nanofibers [133,134]. Recent advances in bioprinting are multiplying the paths for creating complex perfused systems, and the discovery of new materials that accomplish the full requirements of native living cells is still under extensive development.

3. Kidney-on-Chip

The kidney is the second major target of drugs and chemicals after the liver, receiving 25% of the cardiac output and being responsible for drug and metabolite excretion and fluid–electrolyte balance. Thus, exposure of the kidney to high concentrations of xenobiotics and drugs metabolites leads to drug-induced nephrotoxicity, and furthermore to (i) acute kidney injury (AKI), which is often associated with increased morbidity and mortality; (ii) chronic kidney disease (CKD); and (iii) end-stage renal disease (ESRD) in the general population [135]. The physiological function of the kidney is supported by a very complex architecture of the renal tissue that consists of a 3D distribution of more than ten cell types within an ECM and an elaborated vasculature system with high blood flow [2]. The kidneys are responsible for maintaining the balance of the body fluids and also discarding metabolites regulating filtration, reabsorption, and secretion processes through the nephron, the kidney's functional unit. The nephron consists of (i) the renal corpuscle, which includes the Bowman's space, encapsulating the capillaries (glomerulus); (ii) the proximal renal tubule, which reabsorbs sodium chloride and sodium bicarbonate, completes the reabsorption of glucose, and also is the unique site of amino-acid and anion transport; and finally (iii) the renal distal convoluted tubule, which plays a critical role in sodium, potassium, and divalent cation homeostasis [136,137].

The mechanisms through which various drugs are nephrotoxic are diverse and can be selective or non-selective, depending on whether they injure a specific type or multiple types of cells. Moreover, the nephrotoxic effect can be direct, many drugs having an affinity for specific transporters throughout the nephron and causing cell death in the corresponding segments through a variety of mechanisms, or indirect, secondary to the osmotic effect and hemodynamic changes (ex: iodinated contrast agents), drug-induced nephrolithiasis or ischemia.

3.1. *In Vitro 2D/3D Models for Kidney Disease Modeling*

Current approaches to further testing drugs before entering clinical trials to assess nephrotoxicity are using in vitro models, such as 2D, 3D, and microfluidic models—kidney-on-chip and the more recently introduced approaches based on stem cells [3].

The majority of the 2D models use cell lines distributed in tight monolayers, such as porcine LLC-PK1 cells and Madin–Darby canine kidneys. Human cells such as the renal tubular cell line HK-2 have also been used in vitro, albeit with limited applicability due to their poor expression of the SLC2 transporter family. Moreover, due to the kidney's complex structure sustaining its function, the 2D models (despite being the gold standard for a long time now) are a poor representation of the in vivo environment, including low expression of OATs and OAT2 transporters and poor apical-basal polarization [8]; 3D in vitro models known as organoids, are challenging to develop and are not currently validated for prediction of drugs nephrotoxicity. Most of them mimic only a part of the nephron (glomerulus or tubules) and are tested for a limited number of drugs. One of the challenges in recreating an in vitro nephron model is to step forward from traditional 2D models characterized by rather static conditions to mimic the dynamic conditions such as blood and urinary flow, exposing tubular epithelial cells to fluid shear stress facilitating transepithelial osmotic gradient [138]. For kidney-tissue engineering, the most challenging part is recreating the vascular network. Xuan Mu et al. tried to mimic a nephron by combining fibrillogenesis and liquid molding to build a 3D vascular network in the hydrogel. They succeeded in obtaining a cytocompatible, mechanically stable vascular network that could mimic passive diffusion of organic molecules and is showing promise in simulating physiological functions based on active mass transfer in hydrogels. This would benefit not only the drug screening for nephrotoxicity but also other vasculature-rich organ-related research [139].

3.2. *Kidney-on-Chip Models*

In this view, the kidney-on-chip approach tries to mimic the dynamic, flowing environment more closely and consists of a 2D/3D cellular model developed in a microfluidic device. This whole system finally consists of a 3D architecture built up by renal cells grown on an ECM interface or next to perfusable microchannels where the media and/or other body fluids such as blood or urine can flow across the cell surface [140] (Figure 2). Using these models, studies have shown that fluid shear stress (FSS) has a major influence on renal tubular cells' phenotype [141], hence their importance in nephrotoxicity and other renal-function-related studies. FSS can induce inflammation through immune-cell-mediated activation and monocyte adhesion [142]; it increases the expression of certain genes such as ABCG2, RBP4 (a marker for tubular function loss) [143], CYP1A1, and SLC47A1, affects the cytoskeleton organization, and upregulates the formation of tight junctions [144]. Consequently, kidney-on-chip models are designed for high-throughput screening of drug toxicity by delivering essential output data regarding the glomerular filtration processes, the drugs pharmacokinetics with impact on drugs validation, and relevant dose determination. In this view, several types of kidney-on-chip models have been developed targeting different segments of the renal unit.

3.2.1. Glomerulus-on-Chip

The glomerulus is the filtering unit of the kidney, consisting of a capillary network and podocytes, highly differentiated epithelial cells that are responsible for the actual filtration process of the blood. Considering this, there is an obvious interest in developing a functional system recapitulating the glomerular function to support preclinical stages of new drug development or to sustain research in the kidney disease field, including (but not limited to) kidney neoplasms. Despite the efforts made in the past few years to recreate a glomerulus-on-chip system, the main challenge in developing an in vitro glomerulus model is the lack of functional human podocytes. Various studies have been performed to create a glomerulus-on-chip model by obtaining human-induced pluripotent stem-cell (iPSCs)-derived podocytes and placing them in a microfluidic device together with endothelial cells (primary or secondary iPSC-derived). These microfluidic devices were designed to create the two compartments separated by a porous membrane. Musah et al. showed that podocytes extended their processes through the membrane when exposed to constant flow and mechanical strain, and selective filtration was proven by the presence of excreted inulin and retention of albumin on each side of the membrane [145]. Roye et al. also tried to build a glomerulus-on-chip with hPSC-derived podocyte and endothelial cells, and succeeded in obtaining essential functionality and structure [146]. Both models were carried out to study the effect of adryamicin as a chemotherapeutic agent, and showed that the treatment induced podocyte detachment and endothelial-barrier disruption, leading to albuminuria. This glomerulus-on-chip model could be a promising start for an in vitro study of proteinuria and glomerular kidney disease and chemotherapeutic drug-nephrotoxicity assessment. Moreover, in 2017 Wang et al. developed a glomerulus-on-chip model using rat cells to create a diabetic nephropathy model that also served to assess the hyperglycemic pathological response [147].

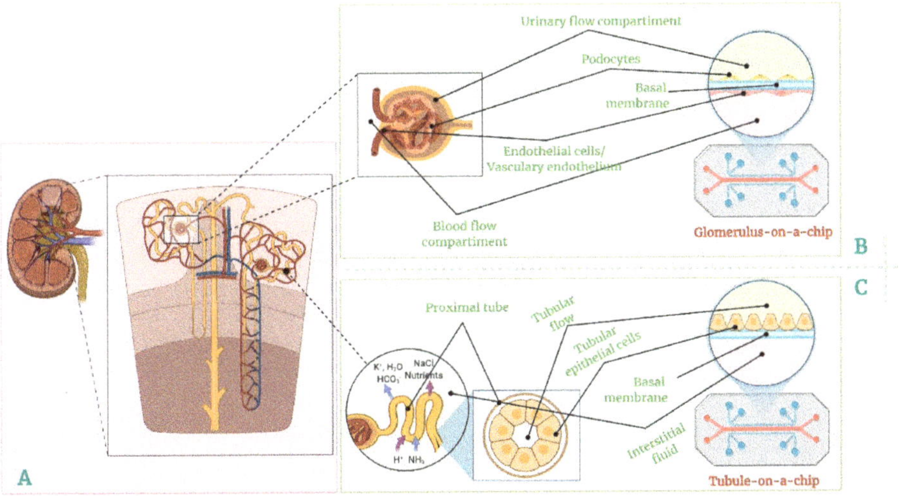

Figure 2. Complex kidney-on-a-chip for personalized medicine: (**A**) kidney and nephron, (**B**) glomerulus-on-a-chip, and (**C**) tubule-on-a-chip.

3.2.2. Tubule-on-Chip

Most kidney-on-chip studies use proximal tubule cells, as this part of the nephron is the primary site of drug clearance and a critical target for drug-induced nephrotoxicity. Therefore, the in vitro reproduction of proximal tubule function is of major interest in the preclinical assessment of candidate compounds. For this, hollow fibers are used as scaffolds to design a tubular model that resembles the structure of the proximal tubule. Such models were developed by coating the hollow fibers with a hydrogel that served as ECM for human proximal-tube endothelial cells (hPTECs) [148]. This approach enabled the observation of the secretory clearance of albumin-bound uremic toxins and albumin reabsorption [149]. Other studies developed a proximal tubule-on-chip using a polyester membrane to split the main compartment of the device into a luminal-like channel and a basal interstitial-like channel that were populated with PTECs seeded on an ECM coating [90]. This study reports significant changes in PTEC physiology and morphology, such as polarization, display of columnar shape, primary cilia, and transporters due to the simulated FSS induced within the system, and highlights the impossibility to achieve all these under static-culture conditions. These observations were also confirmed by other studies [150,151]. Another strategy in achieving proximal tubule-on-chip systems is bioprinting of tubular 3D architectures [152]. Homan et al. used this technique to obtain functional proximal tubule-on-chips by growing PTECs on a fibrinogen ECM coating the lumen of a structure printed with Pluronic ink [153].

One of the first kidney-on-chip models for nephrotoxicity screening used primary renal tubular cells to assess cisplatin toxicity, and the results most resembled in vivo models [154]. Other nephrotoxicity studies performed on kidney-on-chip models include the study of ifosfamide and acrolein nephrotoxicity assessment performed by Le Clerc et al. [155], the gentamicin kinetics study using MDCK on a porous membrane coated by fibronectin [156], and the toxicological assessment of polymyxins and their nephrotoxic potential on a 3D MPS human model [151].

Lastly, one interesting approach in nephrotoxicity testing is using versatile high-throughput screening platforms such as the OrganoPlate™ system powered by Mimetas for robust and reproducible results in preclinical studies. In this view, several studies report the expression and function of renal transporters and cell-polarization response to cisplatin [157] by modeling the proximal tubule functions in independent chips placed in a standard microtiter plate [158].

3.3. *Multiorgan-on-Chip Models That Include Kidney-on-Chip*

Taking the use of OOC models even to the next level, recent studies report the development of multiorgan-on-chip or body-on-chip models as platforms enabling the screening of multiorgan toxicity. In particular, integrating a kidney-on-chip device into a multiorgan system would bring valuable insights regarding secondary toxicity of drugs beyond systemic toxicity and the inflammatory response. Maschmayer et al. created a four tissue coculture (skin, small intestine, kidney, and liver), with 28 days of reproducible capacity, on a microphysiological four-organ-chip model, which is a promising line of research for pharmacodynamic and pharmacokinetic drug parameters and toxic profile [159]. Chang et al. used a combination of liver and kidney cells on a chip platform to study the nephrotoxic effect of aristolochic acid, a nephrotoxic agent that first needs to be activated in the liver [160].

4. Bladder-on-Chip

The urinary bladder is a hollow organ located in the lower abdomen that acts as a reservoir for the temporary storage of urine received from kidneys, further expelled during micturition through the urethra [161]. As a structure, the urinary bladder is composed primarily of smooth muscle, collagen, and elastin [162]. Microscopically, the urinary bladder has a stratified architecture, organized in the following layers: lining epithelium (urothelium), lamina propria, muscularis propria, and serosa, the urothelium acting therefore as a barrier that is exposed to potential carcinogens [163]. Among urinary bladder diseases, cystitis and cancer are the most common; thus, the existing OOC platforms developed until now address these pathologies, with a significantly limited number of published studies available that show the OOC platforms as tools for bladder research are yet to be explored. The use of bladder-on-chip for disease modeling could pave the way for a better knowledge of the disease physiopathology and could advance research by accelerating the discovery of novel drugs, as well as by significantly improving drug-efficacy studies.

4.1. *Modeling Bladder Cancer*

Bladder cancer (BC) is the most common cancer of the urinary system and ranks globally as the 10th most common type of cancer, a pathology that affects men more than women [164,165]. BC is defined as a carcinoma of the urothelial cells and can be divided by the tumor depth of bladder-wall invasion in non-muscle-invasive BC (NMIBC), muscle-invasive BC (MIBC), and metastatic BC, subtypes that are characterized by different molecular signatures, BC being one of the most frequently mutated human cancers [166,167]. While the approval of immunotherapy for both NMIBC and MIBC, as well as targeted therapy (erdafitinib) and antibody-drug conjugate therapy (enfortumab vedotin) significantly improved the therapeutical management of BC, the alarming statistics regarding the disease recurrence and overall survival of BC patients highlight the emerging need for developing better preclinical study models [168]. At the moment, the most intensively used preclinical models for BC research are in vitro 2D cell cultures (cell lines, conditionally reprogrammed cell cultures) and 3D organoids, and in vivo carcinogen-induced mouse models, genetically engineered mouse models, and patient-derived xenograft. However, each preclinical model presents unique features, together with different disadvantages such as failure to mimic the 3D tumor microarchitecture, microenvironment, and tumor heterogeneity; and lack of immune system, thus failing to ease the transition between preclinical models and clinics [169–172]. For example, a highly neglected aspect of many preclinical models is the presence of the tumor microenvironment consisting of malignant and nonmalignant cells (cancer-associated fibroblasts, cancer-associated immune cells), ECM, and blood vessels, an entity that orchestrates cancer progression and modulates therapeutic response [20].

In this view, Liu and colleagues [173] aimed to reconstruct the bladder microenvironment by coculturing four types of cells into a two-layer microfluidic device made of a PDMS piece and a glass slide. The fabricated microfluidic device connected to perfusion equipment featured four indirectly connected cell-culture chambers (BC cells, fibroblasts, macrophages, endothelial cells), ECM channel units (Matrigel), and culture-medium channels. Therefore, using this microfluidic device, four types of cells were allowed to simultaneously interact through soluble biological factors and metabolites that proved to diffuse through the ECM channel units between cell-culture chambers, in a dynamic setup provided by continuous medium perfusion. The validated microfluidic system proved to be a good platform for cell-motility patterns and phenotypic alteration of stromal cells, as well by generation of reticular structures based on BC cells, and opened the perspective of OOC implementation for precision medicine, as BC cells treated with different clinical neo-adjuvant chemotherapy showed different treatment responses, revealing the drug sensitivity of tumor cells in this experimental setup. In another study, a microfluidic chip was designed to coculture BC cells and fibroblasts and further analyze changes in mitochondrial-related protein expression of these cells and their characteristics of energy metabolism [174].

Therefore, to investigate tumor metabolism, the microfluidic chip fabricated was composed of four cell-culture pools, two for BC cells and two for fibroblasts, interconnected through two ECM microchannels (Matrigel) and two microchannels with the outside, which were continuously fed with cell-culture medium injected through the peripheral perfusion channel. By comparing the coculture system with individual cultures of BC cells and fibroblasts, significant differences in lactic-acid concentration and mitochondrial-related protein expression were observed; results that revealed that cells conduct glycolysis more efficiently under coculture conditions, and show enhanced overall mitochondrial activity and protein expression, highlighting the importance of using OOC in bladder tumor energy-metabolism studies. Finally, Lee et al. [175] designed a simple 3D microfluidic device based on PDMS and Matrigel to culture MIBC cells and NMIBC cells, for modeling metastasis. The study showed an increased expression of CD44 and RT4 after 2 weeks of culture in MIBC cells as compared with NMIBC cells, associated with a significant increase of MMP-9 gelatin degradation, showing that the OOC system could be further employed for migration and metastasis studies.

4.2. Modeling Urinary Tract Infections

On the other hand, urinary tract infections (UTI) are among the most common bacterial infections, divided into uncomplicated and complicated infections, isolated infections of the bladder being referred to as cystitis and treated subsequently with antibiotics [176]. However, despite completing the antibiotic-based treatment regimen, cystitis is characterized by a high recurrence frequency, which involves readministration of antibiotics [177]. UTI is most frequently caused by uropathogenic *E. coli* (UPEC) bacteria, which underlies more than 80% of the diagnosed cases [178]. Once entering the urinary bladder, the *E. coli* can float freely in the urine and be eliminated through micturition or form intracellular bacterial communities (IBC) that the bladder struggles to clear [179]. Therefore, to reveal the insights of UPEC infection and IBC formation, Sharma and colleagues [180] used a commercially available OOC purchased from Emulate to model UPEC infection and mimic the interface between the blood vessels and the tissue layers of the human bladder. To reconstruct the bladder native architecture, human bladder microvascular endothelial cells were cocultured with human bladder epithelial cells and exposed dynamically to urine and nutritive cell-culture media, while using the application and release of linear strain to recreate micturition. By infecting the epithelial layer underflow with UPEC and monitoring OOC by microscopy, the bacteria motility, interaction with cells, and IBC formation could be monitored. Moreover, the addition of human-blood isolated neutrophils into the OOC system revealed that their diapedesis to sites of infection on the epithelial side can lead to the formation of neutrophil swarms and neutrophil extracellular traps (NETs), and that IBCs offer substantial protection to bacteria from antibiotic clearance. Administration of antibiotics in the developed bladder-on-a-chip model through urine revealed their potential to kill bacteria floating freely in the urine much faster than bacteria residing in bladder cells, as well as the increased resistance of IBCs to treatment, aspect that highlight the importance of completely eradicating IBCs to avoid infection recurrence. No doubt, this complex study shows the potential of the bladder-on-chip as relevant platforms for modeling infections, as well as their use for drug screening of multiple antibiotics or novel drugs in a physiologically relevant manner.

5. Prostate-on-Chip

The prostate is an exocrine gland that secretes sperm-nourishing and protective fluid. It is located beneath the urinary bladder and part of the male reproductive system [181]. The prostate-tissue architecture consists of ducts and acini lined with glandular or secretory cells on top of basal cells and surrounded by a fibromuscular stroma [182,183]. Consequently, the prostate displays two major components: one is the epithelium, containing neuroendocrine cells and basal cells that express integrins and hold the differentiation potential toward luminal (secretory) cells that will express androgen receptor protein [184,185]; the other component of the prostate tissue is the stroma, which is separated from the epithelium by a basement membrane. Prostate cancer is the most common cancer diagnosed worldwide in men and is characterized by molecular changes caused by genetic and epigenetic modifications leading to malignant transformation of the cells [186,187]. Androgens regulate prostate development and cell differentiation from embryonic development to adults, and thus, a better understanding of early interactions between prostate cells leading to development would be essential in unraveling the mechanisms underlying prostate cancer [185]. Many studies in this field have been developed, starting with 2D conventional cultures and 3D models, and ending with versatile proposals of prostate-on-chip systems. Likewise, for other previously discussed pathologies, in prostate cancer, 2D cell cultures fail to provide the proper tissue-level complexity, not only in their limitation to one cell type but also due to their missing crucial TME aspects, as well as FSS simulation [188]. The schematic representation of the prostate-on-chip model is presented in Figure 3.

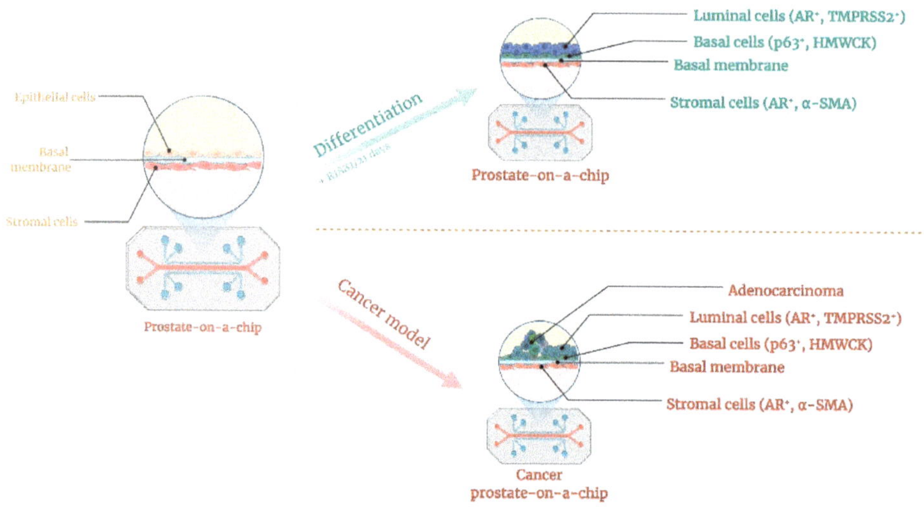

Figure 3. Prostate-on-chip.

On the other hand, animal models that do display the in vivo tissue-level complexity fail to mimic human anatomy or physiology [189]. To address this issue, various engineered microfluidic devices have been proposed for concomitant multiple cell-culture types. For example, Picollet-D'hahan, N. et al. used polyelectrolyte nanofilms to create a model of prostate epithelial cancer cells (PC3) and normal prostate epithelial cells (PNT-2) [190]. Furthermore, Jiang et al. [191] reported the development of a human prostate gland model for a better understanding of the epithelium/stroma interaction, both in terms of unrevealing the R1881-mediated epithelial cell (PrEC)-differentiation mechanism towards functional secretory cells and coculturing of these cells with prostate stromal cells (BHPrS1). For this, they designed a microfluidic device fabricated by soft-lithography techniques from Polydimethylsiloxane (PDMS), which incorporates a commercial polyester membrane of 0.8 μm-diameter pores, 1% porosity, and a thickness of 23 μm. The major advantage of this device is its ability to overcome the impossibility of establishing long-term 2D static cocultures of these cell types due to their different culture-media requirements [191]. Moreover, this device was able to sustain an in vivo-like behavior of the cells that displayed groups forming gland-like buds.

6. Benefits and Challenges of OOCs/TOCs in Urology

Experimental research demonstrated that microfluidic OOC technology might function to screen recently developed anticancer compounds, cellular and nanotechnology-based treatments, improve therapy conditions, and establish the effects (or side effects) of combined therapies in in vivo-like TME (Figure 4).

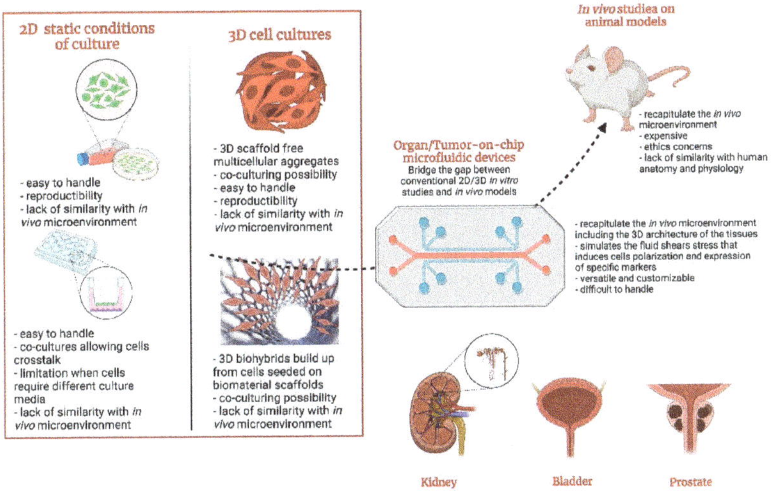

Figure 4. Schematic representation of the available in vitro and in vivo models for the urology-associated pathology study, highlighting their advantages and disadvantages and revealing the potential key role of the organ/tumor-on-chip devices to bridge the gap between conventional 2D/3D culture systems and animal models in preclinical studies.

Imitating organ-level pathophysiology found in vivo is the key factor in developing successful OOC models and requires clear efficacy validation. One of the main challenges in OOC fabrication and usage remains the straightforward optimization to functionally connect multiple organ systems in order to collect media from the output of the microdevice and feed it into another to accurately mimic the sequential adsorption, distribution, metabolism, and excretion of the compounds. The microfluidic technology allows perfusion of cell-culture medium through or across tissue structures and maintains biomechanical stimuli in a controlled manner (for instance, continuous, intermittent, or cyclic). Moreover, during an experiment, the easy access to cells or media for dosing, sampling, and analysis with small-molecule compounds (antibodies, hormones, drugs, etc.) offers an unquestionable advantage over the animal and other previous static models, including spheroids and organoid cultures. Previous studies already showed that TOCs mimic the native environment of cells, including 3D topology or physical stretch and strain, and they are able to maintain precise control over inter-and intra-organ flow rates, forming a miniaturized version of the human body. Additionally, bladder-on-chip models allow modeling of urinary infections by simulating *E. coli* bacterial infection of the urinary tract cells. These insights are crucial in the framework of studies that focus on in vivo processes such as angiogenesis, tumorigenesis, invasion, metastasis, and infection, enabling in vitro analysis of how local microenvironmental signals and chemical gradients influence these processes [192]. The existence of a perfused endothelium-lined capillarity endorses the superiority of OOCs over static models by providing precise control over cell-culture conditions and over platform pharmacokinetics and drug toxicity, respectively. The ability to adjust key parameters such as concentration gradients, cell patterning, tissue–organ interactions, and to replicate an organ-like mechanical microenvironment that exists in human living TME placed OOCs in the "Top 10 Emerging Technologies" at the World Economic Forum [193].

The multiple benefits of OOCs have made them indisputable alternatives for animal studies or 3D culture assays. However, clinical trials require time, and OOCs are still competing with animal testing to predict clinical responses. Moreover, OOCs are not considered as easy to be handled as conventional cell models. Although they allow long-term experiment and minimal user input, the technical robustness remains a challenge, as the compact and complex microfluidic devices are engaging multiple parameters that simultaneously run to achieve optimal functionality. Air-bubble formation or unintended infection risks at connection points are the most common factors that may lead to experiment failure. Shourabi et al. [194] recently designed an optimized integrated microfluidic gradient generator for mechanical stimulation and drug delivery by dynamic culturing of human lung cancer cells (A549 cell line) made from two PDMS-layer microfluidic chips with a porous membrane interfacing in between. The key feature of this system consists in two bubble trappers designed to remove the unwanted bubbles that could enhance the shear stress on cells. In this way, the concentration gradient generator's performance is guaranteed, and precise control of fluid shear stress on cells is obtained. The results showed that the chip exhibits a high cell-viability rate (95%) and will be employed to study the toxicity effect of different concentrations of cisplatin on renal cells.

Microfluidic OOC microdevices can be also employed for studying the role of cancer stem cells (CSCs) in tumor initiation and cancer relapse [195]. As the chemotherapeutic regiments kill tumor cells but are ineffective on CSCs, the microfluidic OOCs have the potential to give insights on tumor cell–CSC interaction, as well as accelerating drug development to also target this subpopulation of cells [196]. By employing a biomimetic in vitro model such as OOCs for CSC research, various environmental parameters such as oxygen gradient, glucose, and fluid shear stress can be mimicked, offering valuable insights on CSCs' multicellular interaction and drug-resistance mechanisms to conventional therapies [197].

However, there is still a place for developing new materials that facilitates the non-invasive assessment of drug effects and meets the industrial-scale production demands. Novel materials based on biopolymers have been developed in the last few years to overcome the main drawbacks of PDMS-based OOC microdevices, and significant attempts to produce scalable systems have been reported [198]. Therefore, as the technology has evolved in the last years and the introduction of commercial prefabricated tumor chips that are user-friendly and custom-made for culture and fluidic control is expanding [199], it will undoubtedly facilitate the possibilities of medical research for oncology treatments.

7. Conclusions and Future Considerations

Human OOCs/TOCs offer a promising alternative to conventional 2D/3D static cultures, and even to in vivo models that lack similarity with human anatomy and/or physiology. These microfluidic devices hold the potential to sustain relevant preclinical studies for developing new therapies, including anticancer drugs, and bring new insights for a better understanding of pathogenesis mechanisms. Replicating the physiological TME, TOC models are seen as promising and more realistic alternatives for investigating the metastasis, distribution, and mechanism of tumor propagation. Involving microfluidic technologies, OOC models can mimic the complexity of in vivo tumors and can be more accurately used to predict therapeutic efficacy and drug toxicity or side effects. Although OOCs closely mimic the native organ functions, one must consider that most of the cell cultures interact with polymeric substrates and porous polymeric membranes to replicate the physiological microenvironments. Thus, the polymer material properties such as surface roughness, wettability, and mechanical properties significantly affect cell adhesion and proliferation.

Moreover, size porosity and biodegradability influence cell migration and viability. While most of the studies present the use of common synthetic polymers beneficial for OOCs fabrication, other studies employ less common polymeric structures with improved cytocompatibility, wettability, or mechanical stability. Although there is a place for deeper investigations to optimize the materials' properties and fabrication methods, the OOC/TOC models are adding necessary steps to personalized medicine for creating high-precision remedies with the possibility to use biopsy from the patient that could be expanded in vitro and evaluated for cost-effective screening of treatments that are particularly efficient for that patient.

Author Contributions: B.G., A.H., M.C. and E.S. reviewed the literature with respect to kidney-on-chip models, bladder-on-chip models, and prostate-on-chip models; wrote Sections 3–5, contributed to Sections 1, 6 and 7; and illustrated Figures 2–4. E.I.B., H.I., R.-C.P. and C.Z. reviewed the literature regarding OOC/TOC fabrication; wrote Sections 1, 2, 6 and 7; and illustrated Figure 1. V.J. coordinated and edited the entire work. All authors have read and agreed to the published version of the manuscript.

Funding: This work was supported by a grant of the Romanian Ministry of Education and Research, CNCS-UEFISCDI, project number PN-III-P1-1.1-PD-2019-0955, within PNCDI III.

Institutional Review Board Statement: Not applicable.

Informed Consent Statement: Not applicable.

Data Availability Statement: Data are contained within the article.

Conflicts of Interest: The authors declare no conflict of interest.

References

1. Faria, J.; Ahmed, S.; Gerritsen, K.G.; Mihaila, S.M.; Masereeuw, R. Kidney-Based in Vitro Models for Drug-Induced Toxicity Testing. *Arch. Toxicol.* **2019**, *93*, 3397–3418. [CrossRef] [PubMed]
2. Zanetti, F. Kidney-on-a-Chip. In *Organ-on-a-Chip*; Elsevier: Amsterdam, The Netherlands, 2020; pp. 233–253.
3. Tiong, H.Y.; Huang, P.; Xiong, S.; Li, Y.; Vathsala, A.; Zink, D. Drug-Induced Nephrotoxicity: Clinical Impact and Preclinical in Vitro Models. *Mol. Pharm.* **2014**, *11*, 1933–1948. [CrossRef] [PubMed]
4. Day, C.-P.; Merlino, G.; Van Dyke, T. Preclinical Mouse Cancer Models: A Maze of Opportunities and Challenges. *Cell* **2015**, *163*, 39–53. [CrossRef] [PubMed]
5. Berlin, J.A.; Glasser, S.C.; Ellenberg, S.S. Adverse Event Detection in Drug Development: Recommendations and Obligations beyond Phase 3. *Am. J. Public Health* **2008**, *98*, 1366–1371. [CrossRef] [PubMed]
6. Sun, W.; Luo, Z.; Lee, J.; Kim, H.; Lee, K.; Tebon, P.; Feng, Y.; Dokmeci, M.R.; Sengupta, S.; Khademhosseini, A. Organ-on-a-chip for Cancer and Immune Organs Modeling. *Adv. Healthc. Mater.* **2019**, *8*, 1801363. [CrossRef] [PubMed]
7. Taylor, D. The Pharmaceutical Industry and the Future of Drug Development. In *Pharmaceuticals in the Environment*; Royal Society of Chemistry: London, UK, 2015; pp. 1–33. [CrossRef]
8. Soo, J.Y.-C.; Jansen, J.; Masereeuw, R.; Little, M.H. Advances in Predictive in Vitro Models of Drug-Induced Nephrotoxicity. *Nat. Rev. Nephrol.* **2018**, *14*, 378–393. [CrossRef] [PubMed]

9. Akhtar, A. The Flaws and Human Harms of Animal Experimentation. *Camb. Q. Healthc. Ethics* **2015**, *24*, 407–419. [CrossRef]
10. Bracken, M.B. Why Animal Studies Are Often Poor Predictors of Human Reactions to Exposure. *J. R. Soc. Med.* **2009**, *102*, 120–122. [CrossRef] [PubMed]
11. Baumans, V. Use of Animals in Experimental Research: An Ethical Dilemma? *Gene Ther.* **2004**, *11*, S64–S66. [CrossRef]
12. Festing, S.; Wilkinson, R. The Ethics of Animal Research: Talking Point on the Use of Animals in Scientific Research. *EMBO Rep.* **2007**, *8*, 526–530. [CrossRef] [PubMed]
13. European Parliament. Available online: Https://Www.Europarl.Europa.Eu/Plenary/En/Vod.Html?Mode=chapter&vodLanguage=EN&vodId=6ea360e5-0dd3-Decd-2a72-642c028c0a34&date=20210708# (accessed on 25 February 2022).
14. American Institute for Cancer Research. Available online: https://www.wcrf.org/dietandcancer/worldwide-cancer-data/2020 (accessed on 25 February 2022).
15. World Health Organization. Available online: https://gco.iarc.fr/tomorrow/en/dataviz/bars?mode=cancer&group_populations=1&multiple_cancers=1&cancers=39_27_30_29&key=total&show_bar_mode_prop=1&populations=903_904_905_908_909_935&years=2040&sexes=1&types=02020 (accessed on 25 February 2022).
16. Trujillo-de Santiago, G.; Flores-Garza, B.G.; Tavares-Negrete, J.A.; Lara-Mayorga, I.M.; González-Gamboa, I.; Zhang, Y.S.; Rojas-Martínez, A.; Ortiz-López, R.; Álvarez, M.M. The Tumor-on-Chip: Recent Advances in the Development of Microfluidic Systems to Recapitulate the Physiology of Solid Tumors. *Materials* **2019**, *12*, 2945. [CrossRef] [PubMed]
17. Chaicharoenaudomrung, N.; Kunhorm, P.; Noisa, P. Three-Dimensional Cell Culture Systems as an in Vitro Platform for Cancer and Stem Cell Modeling. *World J. Stem Cells* **2019**, *11*, 1065. [CrossRef]
18. Przekwas, A.; Somayaji, M.R. Computational Pharmacokinetic Modeling of Organ-on-Chip Devices and Microphysiological Systems. *Organ-A-Chip* **2020**, 311–361. [CrossRef]
19. Sontheimer-Phelps, A.; Hassell, B.A.; Ingber, D.E. Modelling Cancer in Microfluidic Human Organs-on-Chips. *Nat. Rev. Cancer* **2019**, *19*, 65–81. [CrossRef] [PubMed]
20. Whiteside, T. The Tumor Microenvironment and Its Role in Promoting Tumor Growth. *Oncogene* **2008**, *27*, 5904–5912. [CrossRef] [PubMed]
21. Hoarau-Véchot, J.; Rafii, A.; Touboul, C.; Pasquier, J. Halfway between 2D and Animal Models: Are 3D Cultures the Ideal Tool to Study Cancer-Microenvironment Interactions? *Int. J. Mol. Sci.* **2018**, *19*, 181. [CrossRef] [PubMed]
22. Liu, X.; Fang, J.; Huang, S.; Wu, X.; Xie, X.; Wang, J.; Liu, F.; Zhang, M.; Peng, Z.; Hu, N. Tumor-on-a-Chip: From Bioinspired Design to Biomedical Application. *Microsyst. Nanoeng.* **2021**, *7*, 50. [CrossRef] [PubMed]
23. Low, L.A.; Mummery, C.; Berridge, B.R.; Austin, C.P.; Tagle, D.A. Organs-on-Chips: Into the next Decade. *Nat. Rev. Drug Discov.* **2021**, *20*, 345–361. [CrossRef]
24. Ingber, D.E. Developmentally Inspired Human 'Organs on Chips. *Development* **2018**, *145*, dev156125. [CrossRef]
25. Brill-Karniely, Y.; Dror, D.; Duanis-Assaf, T.; Goldstein, Y.; Schwob, O.; Millo, T.; Orehov, N.; Stern, T.; Jaber, M.; Loyfer, N. Triangular Correlation (TrC) between Cancer Aggressiveness, Cell Uptake Capability, and Cell Deformability. *Sci. Adv.* **2020**, *6*, eaax2861. [CrossRef]
26. Rahmanuddin, S.; Korn, R.; Cridebring, D.; Borazanci, E.; Brase, J.; Boswell, W.; Jamil, A.; Cai, W.; Sabir, A.; Motarjem, P. Role of 3D Volumetric and Perfusion Imaging for Detecting Early Changes in Pancreatic Adenocarcinoma. *Front. Oncol.* **2021**, *11*, 678617. [CrossRef] [PubMed]
27. Gilkes, D.M.; Semenza, G.L.; Wirtz, D. Hypoxia and the Extracellular Matrix: Drivers of Tumour Metastasis. *Nat. Rev. Cancer* **2014**, *14*, 430–439. [CrossRef] [PubMed]
28. Ehsan, S.M.; Welch-Reardon, K.M.; Waterman, M.L.; Hughesbce, C.C.; George, S.C. A three-dimensional in vitro model of tumor cell intravasation. *Integr. Biol (Camb).* **2014**, *6*, 603–610. [CrossRef] [PubMed]
29. Chen, Y.-A.; King, A.D.; Shih, H.-C.; Peng, C.-C.; Wu, C.-Y.; Liao, W.-H.; Tung, Y.-C. Generation of Oxygen Gradients in Microfluidic Devices for Cell Culture Using Spatially Confined Chemical Reactions. *Lab Chip* **2011**, *11*, 3626–3633. [CrossRef] [PubMed]
30. Shih, H.-C.; Lee, T.-A.; Wu, H.-M.; Ko, P.-L.; Liao, W.-H.; Tung, Y.-C. Microfluidic Collective Cell Migration Assay for Study of Endothelial Cell Proliferation and Migration under Combinations of Oxygen Gradients, Tensions, and Drug Treatments. *Sci. Rep.* **2019**, *9*, 8234. [CrossRef]
31. Bhatia, S.N.; Ingber, D.E. Microfluidic Organs-on-Chips. *Nat. Biotechnol.* **2014**, *32*, 760–772. [CrossRef]
32. Barrile, R.; van der Meer, A.D.; Park, H.; Fraser, J.P.; Simic, D.; Teng, F.; Conegliano, D.; Nguyen, J.; Jain, A.; Zhou, M. Organ-on-chip Recapitulates Thrombosis Induced by an Anti-CD154 Monoclonal Antibody: Translational Potential of Advanced Microengineered Systems. *Clin. Pharmacol. Ther.* **2018**, *104*, 1240–1248. [CrossRef]
33. Ferrari, E.; Nebuloni, F.; Rasponi, M.; Occhetta, P. Photo and Soft Lithography for Organ-on-Chip Applications. In *Organ-on-a-Chip*; Springer: Berlin/Heidelberg, Germany, 2022; pp. 1–19.
34. Puryear, J.R., III; Yoon, J.-K.; Kim, Y. Advanced Fabrication Techniques of Microengineered Physiological Systems. *Micromachines* **2020**, *11*, 730. [CrossRef]
35. Sticker, D.; Rothbauer, M.; Lechner, S.; Hehenberger, M.-T.; Ertl, P. Multi-Layered, Membrane-Integrated Microfluidics Based on Replica Molding of a Thiol–Ene Epoxy Thermoset for Organ-on-a-Chip Applications. *Lab Chip* **2015**, *15*, 4542–4554. [CrossRef]
36. Tajeddin, A.; Mustafaoglu, N. Design and Fabrication of Organ-on-Chips: Promises and Challenges. *Micromachines* **2021**, *12*, 1443. [CrossRef]

37. Shrestha, J.; Ghadiri, M.; Shanmugavel, M.; Razavi Bazaz, S.; Vasilescu, S.; Ding, L. A Rapidly Prototyped Lung-on-a-Chip Model Using 3D-Printed Molds. *Organs-A-Chip* **2019**, *1*, 100001. [CrossRef]
38. Yang, Q.; Lian, Q.; Xu, F. Perspective: Fabrication of Integrated Organ-on-a-Chip via Bioprinting. *Biomicrofluidics* **2017**, *11*, 031301. [CrossRef] [PubMed]
39. Kasi, D.G.; de Graaf, M.N.; Motreuil-Ragot, P.A.; Frimat, J.-P.; Ferrari, M.D.; Sarro, P.M.; Mastrangeli, M.; van den Maagdenberg, A.M.; Mummery, C.L.; Orlova, V.V. Rapid Prototyping of Organ-on-a-Chip Devices Using Maskless Photolithography. *Micromachines* **2022**, *13*, 49. [CrossRef]
40. Christoffersson, J.; Mandenius, C.-F. Fabrication of a Microfluidic Cell Culture Device Using Photolithographic and Soft Lithographic Techniques. In *Cell-Based Assays Using iPSCs for Drug Development and Testing*; Springer: Berlin/Heidelberg, Germany, 2019; pp. 227–233.
41. Ferreira, D.A.; Rothbauer, M.; Conde, J.P.; Ertl, P.; Oliveira, C.; Granja, P.L. A Fast Alternative to Soft Lithography for the Fabrication of Organ-on-a-Chip Elastomeric-Based Devices and Microactuators. *Adv. Sci.* **2021**, *8*, 2003273. [CrossRef] [PubMed]
42. Huh, D.; Hamilton, G.; Ingber, D. From 3D cell culture to organs-on-chips. *Trends Cell Biol.* **2011**, *21*, 745–754. [CrossRef]
43. Chiadò, A.; Palmara, G.; Chiappone, A.; Tanzanu, C.; Pirri, C.F.; Roppolo, I.; Frascella, F. A Modular 3D Printed Lab-on-a-Chip for Early Cancer Detection. *Lab Chip* **2020**, *20*, 665–674. [CrossRef] [PubMed]
44. Samadian, H.; Jafari, S.; Sepand, M.; Alaei, L.; Malvajerd, S.S.; Jaymand, M.; Ghobadinezhad, F.; Jahanshahi, F.; Hamblin, M.; Derakhshankhah, H. 3D Bioprinting Technology to Mimic the Tumor Microenvironment: Tumor-on-a-Chip Concept. *Mater. Today Adv.* **2021**, *12*, 100160. [CrossRef]
45. Lantada, A.D.; Pfleging, W.; Besser, H.; Guttmann, M.; Wissmann, M.; Plewa, K.; Smyrek, P.; Piotter, V.; García-Ruíz, J.P. Research on the Methods for the Mass Production of Multi-Scale Organs-on-Chips. *Polymers* **2018**, *10*, 1238. [CrossRef] [PubMed]
46. Gröger, M.; Dinger, J.; Kiehntopf, M.; Peters, F.T.; Rauen, U.; Mosig, A.S. Preservation of Cell Structure, Metabolism, and Biotransformation Activity of Liver-on-chip Organ Models by Hypothermic Storage. *Adv. Healthc. Mater.* **2018**, *7*, 1700616. [CrossRef] [PubMed]
47. Shaegh, S.A.M.; Pourmand, A.; Nabavinia, M.; Avci, H.; Tamayol, A.; Mostafalu, P.; Ghavifekr, H.B.; Aghdam, E.N.; Dokmeci, M.R.; Khademhosseini, A. Rapid Prototyping of Whole-Thermoplastic Microfluidics with Built-in Microvalves Using Laser Ablation and Thermal Fusion Bonding. *Sens. Actuators B Chem.* **2018**, *255*, 100–109. [CrossRef]
48. Thakare, K.; Jerpseth, L.; Pei, Z.; Elwany, A.; Quek, F.; Qin, H. Bioprinting of Organ-on-Chip Systems: A Literature Review from a Manufacturing Perspective. *J. Manuf. Mater. Process.* **2021**, *5*, 91. [CrossRef]
49. Cao, X.; Ashfaq, R.; Cheng, F.; Maharjan, S.; Li, J.; Ying, G.; Hassan, S.; Xiao, H.; Yue, K.; Zhang, Y.S. A Tumor-on-a-chip System with Bioprinted Blood and Lymphatic Vessel Pair. *Adv. Funct. Mater.* **2019**, *29*, 1807173. [CrossRef] [PubMed]
50. O'Cearbhaill, E. 3D Bioprinting Chips Away at Glioblastomal Resistance. *Sci. Transl. Med.* **2019**, *11*, eaax1724. [CrossRef]
51. Bishop, E.S.; Mostafa, S.; Pakvasa, M.; Luu, H.H.; Lee, M.J.; Wolf, J.M.; Ameer, G.A.; He, T.-C.; Reid, R.R. 3-D Bioprinting Technologies in Tissue Engineering and Regenerative Medicine: Current and Future Trends. *Genes Dis.* **2017**, *4*, 185–195. [CrossRef] [PubMed]
52. Tan, B.; Gan, S.; Wang, X.; Liu, W.; Li, X. Applications of 3D Bioprinting in Tissue Engineering: Advantages, Deficiencies, Improvements, and Future Perspectives. *J. Mater. Chem. B* **2021**, *9*, 5385–5413. [CrossRef] [PubMed]
53. Papaioannou, T.G.; Manolesou, D.; Dimakakos, E.; Tsoucalas, G.; Vavuranakis, M.; Tousoulis, D. 3D Bioprinting Methods and Techniques: Applications on Artificial Blood Vessel Fabrication. *Acta Cardiol. Sin.* **2019**, *35*, 284.
54. Colosi, C.; Shin, S.R.; Manoharan, V.; Massa, S.; Costantini, M.; Barbetta, A.; Dokmeci, M.R.; Dentini, M.; Khademhosseini, A. Microfluidic Bioprinting of Heterogeneous 3D Tissue Constructs Using Low-viscosity Bioink. *Adv. Mater.* **2016**, *28*, 677–684. [CrossRef]
55. Keenan, T.M.; Folch, A. Biomolecular Gradients in Cell Culture Systems. *Lab Chip* **2008**, *8*, 34–57. [CrossRef]
56. Nagy, J.; Chang, S.H.; Dvorak, A.M.; Dvorak, H.F. Why Are Tumour Blood Vessels Abnormal and Why Is It Important to Know? *Br. J. Cancer* **2009**, *100*, 865–869. [CrossRef]
57. Han, S.; Kim, S.; Chen, Z.; Shin, H.K.; Lee, S.-Y.; Moon, H.E.; Paek, S.H.; Park, S. 3D Bioprinted Vascularized Tumour for Drug Testing. *Int. J. Mol. Sci.* **2020**, *21*, 2993. [CrossRef] [PubMed]
58. Hwang, D.G.; Choi, Y.; Jang, J. 3D Bioprinting-Based Vascularized Tissue Models Mimicking Tissue-Specific Architecture and Pathophysiology for in Vitro Studies. *Front. Bioeng. Biotechnol.* **2021**, *9*, 685507. [CrossRef]
59. Datta, P.; Dey, M.; Ataie, Z.; Unutmaz, D.; Ozbolat, I.T. 3D Bioprinting for Reconstituting the Cancer Microenvironment. *NPJ Precis. Oncol.* **2020**, *4*, 18. [CrossRef] [PubMed]
60. Park, J.Y.; Ryu, H.; Lee, B.; Ha, D.-H.; Ahn, M.; Kim, S.; Kim, J.Y.; Jeon, N.L.; Cho, D.-W. Development of a Functional Airway-on-a-Chip by 3D Cell Printing. *Biofabrication* **2018**, *11*, 015002. [CrossRef] [PubMed]
61. Kim, W.; Kim, G. Intestinal Villi Model with Blood Capillaries Fabricated Using Collagen-Based Bioink and Dual-Cell-Printing Process. *ACS Appl. Mater. Interfaces* **2018**, *10*, 41185–41196. [CrossRef] [PubMed]
62. Ali, M.; Pr, A.K.; Yoo, J.J.; Zahran, F.; Atala, A.; Lee, S.J. A Photo-crosslinkable Kidney ECM-derived Bioink Accelerates Renal Tissue Formation. *Adv. Healthc. Mater.* **2019**, *8*, 1800992. [CrossRef]
63. Fransen, M.F.; Addario, G.; Bouten, C.V.; Halary, F.; Moroni, L.; Mota, C. Bioprinting of Kidney in Vitro Models: Cells, Biomaterials, and Manufacturing Techniques. *Essays Biochem.* **2021**, *65*, 587–602.

64. Sun, M.; Sun, X.; Wang, Z.; Guo, S.; Yu, G.; Yang, H. Synthesis and Properties of Gelatin Methacryloyl (GelMA) Hydrogels and Their Recent Applications in Load-Bearing Tissue. *Polymers* **2018**, *10*, 1290. [CrossRef] [PubMed]
65. Sarkar, J.; Kamble, S.C.; Kashikar, N.C. Polymeric Bioinks for 3D Hepatic Printing. *Chemistry* **2021**, *3*, 164–181. [CrossRef]
66. Zhang, Y.S.; Arneri, A.; Bersini, S.; Shin, S.-R.; Zhu, K.; Goli-Malekabadi, Z.; Aleman, J.; Colosi, C.; Busignani, F.; Dell'Erba, V. Bioprinting 3D Microfibrous Scaffolds for Engineering Endothelialized Myocardium and Heart-on-a-Chip. *Biomaterials* **2016**, *110*, 45–59. [CrossRef] [PubMed]
67. Lind, J.U.; Busbee, T.A.; Valentine, A.D.; Pasqualini, F.S.; Yuan, H.; Yadid, M.; Park, S.-J.; Kotikian, A.; Nesmith, A.P.; Campbell, P.H. Instrumented Cardiac Microphysiological Devices via Multimaterial Three-Dimensional Printing. *Nat. Mater.* **2017**, *16*, 303–308. [CrossRef] [PubMed]
68. Burkholder-Wenger, A.C.; Golzar, H.; Wu, Y.; Tang, X.S. Development of a Hybrid Nanoink for 3D Bioprinting of Heterogeneous Tumor Models. *ACS Biomater. Sci. Eng.* **2022**, *8*, 777–785. [CrossRef] [PubMed]
69. Wu, Y.; Wenger, A.; Golzar, H.; Tang, X.S. 3D Bioprinting of Bicellular Liver Lobule-Mimetic Structures via Microextrusion of Cellulose Nanocrystal-Incorporated Shear-Thinning Bioink. *Sci. Rep.* **2020**, *10*, 20648. [CrossRef] [PubMed]
70. Lampin, M.; Warocquier-Clérout, R.; Legris, C.; Degrange, M.; Sigot-Luizard, M. Correlation between Substratum Roughness and Wettability, Cell Adhesion, and Cell Migration. *J. Biomed. Mater. Res. Off. J. Soc. Biomater. Jpn. Soc. Biomater.* **1997**, *36*, 99–108. [CrossRef]
71. Majhy, B.; Priyadarshini, P.; Sen, A. Effect of Surface Energy and Roughness on Cell Adhesion and Growth–Facile Surface Modification for Enhanced Cell Culture. *RSC Adv.* **2021**, *11*, 15467–15476. [CrossRef]
72. Chen, L.; Yan, C.; Zheng, Z. Functional Polymer Surfaces for Controlling Cell Behaviors. *Mater. Today* **2018**, *21*, 38–59. [CrossRef]
73. Patel, S.; Thakar, R.G.; Wong, J.; McLeod, S.D.; Li, S. Control of Cell Adhesion on Poly (Methyl Methacrylate). *Biomaterials* **2006**, *27*, 2890–2897. [CrossRef] [PubMed]
74. Riau, A.K.; Mondal, D.; Yam, G.H.; Setiawan, M.; Liedberg, B.; Venkatraman, S.S.; Mehta, J.S. Surface Modification of PMMA to Improve Adhesion to Corneal Substitutes in a Synthetic Core–Skirt Keratoprosthesis. *ACS Appl. Mater. Interfaces* **2015**, *7*, 21690–21702. [CrossRef]
75. Jastrzebska, E.; Zuchowska, A.; Flis, S.; Sokolowska, P.; Bulka, M.; Dybko, A.; Brzozka, Z. Biological Characterization of the Modified Poly (Dimethylsiloxane) Surfaces Based on Cell Attachment and Toxicity Assays. *Biomicrofluidics* **2018**, *12*, 044105. [CrossRef] [PubMed]
76. Premnath, P.; Tavangar, A.; Tan, B.; Venkatakrishnan, K. Tuning Cell Adhesion by Direct Nanostructuring Silicon into Cell Repulsive/Adhesive Patterns. *Exp. Cell Res.* **2015**, *337*, 44–52. [CrossRef]
77. Azadbakht, M.; Madaeni, S.S.; Sahebjamee, F. Biocompatibility of Polyethersulfone Membranes for Cell Culture Systems. *Eng. Life Sci.* **2011**, *11*, 629–635. [CrossRef]
78. Ruben, B.; Elisa, M.; Leandro, L.; Victor, M.; Gloria, G.; Marina, S.; Pandiyan, R.; Nadhira, L. Oxygen Plasma Treatments of Polydimethylsiloxane Surfaces: Effect of the Atomic Oxygen on Capillary Flow in the Microchannels. *Micro Nano Lett.* **2017**, *12*, 754–757. [CrossRef]
79. Ribas, J.; Sadeghi, H.; Manbachi, A.; Leijten, J.; Brinegar, K.; Zhang, Y.S.; Ferreira, L.; Khademhosseini, A. Cardiovascular Organ-on-a-Chip Platforms for Drug Discovery and Development. *Appl. Vitro Toxicol.* **2016**, *2*, 82–96. [CrossRef] [PubMed]
80. Kim, S.; Takayama, S. Organ-on-a-Chip and the Kidney. *Kidney Res. Clin. Pract.* **2015**, *34*, 165–169. [CrossRef] [PubMed]
81. Deng, J.; Wei, W.; Chen, Z.; Lin, B.; Zhao, W.; Luo, Y.; Zhang, X. Engineered Liver-on-a-Chip Platform to Mimic Liver Functions and Its Biomedical Applications: A Review. *Micromachines* **2019**, *10*, 676. [CrossRef]
82. Zamprogno, P.; Wüthrich, S.; Achenbach, S.; Thoma, G.; Stucki, J.D.; Hobi, N.; Schneider-Daum, N.; Lehr, C.-M.; Huwer, H.; Geiser, T. Second-Generation Lung-on-a-Chip with an Array of Stretchable Alveoli Made with a Biological Membrane. *Commun. Biol.* **2021**, *4*, 168. [CrossRef] [PubMed]
83. Ogończyk, D.; Jankowski, P.; Garstecki, P. A Method for Simultaneous Polishing and Hydrophobization of Polycarbonate for Microfluidic Applications. *Polymers* **2020**, *12*, 2490. [CrossRef]
84. He, Y.; Guo, Y.; He, R.; Jin, T.; Chen, F.; Fu, Q.; Zhou, N.; Shen, J. Towards High Molecular Weight Poly (Bisphenol a Carbonate) with Excellent Thermal Stability and Mechanical Properties by Solid-State Polymerization. *Chin. J. Polym. Sci.* **2015**, *33*, 1176–1185. [CrossRef]
85. Aazmi, A.; Zhou, H.; Li, Y.; Yu, M.; Xu, X.; Wu, Y.; Ma, L.; Zhang, B.; Yang, H. Engineered Vasculature for Organ-on-a-Chip Systems. *Engineering* **2021**, *9*, 131–147. [CrossRef]
86. Ma, Y.; Pan, J.-Z.; Zhao, S.-P.; Lou, Q.; Zhu, Y.; Fang, Q. Microdroplet Chain Array for Cell Migration Assays. *Lab Chip* **2016**, *16*, 4658–4665. [CrossRef] [PubMed]
87. Gotoh, K.; Yasukawa, A.; Kobayashi, Y. Wettability Characteristics of Poly (Ethylene Terephthalate) Films Treated by Atmospheric Pressure Plasma and Ultraviolet Excimer Light. *Polym. J.* **2011**, *43*, 545–551. [CrossRef]
88. Bin, Y.; Oishi, K.; Yoshida, K.; Matsuo, M. Mechanical Properties of Poly (Ethylene Terephthalate) Estimated in Terms of Orientation Distribution of Crystallites and Amorphous Chain Segments under Simultaneous Biaxially Stretching. *Polym. J.* **2004**, *36*, 888–898. [CrossRef]
89. Xiang, Y.; Wen, H.; Yu, Y.; Li, M.; Fu, X.; Huang, S. Gut-on-Chip: Recreating Human Intestine in Vitro. *J. Tissue Eng.* **2020**, *11*, 2041731420965318. [CrossRef] [PubMed]

90. Jang, K.-J.; Mehr, A.P.; Hamilton, G.A.; McPartlin, L.A.; Chung, S.; Suh, K.-Y.; Ingber, D.E. Human Kidney Proximal Tubule-on-a-Chip for Drug Transport and Nephrotoxicity Assessment. *Integr. Biol.* **2013**, *5*, 1119–1129. [CrossRef] [PubMed]
91. Yu, F.; Deng, R.; Hao Tong, W.; Huan, L.; Chan Way, N.; IslamBadhan, A.; Iliescu, C.; Yu, H. A Perfusion Incubator Liver Chip for 3D Cell Culture with Application on Chronic Hepatotoxicity Testing. *Sci. Rep.* **2017**, *7*, 14528. [CrossRef] [PubMed]
92. Laput, O.; Vasenina, I.; Salvadori, M.C.; Savkin, K.; Zuza, D.; Kurzina, I. Low-Temperature Plasma Treatment of Polylactic Acid and PLA/HA Composite Material. *J. Mater. Sci.* **2019**, *54*, 11726–11738. [CrossRef]
93. Ko, H.-S.; Lee, S.; Lee, D.; Jho, J.Y. Mechanical Properties and Bioactivity of Poly (Lactic Acid) Composites Containing Poly (Glycolic Acid) Fiber and Hydroxyapatite Particles. *Nanomaterials* **2021**, *11*, 249. [CrossRef]
94. Pensabene, V.; Costa, L.; Terekhov, A.Y.; Gnecco, J.S.; Wikswo, J.P.; Hofmeister, W.H. Ultrathin Polymer Membranes with Patterned, Micrometric Pores for Organs-on-Chips. *ACS Appl. Mater. Interfaces* **2016**, *8*, 22629–22636. [CrossRef] [PubMed]
95. Janvikul, W.; Uppanan, P.; Thavornyutikarn, B.; Kosorn, W.; Kaewkong, P. Effects of Surface Topography, Hydrophilicity and Chemistry of Surface-Treated PCL Scaffolds on Chondrocyte Infiltration and ECM Production. *Procedia Eng.* **2013**, *59*, 158–165. [CrossRef]
96. Dziadek, M.; Menaszek, E.; Zagrajczuk, B.; Pawlik, J.; Cholewa-Kowalska, K. New Generation Poly (ε-Caprolactone)/Gel-Derived Bioactive Glass Composites for Bone Tissue Engineering: Part I. Material Properties. *Mater. Sci. Eng. C* **2015**, *56*, 9–21. [CrossRef] [PubMed]
97. Pensabene, V.; Crowder, S.W.; Balikov, D.A.; Lee, J.B.; Sung, H.-J. *Optimization of Electrospun Fibrous Membranes for in Vitro Modeling of Blood-Brain Barrier*; IEEE: Piscataway, NJ, USA, 2016; pp. 125–128.
98. Wloch, J.; Terzyk, A.P.; Wisniewski, M.; Kowalczyk, P. Nanoscale Water Contact Angle on Polytetrafluoroethylene Surfaces Characterized by Molecular Dynamics-Atomic Force Microscopy Imaging. *Langmuir* **2018**, *34*, 4526–4534. [CrossRef] [PubMed]
99. Uehara, H.; Jounai, K.; Endo, R.; Okuyama, H.; Kanamoto, T.; Porter, R.S. High Modulus Films of Polytetrafluoroethylene Prepared by Two-Stage Drawing of Reactor Powder. *Polym. J.* **1997**, *29*, 198–200. [CrossRef]
100. Ahadian, S.; Civitarese, R.; Bannerman, D.; Mohammadi, M.H.; Lu, R.; Wang, E.; Davenport-Huyer, L.; Lai, B.; Zhang, B.; Zhao, Y. Organ-on-a-chip Platforms: A Convergence of Advanced Materials, Cells, and Microscale Technologies. *Adv. Healthc. Mater.* **2018**, *7*, 1700506. [CrossRef] [PubMed]
101. Grant, J.; Özkan, A.; Oh, C.; Mahajan, G.; Prantil-Baun, R.; Ingber, D.E. Predicting Drug Concentrations in PDMS Microfluidic Organ Chips. *Lab Chip* **2021**, *21*, 3509–3519. [CrossRef] [PubMed]
102. Wang, M.; Duan, B. Materials and Their Biomedical Applications. In *Encyclopedia of Biomedical Engineering*; Elsevier: Amsterdam, The Netherlands, 2019; pp. 135–152.
103. Wu, Q.; Liu, J.; Wang, X.; Feng, L.; Wu, J.; Zhu, X.; Wen, W.; Gong, X. Organ-on-a-Chip: Recent Breakthroughs and Future Prospects. *Biomed. Eng. Online* **2020**, *19*, 9. [CrossRef]
104. Osório, L.A.; Silva, E.; Mackay, R.E. A Review of Biomaterials and Scaffold Fabrication for Organ-on-a-Chip (OOAC) Systems. *Bioengineering* **2021**, *8*, 113. [CrossRef] [PubMed]
105. Arif, U.; Haider, S.; Haider, A.; Khan, N.; Alghyamah, A.A.; Jamila, N.; Khan, M.I.; Almasry, W.A.; Kang, I.-K. Biocompatible Polymers and Their Potential Biomedical Applications: A Review. *Curr. Pharm. Des.* **2019**, *25*, 3608–3619. [CrossRef] [PubMed]
106. Raj, M.K.; Chakraborty, S. PDMS Microfluidics: A Mini Review. *J. Appl. Polym. Sci.* **2020**, *137*, 48958. [CrossRef]
107. Fujii, T. PDMS-Based Microfluidic Devices for Biomedical Applications. *Microelectron. Eng.* **2002**, *61*, 907–914. [CrossRef]
108. Carter, S.-S.D.; Atif, A.-R.; Kadekar, S.; Lanekoff, I.; Engqvist, H.; Varghese, O.P.; Tenje, M.; Mestres, G. PDMS Leaching and Its Implications for On-Chip Studies Focusing on Bone Regeneration Applications. *Organs-Chip* **2020**, *2*, 100004. [CrossRef]
109. Bunge, F.; den Driesche, S.V.; Vellekoop, M.J. Microfluidic Platform for the Long-Term on-Chip Cultivation of Mammalian Cells for Lab-on-a-Chip Applications. *Sensors* **2017**, *17*, 1603. [CrossRef] [PubMed]
110. Gezer, P.G.; Brodsky, S.; Hsiao, A.; Liu, G.L.; Kokini, J.L. Modification of the Hydrophilic/Hydrophobic Characteristic of Zein Film Surfaces by Contact with Oxygen Plasma Treated PDMS and Oleic Acid Content. *Colloids Surf. B Biointerfaces* **2015**, *135*, 433–440. [CrossRef] [PubMed]
111. Zhang, W.; Choi, D.S.; Nguyen, Y.H.; Chang, J.; Qin, L. Studying Cancer Stem Cell Dynamics on PDMS Surfaces for Microfluidics Device Design. *Sci. Rep.* **2013**, *3*, 2332. [CrossRef] [PubMed]
112. Chuah, Y.J.; Kuddannaya, S.; Lee, M.H.A.; Zhang, Y.; Kang, Y. The Effects of Poly (Dimethylsiloxane) Surface Silanization on the Mesenchymal Stem Cell Fate. *Biomater. Sci.* **2015**, *3*, 383–390. [CrossRef] [PubMed]
113. Pitingolo, G.; Riaud, A.; Nastruzzi, C.; Taly, V. Gelatin-Coated Microfluidic Channels for 3d Microtissue Formation: On-Chip Production and Characterization. *Micromachines* **2019**, *10*, 265. [CrossRef] [PubMed]
114. Steinfeld, B.; Scott, J.; Vilander, G.; Marx, L.; Quirk, M.; Lindberg, J.; Koerner, K. The Role of Lean Process Improvement in Implementation of Evidence-Based Practices in Behavioral Health Care. *J. Behav. Health Serv. Res.* **2015**, *42*, 504–518. [CrossRef]
115. Khetani, S.; Yong, K.W.; Ozhukil Kollath, V.; Eastick, E.; Azarmanesh, M.; Karan, K.; Sen, A.; Sanati-Nezhad, A. Engineering Shelf-Stable Coating for Microfluidic Organ-on-a-Chip Using Bioinspired Catecholamine Polymers. *ACS Appl. Mater. Interfaces* **2020**, *12*, 6910–6923. [CrossRef]
116. Park, S.E.; Georgescu, A.; Oh, J.M.; Kwon, K.W.; Huh, D. Polydopamine-Based Interfacial Engineering of Extracellular Matrix Hydrogels for the Construction and Long-Term Maintenance of Living Three-Dimensional Tissues. *ACS Appl. Mater. Interfaces* **2019**, *11*, 23919–23925. [CrossRef]

117. Mikhail, A.S.; Ranger, J.J.; Liu, L.; Longenecker, R.; Thompson, D.B.; Sheardown, H.D.; Brook, M.A. Rapid and Efficient Assembly of Functional Silicone Surfaces Protected by PEG: Cell Adhesion to Peptide-Modified PDMS. *J. Biomater. Sci. Polym. Ed.* **2010**, *21*, 821–842. [CrossRef]
118. Henry, O.Y.; Villenave, R.; Cronce, M.J.; Leineweber, W.D.; Benz, M.A.; Ingber, D.E. Organs-on-Chips with Integrated Electrodes for Trans-Epithelial Electrical Resistance (TEER) Measurements of Human Epithelial Barrier Function. *Lab Chip* **2017**, *17*, 2264–2271. [CrossRef]
119. Wang, Y.I.; Abaci, H.E.; Shuler, M.L. Microfluidic Blood–Brain Barrier Model Provides in Vivo-like Barrier Properties for Drug Permeability Screening. *Biotechnol. Bioeng.* **2017**, *114*, 184–194. [CrossRef]
120. Domansky, K.; Sliz, J.D.; Wen, N.; Hinojosa, C.; Thompson, G.; Fraser, J.P.; Hamkins-Indik, T.; Hamilton, G.A.; Levner, D.; Ingber, D.E. SEBS Elastomers for Fabrication of Microfluidic Devices with Reduced Drug Absorption by Injection Molding and Extrusion. *Microfluid. Nanofluidics* **2017**, *21*, 107. [CrossRef]
121. Domansky, K.; Leslie, D.C.; McKinney, J.; Fraser, J.P.; Sliz, J.D.; Hamkins-Indik, T.; Hamilton, G.A.; Bahinski, A.; Ingber, D.E. Clear Castable Polyurethane Elastomer for Fabrication of Microfluidic Devices. *Lab Chip* **2013**, *13*, 3956–3964. [CrossRef] [PubMed]
122. Ding, C.; Chen, X.; Kang, Q.; Yan, X. Biomedical Application of Functional Materials in Organ-on-a-Chip. *Front. Bioeng. Biotechnol.* **2020**, *8*, 823. [CrossRef] [PubMed]
123. Pasman, T.; Grijpma, D.; Stamatialis, D.; Poot, A. Flat and Microstructured Polymeric Membranes in Organs-on-Chips. *J. R. Soc. Interface* **2018**, *15*, 20180351. [CrossRef] [PubMed]
124. Busek, M.; Nøvik, S.; Aizenshtadt, A.; Amirola-Martinez, M.; Combriat, T.; Grünzner, S.; Krauss, S. Thermoplastic Elastomer (TPE)–Poly (Methyl Methacrylate)(PMMA) Hybrid Devices for Active Pumping PDMS-Free Organ-on-a-Chip Systems. *Biosensors* **2021**, *11*, 162. [CrossRef] [PubMed]
125. Nguyen, T.; Jung, S.H.; Lee, M.S.; Park, T.-E.; Ahn, S.; Kang, J.H. Robust Chemical Bonding of PMMA Microfluidic Devices to Porous PETE Membranes for Reliable Cytotoxicity Testing of Drugs. *Lab Chip* **2019**, *19*, 3706–3713. [CrossRef]
126. Echave, M.C.; Burgo, L.S.; Pedraz, J.L.; Orive, G. Gelatin as Biomaterial for Tissue Engineering. *Curr. Pharm. Des.* **2017**, *23*, 3567–3584. [CrossRef]
127. Santoro, M.; Tatara, A.M.; Mikos, A.G. Gelatin Carriers for Drug and Cell Delivery in Tissue Engineering. *J. Control. Release* **2014**, *190*, 210–218. [CrossRef]
128. Lee, H.; Cho, D.-W. One-Step Fabrication of an Organ-on-a-Chip with Spatial Heterogeneity Using a 3D Bioprinting Technology. *Lab Chip* **2016**, *16*, 2618–2625. [CrossRef]
129. Lam, T.; Dehne, T.; Krüger, J.P.; Hondke, S.; Endres, M.; Thomas, A.; Lauster, R.; Sittinger, M.; Kloke, L. Photopolymerizable Gelatin and Hyaluronic Acid for Stereolithographic 3D Bioprinting of Tissue-engineered Cartilage. *J. Biomed. Mater. Res. B Appl. Biomater.* **2019**, *107*, 2649–2657. [CrossRef] [PubMed]
130. Nawroth, J.C.; Scudder, L.L.; Halvorson, R.T.; Tresback, J.; Ferrier, J.P.; Sheehy, S.P.; Cho, A.; Kannan, S.; Sunyovszki, I.; Goss, J.A. Automated Fabrication of Photopatterned Gelatin Hydrogels for Organ-on-Chips Applications. *Biofabrication* **2018**, *10*, 025004. [CrossRef] [PubMed]
131. Yu, F.; Hunziker, W.; Choudhury, D. Engineering Microfluidic Organoid-on-a-Chip Platforms. *Micromachines* **2019**, *10*, 165. [CrossRef]
132. Benton, G.; Arnaoutova, I.; George, J.; Kleinman, H.K.; Koblinski, J. Matrigel: From Discovery and ECM Mimicry to Assays and Models for Cancer Research. *Adv. Drug Deliv. Rev.* **2014**, *79*, 3–18. [CrossRef]
133. Li, Y.; Wang, S.; Huang, R.; Huang, Z.; Hu, B.; Zheng, W.; Yang, G.; Jiang, X. Evaluation of the Effect of the Structure of Bacterial Cellulose on Full Thickness Skin Wound Repair on a Microfluidic Chip. *Biomacromolecules* **2015**, *16*, 780–789. [CrossRef]
134. Anton-Sales, I.; Beekmann, U.; Laromaine, A.; Roig, A.; Kralisch, D. Opportunities of Bacterial Cellulose to Treat Epithelial Tissues. *Curr. Drug Targets* **2019**, *20*, 808–822. [CrossRef] [PubMed]
135. Wu, H.; Huang, J. Drug-Induced Nephrotoxicity: Pathogenic Mechanisms, Biomarkers and Prevention Strategies. *Curr. Drug Metab.* **2018**, *19*, 559–567. [CrossRef] [PubMed]
136. Zhuo, J.L.; Li, X.C. Proximal Nephron. *Compr. Physiol.* **2013**, *3*, 1079.
137. McCormick, J.; Ellison, D. Distal Convoluted Tubule. *Compr. Physiol.* **2015**, *5*, 45–98. [PubMed]
138. Nolin, T.D.; Himmelfarb, J. Mechanisms of Drug-Induced Nephrotoxicity. *Advers. Drug React.* **2010**, *196*, 111–130.
139. Mu, X.; Zheng, W.; Xiao, L.; Zhang, W.; Jiang, X. Engineering a 3D Vascular Network in Hydrogel for Mimicking a Nephron. *Lab Chip* **2013**, *13*, 1612–1618. [CrossRef] [PubMed]
140. Wilmer, M.J.; Ng, C.P.; Lanz, H.L.; Vulto, P.; Suter-Dick, L.; Masereeuw, R. Kidney-on-a-Chip Technology for Drug-Induced Nephrotoxicity Screening. *Trends Biotechnol.* **2016**, *34*, 156–170. [CrossRef] [PubMed]
141. Raghavan, V.; Rbaibi, Y.; Pastor-Soler, N.M.; Carattino, M.D.; Weisz, O.A. Shear Stress-Dependent Regulation of Apical Endocytosis in Renal Proximal Tubule Cells Mediated by Primary Cilia. *Proc. Natl. Acad. Sci. USA* **2014**, *111*, 8506–8511. [CrossRef] [PubMed]
142. Miravète, M.; Dissard, R.; Klein, J.; Gonzalez, J.; Caubet, C.; Pecher, C.; Pipy, B.; Bascands, J.-L.; Mercier-Bonin, M.; Schanstra, J.P. Renal Tubular Fluid Shear Stress Facilitates Monocyte Activation toward Inflammatory Macrophages. *Am. J. Physiol.-Ren. Physiol.* **2012**, *302*, F1409–F1417. [CrossRef] [PubMed]

143. Van de Water, B.; Jaspers, J.; Maasdam, D.H.; Mulder, G.J.; Nagelkerke, J.F. In Vivo and in Vitro Detachment of Proximal Tubular Cells and F-Actin Damage: Consequences for Renal Function. *Am. J. Physiol.-Ren. Physiol.* **1994**, *267*, F888–F899. [CrossRef] [PubMed]
144. Ross, E.J.; Gordon, E.R.; Sothers, H.; Darji, R.; Baron, O.; Haithcock, D.; Prabhakarpandian, B.; Pant, K.; Myers, R.M.; Cooper, S.J. Three Dimensional Modeling of Biologically Relevant Fluid Shear Stress in Human Renal Tubule Cells Mimics in Vivo Transcriptional Profiles. *Sci. Rep.* **2021**, *11*, 14053. [CrossRef] [PubMed]
145. Musah, S.; Mammoto, A.; Ferrante, T.C.; Jeanty, S.S.; Hirano-Kobayashi, M.; Mammoto, T.; Roberts, K.; Chung, S.; Novak, R.; Ingram, M. Mature Induced-Pluripotent-Stem-Cell-Derived Human Podocytes Reconstitute Kidney Glomerular-Capillary-Wall Function on a Chip. *Nat. Biomed. Eng.* **2017**, *1*, 0069. [CrossRef] [PubMed]
146. Roye, Y.; Bhattacharya, R.; Mou, X.; Zhou, Y.; Burt, M.A.; Musah, S. A Personalized Glomerulus Chip Engineered from Stem Cell-Derived Epithelium and Vascular Endothelium. *Micromachines* **2021**, *12*, 967. [CrossRef] [PubMed]
147. Wang, L.; Tao, T.; Su, W.; Yu, H.; Yu, Y.; Qin, J. A Disease Model of Diabetic Nephropathy in a Glomerulus-on-a-Chip Microdevice. *Lab Chip* **2017**, *17*, 1749–1760. [CrossRef]
148. Ng, C.P.; Zhuang, Y.; Lin, A.W.H.; Teo, J.C.M. A Fibrin-Based Tissue-Engineered Renal Proximal Tubule for Bioartificial Kidney Devices: Development, Characterization and in Vitro Transport Study. *Int. J. Tissue Eng.* **2012**, *2013*, 319476. [CrossRef]
149. Jansen, J.; Fedecostante, M.; Wilmer, M.; Peters, J.; Kreuser, U.; Van Den Broek, P.; Mensink, R.; Boltje, T.; Stamatialis, D.; Wetzels, J. Bioengineered Kidney Tubules Efficiently Excrete Uremic Toxins. *Sci. Rep.* **2016**, *6*, 26715. [CrossRef] [PubMed]
150. Sciancalepore, A.G.; Sallustio, F.; Girardo, S.; Gioia Passione, L.; Camposeo, A.; Mele, E.; Di Lorenzo, M.; Costantino, V.; Schena, F.P.; Pisignano, D. A Bioartificial Renal Tubule Device Embedding Human Renal Stem/Progenitor Cells. *PLoS ONE* **2014**, *9*, e87496. [CrossRef] [PubMed]
151. Weber, E.J.; Chapron, A.; Chapron, B.D.; Voellinger, J.L.; Lidberg, K.A.; Yeung, C.K.; Wang, Z.; Yamaura, Y.; Hailey, D.W.; Neumann, T. Development of a Microphysiological Model of Human Kidney Proximal Tubule Function. *Kidney Int.* **2016**, *90*, 627–637. [CrossRef] [PubMed]
152. Sochol, R.D.; Gupta, N.R.; Bonventre, J.V. A Role for 3D Printing in Kidney-on-a-Chip Platforms. *Curr. Transplant. Rep.* **2016**, *3*, 82–92. [CrossRef]
153. Homan, K.A.; Kolesky, D.B.; Skylar-Scott, M.A.; Herrmann, J.; Obuobi, H.; Moisan, A.; Lewis, J.A. Bioprinting of 3D Convoluted Renal Proximal Tubules on Perfusable Chips. *Sci. Rep.* **2016**, *6*, 34845. [CrossRef] [PubMed]
154. Odijk, M.; van der Meer, A.D.; Levner, D.; Kim, H.J.; van der Helm, M.W.; Segerink, L.I.; Frimat, J.-P.; Hamilton, G.A.; Ingber, D.E.; van den Berg, A. Measuring Direct Current Trans-Epithelial Electrical Resistance in Organ-on-a-Chip Microsystems. *Lab Chip* **2015**, *15*, 745–752. [CrossRef] [PubMed]
155. Snouber, L.C.; Letourneur, F.; Chafey, P.; Broussard, C.; Monge, M.; Legallais, C.; Leclerc, E. Analysis of Transcriptomic and Proteomic Profiles Demonstrates Improved Madin–Darby Canine Kidney Cell Function in a Renal Microfluidic Biochip. *Biotechnol. Prog.* **2012**, *28*, 474–484. [CrossRef]
156. Kim, S.; LesherPerez, S.C.; Yamanishi, C.; Labuz, J.M.; Leung, B.; Takayama, S. Pharmacokinetic Profile That Reduces Nephrotoxicity of Gentamicin in a Perfused Kidney-on-a-Chip. *Biofabrication* **2016**, *8*, 015021. [CrossRef] [PubMed]
157. Vormann, M.K.; Gijzen, L.; Hutter, S.; Boot, L.; Nicolas, A.; van den Heuvel, A.; Vriend, J.; Ng, C.P.; Nieskens, T.T.; van Duinen, V. Nephrotoxicity and Kidney Transport Assessment on 3D Perfused Proximal Tubules. *AAPS J.* **2018**, *20*, 90. [CrossRef] [PubMed]
158. Trietsch, S.J.; Israëls, G.D.; Joore, J.; Hankemeier, T.; Vulto, P. Microfluidic Titer Plate for Stratified 3D Cell Culture. *Lab Chip* **2013**, *13*, 3548–3554. [CrossRef] [PubMed]
159. Maschmeyer, I.; Lorenz, A.K.; Schimek, K.; Hasenberg, T.; Ramme, A.P.; Hübner, J.; Lindner, M.; Drewell, C.; Bauer, S.; Thomas, A. A Four-Organ-Chip for Interconnected Long-Term Co-Culture of Human Intestine, Liver, Skin and Kidney Equivalents. *Lab Chip* **2015**, *15*, 2688–2699. [CrossRef]
160. Chang, S.-Y.; Weber, E.J.; Sidorenko, V.S.; Chapron, A.; Yeung, C.K.; Gao, C.; Mao, Q.; Shen, D.; Wang, J.; Rosenquist, T.A. Human Liver-Kidney Model Elucidates the Mechanisms of Aristolochic Acid Nephrotoxicity. *JCI Insight* **2017**, *2*, e95978. [CrossRef] [PubMed]
161. Brooks, J.D. Anatomy of the Lower Urinary Tract and Male Genitalia. In *Campbell-Walsh Urology*; Saunders/Elsevier: Philadelphia, PA, USA, 2007.
162. de Groat, W.C.; Yoshimura, N. Anatomy and Physiology of the Lower Urinary Tract. *Handb. Clin. Neurol.* **2015**, *130*, 61–108.
163. Bolla, S.R.; Odeluga, N.; Jetti, R. Histology, Bladder. In *StatPearls*; StatPearls Publishing: Treasure Island, FL, USA, 2021.
164. Sung, H.; Ferlay, J.; Siegel, R.L.; Laversanne, M.; Soerjomataram, I.; Jemal, A.; Bray, F. Global Cancer Statistics 2020: GLOBOCAN Estimates of Incidence and Mortality Worldwide for 36 Cancers in 185 Countries. *CA Cancer J. Clin.* **2021**, *71*, 209–249. [CrossRef] [PubMed]
165. Bray, F.; Ferlay, J.; Soerjomataram, I.; Siegel, R.L.; Torre, L.A.; Jemal, A. Global Cancer Statistics 2018: GLOBOCAN Estimates of Incidence and Mortality Worldwide for 36 Cancers in 185 Countries. *CA Cancer J. Clin.* **2018**, *68*, 394–424. [CrossRef] [PubMed]
166. Cumberbatch, M.G.K.; Jubber, I.; Black, P.C.; Esperto, F.; Figueroa, J.D.; Kamat, A.M.; Kiemeney, L.; Lotan, Y.; Pang, K.; Silverman, D.T. Epidemiology of Bladder Cancer: A Systematic Review and Contemporary Update of Risk Factors in 2018. *Eur. Urol.* **2018**, *74*, 784–795. [CrossRef]
167. Lenis, A.T.; Lec, P.M.; Chamie, K. Bladder Cancer: A Review. *Jama* **2020**, *324*, 1980–1991. [CrossRef]

168. Zhu, S.; Zhu, Z.; Ma, A.-H.; Sonpavde, G.P.; Cheng, F.; Pan, C. Preclinical Models for Bladder Cancer Research. *Hematol. Clin.* **2021**, *35*, 613–632. [CrossRef] [PubMed]
169. Suh, Y.S.; Jeong, K.-C.; Lee, S.-J.; Seo, H.K. Establishment and Application of Bladder Cancer Patient-Derived Xenografts as a Novel Preclinical Platform. *Transl. Cancer Res.* **2017**, *6*, S733–S743. [CrossRef]
170. Bernardo, C.; Costa, C.; Palmeira, C.; Pinto-Leite, R.; Oliveira, P.; Freitas, R.; Amado, F.; Santos, L.L. What We Have Learned from Urinary Bladder Cancer Models. *J. Cancer Metastasis Treat.* **2016**, *2*, 51–58.
171. Arantes-Rodrigues, R.; Colaco, A.; Pinto-Leite, R.; Oliveira, P.A. In Vitro and in Vivo Experimental Models as Tools to Investigate the Efficacy of Antineoplastic Drugs on Urinary Bladder Cancer. *Anticancer Res.* **2013**, *33*, 1273–1296.
172. DeGraff, D.J.; Robinson, V.L.; Shah, J.B.; Brandt, W.D.; Sonpavde, G.; Kang, Y.; Liebert, M.; Wu, X.-R.; Taylor, J.A. Current Preclinical Models for the Advancement of Translational Bladder Cancer Research. *Mol. Cancer Ther.* **2013**, *12*, 121–130. [CrossRef] [PubMed]
173. Liu, P.; Cao, Y.; Zhang, S.; Zhao, Y.; Liu, X.; Shi, H.; Hu, K.; Zhu, G.; Ma, B.; Niu, H. A Bladder Cancer Microenvironment Simulation System Based on a Microfluidic Co-Culture Model. *Oncotarget* **2015**, *6*, 37695. [CrossRef]
174. Xu, X.-D.; Shao, S.-X.; Cao, Y.-W.; Yang, X.-C.; Shi, H.-Q.; Wang, Y.-L.; Xue, S.-Y.; Wang, X.-S.; Niu, H.-T. The Study of Energy Metabolism in Bladder Cancer Cells in Co-Culture Conditions Using a Microfluidic Chip. *Int. J. Clin. Exp. Med.* **2015**, *8*, 12327. [PubMed]
175. Lee, E.; Kwon, C.; Han, H.; Park, J.; Kim, Y.-C.; Ok, M.-R.; Seok, H.-K.; Jeon, H. Bladder Cancer-on-a-Chip for Analysis of Tumor Transition Mechanism. In Proceedings of the 10th World Biomaterials Congress, Montréal, QC, Canada, 17–22 May 2016. [CrossRef]
176. Hooton, T.M. Uncomplicated Urinary Tract Infection. *N. Engl. J. Med.* **2012**, *366*, 1028–1037. [CrossRef]
177. Sen, A. Recurrent Cystitis in Non-Pregnant Women. *BMJ Clin. Evid.* **2008**, *2008*, 0801. [PubMed]
178. Foxman, B. The Epidemiology of Urinary Tract Infection. *Nat. Rev. Urol.* **2010**, *7*, 653–660. [CrossRef]
179. Kudinha, T. The Pathogenesis of Escherichia Coli Urinary Tract Infection. In *Escherichia coli-Recent Advances on Physiology, Pathogenesis and Biotechnological Applications*; IntechOpen: London, UK, 2017; pp. 45–61.
180. Sharma, K.; Dhar, N.; Thacker, V.V.; Simonet, T.M.; Signorino-Gelo, F.; Knott, G.W.; McKinney, J.D. Dynamic Persistence of UPEC Intracellular Bacterial Communities in a Human Bladder-Chip Model of Urinary Tract Infection. *Elife* **2021**, *10*, e66481. [CrossRef] [PubMed]
181. Aaron, L.; Franco, O.E.; Hayward, S.W. Review of Prostate Anatomy and Embryology and the Etiology of Benign Prostatic Hyperplasia. *Urol. Clin.* **2016**, *43*, 279–288. [CrossRef]
182. Jiang, L.; Ivich, F.; Tran, M.; Frank, S.B.; Miranti, C.K.; Zohar, Y. *Development of a Microfluidic-Based Model of a Human Prostate Gland*; Chemical and Biological Microsystems Society: Palm Springs, CA, USA, 2018; pp. 1625–1627.
183. Ivich, F.; Tran, M.; Tahsin, S.; Frank, S.B.; Kraft, A.; Miranti, C.K.; Zohar, Y.; Jiang, L. *Application of a Microfluidic-Based Model of a Human Prostate Gland for Cancer Research*; IEEE: Piscataway, NJ, USA, 2018; pp. 109–112.
184. Frank, S.B.; Miranti, C.K. Disruption of Prostate Epithelial Differentiation Pathways and Prostate Cancer Development. *Front. Oncol.* **2013**, *3*, 273. [CrossRef]
185. Lamb, L.E.; Knudsen, B.S.; Miranti, C.K. E-Cadherin-Mediated Survival of Androgen-Receptor-Expressing Secretory Prostate Epithelial Cells Derived from a Stratified in Vitro Differentiation Model. *J. Cell Sci.* **2010**, *123*, 266–276. [CrossRef]
186. Pomerantz, M.M.; Qiu, X.; Zhu, Y.; Takeda, D.Y.; Pan, W.; Baca, S.C.; Gusev, A.; Korthauer, K.D.; Severson, T.M.; Ha, G. Prostate Cancer Reactivates Developmental Epigenomic Programs during Metastatic Progression. *Nat. Genet.* **2020**, *52*, 790–799. [CrossRef] [PubMed]
187. Wang, G.; Zhao, D.; Spring, D.J.; DePinho, R.A. Genetics and Biology of Prostate Cancer. *Genes Dev.* **2018**, *32*, 1105–1140. [CrossRef] [PubMed]
188. Ivich, F. Development of a Microfluidic Model of a Human Prostate Gland for Cancer Research. Ph.D. Thesis, The University of Arizon, Tucson, AZ, USA, 2019.
189. Liu, Y.; Gill, E.; Huang, Y.Y.S. Microfluidic on-chip biomimicry for 3D cell culture: A fit-for-purpose investigation from the end user standpoint. *Future Sci. OA* **2017**, *3*, pFSO173. [CrossRef] [PubMed]
190. Picollet-d'Hahan, N.; Gerbaud, S.; Kermarrec, F.; Alcaraz, J.-P.; Obeid, P.; Bhajun, R.; Guyon, L.; Sulpice, E.; Cinquin, P.; Dolega, M.E. The Modulation of Attachment, Growth and Morphology of Cancerous Prostate Cells by Polyelectrolyte Nanofilms. *Biomaterials* **2013**, *34*, 10099–10108. [CrossRef]
191. Jiang, L.; Ivich, F.; Tahsin, S.; Tran, M.; Frank, S.; Miranti, C.; Zohar, Y. Human Stroma and Epithelium Co-Culture in a Microfluidic Model of a Human Prostate Gland. *Biomicrofluidics* **2019**, *13*, 064116. [CrossRef]
192. Imparato, G.; Urciuolo, F.; Netti, P.A. Organ on Chip Technology to Model Cancer Growth and Metastasis. *Bioengineering* **2022**, *9*, 28. [CrossRef]
193. Top 10 Emerging Technologies. Available online: https://www.weforum.org/agenda/2016/06/top-10-emerging-technologies-2016/2016 (accessed on 25 February 2022).
194. Shourabi, A.Y.; Kashaninejad, N.; Saidi, M.S. An Integrated Microfluidic Concentration Gradient Generator for Mechanical Stimulation and Drug Delivery. *J. Sci. Adv. Mater. Devices* **2021**, *6*, 280–290. [CrossRef]
195. Adamowicz, J.; Pokrywczyńska, M.; Tworkiewicz, J.; Wolski, Z.; Drewa, T. The Relationship of Cancer Stem Cells in Urological Cancers. *Cent. Eur. J. Urol.* **2013**, *66*, 273. [CrossRef] [PubMed]

196. Chen, W.-Y.; Evangelista, E.A.; Yang, J.; Kelly, E.J.; Yeung, C.K. Kidney Organoid and Microphysiological Kidney Chip Models to Accelerate Drug Development and Reduce Animal Testing. *Front. Pharmacol.* **2021**, *12*, 695920. [CrossRef] [PubMed]
197. Barisam, M.; Niavol, F.R.; Kinj, M.A.; Saidi, M.S.; Ghanbarian, H.; Kashaninejad, N. Enrichment of Cancer Stem-like Cells by Controlling Oxygen, Glucose and Fluid Shear Stress in a Microfluidic Spheroid Culture Device. *J. Sci. Adv. Mater. Devices* **2022**, *7*, 100439. [CrossRef]
198. Radisic, M.; Loskill, P. Beyond PDMS and Membranes: New Materials for Organ-on-a-Chip Devices. *ACS Biomater. Sci. Eng.* **2021**, *7*, 2861–2863. [CrossRef] [PubMed]
199. Zhang, B.; Radisic, M. Organ-on-a-Chip Devices Advance to Market. *Lab Chip* **2017**, *17*, 2395–2420. [CrossRef] [PubMed]

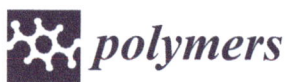

polymers MDPI

Article

pH and Thermoresponsive PNIPAm-co-Polyacrylamide Hydrogel for Dual Stimuli-Responsive Controlled Drug Delivery

Kokila Thirupathi [1,*], Thi Tuong Vy Phan [2,3,†], Madhappan Santhamoorthy [4], Vanaraj Ramkumar [4,†] and Seong-Cheol Kim [4,*]

1 Department of Physics, Sri Moogambigai College of Arts and Science for Women, Palacode 636808, India
2 Center for Advanced Chemistry, Institute of Research and Development, Duy Tan University, 03 Quang Trung, Hai Chau, Danang 550000, Vietnam
3 Faculty of Environmental and Chemical Engineering, Duy Tan University, 03 Quang Trung, Hai Chau, Danang 550000, Vietnam
4 School of Chemical Engineering, Yeungnam University, Gyeongsan 38541, Republic of Korea
* Correspondence: kokila66@gmail.com (K.T.); sckim07@ynu.ac.kr (S.-C.K.)
† These authors contributed equally to this paper.

Abstract: The therapeutic delivery system with dual stimuli-responsiveness has attracted attention for drug delivery to target sites. In this study, we used free radical polymerization to develop a temperature and pH-responsive poly(N-isopropyl acrylamide)-co-poly(acrylamide) (PNIPAM-co-PAAm). PNIPAm-co-PAAm copolymer by reacting with N-isopropyl acrylamide (NIPAm) and acrylamide (Am) monomers. In addition, the synthesized melamine-glutaraldehyde (Mela-Glu) precursor was used as a cross-linker in the production of the melamine cross-linked PNIPAm-co-PAAm copolymer hydrogel (PNIPAm-co-PAAm-Mela HG) system. The temperature-responsive phase transition characteristics of the resulting PNIPAM-co-PAAm-Mela HG systems were determined. Furthermore, the pH-responsive drug release efficiency of curcumin was investigated under various pH and temperature circumstances. Under the combined pH and temperature stimuli (pH 5.0/45 °C), the PNIPAm-co-PAAm-Mela HG demonstrated substantial drug loading (74%), and nearly complete release of the loaded drug was accomplished in 8 h. Furthermore, the cytocompatibility of the PNIPAm-co-PAAm-Mela HG was evaluated on a human liver cancer cell line (HepG2), and the findings demonstrated that the prepared PNIPAm-co-PAAm-Mela HG is biocompatible. As a result, the PNIPAm-co-PAAm-Mela HG system might be used for both pH and temperature-stimuli-responsive drug delivery.

Keywords: thermoresponsive copolymer; curcumin; pH-stimuli; cytocompatibility; drug delivery

Citation: Thirupathi, K.; Phan, T.T.V.; Santhamoorthy, M.; Ramkumar, V.; Kim, S.-C. pH and Thermoresponsive PNIPAm-co-Polyacrylamide Hydrogel for Dual Stimuli-Responsive Controlled Drug Delivery. *Polymers* **2023**, *15*, 167. https://doi.org/10.3390/polym15010167

Academic Editors: Ariana Hudita and Bianca Gălăţeanu

Received: 18 November 2022
Revised: 21 December 2022
Accepted: 23 December 2022
Published: 29 December 2022

1. Introduction

The controlled delivery of therapeutic drugs has been considered an effective technique for maintaining therapeutic effectiveness [1–3]. Controlled drug delivery carriers enable therapeutic concentrations to be maintained while also protecting drugs from enzymatic degradation, improving drug solubility, decreasing adverse effects, and extending release time [4]. Although drug delivery carriers are advantageous, they do have significant clinical limits [5]. As a result, self-regulating drug delivery systems that deliver drugs based on changes in physiological parameters are required [6]. As drug carrier systems, a variety of drug delivery methods, such as nanoparticles, polymeric materials, and lipids, have been employed [7–9].

Hydrogels are cross-linked polymeric networks having the ability to expand in an aqueous medium or biological fluid and hold a considerable amount of water [10,11]. The hydrogels mimic adjacent tissues due to their biocompatibility and high-water content.

Because of their swelling–deswelling capabilities, hydrogels are widely used in biotechnological and biomedical applications such as tissue engineering, regenerative medicine, and diagnostic biosensor fields [12,13]. Chemically cross-linked hydrogel polymers are currently attracting much research attention as prospective drug delivery matrices. The drug release behavior might be influenced by internal or external stimuli such as pH, temperature, light, and ultrasound [14–19]. Dual-stimuli materials, such as temperature and pH-sensitive hydrogel materials, are frequently used in biomedical fields because these parameters may be easily controlled in vitro and in vivo [20].

However, for practical applications, homogenous dispersion of therapeutic molecules in hydrogels is essential, which is one of the key disadvantages of present hydrogel materials. Because of the enhanced swelling capacity of hydrogels, the drug is released quickly. The selective and controlled release of a loaded drug is considered to be an efficient strategy for minimizing undesirable side effects on normal cells and tissues [21]. As a result, the development of dual-stimuli controlled delivery systems, such as pH and temperature-responsive systems, is thought to be important for regulated and selective drug delivery in target sites. Because of their ability to respond to temperature changes in the presence of biological fluid, thermosensitive hydrogels are gaining popularity in drug delivery and tissue encapsulation [22]. Among them, thermoresponsive poly(N-isopropyl acrylamide) (PNIPAm) is the most well-investigated polymer due to its distinctive phase transition from an extended hydrophilic state to a collapsed hydrophobic state in water at around 32 °C. The PNIPAm-based hydrogels undergo an abrupt phase transition both below and above LCST (about 32 °C). In the presence of temperature stimuli, the drug-loaded PNIPAm hydrogels undergo globular structural change, releasing the loaded drugs from the hydrogel network [23]. At swelling temperatures below LCST, the cross-linked hydrogel formed with PNIPAm polymer is hydrophilic. They, on the other hand, acquire a collapsed state whenever the temperature in the aqueous medium is above the LSCT.

Aside from thermoresponsive hydrogels, dual-stimuli responsive hydrogels are produced by copolymerizing two or more appropriate monomers. These pH-responsive hydrogels are a type of stimuli-responsive material used in biological applications. In response to pH variations, such hydrogel systems regulate drug release from the drug-loaded hydrogel. These pH-responsive hydrogels are often synthesized from polymers with weakly basic (-NH_2) or weakly acidic (-COOH) functional units [16,24]. The drug release can be accomplished by the protonation and deprotonation of these functional groups present in hydrogels. Both pH and temperature-responsive hydrogel systems are required for specific biomedical applications such as synergistic chemo-photothermal therapy and magnetic hyperthermia-induced drug delivery applications. Copolymerization of NIPAm with appropriate pH-sensitive monomers bearing basic functional groups in the presence of a limited number of cross-linkers results in dual pH and temperature-stimuli-responsive hydrogels [25].

We developed a thermo- and pH-responsive PNIPAm-based copolymer by combining polyacrylamide (PAAm) units cross-linked with melamine (Mel) units to produce PNIPAm-co-PAAm-Mel copolymer hydrogel (HG). The thermoresponsive PNIPAm has used swelling–deswelling behavior under temperature changes to achieve its sharp phase transition. Because of its basic amine functional groups, PAAm has a pH-responsive segment that can act as a drug-binding site. Additionally, Mel units have been employed to increase the cross-linkage of the copolymer network as well as the number of drug-binding sites. In order to characterize the synthesized PNIPAm-co-PAAm-Mel HG material, various experimental methods such as ^{1}H NMR, FTIR, SEM, and zeta potential were used. Curcumin (Cur) was used as a model cargo to assess in vitro drug loading and pH and temperature-responsive release characteristics. The in vitro drug release study performance at 8 h revealed that under combined pH and temperature stimuli, nearly complete release of Cur occurs at 45 °C and pH 5.0, compared to approximately 75% release only at pH stimuli (pH 5.0) or approximately 58% release only at temperature stimuli (45 °C), respectively. As a result, the prepared dual-stimuli responsive PNIPAm-co-PAAm-Mel HG system may be used for selective drug delivery to the target region.

2. Materials and Methods

2.1. Reagents and Chemicals

Acrylamide (AAm, 99%), melamine (Mel, 99%), 2,2-azobisisobutyronitrile (AIBN, 12 wt.% in acetone), N-Isopropylacrylamide (NIPAm, 97%), ethanol (99%), tetrahydrofuran (THF, 99.9%), diethyl ether (99.7%) and curcumin (98%), were purchased from Sigma Aldrich Chemical Co., Saint Louis, MO, USA, and used as received.

2.2. Synthesis of PNIPAm-co-PAAm Copolymer

The PNIPAm-co-PAAm copolymer was synthesized via a free radical polymerization process using AIBN as the initiator [26]. 2.0 g (17.5 mmol) NIPAm and 1.38 g (17.7 mmol) AAm were solubilized in a 50 mL two-necked round bottom flask containing 15 mL dry THF solvent for this experiment. The reaction flask was then continuously purged with nitrogen gas for 45 min before adding around 0.05 g of AIBN in THF (0.5 mL) and performing the reaction at 68 °C for 24 h under inert conditions. The resulting viscous mass was then precipitated in hexane (100 mL). This precipitation procedure was repeated five times to remove the unreacted monomer, and the white precipitate that formed was vacuum dried at room temperature. The copolymer was named PNIPAm-co-PAAm copolymer (Scheme 1A).

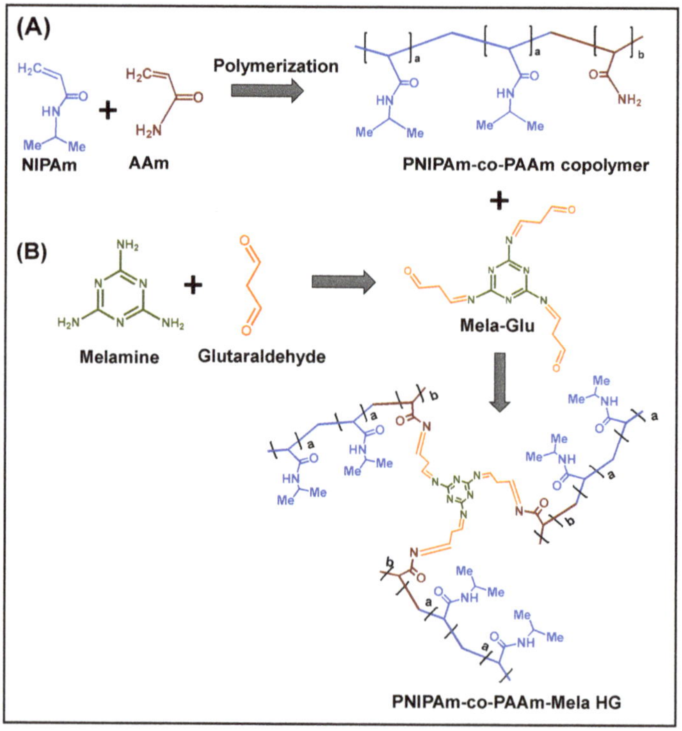

Scheme 1. The scheme represents the (**A**) preparation of PNIPAm-co-PAAm copolymer; (**B**) Mela-Glu cross-linked PNIPAm-co-PAAm HG system.

2.3. Synthesis of Glutaraldehyde-Modified Melamine Precursor

Approximately 1.0 g (7.9 mmol) melamine was mixed in 25 mL of water: ethanol (40:60 vol/vol) mixture for this reaction. In the presence of an acetic acid catalyst, 2.38 mL

(2.3 mmol) of glutaraldehyde was added. The reaction was carried out at 90 °C for 48 h. The product was purified and recrystallized from hot methanol after the reaction mixture was concentrated using a rotary evaporator. The final product was designated as Mela-Glu precursor (Scheme 1B).

2.4. *PNIPAm-co-PAAm Copolymer Cross-Linked with Mela-Glu Precursor*

The PNIPAm-co-PAAm copolymer was cross-linked with Mela-Glu precursor as follows. Approximately 1.0 g of PNIPAm-co-PAAm copolymer was dissolved in 25 mL ethanol. Next, 0.3 g of Mela-Glu precursor in ethanol (5 mL) was slowly introduced, allowing for a cross-linking reaction between the amine groups of PAAm segments and the aldehyde groups of Mela-Glu units through Schiff base reaction in the presence of an acetic acid catalyst [27]. The obtained hydrogel sample was placed in a dialysis membrane tubing with molecular weight cutoff (MWCO) at 3.5 kDa and placed in a beaker containing 200 mL of ethanol-water (1:1 *v*/*v*) mixture. The surrounding solvent was exchanged every 6 h, and a fresh 200 mL of ethanol-water was used. Finally, the purified sample was separated and dried at room temperature under vacuum for 12 h. The Mela-Glu precursor cross-linked hydrogel obtained was designated as PNIPAm-co-PAAm-Mela HG (Scheme 1).

2.5. *Characterization*

^{1}H-NMR analysis of the PNIPAm-co-PAAm copolymer, Mela-Glu precursor, and PNIPAm-co-PAAm-Mela HG sample was carried out using the OXFORD instrument (600 MHz). Scanning electron microscopy (SEM, JEOL 6400 instrument) at 10 kV was applied to measure the surface structure of the prepared copolymer sample. X-ray photoelectron spectroscopy (XPS) analysis was performed on the XPS instrument (Tucson, AZ, USA 85706). Fourier-transform infrared (FTIR) analysis was carried out using JASCO FTIR 4100 instrument. Particle size and the zeta potential measurements were performed on the Malvern Zetasizer Nano-ZS. UV-vis spectral analysis was performed using an Agilent Inc. UV-Vis spectrophotometer.

2.6. *Turbidity Measurement*

For this experiment, about 25 mg/mL of the PNIPAm-co-PAAm-Mela HG sample was dissolved in 5 mL distilled water to measure the turbidity of the synthesized PNIPAm-co-PAAm-Mela HG sample. The absorbance of the obtained polymer solution was measured at various temperatures ranging from 25 to 80 °C. The absorbance of the sample at each temperature was measured.

2.7. *Drug Loading into PNIPAm-co-PAAm-Mela HG*

A model drug, Cur, was used for drug loading and release experiments in vitro. The swelling diffusion method was used to load Cur into the PNIPAm-co-PAAm-Mela HG system at a polymer: drug *w*/*w* ratio of (10:3) [28]. In 3 mL of water, 0.1 g of PNIPAm-co-PAAm-Mela HG sample was dissolved, and 30 mg of Cur drug was combined. For 24 h, the suspension mixture was stirred at room temperature. The drug loading content into the PNIPAm-co-PAAm-Mela HG was determined using a UV-Vis spectrophotometer at 425 nm after the Cur encapsulated copolymer was separated by centrifugation at 50 °C. The drug-loaded sample was labeled as PNIPAm-co-PAAm-Mela@Cur HG (Scheme 2). The drug loading percentage was calculated to be around 78% using the following equation.

$$\text{Drug loading (\%)} = (\text{Wt. of the drug in sample}/\text{Wt. of the sample}) \times 100 \quad (1)$$

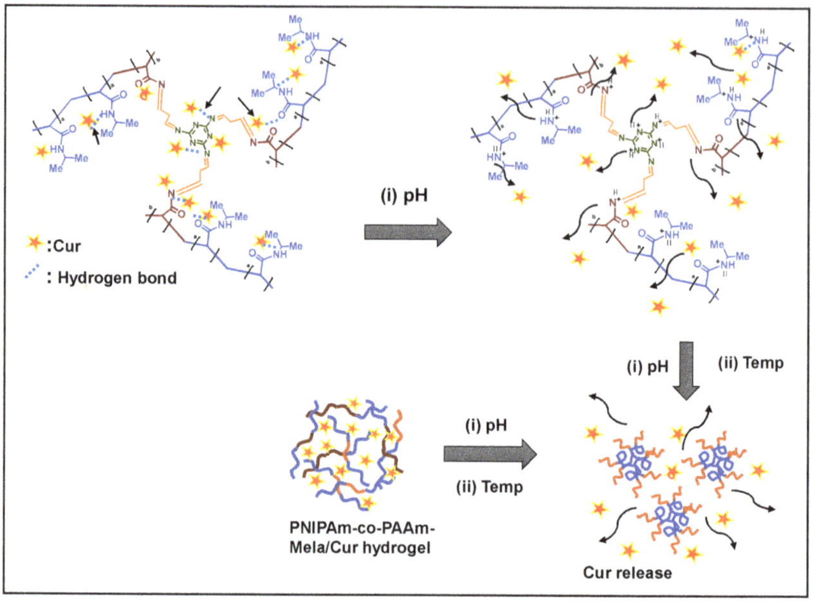

Scheme 2. A proposed schematic representation of drug loading into the PNIPAm-co-PAAm-Mela HG system. The pH stimuli and the combined pH and temperature-stimuli-responsive release behavior of the PNIPAm-co-PAAm-Mela HG system.

2.8. pH and Temperature-Responsive Drug Release Experiments

The PNIPAm-co-PAAm-Mela/Cur HG was evaluated in vitro under various circumstances, including (i) varied pH (pH 7.4 and 5.0); (ii) different temperatures (25 °C and 45 °C); and (iii) the combination of pH + temperature (pH 7.4/45 °C and pH 5/45 °C, respectively). For these experiments, approximately 25 mg/mL of drug-loaded sample PNIPAm-co-PAAm-Mela@Cur HG was placed in a dialysis bag (Mol. wt. cut off 5000 kDa), and the bag was immersed in a beaker containing 25 mL of phosphate-buffered saline (PBS) solution at different pH and temperature under gentle magnetic stirring. At the set time, about 1 mL of release media was removed, and the released Cur was measured using a UV-Vis spectrophotometer at 425 nm. The calibration curve plot was used to calculate the released Cur. The cumulative drug release was calculated using the equation below. Drug release (%) = (Amount of Cur released at time t/Total amount of Cur in the HG sample) × 100.

2.9. Cytocompatibility Study

The synthesized PNIPAm-co-PAAm-Mela HG, Cur loaded PNIPAm-co-PAAm-Mela@Cur HG, and pure Cur samples were tested for in vitro cytocompatibility utilizing the 3-(4,5-dimethylthiazol-2-yl)-2,5-diphenyl tetrazolium bromide (MTT) assay. HepG2 cells (2×10^4 cells/well) were grown in a 96-well plate for 24 h at 37 °C for this experiment. The existing medium was replaced with new media containing varying concentrations of PNIPAm-co-PAAm-Mela HG, PNIPAm-co-PAAm-Mela@Cur HG, and pure Cur. After 4 h, the MTT solution was added to each well and incubated for another 4 h. Following that, about 20 µL of DMSO was added to dissolve the existing formazan crystals, and the absorbance at 592 nm was measured using a microplate reader.

2.10. Statistical Analysis

All results, expressed as the mean ± SD, were analyzed using a two-tailed Student's t-test or one-way analysis of variance (ANOVA). The acceptable level of significance was $p < 0.05$.

3. Results and Discussion

3.1. Structural Study of PNIPAm-co-PAAm-Mela HG System

The ^{1}H NMR spectrum analysis confirmed the structure of the prepared PNIPAm-co-PAAm-Mela HG. The ^{1}H NMR spectra of the hydrogel system in $CDCl_3$ solvent are shown in Figure 1. Because of the solubility of hydrophilic PAAm and hydrophobic PNIPAm and Mela segments, all of the relevant proton signals for PNIPAm-co-PAAm-Mela HG block copolymer are presented in Figure 1a. The resonance peaks for methyl (-CH_2) and -OCH_2- groups were δ0.7 ppm, δ1.2 ppm and δ1.8 ppm, respectively. The significant resonance signal at δ3.2 ppm indicates that the copolymer included the PNIPAm polymer segment in the PNIPAm-co-PAAm copolymer HG. The resonance peaks of methyl protons in isopropyl units are indicated by the peak at δ3.6 ppm. Furthermore, after Mela groups cross-linking, the resonance signals at δ2.4 ppm and δ1.82 ppm indicated that the melamine groups were cross-linked via glutaraldehyde units Figure 1b.

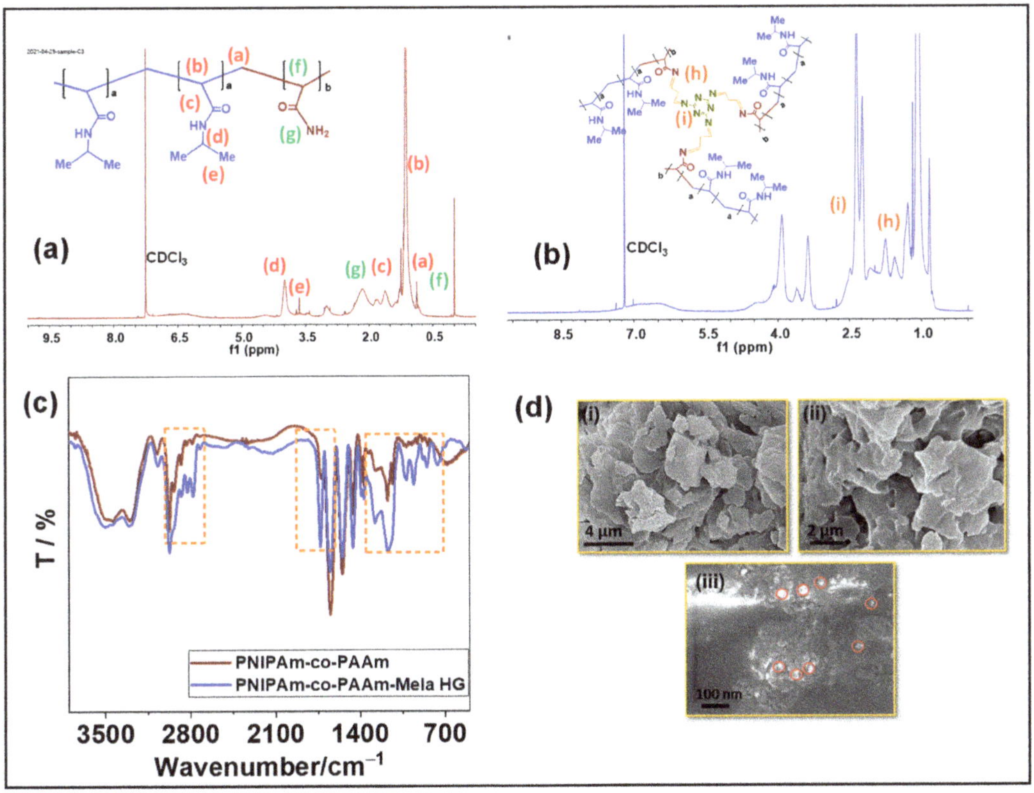

Figure 1. Characterization of PNIPAm-co-PAAm copolymer samples. (**a**,**b**) ^{1}H NMR spectrum; and (**c**) FT-IR analysis of PNIPAm-co-PAAm copolymer; and PNIPAm-co-PAAm-Mela HG samples, respectively. (**d**) SEM images of the (**i**) PNIPAm-co-PAAm copolymer; (**ii**) PNIPAm-co-PAAm-Mela HG; (**iii**) PNIPAm-co-PAAm-Mela/Cur HG samples, respectively. The red circles in figure (**d**) (**iii**) indicate the presence of Cur molecules in the hydrogel samples.

The structure of the PNIPAm-co-PAAm-Mela HG system was determined using FT-IR spectroscopy. The FTIR spectrum is shown in Figure 1c, with the characteristic vibration peak at 1542 cm^{-1} corresponding to the imine group (-C=N-) vibration, the peak at 1453 cm^{-1} indicating the stretching bands of C-N groups, and the sharp band at 1417 cm^{-1} indicating the N-H groups of NIPAm units in the PNIPAm-co-PAAm copolymer sample. The stretching mode of C=O groups in PNIPAm and PAAm segments was attributed to the absorption peak at 1727 cm^{-1}. Additionally, the intense stretching peak appeared at 2836 cm^{-1} and 2912 cm^{-1} of the PNIPAM-co-PAAM-Mela copolymer system's alkyl carbon C-C stretching modes.

SEM analysis was used to examine the surface morphology of the prepared copolymer samples. SEM analysis was carried out on dried powder samples of PNIPAm-co-PAAm copolymer, PNIPAm-co-PAAm-Mela HG, and Cur loaded PNIPAm-co-PAAm-Mela/Cur HG, respectively. The PNIPAm-co-PAAm copolymer exhibited flake-like particles with an average particle size of around 2–3 µm, as shown in Figure 1d(i). In contrast, the Mela-Glu precursor cross-linked PNIPAm-co-PAAm HG sample (Figure 1d(ii)) exhibited slight aggregation with interconnected particles with appropriate pores with an average diameter of about 1 µm, which could be attributed to the cross-linking of PNIPAm-co-PAAm polymer chains via Mela-Glu precursor. Furthermore, the Cur drug-loaded PNIPAm-co-PAAm-Mela/Cur HG sample exhibited a similar aggregated morphology with micropores (Figure 1d(iii)).

The composition of PNIPAm-co-PAAm-Mela HG was determined by XPS analysis. As seen in Figure 2, the XPS spectra with prominent signals for carbon (C1s), nitrogen (N1s), and oxygen (O1s) peaks suggested that the presence of N-isopropyl acrylamide and acrylamide monomer segments existed in the PNIPAm-co-PAAm-Mela HG system. The high-resolution C1s spectra of PNIPAm-co-PAAm-Mela HG were shown in Figure 2a, with the peak at 284.5 eV and −289.2 mV, representing the aliphatic C-C binding mode. The nitrogen peak at 403.3 eV indicates the C-N groups of N-isopropyl acrylamide and acrylamide monomers in the PNIPAm-co-PAAm-Mela HG. The N1s spectra in Figure 2b revealed two peaks at 398.4 eV and 399.25 eV, evidenced that the C-N and C=N bonds of the amide and imine groups of N-isopropyl acrylamide and acrylamide monomer, as well as the melamine units of PNIPAm-co-PAAm-Mela HG [29]. Furthermore, the O1s signal at 402.9 eV represents the C=O groups of N-isopropyl acrylamide and acrylamide monomer in the PNIPAm-co-PAAm-Mela HG (Figure 2c). This supports that the PNIPAm-co-PAAm-Mela copolymer material was successfully synthesized [30].

3.2. Physicochemical Properties of PNIPAm-co-PAAm-Mela HG System

The surface charge of the synthesized PNIPAm-co-PAAm-Mela HG was evaluated using zeta potential analysis. The presence of amide (-N-H), imine (-C=N-), and carbonyl (-C=O) groups in the copolymer segments increased the pH sensitivity of the PNIPAm-co-PAAm-Mela HG. The zeta potential of the PNIPAm-co-PAAm-Mela HG samples was measured at 25 °C and 45 °C, respectively. As shown in Figure 3a, the zeta potential value decreased from +19 mV to −13 mV and from +16 mV to −9 mV for samples evaluated at 25 °C and 45 °C, respectively, as the pH increased from 2 to 10. At low pH and higher temperature (45 °C), the hydrophobic PNIPAm segments combine to form the hydrophobic micelle core, whereas the hydrophilic PAAm forms the outer shell structure. The PNIPAm-co-PAAm-Mela HG, on the other hand, enhances hydrophilicity at higher pH, allowing them to transit the sol phase. The zeta potential values are slightly lower at higher temperatures (45 °C) than at lower temperatures (25 °C), which may be attributed to the development of aggregated micelles above LCST. The dynamic light scattering (DLS) technique was used to determine the linear to globule phase transition at different temperatures, such as 25 °C and 45 °C. As seen in Figure 3b, the DLS intensity remained consistent at 25 °C for the sample concentration of 25 mg/mL. Moreover, the solution temperature above LCST, the PNIPAm-co-PAAm-Mela HG showed increased turbidity (gel phase) and showed appropriate particle size when the solution temperature increased from 25 °C to 45 °C. On the other hand, when cooling the sample to about 25 °C, the sample did not show any considerable particle size and appeared to be a clear transparent solution. The DLS measurement under repeated heating and cooling conditions was examined at 45 °C and 25 °C, respectively (Figure 3b). DLS methods were also used to investigate the temperature responsiveness of the PNIPAm-co-PAAm-Mela HG polymer. Figure 3c shows that at 25 °C, no significant particles are formed. However, tiny particles are repeatedly formed at 45 °C. This work demonstrated that the PNIPAm-co-PAAm-Mela HG polymer undergoes a significant phase transition at temperatures above the LCST.

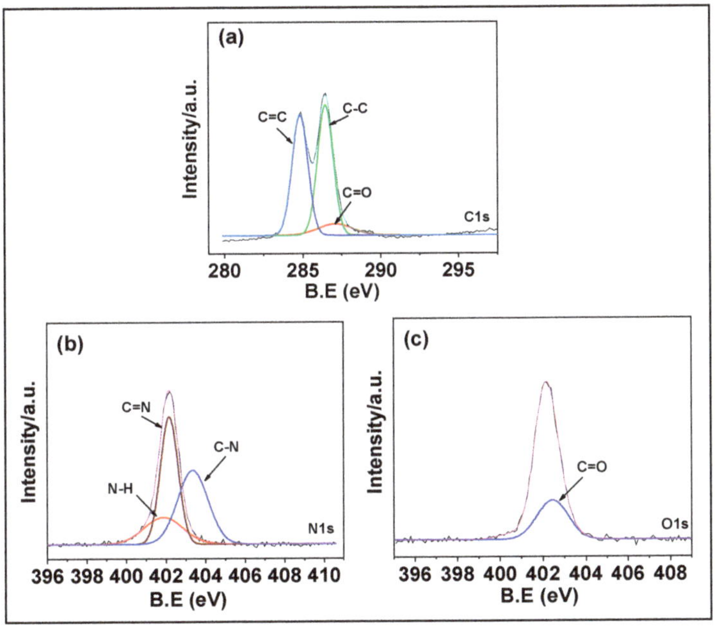

Figure 2. XPS analysis of PNIPAm-co-PAAm-Mela HG. The core level spectra of (**a**) C 1s; (**b**) N 1s; and (**c**) O 1s, respectively.

3.3. Phase Transition Mechanism of PNIPAm-co-PAAm-Mela HG System

The relative turbidity of PNIPAm-co-PAAm-Mela HG polymer (25 mg/mL) was investigated from 25 to 80 °C (Figure 4a). At a medium temperature of less than 40 °C, the sample displayed a transparent and homogenous clear solution, as illustrated in the graph (Figure 4a) and the sample vial. The PNIPAm-co-PAAm-Mela HG polymer absorbs water and becomes hydrated, swelling at this low LCST (linear structure). At temperatures above 40 °C, however, the clear solution became turbid, and the relative turbidity in the solution medium increased with increasing temperature to above 45 °C, indicating that the linear copolymer chain collapses into a hydrophobic globule micelle structure [31]. UV-Vis absorption and relative turbidity were used to study the swelling and deswelling properties of the prepared PNIPAm-co-PAAm-Mela HG at various solution temperatures. At 25 °C and 45 °C, UV-Vis absorbance of the PNIPAm-co-PAAm-Mela HG sample (25 mg/mL) was determined. As shown in Figure 4b, a significant weak absorption peak was seen at 25 °C, which might be attributed to the linear polymer structure and clear solution medium. At 45 °C, however, the solution medium becomes turbid due to temperature-induced micelle formation caused by the transition of the PNIPAm-co-PAAm-Mela HG's hydrophilic linear to hydrophobic globule structure above LCST (Figure 4b).

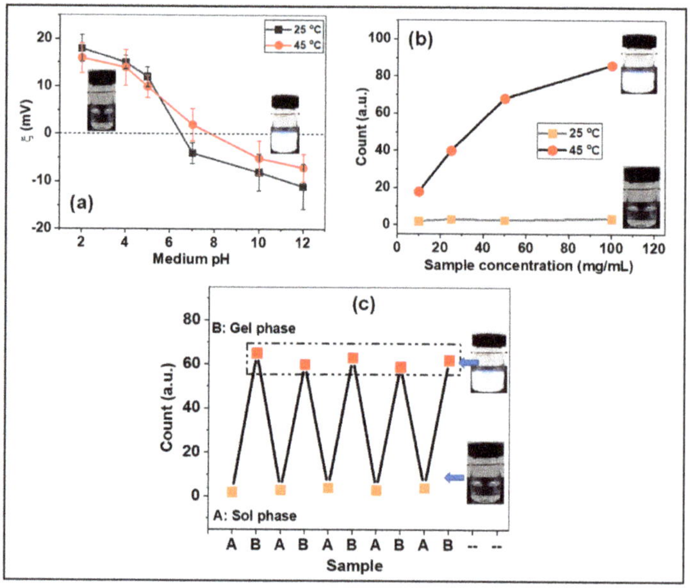

Figure 3. (**a**) Zeta potential and (**b**) particle size of PNIPAm−co−PAAm-Mela HG sample. Mean with error bar $n = 3$. (**c**) DLS analysis of PNIPAm−co−PAAm-Mela HG at below and above LCST.

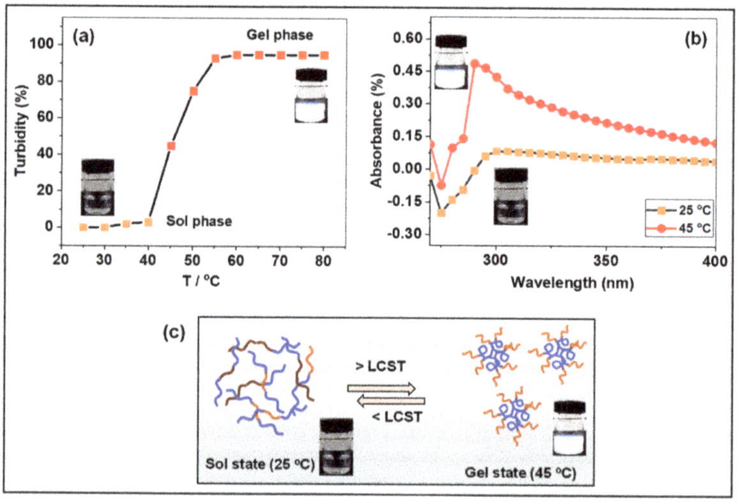

Figure 4. (**a**) Relative turbidity of PNIPAm−co−PAAm-Mela HG at 25 °C to 80 °C. (**b**) UV-vis absorption of PNIPAm-co-PAAm-Mela HG sample at 25°C and 45°C. (**c**) Illustrates the phase transition of PNIPAm−co−PAAm−Mela HG below and above LCST. The violet color indicates the hydrophobic PNIPAm domain, and the brown color indicates the hydrophilic PAAm segments in the presence of water.

The unique properties of dual-stimuli-responsive hydrogels include the ability to undergo noticeable phase transitions in response to physical stimuli rather than chemical or mechanical stimuli. At 25 °C, the PNIPAm-co-PAAm-Mela HG polymer dissolves readily in water and forms a non-cross-linked homogenous solution. The copolymer segments of PNIPAm-co-PAAm-Mela HG comprise hydrophilic amide (-CO-NH-) groups and hydrophobic isopropyl (-CH(CH_3)$_2$) groups. In deionized water, the PNIPAm-co-PAAm-Mela HG polymer undergoes a sharp phase transition and exists in solution as a linear hydrophilic polymer chain below LCST. In contrast, above LCST, the PNIPAm-co-PAAm-Mela HG transformed into a hydrophobic globule coil shape (Figure 4c).

3.4. Cur Loading and pH-Responsive Delivery from PNIPAm-co-PAAm-Mela/Cur HG System

Because the PNIPAm-co-PAAm-Mela HG has a higher LCST than human body temperature, fast micelle formation may be prevented when it is injected into the body. By fine-tuning the temperature stimuli, it is possible to maintain selective and controlled drug release to the target sites. Cur, an anticancer drug, may be encapsulated into the PNIPAm-co-PAAm-Mela HG by mixing them with a polymer solution at low temperatures. At low temperatures, the loaded Cur in the PNIPAm-co-PAAm-Mela HG can be protected against denaturation. The drug molecules can be interacted with the PNIPAm-co-PAAm-Mela HG by hydrogen bonding or electrostatic interactions (Scheme 2). The amide, imine, and carbonyl functional groups in the PNIPAm-co-PAAm-Mela HG system act as drug binding sites as well as protonation centers, ionize in low pH conditions and promote the release of loaded drugs from the PNIPAm-co-PAAm-Mela HG system (Scheme 2).

The pH and temperature-responsive drug release behavior of the prepared Cur loaded PNIPAm-co-PAAm-Mela/Cur HG has been evaluated at different pH and temperature conditions, specifically at different pH (pH 7.4 and 5.0); at different temperatures (25 °C and 45 °C); and the combined pH and temperature (pH 7.4/45 °C, pH 7.4/45 °C, respectively). First, the pH-responsive Cur release behavior from the PNIPAm-co-PAAm-Mela/Cur HG was studied.

Figure 5a demonstrated the Cur release behavior at various pHs, with approximately ~30% and ~82% of Cur released in 12 h at pH 7.4 and 5.0, respectively. The enhanced Cur release observed at pH 5.0 might be attributed to acid-induced protonation of the nitrogen part of PNIPAm and cross-linked Mela groups, as well as the loaded Cur molecules (Figure 5a). Second, Figure 5b depicted the temperature-responsive release behavior, which revealed that around ~26% and ~68% of Cur release was observed at 25 °C and 45 °C, respectively, throughout a 12 h release period. The release of physisorbed Cur molecules was responsible for the increase in release at 45 °C (Figure 5b). Third, at pH 7.4/45 °C and pH 5.0/45 °C, respectively, the combined pH and temperature-stimuli-responsive Cur release were determined. As seen in Figure 5c, a gradual release of Cur was detected, with about ~65% released at pH 7.4/45 °C; an almost complete release of Cur was observed at pH 5.0/45 °C, respectively, in a 12 h release period. Under the combined pH and temperature conditions, enhanced Cur release was observed due to the temperature-induced phase transition and pH-induced protonation of the functional groups and Cur molecules, which induce an electrostatic repulsive force. Therefore, the PNIPAm-co-PAAm-Mel HG system showed considerably enhanced Cur release under the combined pH and temperature stimuli conditions.

The drug loading mechanisms could be described as follows. As shown in Scheme 2, the loaded Cur molecules are strongly associated with the amine, imine, and amide groups via H-bonding/electrostatic interactions at pH 7.4. As a result, only a negligible amount of Cur was released from the PNIPAm-co-PAAm-Mela/Cur HG system at pH 7.4. The enhanced Cur release observed at the combined acidic pH and temperature (pH 5.0/45 °C) might be attributed to the temperature-induced phase change and acid-induced protonation of drug-binding functional sites and drug molecules, both of which force out the Cur molecules from the PNIPAm-co-PAAm-Mela/Cur HG system (Scheme 2).

The experiment results showed that combining pH and temperature stimuli resulted in greater Cur release efficiency from the PNIPAm-co-PAAm-Mela/Cur HG system than single stimuli, such as only pH or temperature stimuli (Tables 1 and 2). The majority of thermoresponsive polymers reported in the literature [23,32–34] primarily focused on temperature (Table 3), but in this work, our proposed melamine cross-linked PNIPAm-co-PAAm-Mela/Cur HG system has advantages such as enhanced drug loading and dual-stimuli-responsive drug release to the target sites.

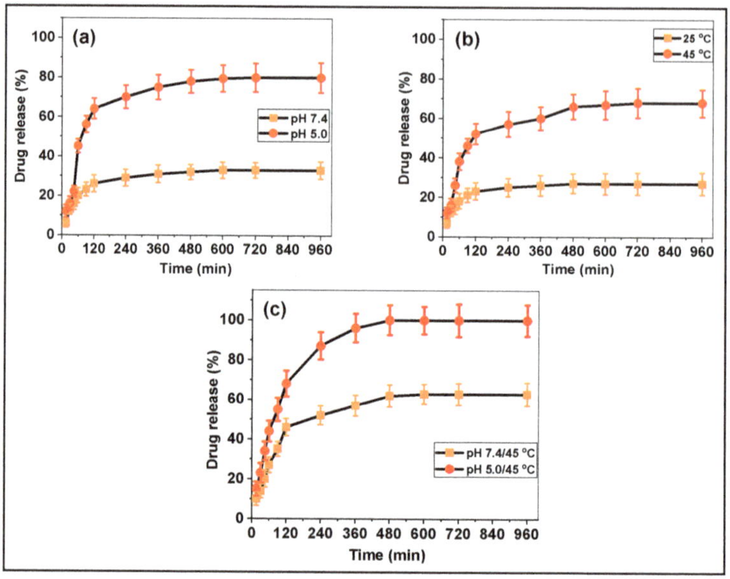

Figure 5. In vitro Cur delivery of PNIPAm-co-PAAm-Mela/Cur HG system. (**a**) Cur release at different pH, (**b**) Cur release at different temperature stimuli, and (**c**) Cur release with the combined pH and temperature conditions, respectively. Mean with error bar $n = 3$.

Table 1. Cur release efficiency from PNIPAm-co-PAAm-Mela HG system at different temperatures.

Sample	LCST (°C)	Release Efficiency (%)	
		25 °C	45 °C
PNIPAm-co-PAAm-Mela HG/Cur	40	26	68

Table 2. Cur release efficiency from PNIPAm-co-PAAm-Mel HG system at pH/temperatures.

pH	Release Efficiency (%)
7.4	30
5.0	82
7.4/45 °C	65
5.0/45 °C	100

Table 3. Various PNIPAM-based copolymer hydrogels for Cur delivery.

Polymers	Stimuli	Refs.
Quaternized triblock terpolymers	Temperature	[32]
pNIPAM grafted chitosan hydrogels	Temperature	[33]
Peptide-PNIPAM hydrogel	Temperature	[23]
Rechargeable pNIPAM hydrogel	Temperature	[34]
PNIPAm-co-PAAm-Mela HG	pH and temperature	This work

3.5. Cytocompatibility

The cytocompatibility of the prepared PNIPAm-co-PAAm copolymer, PNIPAm-co-PAAm-Mela HG, Cur-loaded PNIPAm-co-PAAm-Mela/Cur HG system, and pure Cur was tested in vitro at 37 °C using the HepG2 cell line. The cytocompatibility of control HepG2 cells and different concentrations of synthesized PNIPAm-co-PAAm copolymer samples are shown in Figure 6A. As shown in Figure 6A, the synthesized PNIPAm-co-PAAm copolymer exhibits about ~90% cell viability even at a sample concentration of 200 µg/mL, indicating that the PNIPAm-co-PAAm copolymer is biocompatible in nature [35,36]. On the other hand, Figure 6B shows that the PNIPAm-co-PAAm-Mela HG system without Cur loading demonstrated ~90% cell survival in all investigated sample concentrations, demonstrating that the prepared PNIPAm-co-PAAm-Mela HG also shows excellent biocompatibility to the HepG2 cells. In contrast, the Cur-loaded PNIPAm-co-PAAm-Mela/Cur HG system was shown to be toxic to HepG2 cells at all sample concentrations. It was observed that cells treated with a sample concentration of 200 µg/mL demonstrated nearly complete cell killing efficiency, implying that a concentration of 200 µg/mL of PNIPAm-co-PAAm-Mela HG is sufficient for complete cell killing [37]. This in vitro study indicates that the prepared PNIPAm-co-PAAm-Mela HG might be used for drug loading and pH stimuli-responsive drug release to specific sites.

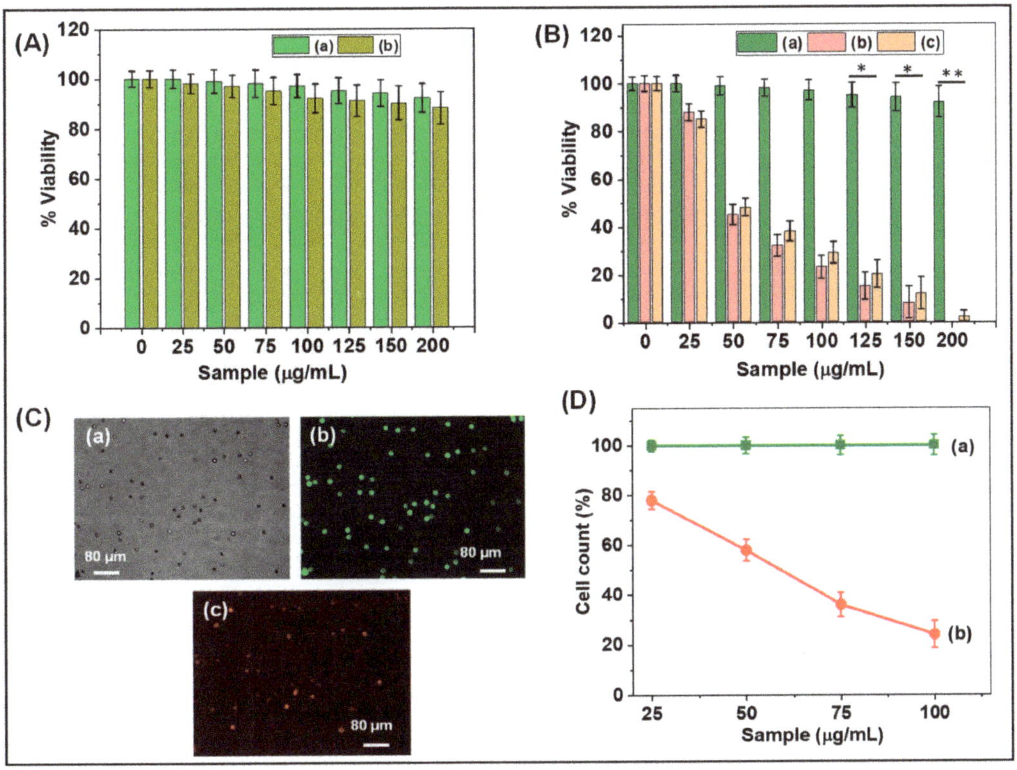

Figure 6. (**A**) In vitro cytocompatibility of (**a**) control HepG2 cells and (**b**) PNIPAm-co-PAAm copolymer; (**B**) In vitro cytocompatibility of (**a**)PNIPAm-co-PAAm-Mela HG; (**b**) PNIPAm-co-PAAm-Mela/Cur HG; and (**c**) pure Cur drug, respectively, at different concentrations. Statistical significance to the cell toxicity with different samples (*, significant $p < 0.05$; **, highly significant $p < 0.01$). (**C**) Fluorescence microscopy images of HepG2 cells represent (**a**) dark-field image; (**b**) PNIPAm-co-PAAm-Mela HG treated cells; and (**c**) Cur loaded PNIPAm-co-PAAm-Mela/Cur HG system, respectively. (**D**) Cell viability of (**a**) PNIPAm-co-PAAm-Mela HG treated cells; and (**b**) Cur loaded PNIPAm-co-PAAm-Mela/Cur HG system, respectively, at different sample concentrations. Mean with error bar $n = 3$.

3.6. Fluorescence Image Analysis

The cell-killing efficacy of the released Cur was determined by staining HepG2 cells with propidium iodide and acridine orange to distinguish between living and dead cells. As seen in Figure 6C, the cells treated with PNIPAm-co-PAAm-Mela HG fluoresced green, indicating that the majority of the cells were alive. The Cur-loaded PNIPAm-co-PAAm-Mela/Cur HG sample treated cells, on the other hand, exhibit red fluorescence, indicating that around >95% of cells were killed [38,39]. This might be due to the released Cur drugs from the PNIPAm-co-PAAm-Mela/Cur HG system, which induced cell death. As seen in Figure 6D, the cell killing efficiency is a function of sample concentration. The cell killing efficiency rose when the PNIPAm-co-PAAm-Mela/Cur HG sample concentration increased, which may be attributed to an increase in released Cur from the PNIPAm-co-PAAm-Mela/Cur HG system, and the released Cur induced more cell death.

4. Conclusions

In this study, we developed a dual-stimulus PNIPAm-co-PAAm-Mela/Cur HG copolymer system for temperature-responsive and pH-induced drug delivery applications. To evaluate the drug delivery behavior of the PNIPAm-co-PAAm-Mela HG copolymer, we employed Cur as a model drug. The PNIPAm-co-PAAm-Mela HG system had much higher Cur loading efficiency (73%) due to the existence of more drug-binding sites of amide, amine, and imine sites in the PNIPAm and Mel cross-linked PAAm segments. The drug release investigation revealed that a nearly complete release of Cur was accomplished in the presence of the combined pH and temperature stimulation conditions (pH 5.0/45 °C Furthermore, the results of the cytocompatibility investigation show that the prepared PNIPAm-co-PAAm-Mela HG system is cytocompatible and that the loaded drug may be released in the intracellular microenvironment. The overall study findings demonstrated that the PNIPAm-co-PAAm-Mela HG system might be used for controlled drug release to specific sites in chemotherapeutic applications.

Author Contributions: Conceptualization, methodology, formal analysis, K.T. and T.T.V.P.; validation, M.S. and V.R.; data curation, K.T.; writing—original draft preparation, T.T.V.P., M.S. and V.R.; writing—review and editing; supervision, project administration, funding acquisition, S.-C.K. All authors have read and agreed to the published version of the manuscript.

Funding: This research was supported by the Basic Science Research Program through the National Research Foundation of Korea (NRF), funded by the Ministry of Education (2020R1I1A3052258). In addition, the work was also supported by the Technology Development Program (S3060516), funded by the Ministry of SMEs and Startups (MSS, Republic of Korea), in 2021.

Institutional Review Board Statement: Not applicable.

Informed Consent Statement: Not applicable.

Data Availability Statement: The data presented in this study are available on request from the corresponding author.

Conflicts of Interest: The authors declare no conflict of interest.

References

1. Hauptstein, S.; Prüfert, F.; Bernkop-Schnürch, A. Self-nanoemulsifying drug delivery systems as novel approach for pDNA drug delivery. *Int. J. Pharm.* **2015**, *487*, 25e31. [CrossRef] [PubMed]
2. Lv, S.; Li, M.; Tang, Z.; Song, W.; Sun, H.; Liu, H.; Chen, X. Doxorubicin-loaded amphiphilic polypeptide-based nanoparticles as an efficient drug delivery system for cancer therapy. *Acta Biomater.* **2013**, *9*, 9330e9342. [CrossRef] [PubMed]
3. Faccia, P.A.; Pardini, F.M.; Amalvy, J.I. Evaluation of pH-sensitive poly(2-hydroxyethyl methacrylate-co-2-(diisopropylamino)ethyl methacrylate) copolymers as drug delivery systems for potential applications in ophthalmic therapies/ocular delivery of drugs. *Express Polym. Lett.* **2015**, *9*, 554e566. [CrossRef]
4. He, C.; Kim, S.W.; Lee, D.S. In Situ Gelling Stimuli-Sensitive Block Copolymer Hydrogels for Drug Delivery. *J. Control. Release* **2008**, *127*, 189–207. [CrossRef] [PubMed]
5. Zhang, Y.; Chan, H.F.; Leong, K.W. Advanced materials and processing for drug delivery: The past and the future. *Adv. Drug Deliv. Rev.* **2013**, *65*, 104e120. [CrossRef]

6. Zhao, L.; Xiao, C.; Ding, J.; He, P.; Tang, Z.; Pang, X.; Zhuang, X.; Chen, X. Facile onepot synthesis of glucose-sensitive nanogel via thiol-ene click chemistry for self-regulated drug delivery. *Acta Biomater.* **2013**, *9*, 6535e6543. [CrossRef]
7. Mitchell, M.J.; Billingsley, M.M.; Haley, R.M.; Wechsler, M.E.; Peppas, N.A.; Langer, R. Engineering precision nanoparticles for drug delivery. *Nat. Rev. Drug Discov.* **2021**, *20*, 101–124. [CrossRef]
8. Sung, Y.K.; Kim, S.W. Recent advances in polymeric drug delivery systems. *Biomater. Res.* **2020**, *24*, 12. [CrossRef]
9. Xu, L.; Wang, X.; Liu, Y.; Yang, G.; Falconer, R.J.; Zhao, C.-X. Lipid Nanoparticles for Drug Delivery. *Adv. Nanobiomed. Res.* **2022**, *2*, 2100109. [CrossRef]
10. Vo, T.S.; Vo, T.T.B.C.; Tran, T.T.; Pham, N.D. Enhancement of water absorption capacity and compressibility of hydrogel sponges prepared from gelatin/chitosan matrix with different polyols. *Prog. Nat. Sci. Mater. Int.* **2022**, *32*, 54–62. [CrossRef]
11. Karoyo, A.H.; Wilson, L.D. A Review on the Design and Hydration Properties of Natural Polymer-Based Hydrogels. *Materials* **2021**, *14*, 1095. [CrossRef] [PubMed]
12. Chai, Q.; Jiao, Y.; Yu, X. Hydrogels for Biomedical Applications: Their Characteristics and the Mechanisms behind Them. *Gels* **2017**, *3*, 6. [CrossRef] [PubMed]
13. Parameswar, A.V.; Fitch, K.R.; Bull, D.S.; Duke, V.R.; Goodwin, A.P. Polyacrylamide Hydrogels Produce Hydrogen Peroxide from Osmotic Swelling in Aqueous Media. *Biomacromolecules* **2018**, *19*, 3421–3426. [CrossRef] [PubMed]
14. Rizwan, M.; Yahya, R.; Hassan, A.; Yar, M.; Azzahari, A.D.; Selvanathan, V.; Sonsudin, F.; Abouloula, C.N. pH Sensitive Hydrogels in Drug Delivery: Brief History, Properties, Swelling, and Release Mechanism, Material Selection and Applications. *Polymers* **2017**, *9*, 137. [CrossRef] [PubMed]
15. Bordbar-Khiabani, A.; Gasik, M. Smart Hydrogels for Advanced Drug Delivery Systems. *Int. J. Mol. Sci.* **2022**, *23*, 3665. [CrossRef]
16. Farjadian, F.; Rezaeifard, S.; Naeimi, M.; Ghasemi, S.; Mohammadi-Samani, S.; Welland, M.E.; Tayebi, L. Temperature and pH-responsive nano-hydrogel drug delivery system based on lysine-modified poly (vinylcaprolactam). *Int. J. Nanomed.* **2019**, *14*, 6901–6915. [CrossRef]
17. Bellotti, E.; Schilling, A.L.; Little, S.R.; Decuzzi, P. Injectable thermoresponsive hydrogels as drug delivery system for the treatment of central nervous system disorders: A review. *J. Control. Release* **2021**, *329*, 16–35. [CrossRef]
18. Gulfam, M.; Jo, S.-H.; Jo, S.-W.; Vu, T.T.; Park, S.-H.; Lim, K.T. Highly porous and injectable hydrogels derived from cartilage acellularized matrix exhibit reduction and NIR light dual-responsive drug release properties for application in antitumor therapy. *NPG Asia Mater.* **2022**, *14*, 8. [CrossRef]
19. Yeingst, T.J.; Arrizabalaga, J.H.; Hayes, D.J. Ultrasound-Induced Drug Release from Stimuli-Responsive Hydrogels. *Gels* **2022**, *8*, 554. [CrossRef]
20. Das, D.; Pal, S. Dextrin/Poly (HEMA): pH Responsive Porous Hydrogel for Controlled Release of Ciprofloxacin. *Int. J. Biol. Macromol.* **2015**, *72*, 171–178. [CrossRef]
21. Brandl, F.; Kastner, F.; Gschwind, R.M.; Blunk, T.; Teßmar, J.; Göpferich, A. Hydrogel-Based Drug Delivery Systems: Comparison of Drug Diffusivity and Release Kinetics. *J. Control. Release* **2010**, *142*, 221–228. [CrossRef] [PubMed]
22. Moorthy, M.S.; Phan, T.T.V.; Ramkumar, V.; Raorane, C.J.; Thirupathi, K.; Kim, S.C. Thermo-Sensitive Poly (N-isopropylacrylamide-co-polyacrylamide) Hydrogel for pH-Responsive Therapeutic Delivery. *Polymers* **2022**, *14*, 4128.
23. Cao, M.; Wang, Y.; Hu, X.; Gong, H.; Li, R.; Cox, H.; Zhang, J.; Waigh, T.A.; Xu, H.; Lu, J.R. Reversible Thermoresponsive Peptide-PNIPAM Hydrogels for Controlled Drug Delivery. *Biomacromolecules* **2019**, *20*, 3601–3610. [CrossRef]
24. Cinay, G.E.; Erkoc, P.; Alipour, M.; Hashimoto, Y.; Sasaki, Y.; Akiyoshi, K.; Kizilel, S. Nanogel-Integrated pH-Responsive Composite Hydrogels for Controlled Drug Delivery. *ACS Biomater. Sci. Eng.* **2017**, *3*, 370–380. [CrossRef] [PubMed]
25. Emam, H.E.; Shaheen, T.I. Design of a dual pH and temperature responsive hydrogel based on esterified cellulose nanocrystals for potential drug release. *Carbohydrate Polym.* **2022**, *278*, 118925. [CrossRef]
26. Wan, W.-M.; Baggett, A.W.; Cheng, F.; Lin, H.; Liu, S.-Y.; Jäkle, F. Synthesis by free radical polymerization and properties of BN-polystyrene and BN-poly(vinylbiphenyl). *Chem. Commun.* **2016**, *52*, 13616–13619. [CrossRef]
27. Salem, M.; Mauguen, Y.; Prangé, T. Revisiting glutaraldehyde cross-linking: The case of the Arg–Lys intermolecular doublet. *Acta Crystallogr. Sect. F Struct. Biol. Cryst. Commun.* **2010**, *66*, 225–228. [CrossRef] [PubMed]
28. Bachmeier, B.E.; Killian, P.H.; Melchart, D. The Role of Curcumin in Prevention and Management of Metastatic Disease. *Int. J. Mol. Sci.* **2018**, *19*, 1716. [CrossRef] [PubMed]
29. Oh, Y.; Moorthy, M.S.; Manivasagan, P.; Bharathiraja, S.; Oh, J. Magnetic hyperthermia and pH-responsive effective drug delivery to the sub-cellular level of human breast cancer cells by modified $CoFe_2O_4$ nanoparticles. *Biochimie* **2017**, *133*, 7–19. [CrossRef]
30. Hou, L.; Wu, P. Comparison of LCST-transitions of homopolymer mixture, diblock and statistical copolymers of NIPAM and VCL in water. *Soft Matter.* **2015**, *11*, 2771–2781. [CrossRef]
31. Wang, Y.; García-Peñas, A.; Gómez-Ruiz, S.; Stadler, F.J. Surrounding Interactions on Phase Transition Temperature Promoted by Organometallic Complexes in Functionalized Poly(N-isopropylacrylamide-co-dopamine methacrylamide) Copolymers. *Macromol. Chem. Phys.* **2020**, *221*, 2000035. [CrossRef]
32. Skandalis, A.; Selianitis, D.; Pispas, S. PnBA-b-PNIPAM-b-PDMAEA Thermo-Responsive Triblock Terpolymers and Their Quaternized Analogs as Gene and Drug Delivery Vectors. *Polymers* **2021**, *13*, 2361. [CrossRef] [PubMed]
33. Luckanagul, J.A.; Pitakchatwong, C.; Bhuket, P.R.N.; Muangnoi, C.; Rojsitthisak, P.; Chirachanchai, S.; Wang, Q.; Rojsitthisak, P. Chitosan-based polymer hybrids for thermo-responsive nanogel delivery of curcumin. *Carbohydrate Polym.* **2018**, *181*, 1119–1127. [CrossRef] [PubMed]

34. Ayar, Z.; Shafieian, M.; Mahmoodi, N.; Sabzevari, O.; Hassannejad, Z. A rechargeable drug delivery system based on pNIPAMhydrogel for the local release of curcumin. *J. Appl. Polym. Sci.* **2021**, *138*, e51167. [CrossRef]
35. Moorthy, M.S.; Hoang, G.; Subramanian, B.; Bui, N.Q.; Panchanathan, M.; Mondal, S.; Tuong, V.P.T.; Kim, H.; Oh, J. Prussian blue decorated mesoporous silica hybrid nanocarriers for photoacoustic imaging-guided synergistic chemo-photothermal combination therapy. *J. Mater. Chem. B.* **2018**, *6*, 5220–5233. [CrossRef]
36. Manivasagan, P.; Bharathiraja, S.; Moorthy, M.S.; Mondal, S.; Nguyen, T.P.; Kim, H.; Phan, T.T.V.; Lee, K.D.; Oh, J. Biocompatible chitosan oligosaccharide modified gold nanorods as highly effective photothermal agents for ablation of breast cancer cells. *Polymers* **2018**, *10*, 232. [CrossRef]
37. Moorthy, M.S.; Bharathiraja, S.; Manivasagan, P.; Oh, Y.; Jang, B.; Phan, T.T.V.; Oh, J. Synthesis of urea-pyridyl ligand functionalized mesoporous silica hybrid material for hydrophobic and hydrophilic drug delivery application. *J. Porous. Mater.* **2018**, *25*, 119–128. [CrossRef]
38. Mondal, S.; Manivasagan, P.; Bharathiraja, P.; Moorthy, M.S.; Nguyen, V.T.; Kim, H.H.; Nam, S.Y.; Lee, K.D.; Oh, J. Hydroxyapatite Coated Iron Oxide Nanoparticles: A Promising Nanomaterial for Magnetic Hyperthermia Cancer Treatment. *Nanomaterials* **2017**, *7*, 426. [CrossRef]
39. Santhamoorthy, M.; Thirupathi, K.; Periyasamy, T.; Thirumalai, D.; Ramkumar, V.; Kim, S.-C. Ethidium bromide-bridged mesoporous silica hybrid nanocarriers for fluorescence cell imaging and drug delivery applications. *New J. Chem.* **2021**, *45*, 20641–20648. [CrossRef]

polymers

MDPI

Article

Composite Fibers Based on Polycaprolactone and Calcium Magnesium Silicate Powders for Tissue Engineering Applications

Cristina Busuioc [1], Andrada-Elena Alecu [1], Claudiu-Constantin Costea [1], Mihaela Beregoi [2], Mihaela Bacalum [3], Mina Raileanu [3], Sorin-Ion Jinga [1] and Iuliana-Mihaela Deleanu [1,*]

1 Department of Science and Engineering of Oxide Materials and Nanomaterials, Faculty of Chemical Engineering and Biotechnologies, University Politehnica of Bucharest, RO-060042 Bucharest, Romania
2 National Institute of Materials Physics, RO-077125 Magurele, Romania
3 National Institute of Physics and Nuclear Engineering, RO-077125 Magurele, Romania
* Correspondence: iuliana.deleanu@upb.ro

Abstract: The present work reports the synthesis and characterization of polycaprolactone fibers loaded with particulate calcium magnesium silicates, to form composite materials with bioresorbable and bioactive properties. The inorganic powders were achieved through a sol–gel method, starting from the compositions of diopside, akermanite, and merwinite, three mineral phases with suitable features for the field of hard tissue engineering. The fibrous composites were fabricated by electrospinning polymeric solutions with a content of 16% polycaprolactone and 5 or 10% inorganic powder. The physico-chemical evaluation from compositional and morphological points of view was followed by the biological assessment of powder bioactivity and scaffold biocompatibility. SEM investigation highlighted a significant reduction in fiber diameter, from around 3 μm to less than 100 nm after the loading stage, while EDX and FTIR spectra confirmed the existence of embedded mineral entities. The silicate phases were found be highly bioactive after 4 weeks of immersion in SBF, enriching the potential of the polymeric host that provides only biocompatibility and bioresorbability. Moreover, the cellular tests indicated a slight decrease in cell viability over the short-term, a compromise that can be accepted if the overall benefits of such multifunctional composites are considered.

Keywords: polycaprolactone; diopside; akermanite; merwinite; electrospinning; bone scaffolds

Citation: Busuioc, C.; Alecu, A.-E.; Costea, C.-C.; Beregoi, M.; Bacalum, M.; Raileanu, M.; Jinga, S.-I.; Deleanu, I.-M. Composite Fibers Based on Polycaprolactone and Calcium Magnesium Silicate Powders for Tissue Engineering Applications. *Polymers* **2022**, *14*, 4611. https://doi.org/10.3390/polym14214611

Academic Editors: Ariana Hudita and Bianca Gălățeanu

Received: 3 October 2022
Accepted: 27 October 2022
Published: 30 October 2022

1. Introduction

Tissue engineering (TE) means the development of biological substitutes to "restore, maintain, or improve tissue function" [1]. One of the strategies of this interdisciplinary field is represented by the design of suitable materials to be implemented as scaffolds. Depending on characteristics, a scaffold can offer temporary cell support, space filling, or be utilized as releasing matrix of active molecules [2]. Numerous biocompatible materials have been researched and clinically studied so far, including metals, ceramics, polymers, and composites; however, choosing the most suitable microenvironment to stimulate cellular adhesion, growth, and differentiation is still a challenge [3,4].

Polymer–bioceramic composites, which result from combining two or more distinct phases, can be successfully used as heterogenous functional matrices with the ability to mimic the natural extracellular matrix (ECM), and are currently intensively researched [5]. Depending on their composition, these kinds of materials can combine the advantages of each constituent to offer a biodegradability rate compatible with morphogenesis rate, biocompatibility, non-toxicity, non-immunogenicity, and adequate morphological and mechanical properties [2,5–8]. In the simplest formulation, a polymer–bioceramic composite is fabricated employing a polymer, to offer flexibility, and a bioceramic, to provide proper mechanical properties; this is similar to bone ECM, which is composed of approximately 35% organic matrix (collagen fibrils) and 65% mineral reinforcement (crystallized calcium phosphate) [2,5].

Among bioceramics, calcium magnesium silicates-based systems play an important role in TE due to their proven biocompatibility, excellent mechanical properties, and high versatility, as their chemical composition and structure can be easily redesigned to respond to different clinical conditions [9]. During the degradation process, these systems release bioactive ions at a controlled rate, with major benefits for bone formation and the human body. Specifically, Ca^{2+}, as the main element of bone tissue, is vital for osteogenesis, supporting cell growth and adapting cellular responses to bioceramics; Si^{2+} inhibits osteoporosis and stimulates the metabolic pathways of bone calcification; and Mg^{2+} modulates the degradation rate and mechanical strength of calcium silicates [6,10–15]. Furthermore, it has been found that Si^{2+} and Mg^{2+} are more important than Ca^{2+} in the process of cell differentiation [10]. So, in other words, each element has its role, but the properties of the resulting biomaterials are directly related to the ions' relative concentration [9,14].

Considering all the above, it is easy to acknowledge ternary silicate bioceramics, such as diopside (D), akermanite (A), or merwinite (M), as frequently used materials in bone regeneration applications. Numerous studies previously investigated their properties, including the capacity to inhibit microbial growth, which is particularly critical for bone matrix [9,16,17], and concluded their applicability for TE as pure structures [6,10,18–20], or doped for improved bioactivity [21–24]. Calcium magnesium silicates have been produced in the form of particles (subsequently employed as fillers or decorations), thin or thick coatings, as well as porous scaffolds; most researchers have reported suitable mechanical properties, high bioactivity, and lack of cytotoxicity. CaO–MgO–SiO_2-based bioceramics can be fabricated by wet or dry techniques: coprecipitation, sol–gel, combustion, spray pyrolysis, fusion process, solid-state sintering, etc. [6,13,15,25–27]. In our work, we chose the sol–gel method; although this requires rather expensive reagents and longer reaction times, it allows good control of the composition and structural characteristics of the obtained powders [13,25].

With respect to the main classes of polymers used in TE, natural and synthetic, the latter category offers the advantage of controlled composition and structure, tunable biodegradability, and possibility of functionalization. Polycaprolactone (PCL), as a synthetic material that enables good processability, precise control over degradation, molecular weight, or hydrophobicity [1], is often employed in nanofibrillar form in TE, drug delivery and wound healing [28–36]. It is approved by FDA and its biodegradability, bioresorbability, biocompatibility, and hydrophilicity can be modelled by the addition of inorganic additives [37]. Solution blow spinning, centrifugal spinning, electrospinning, and pressurized gyration have been studied to obtain PCL nanofibers [38]. Each method has advantages and drawbacks. Electrospinning, limited by its low yields relative to the industrial scale, is still one of the most used as it is a cost-effective, easy-to-apply technique in TE [4,39,40].

Thus, in this work, three types of mineral powders, differing in terms of calcium content, were loaded on polycaprolactone fibers by introducing them in the precursor electrospinning solutions. The novelty of this work resides in the achievement of new composites based on calcium magnesium silicates and polycaprolactone, namely the combination of the bioactive properties of the first with the bioresorbability of the second. Since few studies are available on this topic, their morphological and biological characteristics were investigated, and correlations were established with the compositional and processing parameters.

2. Materials and Methods

2.1. Powder Synthesis

The inorganic powders were synthesized by a sol–gel method and characterized from compositional, structural, and morphological points of view [41]. The sol–gel method is a wet-chemistry approach that involves the conversion of salt or alkoxide-type precursors first in a solution or colloidal suspension and then in a gel with high viscosity by hydrolysis and polycondensation/polymerization processes; such a gel consists of an extended network built on bridging oxygen and with high amounts of liquid phase embedded [42]. Briefly, three different powders were processed, having as a starting point the oxide compositions of diopside ($CaMgSi_2O_6$, D), akermanite ($Ca_2MgSi_2O_7$, A), and merwinite ($Ca_3MgSi_2O_8$, M); these were thermally treated at 1000 or 1300 °C. The pulverulent samples employed as mineral loading consist of consolidated blocks with a high percentage of open porosity; the first composition (D) contains diopside as leading crystalline compound, the second (A) is a balanced mixture of diopside, akermanite, and merwinite, and the third (M) has a majority of dicalcium silicate [41].

2.2. Composites Preparation

The composite fibers were fabricated through an electrospinning technique using chloroform (CF, $CHCl_3$, ≥99%, Sigma-Aldich, St. Louis, MO, USA) and N,N-dimethylformamide (DMF, C_3H_7NO, 99.8%, Sigma-Aldich) as solvents, polycaprolactone (PCL, $(C_6H_{10}O_2)_n$, Mw = 80,000 g/mol, Sigma-Aldich) as the polymeric phase, and the previously described powders as the inorganic component (D, A, M). The CF:DMF solvent ratio was set at 4:1. Electrospinning uses electrical forces to shape a jet of polymeric solution with adequate rheological properties that is pushed through a nozzle and afterwards subjected to processes of drying and stretching, until collected on a grounded support [43]. Briefly, the electrospinning solutions were prepared in two stages, as follows: in the first, the mineral powder was dispersed in the solvent mixture by ultrasonication for 30 min, while in the second, the polymer was dissolved in the obtained suspension by magnetic stirring for 24 h. For each type of powder (D, A, M), two suspensions were achieved, with 5 or 10 wt% inorganic content; PCL concentration was 16 wt% for all cases. The final samples were coded as PCL-D-5%, PCL-D-10%, PCL-A-5%, PCL-A-10%, PCL-M-5% and PCL-M-10%, as a function of loading type and concentration (Table 1).

Table 1. Composition of the electrospinning solutions.

No.	Sample Code	PCL (%)	D (%)	A (%)	M (%)
1	PCL	16	-	-	-
2	PCL-D-5%		5	-	-
3	PCL-D-10%		10	-	-
4	PCL-A-5%		-	5	-
5	PCL-A-10%		-	10	-
6	PCL-M-5%		-	-	5
7	PCL-M-10%		-	-	10

2.3. Investigation Techniques

2.3.1. Physico-Chemical Characterization

The morphology was evaluated by scanning electron microscopy coupled with energy-dispersive X-ray spectroscopy (SEM+EDX) with a FEI Quanta Inspect F electron microscope (FEI Company, Hillsboro, OR, USA),20 or 30 kV accelerating voltage, 10 mm working distance, and gold coating by DC magnetron sputtering for 40 s. The vibrational characteristics were investigated by attenuated total reflection Fourier-transform infrared spectroscopy (ATR-FTIR) with a Thermo Scientific Nicolet iS50 spectrophotometer (Thermo Fisher Scientific, Waltham, MA, USA), 400–4000 cm^{-1} wavenumber range, 4 cm^{-1} resolution, and 64 scans/sample.

2.3.2. Biological Evaluation

The powder bioactivity was assessed by immersion in simulated body fluid (SBF) prepared according to Kokubo et al. [44], $pH = 7.3$, at 37 °C, for 4 weeks. It is widely accepted that such in vitro studies represent a standard approach to evaluate the apatite-forming capability of implantable materials, as a first step towards hard tissue bonding through a chemically stable and mechanically appropriate interface; by immersion in SBF solutions that mimic the composition of human plasma and selecting suitable testing parameters, reliable data on specimen bioactivity can be achieved [45]. In our case, the solid to liquid ratio was 1:10.

Mouse fibroblasts (L929 cells) were grown in MEM supplemented with 2 mM L-glutamine, 10% fetal calf serum (FCS), 100 units/mL of penicillin, and 100 µg/mL of streptomycin, at 37 °C, in a humidified incubator, under an atmosphere containing 5% CO_2. All cell cultivation media and reagents were purchased from Biochrom AG (Berlin, Germany).

Cell viability was evaluated using 3-(4,5-dimethylthiazol-2-yl)-2,5-diphenyltetrazolium bromide (MTT) assay at 24 and 48 h after the cells were seeded onto the investigated fibrous composites. Briefly, the surfaces were sterilized in flow with UV light, 15 min on each side. Following the sterilization, 1 cm^2 squares were placed in 24-well plates and seeded with 20.000 cell/well for 24 and 48 h. After the desired time passed, the medium was exchanged with medium containing 1 mg/mL MTT and further incubated for 4 h in the incubator. Finally, the solution was extracted and the formed formazan crystals were dissolved in dimethyl sulfoxide (DMSO). Negative control was represented by cells cultivated on aluminum foil. The percentage of viable cells was obtained using Equation (1).

$$\textit{Cell viability} = [(A_{570} \textit{ of treated cells})/(A_{570} \textit{ of untreated cells})] \times 100\ (\%), \quad (1)$$

Morphological investigation was performed using fluorescence microscopy for cells grown for 24 h on the developed scaffolds. The cells were grown in the same way presented above; after 24 h, they were washed with PBS, fixed in 4% paraformaldehyde dissolved in PBS for 15 min, followed by a washing step. Sequentially, the cells were stained for 15 min with 20 µg/mL Acridine Orange solution and washed again with PBS. Finally, the fluorescence images were taken using a confocal microscope (Andor DSD2 Confocal Unit, Belfast, UK) mounted on an Olympus BX51 epifluorescence microscope, employing a 40× objective. The images were recorded using a suitable filter cube (excitation filter 466/40 nm, dichroic mirror 488 nm, and emission filter 525/54 nm).

3. Results and Discussion

Figure 1 displays SEM images at different magnifications of the pristine and D-loaded PCL fibers. Figure 1a,b indicates non-woven fibers, randomly arranged in several layers, with an average diameter of 3 µm for the polymeric fibers. The length of the fibers cannot be estimated, but it can be stated that they have a great tendency to gather tightly, sometimes even to stick. Their surface is smooth, the diameter relatively constant along the entire length, and the overall aspect is slightly wavy/winding, suggesting high flexibility. Very rarely, fibers with a much smaller diameter or areas with electrospinning defects can be detected.

The addition of mineral powders (D, A, M) at a proportion of 5 or 10% triggers significant modifications in the general appearance, reducing the fiber diameter to a large extent, below 100 nm, most of them being around 30 nm (Figure 1c,d and Figure 2). This behavior highlights the major changes induced by the presence of the powder in the precursor solution. In addition, the degree of interconnection increases, which makes the scaffolds appear in the form of networks with many points of connectivity, the most congested points being supported by the inorganic aggregates. The presence of micrometric entities homogeneously distributed in the volume of the fiber network is not equivalent to the presence of ceramic aggregates, thus indicating the emergence of polymeric beads on the primary fiber. However, a closer look reveals the existence of several bright areas corresponding to the ceramic bodies. The increase in the loading concentration does not result in significant differences to the general look, which suggests that a large proportion of the powder has sedimented in the precursor solution. The only major dissimilarity appears in the case of D, for which 10% concentration was not favorable for the electrospinning process, resulting in an electrosprayed sample (Figure 1e,f). This is probably correlated with the fact that the powder was thermally processed at a lower temperature (1000 °C) and consists of smaller entities that affect the solution viscosity and rheological properties, hindering the flow process and jet stretching.

Figure 1. SEM images of the electrospun samples: (**a**,**b**) PCL, (**c**,**d**) PCL-D-5%, and (**e**,**f**) PCL-D-10%.

Bafandeha et al. [46] fabricated poly (vinyl alcohol)/chitosan/akermanite scaffolds by electrospinning and reported fiber diameters of less than 100 nm, without beads, for 1 wt% akermanite content and with a considerable number of beads for 2 wt% akermanite, which was attributed to the improper viscosity of the electrospinning solution. In our case, 5 or 10% silicate powder was integrated in the precursor system, leading to considerable fiber thinning, of 30 times, as well as large and numerous local diameter enlargements because of the conductivity and viscosity changes occurring after mineral phase addition. An obvious reduction in fiber diameter was also observed for other PCL scaffolds loaded with mineral powders, such as ZnO, TiO_2, and HAp [47].

Otherwise, the inorganic aggregates are caught as if in a spider web, either covered with a small layer of PCL or just attached to the fiber surface (Figure 2), making their outer side available and prone to biomineralization. Thus, the mineral blocks that are coated with polymer will require a longer time for the bioactivity to be displayed, namely the period necessary for polymer degradation and ceramic surface exposure.

Figure 3 integrates the EDX spectra associated with the simple and composite fibers at the concentration of 5% ceramic powder. If, in the case of PCL, the main elements of the polymer (C, O) are visible, in the other three spectra, additional elements specific to the loading composition can be found (Si, Ca, Mg). In this way, the distribution of the mineral aggregates in the polymeric fibrous network is confirmed.

The loading procedure was also validated from the FTIR spectra (Figure 4) recorded for the pristine and powder-containing fibers. According to the scientific literature, the polymer fingerprint is defined by the vibrational bands typical of C–H, C=O, and C–O, occurring at around 2945, 1720, and 1165 cm^{-1}, respectively [48,49]. For the composite fibers, below 1100 cm^{-1}, some additional bands emerge; in the range 800–1100 cm^{-1}, the signals of Si–O bonds and Si–O–Si groups are visible, whereas below 550 cm^{-1}, bands specific to the Ca–O and Mg–O oxide bonds overlap [50–52].

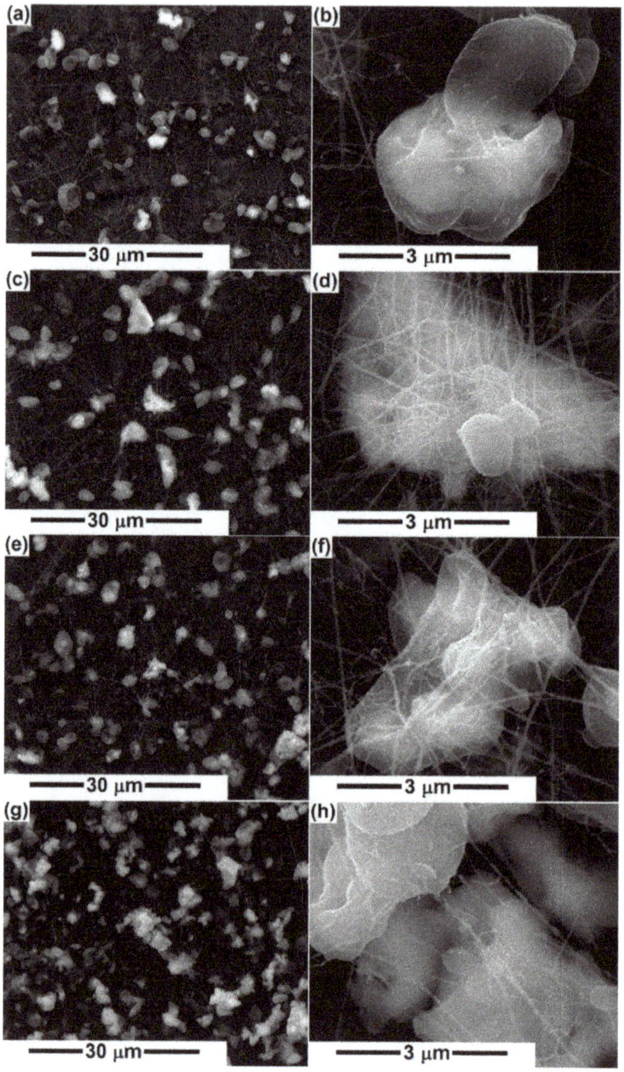

Figure 2. SEM images of the electrospun samples: (**a**,**b**) PCL-A-5%, (**c**,**d**) PCL-A-10%, (**e**,**f**) PCL-M-5%, and (**g**,**h**) PCL-M-10%.

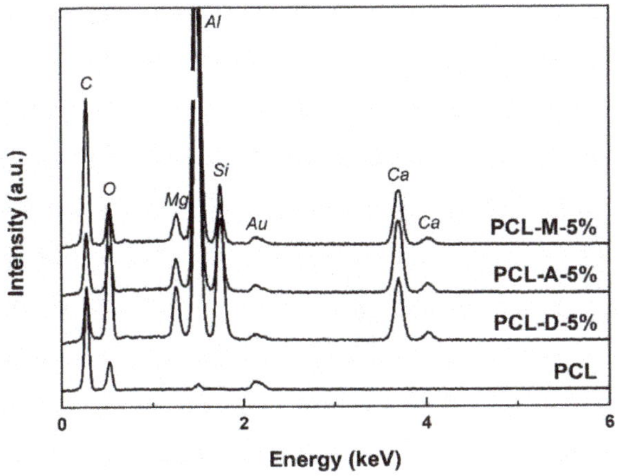

Figure 3. EDX spectra of the electrospun samples with 5% powder loading.

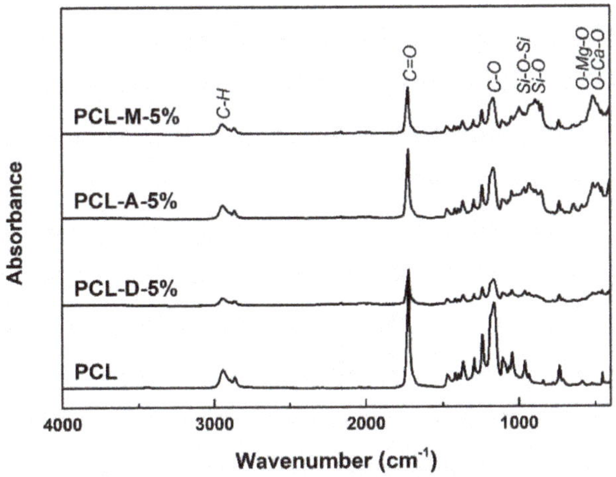

Figure 4. FTIR spectra of the electrospun samples with 5% powder loading.

In the SEM images taken of the thermally treated powders immersed in SBF for 4 weeks (Figure 5), the behavior of D, A and M during the biomineralization process is noticeable. In the case of D and A compositions, a new morphology is highlighted from place to place, namely quasi-spherical, porous structures, similar to a ball of fibers. As Ca concentration increases from D to A, the spheres grow in diameter (from around 100 nm to about 500 nm) and acquire a better-defined morphology, which suggests greater bioactivity. In the case of M, the morphology is different from the other two, and is also specific to apatite, a tangle of fibrillar structures that completely covers the block surface, of 10 nm in diameter for an individual entity. It can be concluded that the increase in Ca content leads to improved bioactivity, speeding up the healing process and subsequently the implant osseointegration. Furthermore, this in vitro assay validates the multifunctional character of the loaded scaffolds, since the bioresorbability of PCL is complemented by the bioactivity of the silicate powders, opening the possibility for the emergence of synergistic effects. Shahrouzifar et al. [23] obtained diopside scaffolds by sintering at 1200 °C and observed some spare nanometric plates of apatite on the surface of the pure scaffold, while F- and Sr-doped scaffolds were generously covered with apatite microspheres after 1 week of immersion in SBF; all these structures displayed a leaf-like morphology. Since our D powder was thermally treated at 1000 °C and the soaking time was longer (1 week vs. 4 weeks), better mineralization was achieved for an undoped material, confirming that the crystallization degree and grain growth represent determining parameters in the process of mineralization.

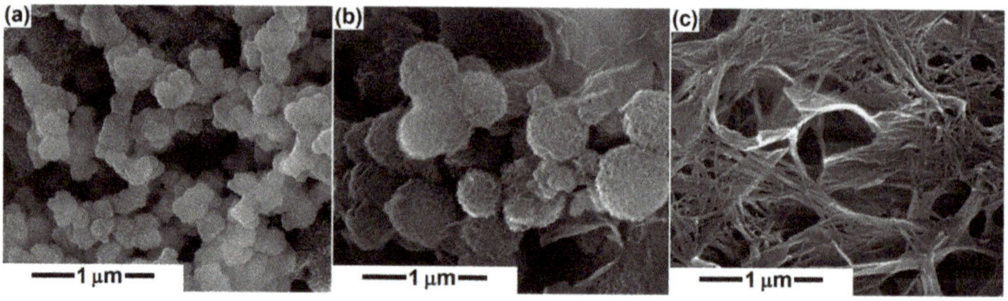

Figure 5. SEM images of the mineral powders after immersion in simulated body fluid for 4 weeks: (**a**) diopside, (**b**) akermanite, and (**c**) merwinite compositions.

When combined with chitosan for the production of scaffolds by 3D-bioprinting, akermanite conferred bioactivity, the surface becoming almost wholly coated with a thick layer of bone-like apatite after 10 days of immersion in SBF; the newly formed precipitate had a sponge-like morphology [10]. Moreover, the soaking period and akermanite concentration were strongly correlated with the degree of coverage, resulting in a dense and homogeneous apatite deposition for 80 wt% akermanite. Thus, it is desirable to have as much akermanite as possible in the fibrous scaffolds, so that a fast and extended mineralization can occur, which will further promote the intimate connection between the living bone and artificial implant, ensuring a stable and lasting bond.

Merwinite has been studied to a lesser extent, but the available data suggest that the rate of apatite growth is extremely high for merwinite compared to akermanite and diopside, following the same trend as CaO content; as observed, the freshly emerged layer displayed a cauliflower-like morphology and completely covered the scaffold surface, the density was extremely favored in the case of calcium resources provided by merwinite [9].

After testing the biocompatibility of the developed composite materials, it was found out that they were not cytotoxic and ensure a favorable platform for fibroblast proliferation. Thus, Figure 6 presents the calculated cell viability, which indicates that a concentration of 10% inorganic powder ensures a slightly better response compared to 5% loading. Moreover, a higher concentration of mineral loading is correlated with a small increase in cell viability with the seeding time.

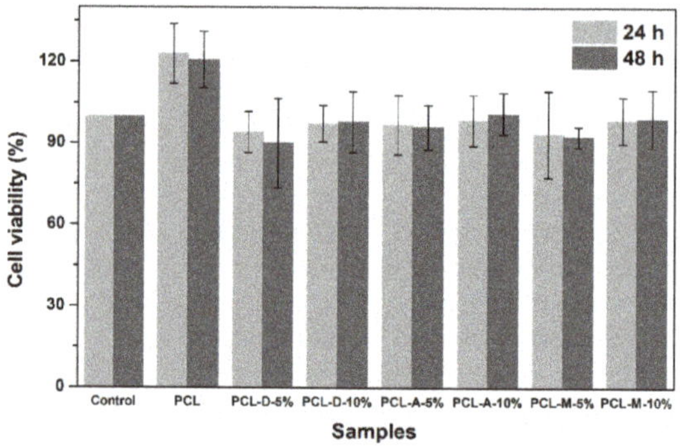

Figure 6. Cell viability for fibroblasts in contact with the electrospun samples for 24 and 48 h.

Indeed, PCL stands out as the sample with the highest cell viability; this result can be explained based on its different morphology compared to all other specimens. In other words, if PCL fibers have a diameter around 3 μm, the size of the powder-containing fibers drops below 100 nm (Figures 1 and 2). Due to its dimensional compatibility at the micrometric scale, the PCL scaffold seems to ensure a better environment for cell adhesion and proliferation. However, the addition of a silicate powder offers an indisputable advantage in terms of bioactivity, which makes this slight decrease in cell viability a compromise we can afford.

The morphological evaluation of L929 cells grown on the investigated materials was performed using fluorescence microscopy; the corresponding images are reported in Figure 7. Similar to the MTT results, the microscopy images reveal the biocompatibility of the electrospun samples, irrespective of the loading type and concentration. The fibroblast cells grown on the control surface show an elongated morphology and an intact oval nucleus, with a normal shape and size (Figure 7A); regardless of the experimental conditions, cell morphology was not altered (Figure 7B–H).

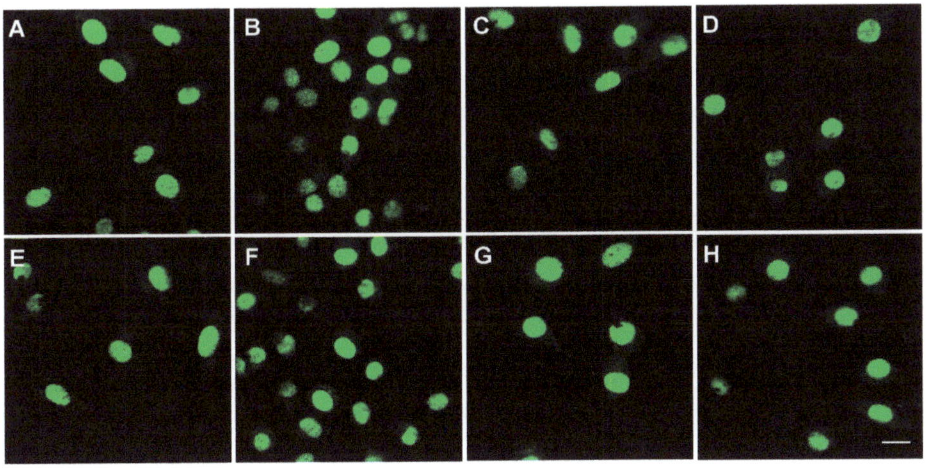

Figure 7. Morphology of fibroblasts in contact with the electrospun samples for 24 h: (**A**) Control, (**B**) PCL, (**C**) PCL-D-5%, (**D**) PCL-D-10%, (**E**) PCL-A-5%, (**F**) PCL-A-10%, (**G**) PCL-M-5%, and (**H**) PCL-M-10%.

The in vitro assays performed in this paper are in concordance with previous studies, which have confirmed that polycaprolactone-based scaffolds, membranes, or gels have good biocompatibility with osteoblasts, bone marrow mesenchymal stem cells, dental pulp cells, or fibroblast-like cells [49,53–60]. The encouraging results obtained for L929 fibroblasts justify additional investigations on the proposed materials. Further studies should focus on more information regarding cell adhesion and migration, as well as a long-term evaluation of cell behavior in the presence of the samples.

4. Conclusions

Calcium has a key role in bone remodeling by modulating the cellular responses to bioceramics, as well as promoting cell growth and osteogenic differentiation. Therefore, three calcium magnesium silicate compositions, corresponding to diopside, akermanite, and merwinite, were processed as thermally treated sol–gel-derived powders and subsequently integrated in polycaprolactone fibers. The fibrous composites were fabricated by electrospinning and evaluated in terms of powder loading degree and particle distribution in the fibers, as well as response to immersion in simulated biological fluid and influence on fibroblast cells. If the inorganic component proved to be bioactive, gradually increasing from diopside to akermanite and then to merwinite (the bioactivity increased with the increase in Ca amount), the loaded fibers represent cell-friendly supports, achieving better cell proliferation when combined with a higher powder concentration.

Author Contributions: Conceptualization, C.B. and I.-M.D.; methodology, C.B. and M.B. (Mihaela Beregoi); validation, C.B.; investigation, C.B., A.-E.A., C.-C.C., M.B. (Mihaela Bacalum) and M.R.; resources, C.B., S.-I.J., M.B. (Mihaela Beregoi), M.B. (Mihaela Bacalum) and M.R.; writing—original draft preparation, C.B., A.-E.A. and I.-M.D.; writing—review and editing, C.B. and I.-M.D.; supervision, C.B. and I.-M.D. All authors have read and agreed to the published version of the manuscript.

Funding: This research received no external funding.

Institutional Review Board Statement: Not applicable.

Data Availability Statement: Not applicable.

Acknowledgments: This work was supported by the grant POCU/993/6/13—153178, co-financed by the European Social Fund within the Sectorial Operational Program Human Capital 2014–2020.

Conflicts of Interest: The authors declare no conflict of interest.

References

1. Langer, R.; Vacanti, J.P. Tissue Engineering. *Science* **1993**, *260*, 920–926. [CrossRef] [PubMed]
2. Costa-Pinto, A.R.; Reis, R.L.; Neves, N.M. Scaffolds Based Bone Tissue Engineering: The Role of Chitosan. *Tissue Eng. Part B Rev.* **2011**, *17*, 331–347. [CrossRef] [PubMed]
3. Gandolfi, M.G.; Zamparini, F.; Degli Esposti, M.; Chiellini, F.; Fava, F.; Fabbri, P.; Taddei, P.; Prati, C. Highly Porous Polycaprolactone Scaffolds Doped with Calcium Silicate and Dicalcium Phosphate Dihydrate Designed for Bone Regeneration. *Mater. Sci. Eng. C* **2019**, *102*, 341–361. [CrossRef]
4. Mohammadalizadeh, Z.; Bahremandi-Toloue, E.; Karbasi, S. Recent Advances in Modification Strategies of Pre- and Post-Electrospinning of Nanofiber Scaffolds in Tissue Engineering. *React. Funct. Polym.* **2022**, *172*, 105202. [CrossRef]
5. Kaur, G.; Kumar, V.; Baino, F.; Mauro, J.C.; Pickrell, G.; Evans, I.; Bretcanu, O. Mechanical Properties of Bioactive Glasses, Ceramics, Glass-Ceramics and Composites: State-of-the-Art Review and Future Challenges. *Mater. Sci. Eng. C* **2019**, *104*, 109895. [CrossRef]
6. Arastouei, M.; Khodaei, M.; Atyabi, S.M.; Jafari Nodoushan, M. Poly Lactic Acid-Akermanite Composite Scaffolds Prepared by Fused Filament Fabrication for Bone Tissue Engineering. *J. Mater. Res. Technol.* **2020**, *9*, 14540–14548. [CrossRef]
7. Salernitano, E.; Migliaresi, C. Composite Materials for Biomedical Applications: A Review. *J. Appl. Biomater. Biomech.* **2003**, *1*, 3–18.
8. Sadeghzade, S.; Emadi, R.; Tavangarian, F.; Doostmohammadi, A. In Vitro Evaluation of Diopside/Baghdadite Bioceramic Scaffolds Modified by Polycaprolactone Fumarate Polymer Coating. *Mater. Sci. Eng. C* **2020**, *106*, 110176. [CrossRef]
9. Collin, M.S.; Venkatraman, S.K.; Sriramulu, M.; Shanmugam, S.; Drweesh, E.A.; Elnagar, M.M.; Mosa, E.S.; Sasikumar, S. Solution Combustion Synthesis of Functional Diopside, Akermanite, and Merwinite Bioceramics: Excellent Biomineralization, Mechanical Strength, and Antibacterial Ability. *Mater. Today Commun.* **2021**, *27*, 102365. [CrossRef]
10. Duman, Ş. Effect of Akermanite Powders on Mechanical Properties and Bioactivity of Chitosan-Based Scaffolds Produced by 3D-Bioprinting. *Ceram. Int.* **2021**, *10*, 13912–13921. [CrossRef]
11. Razavi, M.; Fathi, M.; Savabi, O.; Hashemi Beni, B.; Vashaee, D.; Tayebi, L. Nanostructured Merwinite Bioceramic Coating on Mg Alloy Deposited by Electrophoretic Deposition. *Ceram. Int.* **2014**, *40*, 9473–9484. [CrossRef]
12. Tavangarian, F.; Zolko, C.A.; Davami, K. Synthesis, Characterization and Formation Mechanisms of Nanocrystalline Akermanite Powder. *J. Mater. Res. Technol.* **2021**, *11*, 792–800. [CrossRef]
13. Tavangarian, F. Facile Synthesis and Structural Insight of Nanostructure Akermanite Powder. *Ceram. Int.* **2019**, *7*, 7871–7877. [CrossRef]
14. Praharaj, S.; Venkatraman, S.K.; Vasantharaman, R.; Swamiappan, S. Sol-Gel Combustion Synthesis of Merwinite and Its Biomedical Applications. *Mater. Lett.* **2021**, *300*, 130108. [CrossRef]
15. Sherikar, B.N.; Sahoo, B.; Umarji, A.M. One-Step Synthesis of Diopside (CaMgSi2O6) Ceramic Powder by Solution Combustion Method. *Adv. Powder Technol.* **2020**, *31*, 3492–3499. [CrossRef]
16. Bakhsheshi-Rad, H.R. Coating Biodegradable Magnesium Alloys with Electrospun Poly-L-Lactic Acid-Åkermanite-Doxycycline Nanofibers for Enhanced Biocompatibility, Antibacterial Activity, and Corrosion Resistance. *Coat. Technol.* **2019**, *14*, 124898. [CrossRef]
17. Choudhary, R.; Venkatraman, S.K.; Chatterjee, A.; Vecstaudza, J.; Yáñez-Gascón, M.J.; Pérez-Sánchez, H.; Locs, J.; Abraham, J.; Swamiappan, S. Biomineralization, Antibacterial Activity and Mechanical Properties of Biowaste Derived Diopside Nanopowders. *Adv. Powder Technol.* **2019**, *30*, 1950–1964. [CrossRef]

18. Abdollahi, S.; Paryab, A.; Khalilifard, R.; Anousheh, M.; Malek Khachatourian, A. The Fabrication and Characterization of Bioactive Akermanite/Octacalcium Phosphate Glass-Ceramic Scaffolds Produced via PDC Method. *Ceram. Int.* **2021**, *47*, 6653–6662. [CrossRef]
19. Sayed, M.; Mahmoud, E.M.; Bondioli, F.; Naga, S.M. Developing Porous Diopside/Hydroxyapatite Bio-Composite Scaffolds via a Combination of Freeze-Drying and Coating Process. *Ceram. Int.* **2019**, *45*, 9025–9031. [CrossRef]
20. Goudouri, O.M.; Theodosoglou, E.; Kontonasaki, E.; Will, J.; Chrissafis, K.; Koidis, P.; Paraskevopoulos, K.M.; Boccaccini, A.R. Development of Highly Porous Scaffolds Based on Bioactive Silicates for Dental Tissue Engineering. *Mater. Res. Bull.* **2014**, *49*, 399–404. [CrossRef]
21. Myat-Htun, M. Enhanced Sinterability and in Vitro Bioactivity of Barium-Doped Akermanite Ceramic. *Ceram. Int.* **2020**, *7*, 19062–19068. [CrossRef]
22. Hosseini, Y.; Emadi, R.; Kharaziha, M. Surface Modification of PCL-Diopside Fibrous Membrane via Gelatin Immobilization for Bone Tissue Engineering. *Mater. Chem. Phys.* **2017**, *194*, 356–366. [CrossRef]
23. Shahrouzifar, M.R.; Salahinejad, E.; Sharifi, E. Co-Incorporation of Strontium and Fluorine into Diopside Scaffolds: Bioactivity, Biodegradation and Cytocompatibility Evaluations. *Mater. Sci. Eng. C* **2019**, *103*, 109752. [CrossRef] [PubMed]
24. Mahdy, E.A.; Ahmed, H.Y.; Farag, M.M. Combination of Na-Ca-Phosphate and Fluorapatite in Wollastonite-Diopside Glass-Ceramic: Degradation and Biocompatibility. *J. Non-Cryst. Solids* **2021**, *566*, 120888. [CrossRef]
25. Yamagata, C.; Leme, D.R.; Rodrigues, V.G.; Eretides, G.T.; Dorion Rodas, A.C. Three Routes for the Synthesis of the Bioceramic Powder of the CaO-MgO-SiO_2 System. *Ceram. Int.* **2022**, *48*, 9681–9691. [CrossRef]
26. Feng, J.; Wu, D.; Long, M.; Lei, K.; Sun, Y.; Zhao, X. Diopside Glass-Ceramics Were Fabricated by Sintering the Powder Mixtures of Waste Glass and Kaolin. *Ceram. Int.* **2022**, *48*, 27088–27096. [CrossRef]
27. Srinath, P.; Abdul Azeem, P.; Venugopal Reddy, K.; Chiranjeevi, P.; Bramanandam, M.; Prasada Rao, R. A Novel Cost-Effective Approach to Fabricate Diopside Bioceramics: A Promising Ceramics for Orthopedic Applications. *Adv. Powder Technol.* **2021**, *32*, 875–884. [CrossRef]
28. Siqueira, I.A.W.B.; de Moura, N.K.; de Barros Machado, J.P.; Backes, E.H.; Roberto Passador, F.; de Sousa Trichês, E. Porous Membranes of the Polycaprolactone (PCL) Containing Calcium Silicate Fibers for Guided Bone Regeneration. *Mater. Lett.* **2017**, *206*, 210–213. [CrossRef]
29. Torres, E.; Fombuena, V.; Vallés-Lluch, A.; Ellingham, T. Improvement of Mechanical and Biological Properties of Polycaprolactone Loaded with Hydroxyapatite and Halloysite Nanotubes. *Mater. Sci. Eng. C* **2017**, *75*, 418–424. [CrossRef]
30. Chunyan, Z.; Lan, C.; Jiajia, L.; Dongwei, S.; Jun, Z.; Huinan, L. In Vitro Evaluation of Degradation, Cytocompatibility and Antibacterial Property of Polycaprolactone/Hydroxyapatite Composite Coating on Bioresorbable Magnesium Alloy. *J. Magnes. Alloys* **2021**, *10*, 2252–2265. [CrossRef]
31. Mendes Soares, I.P.; Anselmi, C.; Kitagawa, F.A.; de Ribeiro, R.A.O.; Leite, M.L.; de Souza Costa, C.A.; Hebling, J. Nano-Hydroxyapatite-Incorporated Polycaprolactone Nanofibrous Scaffold as a Dentin Tissue Engineering-Based Strategy for Vital Pulp Therapy. *Dent. Mater.* **2022**, *38*, 960–977. [CrossRef] [PubMed]
32. Surmenev, R.A.; Shkarina, S.; Syromotina, D.S.; Melnik, E.V.; Shkarin, R.; Selezneva, I.I.; Ermakov, A.M.; Ivlev, S.I.; Cecilia, A.; Weinhardt, V.; et al. Characterization of Biomimetic Silicate- and Strontium-Containing Hydroxyapatite Microparticles Embedded in Biodegradable Electrospun Polycaprolactone Scaffolds for Bone Regeneration. *Eur. Polym. J.* **2019**, *113*, 67–77. [CrossRef]
33. Sadeghi, A.; Razavi, S.M.A.; Shahrampour, D. Fabrication and Characterization of Biodegradable Active Films with Modified Morphology Based on Polycaprolactone-Polylactic Acid-Green Tea Extract. *Int. J. Biol. Macromol.* **2022**, *205*, 341–356. [CrossRef] [PubMed]
34. Liu, K.; Wang, J.; Fang, S.; Wang, H.; Bai, Y.; Zhao, Z.; Zhu, Q.; Wang, C.; Chen, G.; Jiang, H.; et al. Effect of Polycaprolactone Impregnation on the Properties of Calcium Silicate Scaffolds Fabricated by 3D Printing. *Mater. Des.* **2022**, *220*, 110856. [CrossRef]
35. Li, J.; Wang, C.; Gao, G.; Yin, X.; Pu, X.; Shi, B.; Liu, Y.; Huang, Z.; Wang, J.; Li, J.; et al. MBG/PGA-PCL Composite Scaffolds Provide Highly Tunable Degradation and Osteogenic Features. *Bioact. Mater.* **2022**, *15*, 53–67. [CrossRef]
36. Ghaziof, S.; Shojaei, S.; Mehdikhani, M.; Khodaei, M.; Jafari Nodoushan, M. Electro-Conductive 3D Printed Polycaprolactone/Gold Nanoparticles Nanocomposite Scaffolds for Myocardial Tissue Engineering. *J. Mech. Behav. Biomed. Mater.* **2022**, *132*, 105271. [CrossRef] [PubMed]
37. El-Morsy, M.A.; Afifi, M.; Ahmed, M.K.; Awwad, N.S.; Ibrahium, H.A.; Alqahtani, M.S. Electrospun Nanofibrous Scaffolds of Polycaprolactone Containing Binary Ions of Pd/Vanadate Doped Hydroxyapatite for Biomedical Applications. *J. Drug Deliv. Sci. Technol.* **2022**, *70*, 103153. [CrossRef]
38. Altun, E.; Ahmed, J.; Onur Aydogdu, M.; Harker, A.; Edirisinghe, M. The Effect of Solvent and Pressure on Polycaprolactone Solutions for Particle and Fibre Formation. *Eur. Polym. J.* **2022**, *173*, 111300. [CrossRef]
39. Salehi, A.O.M.; Keshel, S.H.; Rafienia, M.; Nourbakhsh, M.S.; Baradaran-Rafii, A. Promoting Keratocyte Stem like Cell Proliferation and Differentiation by Aligned Polycaprolactone-Silk Fibroin Fibers Containing Aloe Vera. *Biomater. Adv.* **2022**, *137*, 212840. [CrossRef]
40. Rogina, A. Electrospinning Process: Versatile Preparation Method for Biodegradable and Natural Polymers and Biocomposite Systems Applied in Tissue Engineering and Drug Delivery. *Appl. Surf. Sci.* **2014**, *296*, 221–230. [CrossRef]
41. Alecu, A.E.; Costea, C.C.; Surdu, V.A.; Voicu, G.; Jinga, S.I.; Busuioc, C. Processing of Calcium Magnesium Silicates by the Sol-Gel Route. *Gels* **2022**, *8*, 574. [CrossRef] [PubMed]

42. Klein, L.; Aparicio, M.; Jitianu, A. *Handbook of Sol-Gel Science and Technology: Processing, Characterization and Applications*; Springer: Berlin/Heidelberg, Germany, 2018. [CrossRef]
43. Mitchell, G.R. *Electrospinning: Principles, Practice and Possibilities*; Royal Society of Chemistry: London, UK, 2015. [CrossRef]
44. Kokubo, T.; Kushitani, H.; Sakka, S.; Kitsugi, T.; Yamamuro, T. Solutions Able to Reproduce In Vivo Surface-Structure Changes in Bioactive Glass-Ceramic A-W. *J. Biomed. Mater. Res.* **1990**, *24*, 721–734. [CrossRef] [PubMed]
45. Baino, F.; Yamaguchi, S. The Use of Simulated Body Fluid (SBF) for Assessing Materials Bioactivity in the Context of Tissue Engineering: Review and Challenges. *Biomimetics* **2020**, *5*, 57. [CrossRef] [PubMed]
46. Bafandeha, M.R.; Mojarrabianb, H.M.; Doostmohammadic, A. Poly (Vinyl Alcohol)/Chitosan/Akermanite Nanofibrous Scaffolds Prepared by Electrospinning. *J. Macromol. Sci. B* **2019**, *58*, 749–759. [CrossRef]
47. Jinga, S.I.; Zamfirescu, A.I.; Voicu, G.; Enculescu, M.; Evanghelidis, A.; Busuioc, C. PCL-ZnO/TiO_2/HAp Electrospun Composite Fibres with Applications in Tissue Engineering. *Polymers* **2019**, *11*, 1793. [CrossRef]
48. Abdolmaleki, A.; Mohamadi, Z. Acidic Ionic Liquids Catalyst in Homo and Graft Polymerization of ε-Caprolactone. *Colloid. Polym. Sci.* **2013**, *291*, 1999–2005. [CrossRef]
49. Jinga, S.I.; Costea, C.C.; Zamfirescu, A.I.; Banciu, A.; Banciu, D.D.; Busuioc, C. Composite Fibre Networks Based on Polycaprolactone and Bioactive Glass-Ceramics for Tissue Engineering Applications. *Polymers* **2020**, *12*, 1806. [CrossRef]
50. Negrea, R.; Busuioc, C.; Constantinoiu, I.; Miu, D.; Enache, C.; Iordache, F.; Jinga, S.I. Akermanite Based Coatings Grown by Pulsed Laser Deposition for Metallic Implants Employed in Orthopaedics. *Surf. Coat. Technol.* **2019**, *357*, 1015–1026. [CrossRef]
51. Schitea, R.I.; Nitu, A.; Ciobota, A.A.; Munteanu, A.L.; David, I.M.; Miu, D.; Raileanu, M.; Bacalum, M.; Busuioc, C. Pulsed Laser Deposition Derived Bioactive Glass-Ceramic Coatings for Enhancing the Biocompatibility of Scaffolding Materials. *Materials* **2020**, *13*, 2615. [CrossRef]
52. Prefac, G.A.; Milea, M.L.; Vadureanu, A.M.; Muraru, S.; Dobrin, D.I.; Isopencu, G.O.; Jinga, S.I.; Raileanu, M.; Bacalum, M.; Busuioc, C. CeO_2 Containing Thin Films as Bioactive Coatings for Orthopaedic Implants. *Coatings* **2020**, *10*, 642. [CrossRef]
53. Ho, C.C.; Fang, H.Y.; Wang, B.; Huang, T.H.; Shie, M.Y. The Effects of Biodentine/Polycaprolactone Three-Dimensional-Scaffold with Odontogenesis Properties on Human Dental Pulp Cells. *Int. Endod. J.* **2018**, *51*, e291–e300. [CrossRef] [PubMed]
54. Kang, Y.G.; Wei, J.; Shin, J.V.; Wu, Y.R.; Su, J.; Park, Y.S.; Shin, J.W. Enhanced Biocompatibility and Osteogenic Potential of Mesoporous Magnesium Silicate/Polycaprolactone/Wheat Protein Composite Scaffolds. *Int. J. Nanomedicine* **2018**, *13*, 1107–1117. [CrossRef]
55. Shkarina, S.; Shkarin, R.; Weinhardt, V.; Melnik, E.; Vacun, G.; Kluger, P.J.; Loza, K.; Epple, M.; Ivlev, S.I.; Baumbach, T.; et al. 3D Biodegradable Scaffolds of Polycaprolactone with Silicate-Containing Hydroxyapatite Microparticles for Bone Tissue Engineering: High-Resolution Tomography and in Vitro study. *Sci. Rep.* **2018**, *8*, 8907. [CrossRef] [PubMed]
56. Huang, S.H.; Hsu, T.T.; Huang, T.H.; Lin, C.Y.; Shie, M.Y. Fabrication and Characterization of Polycaprolactone and Tricalcium Phosphate Composites for Tissue Engineering Applications. *J. Dent. Sci.* **2017**, *12*, 33–43. [CrossRef] [PubMed]
57. Jaiswal, A.K.; Chhabra, H.; Soni, V.P.; Bellare, J.R. Enhanced Mechanical Strength and Biocompatibility of Electrospun Polycaprolactone-Gelatin Scaffold with Surface Deposited Nano-Hydroxyapatite. *Mater. Sci. Eng. C* **2013**, *33*, 2376–2385. [CrossRef]
58. Salgado, C.L.; Sanchez, E.M.S.; Zavaglia, C.A.C.; Granja, P.L. Biocompatibility and Biodegradation of Polycaprolactone-Sebacic Acid Blended Gels. *J. Biomed. Mater. Res. A* **2012**, *100A*, 243–251. [CrossRef]
59. Yang, X.; Wang, Y.; Zhou, Y.; Chen, J.; Wan, Q. The Application of Polycaprolactone in Three-Dimensional Printing Scaffolds for Bone Tissue Engineering. *Polymers* **2021**, *13*, 2754. [CrossRef]
60. Coombes, A.G.A.; Rizzi, S.C.; Williamson, M.; Barralet, J.E.; Downes, S.; Wallace, W.A. Precipitation Casting of Polycaprolactone for Applications in Tissue Engineering and Drug Delivery. *Biomaterials* **2004**, *25*, 315–325. [CrossRef]

polymers

MDPI

Article

Cefazolin Loaded Oxidized Regenerated Cellulose/Polycaprolactone Bilayered Composite for Use as Potential Antibacterial Dural Substitute

Arunnee Sanpakitwattana [1], Waraporn Suvannapruk [2], Sorayouth Chumnanvej [3], Ruedee Hemstapat [1,*] and Jintamai Suwanprateeb [2,*]

[1] Department of Pharmacology, Faculty of Science, Mahidol University, Bangkok 10400, Thailand
[2] Biofunctional Materials and Devices Research Group, National Metal and Materials Technology Center (MTEC), Pathum Thani 12120, Thailand
[3] Neurosurgical Unit, Surgery Department, Faculty of Medicine Ramathibodi Hospital, Bangkok 10400, Thailand
* Correspondence: ruedee.hem@mahidol.ac.th (R.H.); jintamai@mtec.or.th (J.S.)

Abstract: Oxidized regenerated cellulose/polycaprolactone bilayered composite (ORC/PCL bilayered composite) was investigated for use as an antibacterial dural substitute. Cefazolin at the concentrations of 25, 50, 75 and 100 mg/mL was loaded in the ORC/PCL bilayered composite. Microstructure, density, thickness, tensile properties, cefazolin loading content, cefazolin releasing profile and antibacterial activity against *S. aureus* were measured. It was seen that the change in concentration of cefazolin loading affected the microstructure of the composite on the rough side, but not on the dense or smooth side. Cefazolin loaded ORC/PCL bilayered composite showed greater densities, but lower thickness, compared to those of drug unloaded composite. Tensile modulus was found to be greater and increased with increasing cefazolin loading, but tensile strength and strain at break were lower compared to the drug unloaded composite. In vitro cefazolin release in artificial cerebrospinal fluid (aCSF) consisted of initial burst release on day 1, followed by a constant small release of cefazolin. The antibacterial activity was observed to last for up to 4 days depending on the cefazolin loading. All these results suggested that ORC/PCL bilayered composite could be modified to serve as an antibiotic carrier for potential use as an antibacterial synthetic dura mater.

Keywords: synthetic dural substitute; drug release; antibiotics carrier; cefazolin

Citation: Sanpakitwattana, A.; Suvannapruk, W.; Chumnanvej, S.; Hemstapat, R.; Suwanprateeb, J. Cefazolin Loaded Oxidized Regenerated Cellulose/Polycaprolactone Bilayered Composite for Use as Potential Antibacterial Dural Substitute. *Polymers* **2022**, *14*, 4449. https://doi.org/10.3390/polym14204449

Academic Editors: Ariana Hudita and Bianca Gălăţeanu

Received: 27 August 2022
Accepted: 20 September 2022
Published: 21 October 2022

1. Introduction

Dura mater is the outermost membrane of meninges which surrounds the brain and spinal cord to retain the cerebrospinal fluid (CSF) inside. During neurosurgical procedures, intracranial lesions might lead to partial dura removal or dural perforation and the restoration of the dura mater to regain a watertight dural closure should be performed. The solitary use of native dura for such reconstruction could struggle to achieve an adequate primary closure in certain situations and might result in post-operative complications, morbidity and even loss of life [1,2]. Several dural substitutes and dural sealants are commercially available and used in place of native dura mater for dural repair and closure restoration. Recently, a new bilayered knitted fabric reinforced composite (ORC/PCL bilayered composite) was developed for potential use as a dural substitute [3–5]. This composite membrane contained two different morphologies including a composite layer which incorporated a relatively faster resorption ORC that could in situ generate pores for tissue ingrowth and a dense nonporous layer of relatively slower resorption PCL on the opposite side which helped provide the load bearing and liquid leakage resistance [6–8]. This type of dural substitute also offered several advantages including low cost, simple production as well as similar mechanical properties to human dura mater.

In general, post-operative CNS infection such as meningitis and post-operative surgical site infections (SSIs) have been reported following craniotomy and neurosurgical procedures [8,9]. Several risk factors associated with these post-operative infections have been identified, including CSF leakage, prolonged operation procedure (>3 h), diabetes, implantation of prosthetic devices such as shunt and cranioplasty materials [9,10]. The predominant pathogens causing these infections are Gram-positive bacteria, particularly *Staphylococcus aureus* (*S. aureus*) and *coagulase-negative staphylococci* (*CoNS*). The incidence rates of post-operative surgical site infections (SSIs) have been reported to decrease when antibiotic prophylaxis strategies were implemented [11,12]. The ideal antimicrobial drugs for prophylaxis should be effective against the most common organisms found at the surgical sites without eliminating other causative organisms [9,13]. Cephalosporins are the most commonly used prophylactic antibiotics in different types of surgical procedure [14]. In particular, cefazolin, the first generation cephalosporin, has been considered as the prophylactic drug of choice for post-neurosurgical infections [11]. This is because it is most effective against Gram-positive bacteria such as *S. aureus* and *CoNS*, which are the major causes of the post-operative SSIs and post-craniotomy infections. Moreover, several advantages of cefazolin have been reported including a low toxicity and low cost [12,15]. Cefazolin is given parenterally via intramuscular or intravenous administration since its absorption from the GI tract is insufficient.

Localized drug delivery is one of the most effective strategies for delivering drugs to the specific targeted site, which led to more preferable therapeutic effects [16]. It has been reported that a localized delivery system showed superiority over the conventional systemic antibiotic delivery [17]. A higher dose reaching directly to the desired site, decreased systemic toxicity, reduced the bacterial resistance as well as improved patient compliance [17–19]. The ideal localized system should exhibit a burst release of a large amount of drug in the initial period and be followed by a continual therapeutic dose release to prevent latent infection [16,19]. Cefazolin has been experimentally loaded into several carriers including a PCL sponge pad or fibers [20,21], PCL/sodium-alginate [22], mesoporous silicon microparticles [23], hydroxyapatite [24] and gelatin nanofiber mats [25,26] for use as localized drug release systems. Initial burst release of cefazolin was typically seen and followed by a sustained release from 24 h to 32 days depending on the type of carrier.

Preventing the incidence of post-neurosurgical infection is clinically crucial during neurosurgical procedures. Since an implant was reported to be one of the risk factors for SSIs, the use of an antibacterial dural substitute with the localized antibiotics delivery would probably help in decreasing this risk. This would be used as an adjuvant or add-on treatment and not as a replacement for systemic antibiotics prophylaxis via an intravenous route. Previously, commercial collagen artificial dura mater was experimentally mixed with cefuroxime sodium, ceftriaxone sodium or norvancomycin aimed to provide sustained drug release directly to the brain surface and it overcame the problem of the blood–brain barrier that lowered drug concentrations in the cerebrospinal fluid compared to the venous blood [27]. However, no antibacterial activity was carried out and only a short release study for up to 72 h was performed, although the release time of 6–7 days was designed. Vancomycin soaked commercial crosslinked collagen based dural substitute was also investigated for possible use in infected and contaminated wounds. However, the study was brief and only colony counting and antibacterial activity at 24 h were measured [28]. Therefore, data of an antibiotic loaded dural substitute that can perform dual functions for dural restoration and local release of antibiotics in the prevention of surgical site infection are still limited and not exhaustively studied.

It was hypothesized that ORC/PCL bilayered composite which was developed and already passed in vitro and in vivo studies [3–5] could be further modified to act as an antibiotic dural substitute. The characteristics and properties of the carriers are important factors which could influence the drug releasing behavior of the material and the incorporation of drug in the carriers would in turn affect the property of the materials. This study was carried out to gain the understanding and knowledge to further develop an antibacterial dural substitute, which has rarely been reported. In this study, cefazolin was chosen as a model antibiotic for investigation. Various concentrations of cefazolin were loaded into previously selected ORC/PCL bilayered composite (P20 formulation) and their effects on physical properties, mechanical properties, drug loading and releasing characteristics and antibacterial activity were investigated for potentially being used as an antibacterial dural substitute in dural closure application.

2. Materials and Methods

2.1. *Raw Materials*

All raw materials including polycaprolactone (PCL, M_W ~ 80,000, Sigma-Aldrich, St. Louis, MI, USA), oxidized regenerated cellulose (ORC, Surgicel®®, Nu-Knit®®, Ethicon Inc., Raritan, NJ, USA) and N-methyl-2-pyrrolidone (NMP, Pharmasolve™, Ashland Inc., Wilmington, DE, USA) were purchased and used in the as-received form. The antibiotic used was cefazolin sodium (Fazolin®®, Siam Bheasach Co., Ltd., Bangkok, Thailand) and was supplied in the powder form.

2.2. *Sample Preparation*

NMP was equally divided into two parts. Cefazolin powders were dissolved in the first part of NMP at the concentrations of 0, 25, 50, 75 and 100 mg/mL and stirred at room temperature until clear solution was obtained while 20 g of PCL was dissolved in the second part of NMP by heating at 60 °C using a hotplate until obtaining a clear viscous PCL solution. After cooling down to a temperature of approximately 40 °C, both parts were then mixed well together, producing 20% *w*/*v* cefazolin loaded PCL solutions. PCL solutions was used to evenly infiltrate both sides of ORC knitted woven fabric (30 × 70 × 1.0 mm^3) and was then additionally recoated on one side of the loaded fabric as described previously [3,4]. The infiltrated and coated fabrics were then submerged in deionized water for 30 s to solidify the PCL and to leach out the solvent and they were then further dried in the oven at 40 °C for 24 h to obtain cefazolin loaded ORC/PCL bilayered composite (designated P20, P20_25, P20_50, P20_75 and P20_100, respectively). Cefazolin loaded PCL was also fabricated for use as comparative samples. The cefazolin loaded PCL solutions, which were prepared in the same manner as described previously, were poured into a cavity of a mold (30 × 70 × 1.0 mm^3). The mold was then submerged in deionized water for 30 s and dried in the oven at 40 °C for 24 h to obtain cefazolin loaded PCL (designated PCL, PCL_25, PCL_50, PCL_75 and PCL_100, respectively). The concentration of cefazolin solution used to infiltrate the ORC/PCL bilayered composite or mix in PCL in this study was selected based on the concentration of cefazolin solution used for typical intravenous or intramuscular injection which ranges from about 20 mg/mL to 250 mg/mL. Table 1 shows the formulations that were employed to prepare the samples.

Table 1. Formulations of the prepared samples.

Samples	Cefazolin (g)	NMP (mL)	PCL (g)	ORC Impregnation
P20	0	100	20	Yes
P20_25	2.5	100	20	Yes
P20_50	5.0	100	20	Yes
P20_75	7.5	100	20	Yes
P20_100	10.0	100	20	Yes
PCL	0	100	20	No
PCL_25	2.5	100	20	No
PCL_50	5.0	100	20	No
PCL_75	7.5	100	20	No
PCL_100	10.0	100	20	No

2.3. Microstructure

Microstructure of the fabricated samples was observed by using a scanning electron microscope (JEOL JSM-5410) using an accelerating voltage of 20 kV. Prior to observation, the samples were gold sputter coated to improve the conductivity and prevent charging of the samples.

2.4. Bulk Density

Weight of the sample and its dimensions (width, length and thickness) were measured by a precision balance (Sartorious) and a vernier caliper (Mitutoyo), respectively. The bulk density was determined by dividing the weight by the calculated volume.

2.5. Tensile Properties

Rectangular specimens (10 mm × 70 mm) were cut from the fabricated samples and were used for tensile testing following ISO 527-4 to determine tensile modulus, tensile strength and tensile strain at break. All the tests were performed by a universal testing machine (Instron 55R 4502) equipped with a 10 kN load cell using a constant crosshead speed of 50 mm/min at 23 °C and 50% RH.

2.6. Total Cefazolin Loading Content

Specimens (10 mm × 10 mm) were immersed in a bottle containing 5 mL of artificial cerebrospinal fluid (aCSF; 6.279 g/L sodium chloride, 0.216 g/L potassium chloride, 0.353 g/L calcium chloride, 0.488 g/L magnesium chloride, 1.932 g/L sodium bicarbonate, 0.358 g/L disodium hydrogen phosphate and 2-[4-(2-hydroxyethyl)piperazin-1-yl] ethanesulfonic acid (HEPES) 11.915 g, pH 7.4 [27]) and incubated for 24 h at 37 °C in an incubator. The specimen was taken out and aCSF was collected and filtered. The absorbance of the collected aCSF was then measured by using a UV–Vis spectrophotometer (Perkin Elmer) at a wavelength of 270 nm and compared with a linear equation of the calibration curve (cefazolin in aCSF, R-squared = 0.9996) to quantify the amount of released cefazolin from the sample. The standard cefazolin solutions ranged from 0.02, 0.05, 0.10, 0.25, 0.5, 1, 5, 10, 15, 20, 25 to 30 μg/mL. The collected sample was further dried at 40 °C and dissolved in 5 mL dimethylformamide (DMF) in a shaking incubator for 24 h at 37 °C, 100 rpm. The DMF solution was then filtered through a silica glass filter and the amount of cefazolin was quantified by using a UV–Vis spectrophotometer in a similar manner as that for aCSF but using the linear calibration curve of cefazolin in DMF solutions (R-squared = 0.9867) instead. The total amount of cefazolin in the samples was calculated by combining the measured cefazolin amount in the aCSF and that in DMF solution. The measurement was performed in triplicate and the result was expressed as the ratio of the total amount of cefazolin (mg) to the mass (g) of each specimen.

2.7. In Vitro Cefazolin Release Study

Specimens (10 mm × 10 mm) were prepared and each sample was placed in a plastic bottle containing 5 mL of aCSF and secured in place with a frame sheet to ensure that both sides of the sample were totally immersed in the solution. This was then incubated in an incubator at 37 °C. Every 24 h, the specimen was removed, slightly blotted with tissue paper and placed in a new bottle containing 5 mL of fresh aCSF. The remaining solution was collected and filtered with a 0.45 μm syringe filter (Corning). The absorbance of cefazolin in the collected solution was then measured by using a UV–Vis spectrophotometer (Perkin Elmer) at a wavelength of 270 nm and compared with a linear equation of the prepared calibration curve (cefazolin in aCSF; R-squared = 0.9996) as previously described to quantify the amount of cefazolin released from the sample.

2.8. Minimum Inhibitory Concentration (MIC)

The MIC value of cefazolin used against Gram-positive *S. aureus* ATCC 25923 was assessed by using the standard broth macrodilution method as described previously [29].

Cefazolin solution was prepared by reconstituting 1 g vial cefazolin with 10 mL sterile aCSF to give a resultant concentration of 100 mg/mL and further diluted to obtain a stock solution of 0.01 mg/mL (10 μg/mL). A series of 2-fold higher concentrations of cefazolin than the final dilution range as well as control was prepared in duplicate as follows: 10, 5.0, 2.5, 1.25, 0.625, 0.3125, 0.1560, 0.0780, 0.0390 and 0.000 μg/mL. In addition, uninoculated wells of antibiotic-free broth were also prepared to ensure the sterility of the cefazolin solution and aCSF. All the wells were then incubated at 35 °C ± 2 °C for 18–20 h. The MIC value was determined as the lowest dilution of cefazolin that completely inhibited the visible growth of the test organism which is evidenced by the absence of turbidity in the well. The bacterial inoculum was prepared according to a standard broth culture method (European Committee for Antimicrobial Susceptibility Testing of the European Society of Clinical and Infectious, 2000). In brief, the bacterial colonies were taken with a sterile loop and transferred to 5 mL of sterile nutrient broth (NB). The broth was then incubated in a shaking incubator at 37 °C until the visible turbidity (4–6 h) was observed. The density of the suspension of bacterial culture was diluted with NB to give a turbidity equivalent to the 0.5 McFarland standard (approximately 1.5×10^8 cfu/mL). This was performed using visual inspection such that the appearance of black lines was compared when observed through the bacterial inoculum and McFarland standard suspension. This inoculum was then transferred to a tube, which was further adjusted with NB to achieve a final organism density of 1.5×10^6 cfu/mL.

2.9. Antibacterial Activity

The antibacterial activity of the samples was evaluated by using the agar disk-diffusion method or Kirby–Bauer test. The *S. aureus* ATCC 25923 (1.5×10^6 cfu/mL) were inoculated on Mueller–Hinton agar plates. Specimens (10 mm × 10 mm) were then gently placed down to ensure even contact to the agar surface. Cefazolin loaded ORC/PCL bilayered composites contained two distinct surfaces, namely rough or porous surface (R) and smooth or dense surface (S), resulting from the processing. Both surfaces of the composite were independently evaluated whereas only one surface of the cefazolin loaded PCL was tested due to indistinguishable appearance between the two sides. All the plates were then incubated at 37 °C for 24 h, in which the samples were transferred to fresh *S. aureus* agar plates every 24 h. This was performed repeatedly until the inhibition (clear) zones were not observed. The size of the clear space around the sample indicated the antibacterial activity of the sample. All dimensional measurement was carried out by using a vernier caliper and used to calculate the inhibition zone values using the following formula:

$$H = (D - d)/2, \tag{1}$$

where H is the inhibition zone (mm);
D is the clear space diameter around the sample (mm);
d is the specimen diameter (mm).

2.10. Statistical Analysis

The data were described as the mean ± standard deviation values. Data were analyzed by using statistical analysis software (GraphPad Prism version 6, GraphPad Software, San Diego, CA, USA). One-way analysis of variance (ANOVA), followed by Tukey's multiple comparisons test, was employed to determine the significant between groups or samples. A p-value < 0.05 was considered statistically significant.

3. Results

3.1. Physical and Mechanical Properties

3.1.1. Microstructure, Bulk Density and Thickness

Figure 1 shows the microstructure of cefazolin loaded samples. Regardless of cefazolin concentrations, all cefazolin loaded PCL generally displayed an open microstructure resulting from the exchange of solvent and water in the fabrication process (Figure 1a,c,e,g). However, the amount of dense area surrounding pores tended to increase with increasing cefazolin content in the PCL solution. Upon examining higher magnification images (Figure 1b,d,f,h), only PCL_25 showed a bicontinuous-like structure having a pore size less than 5 μm while PCL_50, PCL_75 and PCL_100 showed cellular structure having larger pore size of about 10–20 μm. In contrast, cefazolin loaded ORC/PCL bilayered composites exhibited a bilayer structure comprising a nonporous layer and a composite layer. In the case of the nonporous layer side, all formulations similarly displayed dense solid structure of PCL similar to those of cefazolin loaded PCL, but no pores were observed in this case (Figure 1i,j,m,n,q,r,u,v). In the case of the composite layer side, different microstructures were obtained depending on the cefazolin concentrations. At a low cefazolin concentration of 25 mg/mL, the composite layer side consisted of knitted ORC fabric mostly embedded within the continuous PCL matrix with exposed fabric in some areas (Figure 1k). Increasing the cefazolin concentration to 50 mg/mL resulted in a dense PCL layer without any exposed ORC fabric resembling the dense structure of the nonporous layer side (Figure 1o). Further increasing the cefazolin concentration to 75 and 100 mg/mL led to the coating of PCL on top of the ORC fabric since the knitted ORC morphology was still evidenced, but with the thick coating of the PCL layer on top (Figure 1s,w). At high magnification, PCL matrix was observed to be dense for P20_25 and P20_50 (Figure 1l,p) while some pores were observed in the PCL matrix close to the ORC fibers for P20_75 and P20_100 (Figure 1t,x).

Figure 2 shows the bulk density and thickness of cefazolin loaded and drug unloaded samples. The bulk density and thickness of all formulations of cefazolin loaded ORC/PCL bilayered composite did not differ significantly. In comparison to their corresponding drug unloaded samples (P20 or PCL), bulk density of both cefazolin loaded ORC/PCL bilayered composite and cefazolin loaded PCL samples was greater, but the thickness was lower. However, the significant difference ($p < 0.05$) in bulk density between cefazolin loaded and drug unloaded samples was seen for P20_25 and all cefazolin loaded PCL except PCL_100. In contrast, thickness of all cefazolin loaded ORC/PCL bilayered composites was significantly lower than that of P20 while only that of PCL75 was significantly different.

3.1.2. Tensile Properties

Figure 3 shows tensile properties of drug loaded and unloaded samples. Tensile modulus of cefazolin loaded ORC/PCL bilayered composite increased with increasing cefazolin concentration, which reached statistical significance when cefazolin concentration was 50 mg/mL or above (Figure 3a). However, this was not observed in cefazolin loaded PCL, in which the tensile moduli of all samples were seen to be in the same range and did not differ significantly from that of PCL (Figure 3b). Generally, tensile strengths of both cefazolin loaded ORC/PCL bilayered composite and cefazolin loaded PCL were found to be significantly lower compared to unloaded samples ($p < 0.05$). Tensile strength of cefazolin loaded PCL samples tended to decrease when cefazolin concentration was increased while those of cefazolin loaded ORC/PCL bilayered composite remained relatively unchanged (Figure 3c,d). All cefazolin loaded ORC/PCL bilayered composites exhibited similar tensile strain at break regardless of the concentration of cefazolin, but all were significantly lower than that of drug unloaded composite (Figure 3e). In contrast to cefazolin loaded ORC/PCL bilayered composite, tensile strain at break of cefazolin loaded PCL samples tended to decrease with increasing cefazolin concentration, but only reached statistical significance at the highest concentration of cefazolin ($p < 0.05$) (Figure 3f). Comparing between cefazolin loaded ORC/PCL bilayered composite and cefazolin loaded PCL at a similar cefazolin concentration, the composites had greater tensile modulus than those of PCL, but tensile strengths were in similar ranges. Tensile strain at break of cefazolin loaded PCL was much greater than those of cefazolin loaded ORC/PCL bilayered composite.

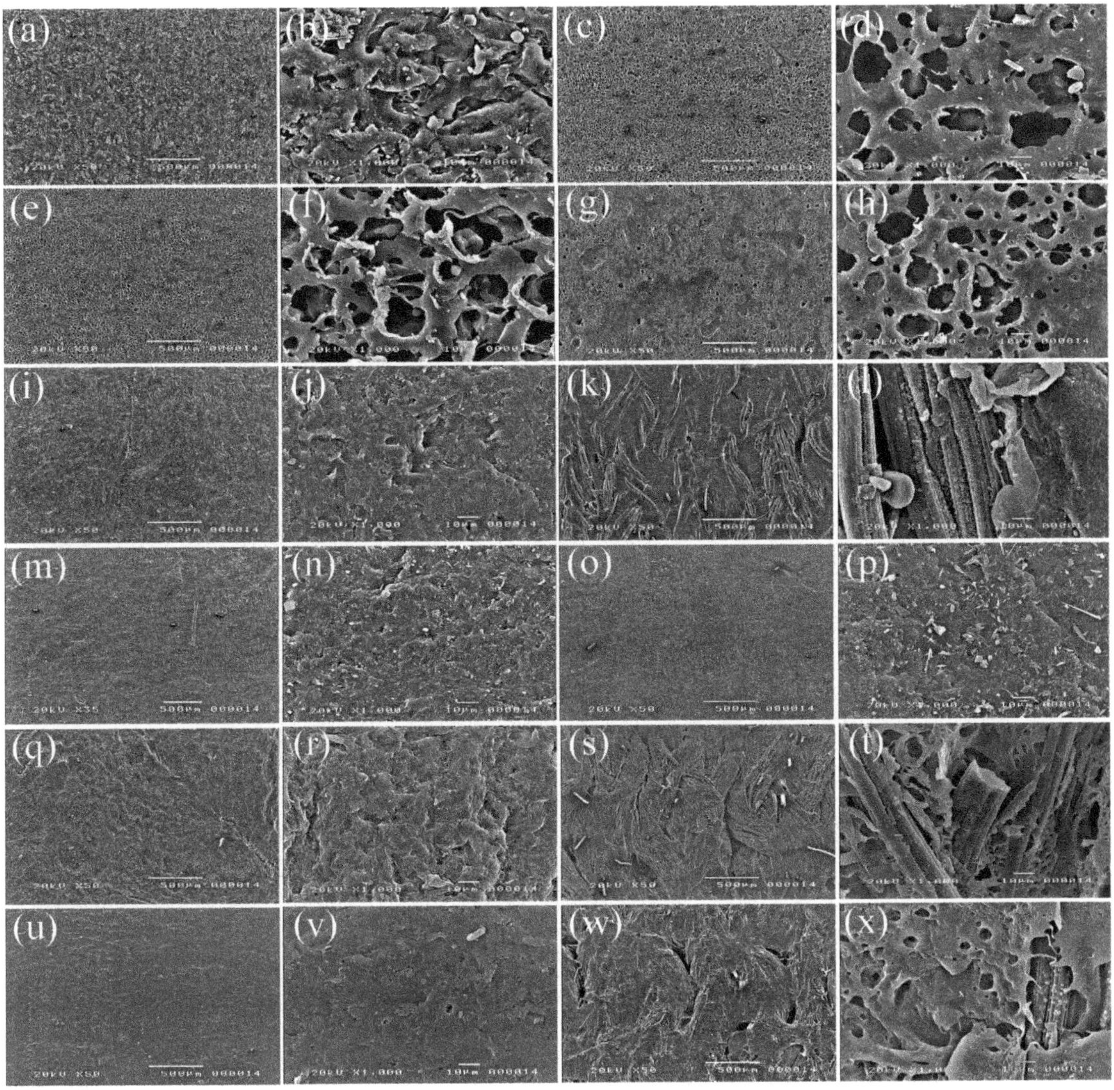

Figure 1. SEM images showing the microstructure of fabricated cefazolin loaded samples: (**a**) PCL_25 at 50×; (**b**) PCL_25 at 1000×; (**c**) PCL_50 at 50×; (**d**) PCL_50 at 1000×; (**e**) PCL_75 at 50×; (**f**) PCL_75 at 1000×; (**g**) PCL_100 at 50×; (**h**) PCL_100 at 1000×; (**i**) P20_25 smooth side at 50×; (**j**) P20_25 smooth side at 1000×; (**k**) P20_25 rough side at 50×; (**l**) P20_25 rough side at 1000×; (**m**) P20_50 smooth side at 50×; (**n**) P20_50 smooth side at 1000×; (**o**) P20_50 rough side at 50×; (**p**) P20_50 rough side at 1000×; (**q**) P20_75 smooth side at 50×; (**r**) P20_75 smooth side at 1000×; (**s**) P20_75 rough side at 50×; (**t**) P20_75 rough side at 1000×; (**u**) P20_100 smooth side at 50×; (**v**) P20_100 smooth side at 1000×; (**w**) P20_100 rough side at 50×; (**x**) P20_100 rough side at 1000×. Magnification 50× (bar = 500 μm) and magnification 1000× (bar = 10 μm).

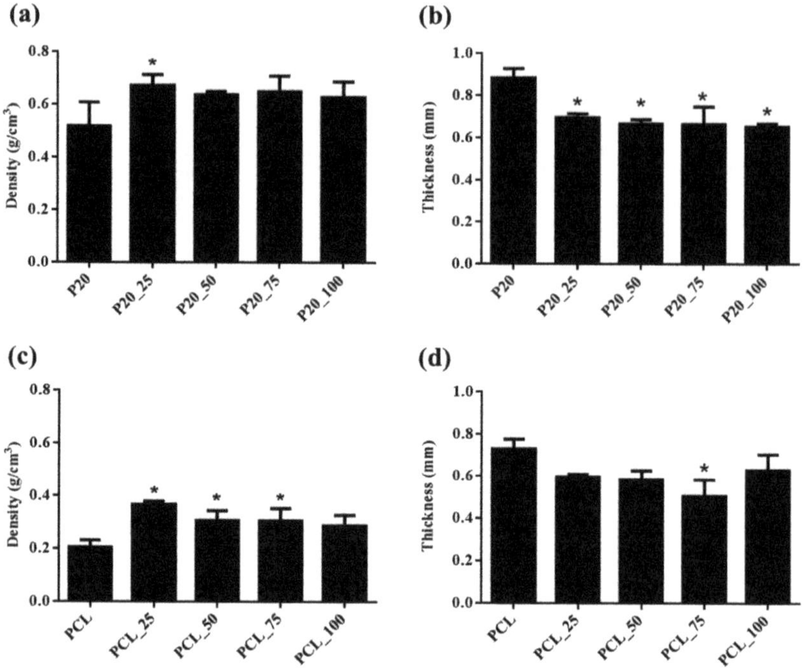

Figure 2. Influence of cefazolin concentration in the ORC/PCL bilayered composite and PCL on density and thickness of fabricated samples: (**a**) Density of cefazolin loaded ORC/PCL bilayered composite; (**b**) thickness of cefazolin loaded ORC/PCL bilayered composite; (**c**) density of cefazolin loaded PCL; (**d**) thickness of cefazolin loaded PCL. Data are expressed as mean ± standard deviation (SD). Significance between cefazolin loaded samples versus control (P20 or PCL) is indicated as * ($p < 0.05$).

3.2. *Total Cefazolin Content*

The measured total cefazolin loading contents in fabricated cefazolin loaded ORC/PCL bilayered composite and cefazolin loaded PCL are shown in Table 2. Increasing cefazolin concentration in the solution significantly increased the uptake of cefazolin in both types of fabricated samples. For a similar cefazolin concentration employed, the loading content of cefazolin in the ORC/PCL bilayered composite was greater than that of cefazolin loaded PCL.

Table 2. Total cefazolin contents loaded in the samples. Data are expressed as mean ± standard deviation (SD).

Samples	Total Cefazolin Content (mg Drug per 100 mg Sample)
P20_25	1.94 ± 0.15
P20_50	2.77 ± 0.17
P20_75	3.61 ± 0.07
P20_100	5.54 ± 0.09
PCL_25	1.74 ± 0.10
PCL-50	2.49 ± 0.13
PCL_75	2.80 ± 0.11
PCL_100	3.28 ± 0.11

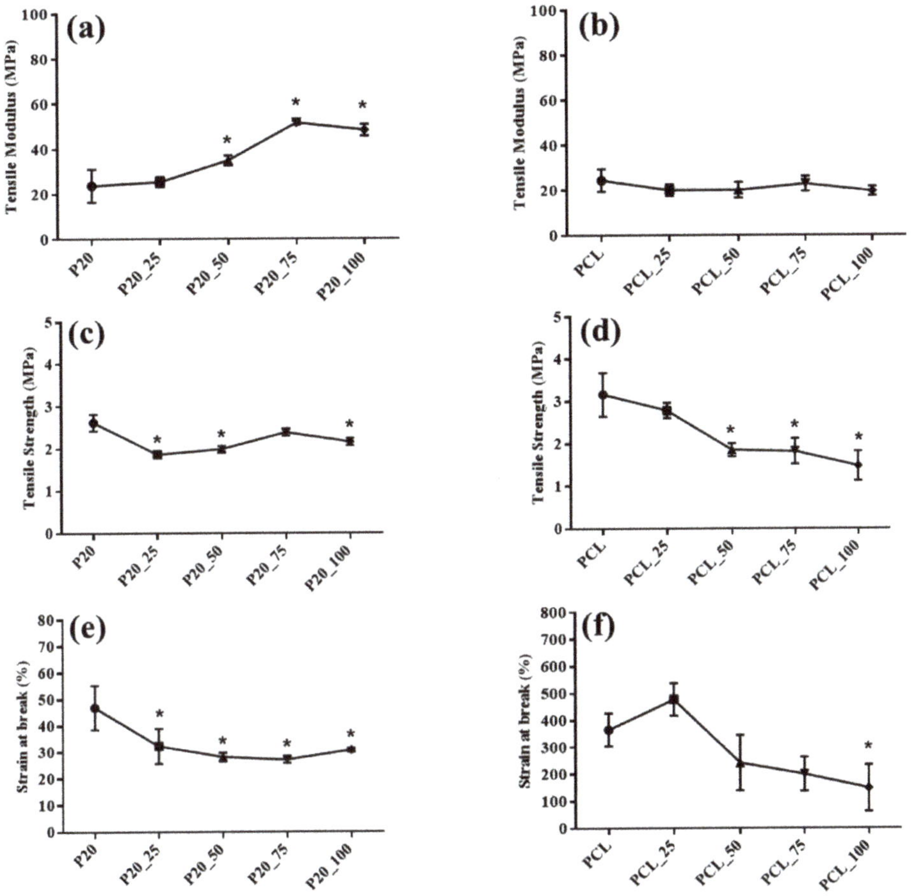

Figure 3. Influence of cefazolin concentration in the ORC/PCL bilayered composite and PCL on tensile properties: (**a**) Tensile modulus of cefazolin loaded ORC/PCL bilayered composite; (**b**) tensile modulus of cefazolin loaded PCL; (**c**) tensile strength of cefazolin loaded ORC/PCL bilayered composite; (**d**) tensile strength of cefazolin loaded PCL; (**e**) tensile strain at break of cefazolin loaded ORC/PCL bilayered composite; (**f**) tensile strain at break of cefazolin loaded PCL. Data are expressed as mean ± standard deviation (SD). Statistical significance between cefazolin loaded samples versus control (P20 or PCL) is indicated as * ($p < 0.05$).

3.3. In Vitro Cefazolin Release

Figure 4 shows the daily concentration of eluted cefazolin (μg/mL) from cefazolin loaded samples for up to 30 days. All samples displayed similar release profiles including a sharp initial burst release of the highest concentration of cefazolin on the first day, followed by a significantly declined release thereafter. While cefazolin was gradually eluted from cefazolin loaded PCL samples for approximately 15 days (Figure 4a), almost all cefazolin was rapidly eluted from cefazolin loaded ORC/PCL bilayered composite during the 4–10 days depending on the cefazolin loading (Figure 4b). Figure 5 shows the cumulative percentage of released cefazolin as a function of time for cefazolin loaded samples over 30 days. The burst and delayed release were observed for cefazolin loaded PCL samples (Figure 5a), where the cumulative percentage of released cefazolin was found to reach approximately 50% in 1 day. In contrast, cefazolin loaded ORC/PCL bilayered composite exhibited a greater burst release, in which most of cefazolin was released in

1 day and followed by constant small release thereafter (Figure 5b). No differences in daily released content or cumulative release among different formulations for both cefazolin loaded PCL and cefazolin loaded ORC/PCL composite were observed.

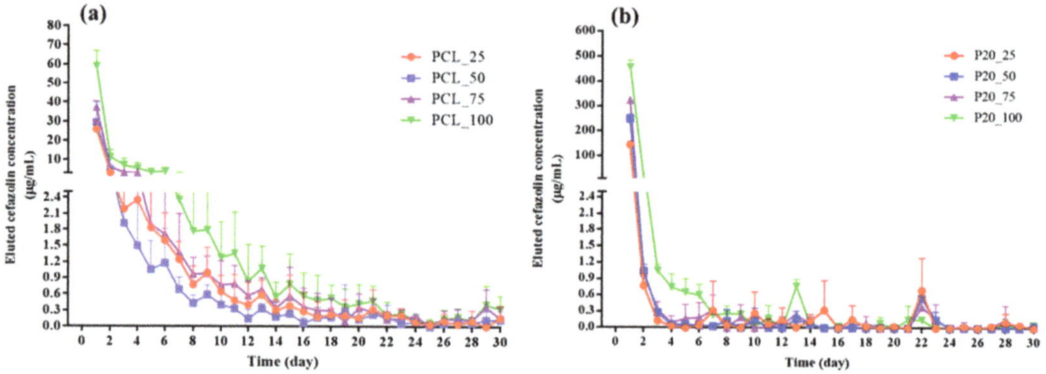

Figure 4. Cefazoline release profile of cefazolin loaded samples in aCSF solution: (**a**) Cefazolin loaded PCL; (**b**) cefazolin loaded ORC/PCL bilayered composites. Data are expressed as mean ± standard deviation (SD).

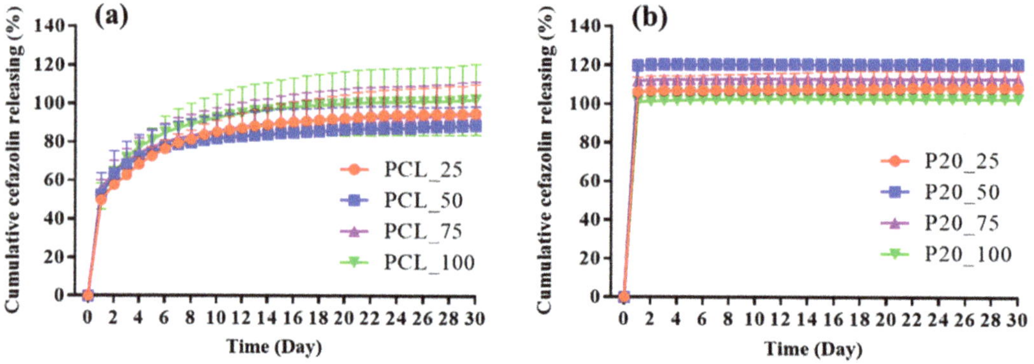

Figure 5. Cumulative cefazoline release profile of cefazolin loaded samples in aCSF solution: (**a**) Cefazolin loaded PCL; (**b**) cefazolin loaded ORC/PCL bilayered composites. Data are expressed as mean ± standard deviation (SD).

3.4. Minimum Inhibitory Concentration (MIC)

Table 3 shows a series of half-decreasing concentrations of cefazolin employed against *S. aureus* ATCC 50923. Based on this result, the MIC value, the lowest concentration of cefazolin that inhibited the bacterial growth, was 0.3150 µg/mL (Well 5).

Table 3. Minimum inhibitory concentration (MIC) value of cefazolin against *S. aureus* ATCC 50923.

Well Number	Cefazolin Concentration (μg/mL)	Bacterial Growth (= No Growth; Clear Solution, ✓= Growth; Turbid Solution)
1	5.0	
2	2.5	
3	1.50	
4	0.650	
5	0.3150	
6	0.1563	✓
7	0.0780	✓
8	0.0390	✓
9	0.0195	✓
10	0.0	✓
11	Nutrient broth (NB), no bacterial inoculum	
12	NB + aCSF, no bacterial inoculum	

3.5. Antibacterial Activity

Figure 6 shows the antimicrobial activity profiles of cefazolin loaded ORC/PCL bilayered composite and cefazolin loaded PCL, which were assessed by using Kirby–Bauer methods for up to 7 days. Cefazolin loaded ORC/PCL bilayered composite showed an antibacterial activity against *S. aureus* for up to 4 days while cefazolin loaded PCL showed a slightly longer activity for up to 5 days. The greatest inhibition zone was observed on day 1, which was relatively similar in both cefazolin loaded ORC/PCL bilayered composite and cefazolin loaded PCL regardless of the concentration of cefazolin. It can be noted that the inhibition zone of cefazolin loaded ORC/PCL bilayered composite tended to display a greater value compared to that of the cefazolin loaded PCL at the same concentration of cefazolin. The inhibition zone of the smooth side of the cefazolin loaded ORC/PCL bilayered composite and cefazolin loaded PCL tended to increase with increasing cefazolin content and the longest inhibition durations (4 and 5 days, respectively) were similarly found at 100 mg/mL (Figure 6b,c). Contrarily, the inhibition zone and antibacterial duration of the rough side of cefazolin loaded ORC/PCL bilayered composite did not correspond with the concentration of cefazolin loading (Figure 6a). In a similar tested period, P20_75 tended to show the greatest inhibition zone while P20_100 showed the second smallest inhibition zone. In the case of antibacterial duration, P20_50 and P20_75 displayed a longer period than P20_50 and P20_100. A drug unloaded composite (P20) also exhibited an antimicrobial activity on day 1 which decreased thereafter, in which the rough side (Figure 6a) exhibited a greater inhibition zone (6.1 ± 0.1 mm) compared to the smooth side (4.1 ± 0.2 mm) (Figure 6b). However, the inhibition zones of P20 on both sides were still smaller than those of cefazolin loaded ORC/PCL bilayered composite. In contrast, the drug unloaded PCL did not show any antibacterial activity.

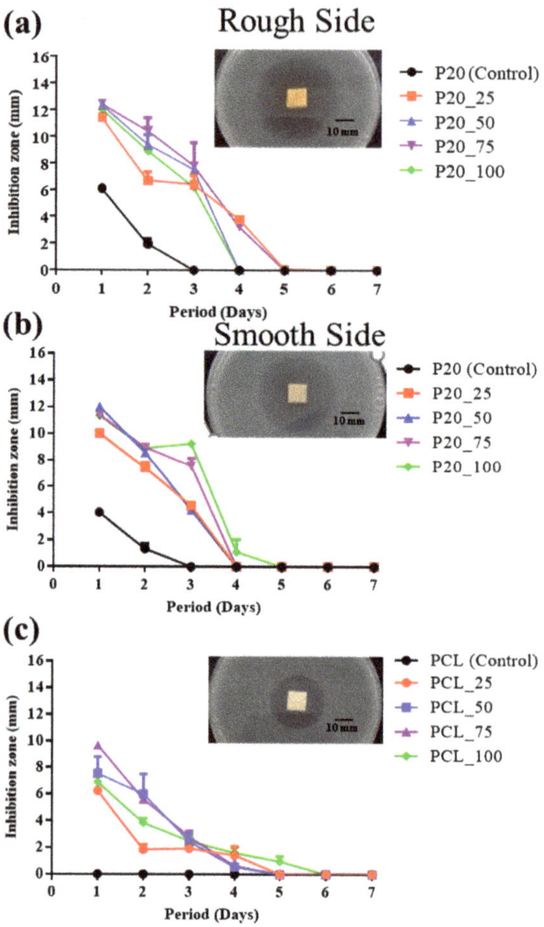

Figure 6. Antibacterial activity of cefazolin loaded samples against *S. aureus* as indicated by inhibition zone: (**a**) Rough side of cefazolin loaded ORC/PCL bilayered composite; (**b**) smooth side of cefazolin loaded ORC/PCL bilayered composite; (**c**) cefazolin loaded PCL. Bar = 10 mm.

4. Discussion

The microstructure of cefazolin loaded PCL resembled the microstructure of porous polymers that were fabricated by a nonsolvent induced phase inversion process wherein the delayed demixing took place when the NMP exchanged into water and precipitation occurred in the PCL solution [30,31]. The microstructure was also observed to be influenced by the cefazolin content which led to an increase in observed viscosity of the PCL solution similar to the increase in polymer content or additives. This caused the insufficiency solvent and nonsolvent exchange to form pores during phase separation and solidification which tended to decrease the porosity of the samples [30,31]. Pore morphology was also affected by cefazolin content. It was previously reported that several pore structures could be obtained by an immersion precipitation process including unconnected latex, nodules, bicontinuous structures, cellular structures or macrovoids depending on the parameters employed in the process, including, but not limit to, solvent and nonsolvent used, polymer solution composition and casting conditions [31,32]. In this study, a bicontinuous-like structure containing interconnected pores was attained at the lowest cefazolin concentration (PCL_25) while cellular structure resulted when using a greater cefazolin concentration.

All cefazolin loaded ORC/PCL bilayered composites retained a bilayered structure which consisted of a composite layer and nonporous PCL layer similarly to those of drug unloaded ORC/PCL bilayered composite, P20 [3]. However, the microstructure of the composite side resembled that of P20 only at the lowest cefazolin loading (P20_25) wherein the viscosity of PCL solution was still comparable and could infiltrate the ORC fabric readily, but not fully cover the surface since the PCL content was low. Using the immediate viscosity solution of P20_50, a fraction of the solution would still be able to infiltrate and fill the gaps among the fibers of ORC and the surplus part that could not infiltrate would spread and coat on the surface instead which resulted in the smooth coating on the surface. Using high viscosity solution of P20_75 and P20_100, limited infiltration was achieved and the majority of the solution would coat on the ORC while preserving the ORC morphology underneath. The coating layer of P20_100 was observed to be thicker than that of P20_75. These changes in solution viscosity and microstructure with increasing cefazolin content were in the same fashion as increasing PCL concentration in fabricating ORC/PCL composite as reported previously [3]. Unlike cefazolin loaded PCL, no effect of such an increase in viscosity was noted on the dense PCL side and all formulations similarly displayed nonporous structure. This was thought to be partly due to the solvent absorption by ORC fabric which concentrated the PCL solution and the easier transport of the solvent though the composite side. These could depress the exchange of solvent and nonsolvent at the dense PCL side, resulting in demixing without pore generation.

Cefazolin loaded ORC/PCL bilayered composite exhibited greater densities, but lower thickness, than drug unloaded composite. These changes after cefazolin loading were also observed in cefazolin loaded PCL. These could all be related to the increase in solution viscosity which decreased the swelling ability of the samples during solvent–water exchange in the solidification step and the decrease in coating layer due to the difficulty to achieve the spreading during the infiltration and recoating process. Despite this, the thickness (0.65–0.69 mm) of the composites was still in the same range as those observed in human dura (0.3–0.8 mm) [33], while densities (0.63–0.67 g/cm^3) were lower than that of human dura (1.03 g/cm^3) [34]. These results suggested that cefazolin loaded composites could still be employed and handled similarly to the drug unloaded ORC/PCL bilayered composite and natural dura mater.

Apart from physical properties, changes in tensile properties were also observed after cefazolin loading. Cefazolin loaded PCL was weaker and less ductile after cefazolin loading since the tensile strength and strain at break decreased with increasing cefazolin concentration, while the tensile modulus was relatively unchanged (Figure 3b,d,f). This was consistent with the previous findings for other types of drug loaded PCL [35–38]. In contrast, tensile strength of cefazolin loaded ORC/PCL bilayered composite was unaffected, but the increase in tensile modulus and the decrease in strain at break with increasing cefazolin concentration were observed instead (Figure 3a,e). Therefore, the underlying mechanism might be different. The changes in tensile properties observed were anticipated to be due to the effect of incorporated cefazolin that could restrict both the deformation of ORC fabric as well as the polymer chain movement of the PCL matrix. In the case of cefazolin loaded PCL, the restriction of PCL chain movement imposed by cefazolin obviously increased the brittleness and decreased the strength of the sample. In the case of ORC/PCL bilayered composite, the reinforcing effect of ORC fabric was relatively greater and might have offset the effect of cefazolin on strength reduction due to PCL chain movement restriction resulting in overall unchanged tensile strength. The restraint of ORC deformation imposed by cefazolin also amplified the stiffness of ORC and gave rise to the increase in tensile modulus. Despite the changes in the tensile properties of all cefazolin loaded ORC/PCL bilayered composites, these tensile values were still within the same range of human dura mater as reported previously, including 11.2–171.5 MPa, 1.3–27.1 MPa and 16.0–49.7% for tensile modulus, tensile strength and strain at break, respectively [34,39,40].

The release profiles of cefazolin loaded PCL were biphasic, comprising the burst release of approximately half of the total cefazolin content and followed by a gradual release for 15 days which was similar to typical burst biphasic release kinetics of drug loaded polymer matrix wherein the initial burst release phase is followed by a power-law phase resulting from Fickian diffusion processes or "anomalous" processes which encompass both diffusion of drugs and swelling of polymers [41,42]. In contrast, the release profile of cefazolin loaded ORC/PCL bilayered composite was found to be mainly burst release where most of the cefazolin was eluted in this phase and followed by relatively constant release thereafter for up to 4–10 days. This is monophasic release kinetics comprising a zero-order release phase that displays a constant rate of drug release via non-Fickian diffusion [42]. The difference in these release profiles between two types of cefazolin loaded samples could be related to the dissimilarity of the composition and microstructure as observed. The cefazolin loaded ORC/PCL bilayered composite contained two layers comprising a composite layer which in situ produced porous channels by relatively faster ORC resorption and the relatively slower degradation of the nonporous PCL layer. These microstructures would aid the diffusion of aCSF solution into the samples, mainly through the composite layer, and caused a rapid dissolution and release of cefazolin resulting in the burst release as observed. Since three different mechanisms of drug release from the polymer matrix were reported, including release of surface loaded drug, diffusion and degradation of the carrier [43,44], the release of antibiotics from dense PCL layer in ORC/PCL bilayered composite could also occur, but contributed less than that of the composite layer. In the case of cefazolin loaded PCL, the diffusion pathway of aCSF into the PCL matrix to dissolve and lead to the release of cefazolin would be more difficult and drug would be slowly released [45]. The initial rapid release of cefazolin loaded PCL might be due to the release of surface loaded cefazolin and part of the drugs which were close to the inner surface of PCL. A subsequent gradual release would be caused by the diffusion and swelling or degradation of PCL. Moreover, both release phases of cefazolin loaded PCL would also be amplified by the open porous microstructure which could act as an additional route for solution diffusion and drug release. Compared to other cefazolin loaded carriers [20–26], cefazolin loaded PCL fabricated in this study exhibited similar biphasic release kinetics to those reported previously, especially when using PCL as a carrier. The rate of elution tended to be faster than that of dense electrospun fiber [21], but comparable to that of PCL sponge which contained microscopic pores [20]. In contrast, ORC/PCL bilayered composite displayed much faster release of cefazolin than cefazolin loaded carriers reported previously and also exhibited different release kinetics. It should be noted that the limitation of using of a UV–Vis spectrophotometer for cefazolin measurement is that it might not specifically detect the cefazolin compared to chromatographic or mass spectrometric methods.

Concerning the MIC of cefazolin against *S. aureus* which was determined to be 0.3150 μg/mL, the duration of the release content of cefazolin from cefazolin loaded ORC/PCL bilayered composite which was still closed to MIC value was approximately 3 days for all samples, except that of P20_100, which was 10 days. This correlated quite well with the results of antibacterial activity which was found to last for 3–4 days. In the case of cefazolin loaded PCL, the duration for which the released cefazolin concentration was still in MIC range was about 14 days which differed from the antibacterial activity durations of 4–5 days. This could be due to the dissimilarity in the solution saturated condition in the release study compared to the moistened agar in antibacterial activity testing which would affect the cefazolin release, in particular from PCL matrix. Interestingly, it was also observed that a drug unloaded ORC/PCL composite (P20) also exhibited antibacterial activity. This was probably due to the known antibacterial property of ORC that was reported to produce an acidic environment which was not suitable for microorganisms to survive [46,47]. The antibacterial activity of ORC in the composites can be further confirmed by observation that a greater inhibition zone was observed from the rough side of the composite where the knitted ORC fabric was presented compared to that of the smooth PCL side. However, no clear difference in the inhibition zone of the cefazolin loaded ORC/PCL composite between the rough side and smooth side was observed. This was possibly because the antibacterial activity of released cefazolin was much greater than that of ORC and dominated the activity. In addition, the greater transport ability of solution through the rough side might not be advantageous over the smooth side in this contact situation compared to full immersion in the release test.

The burst effect can be regarded as a negative consequence when long-term controlled release applications are needed, but a rapid release or high initial rates of delivery may be the optimal mechanism in certain situations [48]. It should be noted that the purpose of antimicrobial prophylaxis is to achieve drug levels in serum and tissue that surpass the MIC of the bacterial organisms likely to be faced during operation and prophylaxis after wound close is unessential [49–51]. Therefore, the burst release of cefazolin loaded ORC/PCL bilayered composite, which could be a problem in some applications which require a sustained release of drug for a long period of time, might not be relevant for an antibacterial dural substitute, which requires a localized release of high concentration of antibiotic prophylaxis to ensure that any bacterial infections (if any) are totally and immediately eliminated at the site of implantation. Additionally, the burst release of a high concentration of cefazolin observed in the present study could be an advantage as this could reduce the risk of antibiotic resistance.

5. Conclusions

The antibacterial synthetic dural substitute was successfully fabricated by incorporating cefazolin into ORC/PCL bilayered composite, which still had the physical and mechanical properties in the range of natural dura mater. Cefazolin released from ORC/PCL bilayered composite was found to be monophasic, comprising mainly a burst release. Among all formulations, P20_25 might be most suitable since it showed relatively similar releasing profile and antibacterial activity to other higher drug loaded formulations and still exhibited comparable microstructure and physical and mechanical properties to the drug unloaded composite. Although cefazolin loaded ORC/PCL bilayered composite was found to exhibit a burst release of antibiotic, in which its antibacterial activity was sustained for up to 4 days, this might not fully predict in vivo performance where the environment and the clearance of the released cefazolin from the site of injury might be different. Further studies in animals are required to confirm its performance prior to translating to clinical study.

Author Contributions: Conceptualization, R.H. and J.S.; methodology, R.H. and J.S.; formal analysis, A.S., W.S., R.H. and J.S.; investigation, A.S. and W.S.; writing—original draft preparation, A.S., R.H. and J.S.; writing—review and editing, A.S., W.S., S.C., R.H. and J.S.; supervision, S.C., R.H. and J.S.; project administration, R.H. and J.S.; funding acquisition, A.S., R.H., S.C. and J.S. All authors have read and agreed to the published version of the manuscript.

Funding: A.S. was financially supported by a scholarship from the Science Achievement Scholarship of Thailand. Collaborative research project grants from the Faculty of Science and the Faculty of Medicine Ramathibodi Hospital, Mahidol University and National Metal and Materials Technology Center (MTEC) are acknowledged for providing financial and technical support for carrying out this research.

Institutional Review Board Statement: Not applicable.

Informed Consent Statement: Not applicable.

Data Availability Statement: The data presented in this study are available on request from the corresponding author. The data are not publicly available due to confidentiality.

Acknowledgments: Phetrung Phanpiriya (MTEC) is acknowledged for assistance in the antibacterial activity test.

Conflicts of Interest: The authors declare no conflict of interest.

References

1. Wang, W.; Ao, Q. Research and application progress on dural substitutes. *J. Neurorestoratol.* **2019**, *7*, 161–170. [CrossRef]
2. Choi, E.H.; Chan, A.Y.; Brown, N.J.; Lien, B.V.; Sahyouni, R.; Chan, A.K.; Roufail, J.; Oh, M.Y. Effectiveness of repair techniques for spinal dural tears: A systematic review. *World Neurosurg.* **2021**, *149*, 140–147. [CrossRef] [PubMed]
3. Suwanprateeb, J.; Luangwattanawilai, T.; Theeranattapong, T.; Suvannapruk, W.; Sorayouth, C.; Hemstapat, W. Bilayer oxidized regenerated cellulose/poly ε-caprolactone knitted fabric-reinforced composite for use as an artificial dural substitute. *J. Mater. Sci. Mater. Med.* **2016**, *27*, 122–134. [CrossRef] [PubMed]
4. Hemstapat, R.; Suvannapruk, W.; Thammarakcharoen, F.; Chumnanvej, S.; Suwanprateeb, J. Performance evaluation of bilayer oxidized regenerated cellulose/poly ε-caprolactone knitted fabric reinforced composites for dural substitution. *Proc. Inst. Mech. Eng. Part H J. Eng. Med.* **2020**, *234*, 854–863. [CrossRef]
5. Chumnanvej, S.; Luangwattanawilai, T.; Rawiwet, V.; Suwanprateeb, J.; Rattanapinyopituk, K.; Huaijantug, S.; Yinharnmingmongkol, C.; Hemstapat, R. In vivo evaluation of bilayer ORC/PCL composites in a rabbit model for using as a dural substitute. *Neurol. Res.* **2020**, *42*, 879–889. [CrossRef]
6. Dwivedi, R.; Kumar, S.; Pandey, R.; Mahajan, A.; Nandana, D.; Katti, D.S.; Mehrotra, D. Polycaprolactone as biomaterial for bone scaffolds: Review of literature. *J. Oral Biol. Craniofac. Res.* **2020**, *10*, 381–388. [CrossRef] [PubMed]
7. Backes, E.H.; Harb, S.V.; Beatrice, C.A.G.; Shimomura, K.M.B.; Passador, F.R.; Costa, L.C.; Pessan, L.A. Polycaprolactone usage in additive manufacturing strategies for tissue engineering applications: A review. *J. Biomed. Mater. Res. B Appl. Biomater.* **2022**, *110*, 1479–1503. [CrossRef] [PubMed]
8. Kleine, J.; Leisz, S.; Ghadban, C.; Hohmann, T.; Prell, J.; Scheller, C.; Strauss, C.; Simmermacher, S.; Dehghani, F. Variants of oxidized regenerated cellulose and their distinct effects on neuronal tissue. *Int. J. Mol. Sci.* **2021**, *22*, 11467. [CrossRef] [PubMed]
9. Bassetti, M.; Righi, E.; Astilean, A.; Corcione, S.; Petrolo, A.; Farina, E.C.; De Rosa, F.G. Antimicrobial prophylaxis in minor and major surgery. *Minerva Anestesiol.* **2015**, *81*, 76–91. [PubMed]
10. Erman, T.; Demirhindi, H.; Göçer, A.İ.; Tuna, M.; İldan, F.; Boyar, B. Risk factors for surgical site infections in neurosurgery patients with antibiotic prophylaxis. *Surg. Neurol.* **2005**, *63*, 107–113. [CrossRef] [PubMed]
11. Bratzler, D.W.; Dellinger, E.P.; Olsen, K.M.; Perl, T.M.; Auwaerter, P.G.; Bolon, M.K.; Fish, D.N.; Napolitano, L.M.; Sawyer, R.G.; Slain, D.; et al. Clinical practice guidelines for antimicrobial prophylaxis in surgery. *Am. J. Health Syst. Pharm.* **2013**, *70*, 195–283. [CrossRef]
12. Liu, W.; Ni, M.; Zhang, Y.; Groen, R.J. Antibiotic prophylaxis in craniotomy: A review. *Neurosurg. Rev.* **2014**, *37*, 407–414. [CrossRef]
13. Gyssens, I.C. Preventing postoperative infections: Current treatment recommendations. *Drugs* **1999**, *57*, 175–185. [CrossRef] [PubMed]
14. Brocard, E.; Reveiz, L.; Régnaux, J.P.; Abdala, V.; Ramón-Pardo, P.; del Rio Bueno, A. Antibiotic prophylaxis for surgical procedures: A scoping review. *Rev. Panam. Salud. Publica* **2021**, *45*, e62. [CrossRef] [PubMed]
15. Kusaba, T. Safety and efficacy of cefazolin sodium in the management of bacterial infection and in surgical prophylaxis. *Clin. Med. Ther.* **2009**, *1*, 1607–1615. [CrossRef]
16. Wassif, R.K.; Elkayal, M.; Shamma, R.N.; Elkheshen, S.A. Recent advances in the local antibiotics delivery systems for management of osteomyelitis. *Drug Deliv.* **2021**, *28*, 2392–2414. [CrossRef] [PubMed]
17. Stebbins, N.D.; Ouimet, M.A.; Uhrich, K.E. Antibiotic-containing polymers for localized, sustained drug delivery. *Adv. Drug Deliv. Rev.* **2014**, *78*, 77–87. [CrossRef]
18. Dash, A.K.; Cudworth, G.C. Therapeutic applications of implantable drug delivery systems. *J. Pharmacol. Toxicol. Methods* **1998**, *40*, 1–12. [CrossRef]
19. Wu, P.; Grainger, D.W. Drug/device combinations for local drug therapies and infection prophylaxis. *Biomaterials* **2006**, *27*, 2450–2467. [CrossRef] [PubMed]
20. Mutsuzaki, H.; Oyane, A.; Sogo, Y.; Sakane, M.; Ito, A. Cefazolin-containing poly(ε-caprolactone) sponge pad to reduce pin tract infection rate in rabbits. Asia-Pac. *J. Sports Med. Arthrosc. Rehabil. Technol.* **2014**, *1*, 54–61.
21. Radisavljevic, A.; Stojanovic, D.B.; Perisic, S.; Djokic, V.; Radojevic, V.; Rajilic-Stojanovic, M.; Uskokovic, P.S. Cefazolin-loaded polycaprolactone fibers produced via different electrospinning methods: Characterization, drug release and antibacterial effect. *Eur. J. Pharm. Sci.* **2018**, *124*, 26–36. [CrossRef]
22. Lee, J.H.; Park, J.K.; Son, K.H.; Lee, J.W. PCL/sodium-alginate based 3D-printed dual drug delivery system with antibacterial activity for osteomyelitis therapy. *Gels* **2022**, *8*, 163. [CrossRef]
23. Yazdi, I.K.; Murphy, M.B.; Loo, C.; Liu, X.; Ferrari, M.; Weiner, B.K.; Tasciotti, E. Cefazolin-loaded mesoporous silicon microparticles show sustained bactericidal effect against *Staphylococcus aureus*. *J. Tissue Eng.* **2014**, *5*, 2041731414536573. [CrossRef] [PubMed]
24. Munir, M.U.; Ihsan, A.; Javed, I.; Ansari, M.T.; Bajwa, S.Z.; Bukhari, S.N.A.; Ahmed, A.; Malik, M.Z.; Khan, W.S. Controllably biodegradable hydroxyapatite nanostructures for cefazolin delivery against antibacterial aesistance. *ACS Omega* **2019**, *4*, 7524–7532. [CrossRef]
25. Rath, G.; Hussain, T.; Chauhana, G.; Garg, T.; Goyall, A.K. Fabrication and characterization of cefazolin-loaded nanofibrous mats for the recovery of post-surgical wound. *Artif. Cells Nanomed. Biotechnol.* **2016**, *44*, 1783–1792. [CrossRef] [PubMed]

26. Rath, G.; Hussain, T.; Chauhana, G.; Garg, T.; Goyall, A.K. Development and characterization of cefazolin loaded zinc oxide nanoparticles composite gelatin nanofiber mats for postoperative surgical wounds. *Mater. Sci. Eng. C* **2016**, *58*, 242–253. [CrossRef]
27. Wang, H.; Dong, H.; Kang, C.G.; Lin, C.; Ye, X.; Zhao, Y.L. Preliminary exploration of the development of a collagenous artificial dura mater for sustained antibiotic release. *Chin. Med. J.* **2013**, *126*, 3329–3333. [PubMed]
28. Kaplan, M.; Akgun, B.; Demirdag, K.; Akpolat, N.; Kozan, S.K.; Cagasar, O.; Yakar, H. Use of antibiotic-impregnated DuraGen ® to reduce the risk of infection in dura repair: An in vitro study. *Cen. Eur. Neurosurg.* **2011**, *72*, 75–77. [CrossRef]
29. Suwanprateeb, J.; Thammarakcharoen, F.; Phanphiriya, P.; Chokevivat, W.; Suvannapruk, W.; Chernchujit, B. Preparation and characterizations of antibiotic impregnated microporous nano-hydroxyapatite for osteomyelitis treatment. *Biomed. Eng. Appl. Basis Commun.* **2014**, *26*, 1450041. [CrossRef]
30. Guillen, G.R.; Pan, Y.; Li, M.; Hoek, E.M.V. Preparation and characterization of membranes formed by nonsolvent induced phase separation: A review. *Ind. Eng. Chem. Res.* **2011**, *50*, 3798–3817. [CrossRef]
31. Tan, X.M.; Rodrigue, D. A review on porous polymeric membrane preparation. Part I: Production techniques with polysulfone and poly (vinylidene fluoride). *Polymers* **2019**, *11*, 1160. [CrossRef] [PubMed]
32. van de Witte, P.; Dijkstra, P.J.; van den Berg, J.W.A.; Feijen, J. Phase separation processes in polymer solutions in relation to membrane formation. *J. Membr. Sci.* **1996**, *117*, 1–31. [CrossRef]
33. Bashkatov, A.N.; Genina, E.A.; Sinichkin, Y.P.; Kochubey, V.I.; Lakodina, N.A.; Tuchin, V.V. Glucose and mannitol diffusion in human dura mater. *Biophys. J.* **2003**, *85*, 3310–3318. [CrossRef]
34. Van Noort, R.; Black, M.M.; Martin, T.R.P.; Meanley, S. A study of the uniaxial mechanical properties of human dura mater preserved in glycerol. *Biomaterials* **1981**, *2*, 41–45. [CrossRef]
35. Opálková Šišková, A.; Bucková, M.; Kroneková, Z.; Kleinová, A.; Nagy, Š.; Rydz, J.; Opálek, A.; Sláviková, M.; Eckstein Andicsová, A. The drug-loaded electrospun poly(ε-caprolactone) mats for therapeutic application. *Nanomaterials* **2021**, *11*, 922. [CrossRef]
36. Gao, Y.; Li, J.; Xu, C.; Hou, Z.; Yang, L. Mechanical properties and drug loading rate of a polycaprolactone 5-fluorouracil controlled drug delivery system. *Mater. Res. Express* **2021**, *8*, 095302. [CrossRef]
37. Glover, K.; Mathew, E.; Pitzanti, G.; Magee, E.; Lamprou, D. 3D bioprinted scaffolds for diabetic wound healing applications. *Drug Deliv. Transl. Res.* **2022**. [CrossRef]
38. Rychter, M.; Baranowska-Korczyc, A.; Milanowski, B.; Jarek, M.; Maciejewska, B.M.; Coy, E.L.; Lulek, J. Cilostazol-loaded poly(ε-caprolactone) electrospun drug delivery System for cardiovascular applications. *Pharm. Res.* **2018**, *35*, 32. [CrossRef]
39. van Noort, R.; Martin, T.R.; Black, M.M.; Barker, A.T.; Montero, C.G. The mechanical properties of human dura mater and the effects of storage media. *Clin. Phys. Physiol. Meas.* **1981**, *2*, 197–203. [CrossRef]
40. Zarzur, E. Mechanical properties of the human lumbar dura mater. *Arq. Neuropsiquiatr.* **1996**, *54*, 455–460. [CrossRef]
41. Kamaraj, N.; Rajaguru, P.Y.; Issac, P.K.; Sundaresan, S. Fabrication, characterization, in vitro drug release and glucose uptake activity of 14-deoxy, 11, 12-didehydroandrographolide loaded polycaprolactone nanoparticles. *Asian J. Pharm. Sci.* **2017**, *12*, 353–362. [CrossRef] [PubMed]
42. Yoo, J.; Won, Y.Y. Phenomenology of the initial burst release of drugs from PLGA microparticles. *ACS Biomater. Sci. Eng.* **2020**, *6*, 6053–6062. [CrossRef]
43. Hu, X.; Liu, S.; Zhou, G.; Huang, Y.; Xie, Z.; Jing, X. Electrospinning of polymeric nanofibers for drug delivery applications. *J. Control. Release* **2014**, *185*, 12–21. [CrossRef] [PubMed]
44. Karuppuswamy, P.; Reddy, V.J.; Navaneethan, B.; Luwang, L.A.; Ramakrishna, S. Polycaprolactone nanofibers for the controlled release of tetracycline hydrochloride. *Mater. Lett.* **2015**, *141*, 180–186. [CrossRef]
45. Liu, D.Q.; Cheng, Z.Q.; Feng, Q.J.; Li, H.J.; Ye, S.F.; Teng, B. Polycaprolactone nanofibres loaded with 20(S)-protopanaxadiol for in vitro and in vivo anti-tumour activity study. *R. Soc. Open Sci.* **2018**, *5*, 180137. [CrossRef]
46. Frantz, V.K. Absorbable Cotton, Paper and Gauze: (Oxidized Cellulose). *Ann. Surg.* **1943**, *118*, 116–126. [CrossRef]
47. Spangler, D.; Rothenburger, S.; Nguyen, K.; Jampani, H.; Weiss, S.; Bhende, S. *In Vitro* Antimicrobial Activity of Oxidized Regenerated Cellulose against Antibiotic-Resistant Microorganisms. *Surg. Infect.* **2003**, *4*, 255–262. [CrossRef]
48. Huang, X.; Brazel, C.S. On the importance and mechanisms of burst release in matrix-controlled drug delivery systems. *J. Control. Release* **2001**, *73*, 121–136. [CrossRef]
49. Page, C.P.; Bohnen, J.M.; Fletcher, J.R.; McManus, A.T.; Solomkin, J.S.; Wittmann, D.H. Antimicrobial prophylaxis for surgical wounds. Guidelines for clinical care. *Arch. Surg.* **1993**, *128*, 79–88. [CrossRef]
50. Patel, S.; Thompson, D.; Innocent, S.; Narbad, V.; Selway, R.; Barkas, K. Risk factors for surgical site infections in neurosurgery. *Ann. R. Coll. Surg. Engl.* **2019**, *101*, 220–225. [CrossRef]
51. Mindermann, T. Empirically adapted or personalized antibiotic prophylaxis in select cranial neurosurgery? *Acta Neurochir.* **2021**, *163*, 365–367. [CrossRef] [PubMed]

polymers

MDPI

Article

Structural Breakdown of Collagen Type I Elastin Blend Polymerization

Nils Wilharm [1,2,*], Tony Fischer [3], Alexander Hayn [3,4] and Stefan G. Mayr [1,2,*]

1 Leibniz-Institut für Oberflächenmodifizierung e.V. (IOM), Permoserstr. 15, 04318 Leipzig, Germany
2 Division of Surface Physics, Department of Physics and Earth Sciences, Leipzig University, Linnéstraße 5, 04103 Leipzig, Germany
3 Biological Physics Division, Department of Physics and Earth Sciences, Leipzig University, Linnéstraße 5, 04103 Leipzig, Germany
4 Division of Hepatology, Department of Medicine II, Leipzig University Medical Center, 04103 Leipzig, Germany
* Correspondence: nils.wilharm@iom-leipzig.de (N.W.); stefan.mayr@iom-leipzig.de (S.G.M.)

Abstract: Biopolymer blends are advantageous materials with novel properties that may show performances way beyond their individual constituents. Collagen elastin hybrid gels are a new representative of such materials as they employ elastin's thermo switching behavior in the physiological temperature regime. Although recent studies highlight the potential applications of such systems, little is known about the interaction of collagen and elastin fibers during polymerization. In fact, the final network structure is predetermined in the early and mostly arbitrary association of the fibers. We investigated type I collagen polymerized with bovine neck ligament elastin with up to 33.3 weight percent elastin and showed, by using a plate reader, zeta potential and laser scanning microscopy (LSM) experiments, that elastin fibers bind in a lateral manner to collagen fibers. Our plate reader experiments revealed an elastin concentration-dependent increase in the polymerization rate, although the rate increase was greatest at intermediate elastin concentrations. As elastin does not significantly change the structural metrics pore size, fiber thickness or 2D anisotropy of the final gel, we are confident to conclude that elastin is incorporated homogeneously into the collagen fibers.

Keywords: elastin; collagen; polymerization; fiber formation

Citation: Wilharm, N.; Fischer, T.; Hayn, A.; Mayr, S.G. Structural Breakdown of Collagen Type I Elastin Blend Polymerization. *Polymers* **2022**, *14*, 4434. https://doi.org/10.3390/polym14204434

Academic Editors: Ariana Hudita and Bianca Gălățeanu

Received: 20 September 2022
Accepted: 18 October 2022
Published: 20 October 2022

1. Introduction

Thermoresponsive hydrogels find widespread applications in medicine where synthetic and protein-based hydrogels are described. The potential applications include drug delivery, tissue engineering or the separation of bio molecules [1–4]. For example, an elastin-like polypeptide sequence, attached to graphene, with a high switching rate was designed, which revealed shrinking/bending upon irradiation with NIR (near infrared) light [5]. Examples of drug delivery have been presented by various groups, such as when an elastin-like peptide (ELP) solution is loaded with an anti-tumor drug. After injection into a tumor, the ELP coacervates due to the body temperature and forms a depot from which the drug is released over some time [6]. Similarly, the loading of ELPs with bone morphogenetic protein was described to enhance mineralization [7]. In another application, ELPs were combined with chitosan to form a multilayer system, which changes its wettability state when heated above 50 °C [8]. This system can assist in fine tuning cellular adhesion.

The temperature-induced contraction has an identical root for synthetic as well as protein-based polymers. All these polymers exhibit a lower critical solution temperature (LCST) upon which they become insoluble. The reason for this behavior is the imbalance between hydrophilic–hydrophobic interactions between a polymer and a solvent. Hydrophobic segments along a polymer chain can reduce their solvent-accessible surface area upon an increased temperature by aggregation, which exerts a pulling force on the non-contracting network segments. In fact, the driving force for the contraction is the entropy gain for the solvent molecules. Water molecules around hydrophobic segments are highly ordered but with an increasing temperature this order is disrupted and the hydrophobic segments can associate and fold. The contraction is then actually induced when the entropy gain by the released water molecules is greater than the enthalpy gain by water binding to the polymer [9]. Poly(N-isopropylacrylamide) (PNIPAM) is one of the most investigated thermoresponsive polymers as its transition temperature is relatively insensitive to environmental conditions and is in the physiological regime (~32 °C) [10]. However, PNIPAM polymers have been shown to reduce cell viability for different polymerization types as well as different cell types [11]. Elastin is, therefore, a prime candidate for bio compatible thermoresponsive hydrogels as it is composed of alternating hydrophilic and hydrophobic segments and has already been shown to exhibit a LCST [12]. Recently, a collagen elastin thermoactuator with a tunable transition temperature in the physiological temperature regime was designed [13]. It was demonstrated that the incorporation of elastin from bovine neck ligament into a 2 mg/mL type I collagen gel resulted in a reversible thermoswitchable system with a transition temperature in the physiological temperature regime. As the system showed a temperature-induced phase transition like a volume contraction, it was argued that this process can be described by Euler buckling, which refers to the buckling of a rod under an axial critical load. In fact, two cases are possible when collagen and elastin are polymerized: the formation of individual fibers between collagen fibers ("perpendicular") and the incorporation of elastin monomers in a parallel manner ("lateral") into a collagen fiber. Although a lateral fiber alignment seemed likely as the buckling behavior was observed, convincing experiments have been lacking so far. We now present insights into the polymerization features of an elastin collagen hydrogel with significant evidence that elastin is laterally incorporated into the collagen fiber. Both elastin and collagen are the main components of connective tissue and exhibit distinct features. While elastin is structurally heterogeneous as the hydrophilic segments contain some α-helix and the hydrophobic segments are mostly of a random coil design, collagen is relatively homogeneous as three single peptide chains form a collagen triple helix [12,14]. The elastin is expressed in the endoplasmic reticulum, transported outside of the cell and then bound to micro fibrils, where, by a complex mechanism involving the crosslinking of lysine residues, an elastin fiber is formed with elastin on the inside and several micro fibrillar proteins on the outside [15]. Collagen fibrils are formed by an association of collagen triple helical monomers via an enzymatic crosslinking of lysine residues, and several fibrils then associate into fibers; however, our preparation steps were devoid of any crosslinking steps, so that the interaction between the collagen and elastin were dominated by an electrostatic interaction [16].

This work aims to present evidence that the structural metrics pore size and fiber diameter of a type I collagen hydrogel are not significantly changed upon the addition of bovine elastin. As we have found evidence supporting our thesis, we conclude that elastin monomers attach in a parallel fashion to the collagen fibers. This agrees with our own earlier studies with circular dichroism experiments, where we saw a systematic decrease in the helical structures in collagen and elastin after polymerization [13].

2. Experiments

2.1. Hydrogel Preparation

Collagen hydrogels for all experiments in this study were prepared using the same protocol as described before [13]. The basis for all the hydrogels was a mixture of collagen I monomers from rat tail (collagen R, 0.4% solution and Cat. No. 47256.01; SERVA Electrophoresis, Heidelberg, Germany) and bovine skin (collagen G, 0.4% solution and Cat. No. L 7213; Biochrom, Berlin, Germany) in a mass fraction of 1:2, respectively. To initiate the polymerization of the monomer mixture solution, a 1 M phosphate buffered solution containing disodium hydrogen phosphate (Cat. No. 71636, Merck KGaA, Darmstadt, Germany), sodium dihydrogen phosphate (Cat. No. 71507, Merck KGaA, Darmstadt, Germany) and ultrapure water was added to produce a final pH value of 7.5, ionic strength of 0.7, and a phosphate concentration of 400 mM. To produce the collagen–elastin hydrogels, appropriate amounts of elastin powder (elastin, Cat. No. 6527, Merck KGaA, Darmstadt, Germany) were added to the buffer solution prior to the polymerization. All solutions were kept on ice. The polymerization of the final solutions was initialized by placing the samples in an incubator at 37 °C.

2.2. UV/VIS Plate Reader Experiments

The experiments were performed on a TECAN infinite® 200 plate reader (absorbance mode, 405 nm, target temperature 37 °C, 25 flashes, and a sampling rate of 1/min; TECAN Trading AG, Männedorf, Switzerland) using flat bottom 96-well plates (Carl-Roth GmbH, Karlsruhe, Germany) for the sample preparation and measurement. For a single experiment, three solutions were prepared directly before the measurement, namely, (I) 1.2 mL of a 2 mg/mL collagen solution (0.4 mL R, 0.8 mL G, each 4 mg/mL of stock solution), (II) 1.2 mL of a 2 mg/mL collagen solution (same as above) with 0.6 mg of elastin and (III) 1.2 mL of a 2 mg/mL collagen solution (same as above) with 1.2 mg of elastin. A 200 µL amount of the final solution was filled in each well, resulting in 6 wells of a 96-well plate per condition and 18 wells for all three conditions. The remaining 78 wells were filled with distilled water to ensure high humidity during the polymerization and to counteract the dehydration of the samples.

2.3. Zeta Potential

A 6 mg amount of elastin was dissolved in a 3 mL phosphate buffer (pH 7.5, 400 mM) at 4 °C. Additionally, 3 mL of collagen (1 mL R and 2 mL G, each 4 mg/mL stock solution) was mixed with 3 mL of the phosphate buffer (pH 7.5, 400 mM) at 4 °C. Both solutions were subjected to zeta potential measurements on a NanoZS (Malvern) zetasizer with a backscatter optical arrangement (173°). Each condition was measured three times by preparing a fresh solution each time. Each sample (1 mL) was left for 120 s in the instrument, which was precooled at 4 °C, to reach the thermal equilibrium. Each sample was measured eight times with twenty runs and a 30 s delay between the measurements.

2.4. Pore size and Fiber Diameter

The collagen and collagen–elastin samples were prepared as described above. A 200 µL amount of the ice-cooled solution was placed in each well of a cooled 24-well µ-plate (µ-Plate 24 Well ibiTreat; Cat. No. 82406; ibidi, Gräfelfing, Germany). Subsequently, the 24-well plate was placed in an incubator at 37 °C to start the polymerization for 2 h. The polymerized hydrogels were washed three times using PBS and fluorescently stained by applying 20µg/mL of 5(6)-Carboxytetramethylrhodamine N-succinimidylester (TAMRA-SE; Cat. No. 21955; Merck KGaA, Darmstadt, Germany) overnight and subsequently washed three times using the PBS. Using an LSM microscope (TCS SP8; Leica, Wetzlar, Germany), three-dimensional image cubes of the fluorescence signal of the TAMRA-SE using a 561 nm excitation laser and HC PL APO CS2 40x/1.10 water immersion objective were recorded. The pore size and fiber diameter were determined as published previously [17,18]. To compensate for the apparent collagen–elastin clusters that would disturb a fiber diameter determination, a custom-built cluster deletion algorithm was used to solely measure the actual fiber diameter sizes.

2.5. Directionality

The above obtained images were also used to quantify the directionality and its standard deviation of the network. The imageJ plugin, "Directionality", with the "Fourier components" method was used. A total of 70 planes of each of 10 different random positions for both gels (collagen and collagen–elastin) were summarized into one image which was then analyzed. The clusters in the collagen–elastin samples were removed prior to analysis using the same, custom-built cluster deletion algorithm as described above.

2.6. Elastin Influence on the Network Structure

To investigate the influence of elastin polymerization on the final hydrogel structure, we used a Col-F collagen binding reagent (Col-F; Cat. No. 6346; ImmunoChemistry Technologies, Bloomington, MN, USA) and a collagen I antibody (Immunotag™ Collagen I Polyclonal Antibody; Cat. No. #ITT5769; G-Biosciences, St. Louis, MO, USA). The collagen–elastin hydrogels were prepared in 24-well μ-plates as described above.

For the collagen I antibody staining, the samples were incubated with 5% BSA solved in PBS for 30 min at room temperature—with aspirate goat serum, and incubate sections with primary antibody (ITT5769) in PBS overnight at 4 °C or 1 h at 37 °C; 3 × 1:1000 (600 μL/well). The samples were washed three times with PBS for 5 min each.

For the Col-F staining, the samples were incubated with 3% BSA solved in PBS for 30 min at room temperature—with aspirate goat serum, and incubate sections with primary antibody (ITT5769) in PBS overnight at 4 °C or 1 h at 37 °C; 3 × 1:200 (300 μL/well). The samples were washed three times with PBS for 5 min each.

Three-dimensional images were recorded using an LSM microscope (TCS SP8; Leica, Wetzlar, Germany) with a 63×/1.20 HC PL APO CS2 water immersion objective and a 488 nm (Col-F) and 561 nm (collagen antibody) excitation laser, respectively. The final image dimensions were 100 μm by 100 μm in x-y and a roughly 30 μm to 50μm z dimension.

2.7. Live Polymerization

The samples were prepared as described above. A 1 mL amount of the cooled solution was placed in a well of a pre-cooled 24-well μ-plate and then placed in a LSM microscope with an incubation chamber (TCS SP8; Leica, Wetzlar, Germany) at 37 °C and 100% relative humidity. Using a HC PL APO CS2 40×/1.10 water immersion objective and a 561 nm laser in the reflection mode, a 1 h recording of the polymerization process and hydrogel network formation was observed and recorded as live imaging videos. The videos had an image size of 1024 × 1024 px with a frame-rate of 1 fps.

2.8. Statistical Methods

The employed statistical methods included the mean, median, standard deviation and a box plot as well as a Mann–Whitney-U test. The methods are named at the relevant position.

3. Results

3.1. Plate Reader

Figure 1 displays the polymerization curves of a collagen solution and two collagen elastin solutions at 37 °C. The heating curve of the collagen is in strong agreement with the literature data, as it highlights the onset of clouding after 30 min as well as no significant changes in the turbidity after 2 h [19]. The clouding curve is generally associated with fiber formation which increasingly contributes to light scattering. The addition of elastin then introduces several features into the polymerization process. The most striking feature is that the final absorption (>2 h) was only slightly increased, although 25% or 30%, respectively, should be expected as this is the net increase in the biomass for each sample. This is a strong indication that the alignment of elastin and collagen monomers must occur in a lateral manner, as the opposite case of a perpendicular alignment would contribute to light absorption and scattering. The small increase in absorption, at times >2 h, might be a consequence of an elevated fiber thickness as elastin monomers attach to the collagen triple helix. The lateral addition of elastin is also likely as circular dichroism experiments on elastin–collagen gels have shown that the addition of elastin leads to a reduced PPII (polyproline II) content, probably due to PPII helix distortion [13]. The addition of low amounts of elastin increases the polymerization rate by a factor of two while the polymerization rate maximum is shifted to an earlier time (37 min instead of 49 min, see Table 1). At these concentrations, the elastin may act as a nucleation center for polymerization. Additional effects which could fasten the assembly might include the burying of hydrophobic domains in collagen but especially in elastin, which has alternating hydrophilic and hydrophobic segments [20,21]. Additionally, a potential entropy gain by a helix distortion, as mirrored in the reduced PPII helix content in collagen after an elastin addition, supports the thesis of a conformation-dependent collagen–elastin interaction [13]. Such an entropy gain by a helix distortion was described for an alpha helix [22]. Taken together, collagen's and especially elastin's propensity to bury their hydrophobic domains, as well as a general increase in the monomer concentration, might contribute to an increase in the polymerization rate. In terms of the type of fiber alignment, we argue that hydrophobic burying implies a parallel alignment, as in the otherwise perpendicular type no significant burying can take place.

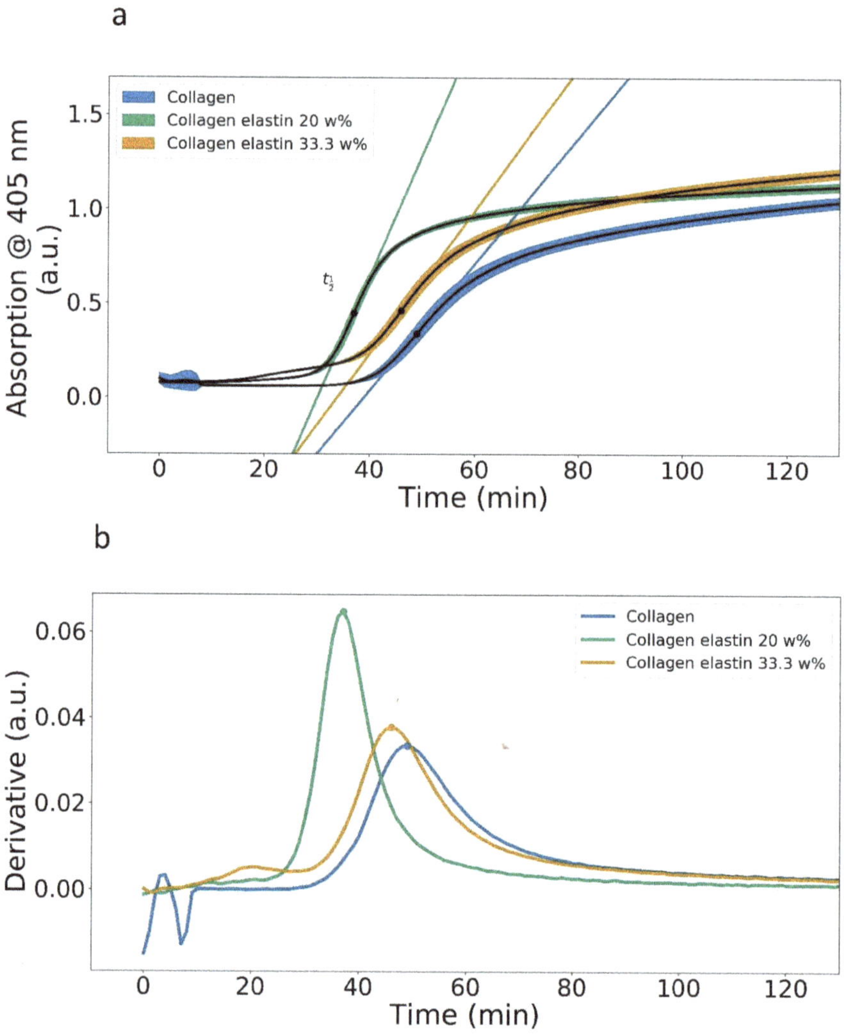

Figure 1. (**a**) Mean polymerization curves at 37 °C for a 2 mg/mL collagen solution as well as two collagen–elastin solutions containing 20 w% (0.6 mg/mL elastin) and 33.3 w% (1.2 mg/mL elastin), respectively. (**b**) Derivative of the mean of the curves in (**a**). Six wells were recorded per sample and the color-coded curves in figure (**a**) denote one standard deviation. Supplementary Figure S3 shows the extended curves. The random spikes in the beginning of the collagen curve result from water condensation and evaporation under the well plate cover.

Table 1. Characteristic values for polymerization. $t_{\frac{1}{2}}$ stands for the time where the derivative of the turbidity curves has the greatest value, i.e., the increase in turbidity is the greatest, while Abs. at t(1/2) (a.u.) stands for the absorption value (turbidity) at the time of $t_{\frac{1}{2}}$.

	Maximum Rate (a.u./Time)	$t_{\frac{1}{2}}$ **(min)**	**Abs. at** $t_{\frac{1}{2}}$ **(a.u.)**	**Abs. after 120 min (a.u.)**
Collagen, 405 nm	0.033	49	0.34	1.01 ± 0.03
Collagen + 0.6 mg Elastin, 405 nm	0.064	37	0.46	1.11 ± 0.02
Collagen + 1.2 mg Elastin, 405 nm	0.037	46	0.45	1.17 ± 0.03

A further addition of elastin, however, decreased the polymerization rate. This was unexpected as the polymerization rate is always proportional to the monomer concentration. Thereby, at relevant concentrations, elastin can be viewed as a perturbation towards polymerization as it may interfere with the proper alignment of collagen triple helical monomers. A fingerprint of this feature was the additional absorption shoulder at 20 min which probably signified a second polymerization process introduced by the elastin. This shoulder is believed to originate from the formation of elastin–collagen clusters which form at elevated elastin concentrations. This is a likely process, as elastin to collagen ratios of more than 0.22 will exceed elastin–collagen equimolarity. In fact, based on the molar masses of collagen (300,000 Da) and elastin monomers (ca. 67,000 Da), 0.5 mg/mL of elastin is sufficient to accommodate 2 mg/mL of collagen in an equimolar manner [23,24]. The plate reader experiments fell well within this consideration, as they showed that 0.6 mg/mL of elastin did not lead to an additional clustering peak at 20 min, while the 1.2 mg/mL sample did so; therefore, the upper limit for an elastin addition seems to lie between these values. As the absorption value in the 33.3 w% curve of the 20 min peak was much smaller than the maximum absorption, it can be argued that most of the biomass was polymerized into the gel. Another interpretation may be that the shoulder at 20 min signified clouding by elastin coacervation, a well-known effect which describes the heat-induced elastin aggregation by an association of hydrophobic elastin segments. However, elastin coacervation is quite fast and usually complete after several minutes; therefore, we can exclude this effect here [25]. It is important to note that a similar experiment was performed by Vazquez-Portalatin et al. by also using collagen type I and bovine neck ligament elastin. They similarly recorded the clouding of elastin–collagen solutions for several elastin–collagen ratios [26]. Opposed to our experiments, they observed an overall increase in the polymerization rate and a shift in the polymerization start to earlier times with an increasing elastin proportion; however, the maximum polymerization rate was around 21 min, which was twice as fast as our observation of around 40 min. Additionally, the turbidity-dependence on the elastin concentration was much lower than in our experiments. This might not only have to do with the fact that they used only rat tail collagen (R collagen), another wavelength (313 nm) and PBS (phosphate-buffered saline) instead of a phosphate buffer. In fact, they used comparable elastin–collagen ratios but with a 1:10 dilution. This gives credit to our above claim of a saturative process during polymerization. Obviously, in our experiments, the addition of elastin at elevated concentrations seemed to induce a second polymerization process apart from the "classical" polymerization which we introduce as a cluster formation, probably because the monomers met more often which also increased the chance that the monomers met without being optimally aligned in the gel. These clusters grew on their own without participating in the classical polymerization. In fact, Paderi et. al. discuss how a perpendicular chain alignment can inhibit collagen polymerization which may, in our case, have been the nucleation center for the cluster formation [27].

3.2. Videos of Polymerization

Videos S1–S3 (supplement) show the fiber formation of a collagen solution and two elastin–collagen solutions, while Video S4 (supplement) shows a comparison video. Video S1 is characterized by early and quick flashes of fibers and nodes which resulted from their diffusion through the focal plane. Small fibers and nodes could be observed as early as five minutes after combining both solutions (the collagen stock and buffer). This was contrasted to the plate reader experiments where no significant changes in the absorbance were observed before 25 min. This was because the plate reader measures absorbance which is quite small for small particles, so that only sufficiently large particles or fibers can contribute to the absorbance. The videos emphasize that the fibers assembled rather quickly while they were still subjected to convection, i.e., liquid flow. The onset of polymerization was characterized by a fiber flow velocity reduction which came to a complete stop as soon as sufficiently large enough fibers had come into contact. The fiber growth occurred in the early stages end to end and was then followed by a fiber thickening, which is in line with the literature claims that axial growth is much faster than lateral growth [28]. The sequence of the axial followed by the lateral fiber growth was retained when the elastin was added, implying that the elastin did not significantly interfere with the fiber assembly process in terms of the network structure. Moreover, when the polymerization sequence of the collagen–elastin solution was identical to the one of the pure collagen polymerization sequence, then the elastin must have been homogeneously incorporated into the collagen system, i.e., laterally. It was further obvious, that the elastin-containing networks polymerized earlier, which was in good agreement with the plate reader experiments. Consequently, elastin seemed to facilitate polymerization as described above, although this effect was concentration-dependent. The maturing collagen network was still drifting through the focal plane as seen in the appearance and disappearance of fibers and nodes. This implies that the network was subjected to density fluctuations during the polymerization. This was contrasted to the elastin containing networks, which did not drift through the focal plane. This might relate to our observations, namely, that the elastin-containing gels appeared to stick to the walls of the petri dish. This effect might limit the z-drift. A final observation was that the elastin-containing networks contained some clusters which were more prominent in the high-elastin concentration sample. Video S4 shows quite nicely how these clusters disappeared after around 30 min. We believe that the clusters sunk either to the bottom of the gel or were bound randomly to the existing fibers, although we could not observe such diffusion to the fibers. We further argue that the presence of these clusters coincided with the presence of the elevated absorbance around 20 min in the elastin-containing turbidity curves (Figure 1); however, further experiments are required to understand the interaction between the elastin and collagen R and G. This question bears some importance, as G collagen is more closely related to the formation of nodes than R collagen [19]. In fact, further applications might demand answering the question of whether elastin is also present in the nodes as a local matrix stiffness can guide the cell migration [19].

3.3. Zeta Potential

The zeta potential measurements of the individual collagen and elastin solutions in the phosphate buffer at pH = 7.5 and 4 °C revealed that all solutions exhibited a zeta potential around −4 mV (Figure 2). Values in this range are optimal for aggregation as values smaller than ± 30 mV are considered to induce aggregation [29]. Although biopolymers such as collagen and elastin have plenty of ionizable groups, zeta potential values around zero indicate a low degree of ionization. This low potential, as described above, favors monomer aggregation in any way, including laterally, as the resulting hydration shell will be small at these values so that the repulsion will effectively play no role. In fact, both collagen assembly and elastin assembly (coacervation) are endothermic and entropy driven at 37 °C, while the loss of an ordered hydration shell is the largest contribution to entropy gain [25,30]. However, a decrease in Gibbs energy is roughly twice as much in collagen than it is in elastin, implying that collagen can more easily lose its hydration shell. Moreover, although the thermodynamics for elastin relate to the effect of coacervation, we did not see this effect in the plate reader experiments, where no early clouding could be detected. Collagen and elastin monomers could, therefore, align in a parallel manner before a temperature increase shifts the Gibbs energy change from positive to negative such that, after a loss of the respective hydration shell, the collagen–elastin association is more favorable than an elastin–elastin association (coacervation). Elsewhere, the lateral merging of a hydration shell of peptides has been described which opens up the possibility of a multistep mechanism of an early elastin–collagen interaction [31]. It is also noted that the employed collagen was already in its triple helical state so that the elastin should not have interfered with the triple helix formation; however, the data of Wilharm et. al. show that the circular dichroism of collagen–elastin is not an ideal superposition for each component and that it lacks some PPII content [13]. This shows how the presence of elastin might impact the collagen helix anyway, probably due to the destabilization of the intricate H-bond equilibrium in collagen, probably in a lateral manner.

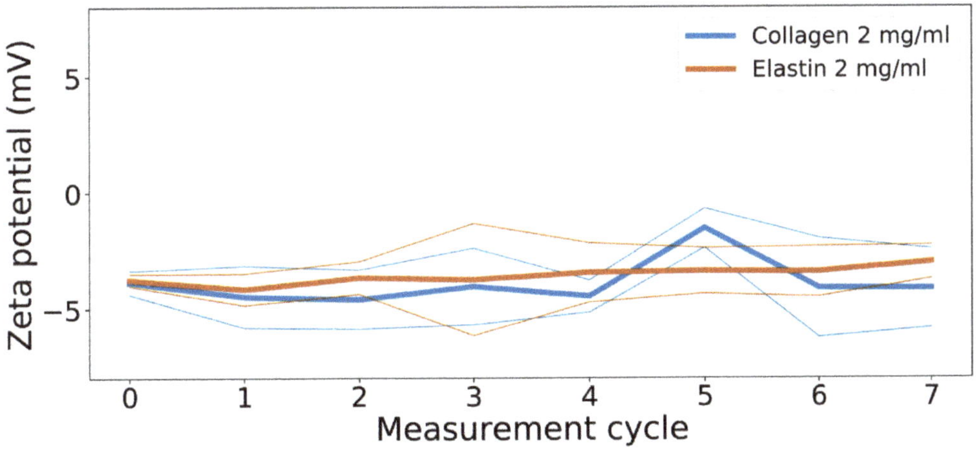

Figure 2. Zeta size measurements of a collagen (2 mg/mL) and an elastin solution (2 mg/mL) in a phosphate buffer at pH 7.5 and 4 °C. Incremented mean (thick lines) and one standard deviation (thin lines) are shown. As data points along the curves have slightly varying total run times from 7 to 10 min (due to variations in temperature regulation by the device), the data points were averaged accordingly and plotted over the increment. Original data can be found in supporting Figure S4.

3.4. Directionality

Figure 3 shows the 2D anisotropy for an exemplary network, while Figure 4 compares the 2D anisotropy of a collagen and a collagen–elastin network. The sections of all samples show two preferred directionalities, one around 65° and another around −80°. The directions seem to be inversely populated by collagen and elastin–collagen. Regarding the origin of this preferred orientation, one explanation might be the gelation condition. In fact, the gels were gelled within an incubator placed on a microscope. The incoming air and humidity induced a mild current which might have oriented the fibers accordingly. This is also visible in the Videos S1–S4 where the material is drifting until the polymerization starts and the flow is restricted. Although this drift was unavoidable when using this experimental approach, this technique was used to specifically prepare oriented gels [32]. While the addition of elastin did not change the network directionality, the standard deviation of the angle distribution might have been slightly increased (Figure 4). This direct comparison between the respective standard deviations across the angle distributions of all 10 position reveals a minor significant difference as the Mann–Whitney U test was $p = 0.08$, which was larger than the generally accepted threshold of 0.05. Under the assumption of this threshold, both distributions would not originate from one set of data, i.e., the addition of elastin would lead to a flattened angle distribution. It can be concluded that elastin's presence interferes with the formation of larger, similarly oriented domains. Mostaço-Guidolin et al. have found the interesting observation that a similar orientation of collagen and elastin fibers in the arterial wall of rabbits was greatest when they were middle-aged and lowest when they were young or old [33]. In the context of our experiments, this might imply that an elastin addition creates less mature networks as it seems to slightly interfere with a proper alignment of collagen fibers. The above-described plate reader experiments support this claim, as they showed an elastin concentration-dependent increase in the polymerization rate. A faster rate means less time for the monomers to perfectly align, such that stacking irregularities can occur. This was already discussed earlier, where a faster rate was suspected to contribute also to the cluster formation. Nonetheless, our analysis of the 2D anisotropy in the collagen and elastin–collagen networks could not conclusively portray a difference in the 2D anisotropy, as the p-value was just slightly larger than 0.05, which supports our claim of a lateral elastin–collagen alignment. In fact, a predominantly perpendicular alignment of the collagen and elastin fibers should significantly increase the standard deviation of the angle distribution. Additionally, the above-mentioned bubbles were removed prior to the 2D anisotropy analysis so that the observed potential increase in the standard deviation might as well have originated from this preprocess, implying that there was no real difference at all between the collagen and elastin in terms of the 2D anisotropy. The above arguments are in line with the narrative that if 2D anisotropy as a network metric does not significantly change upon an elastin addition, then the structure cannot be changed, i.e., elastin is incorporated mostly homogeneously into collagen.

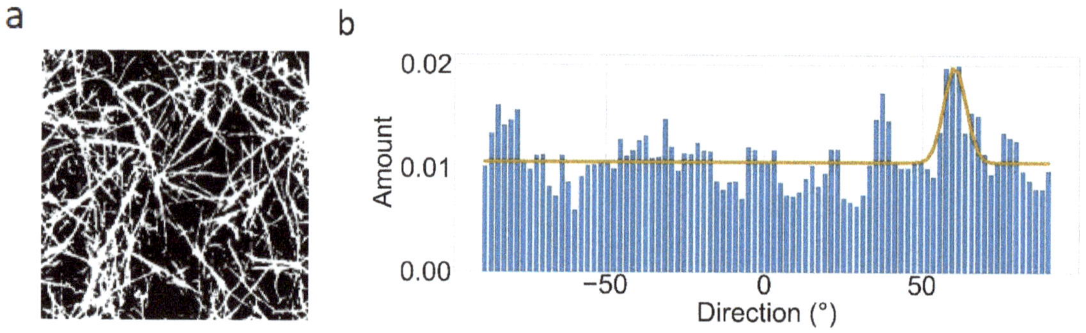

Figure 3. (**a**) Exemplary skeletonized image of a 2 mg/mL collagen network. (**b**): Angle distribution of (**a**). The degree values are given in the mathematical sense, i.e., 0° is pointing to the right. The fit is a Gaussian function, provided by ImageJ.

3.5. Laser Scanning Microscopy (LSM)

LSM recordings of a collagen–elastin hydrogel with primary collagen antibody staining support the above claims of lateral collagen–elastin polymerization (Figure 5). The left image (Figure 5a) represents an exemplary primary collagen antibody-stained sample. In total, images from seven random positions were recorded which all displayed the discussed features.

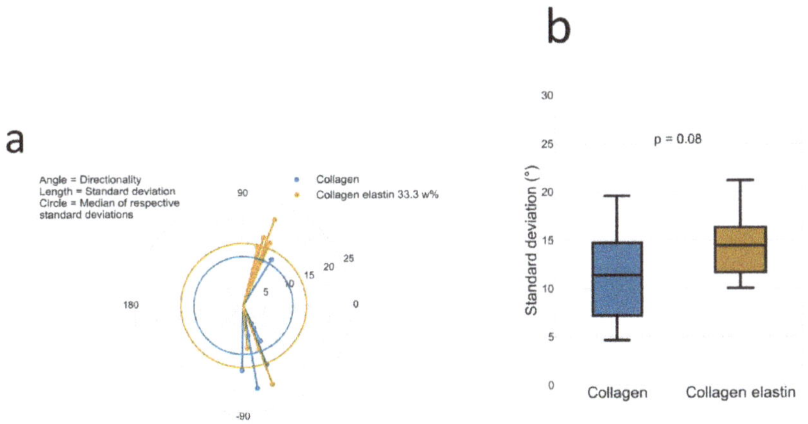

Figure 4. (**a**) Polar plot of the 2D anisotropy analysis of each 10 random positions in one 2 mg/mL collagen network and one collagen–elastin (33.3 w% elastin) network. The coordinates refer to the determined angle (polar angle) and the standard deviation (radial length) while the circles refer to the median values of the distributions of the standard deviations, i.e., elastin increases the standard deviation of the network and, thereby, the 2D anisotropy. (**b**) Comparison between the standard deviation of the angle distribution of the samples already displayed in Figure 4. Significance was tested with the Mann–Whitney U test. This standard deviation is referred to as "2D anisotropy".

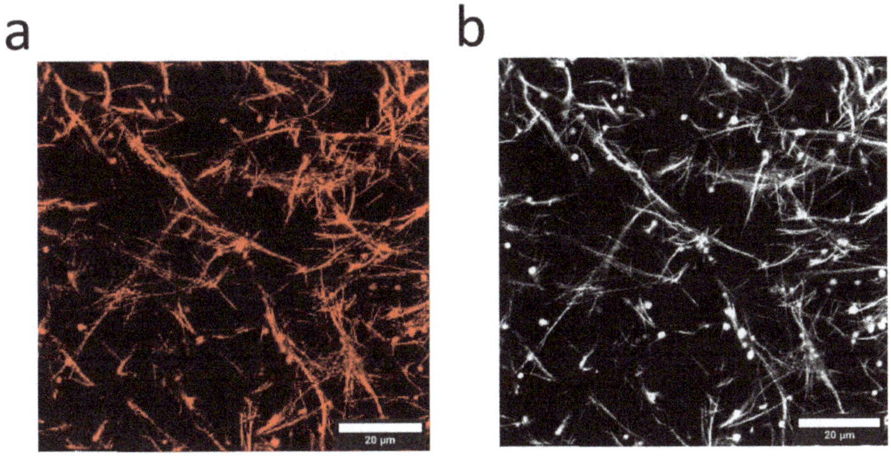

Figure 5. Fluorescence images of a 33.3 w% elastin–collagen gel. (**a**) collagen type I antibody and (**b**) Col-F. The dots in each image are most likely clustered elastin–collagen monomers.

The right image (Figure 5b) shows all the network features (collagen + elastin). A comparison with the primary collagen antibody-stained image reveals identical features in both images while any observable differences must be attributed to the brightness thresholding of the image software. The left image contains the well-known features of the R + G collagen mixture, namely, the nodes and fibers, while the right image does not convey any additional structural features; therefore, a lateral alignment of the elastin and collagen monomers appears likely. This is further plausible, given the architecture of the elastic fiber under physiological conditions. Elastin is synthesized in the endoplasmic reticulum and then transported outside of a cell by binding to an elastin-binding protein. Upon binding of this protein to the galactosugars of micro fibrils outside of a cell, elastin is released from the elastin-binding protein and interacts then with the microfibrils. Elastin is then incorporated in a complex way into the microfibrils resulting finally in a fiber which contains elastin on the inside and a microfibrillar shell outside [15]. Basically, elastin needs a scaffold to be deposited on and several proteins of the fibrillin class as well as MAGP-1 where it is shown to interact with elastin [34]. Furthermore, although the elastin–fibrillin interaction is very complex, it has been shown that elastin binds to a glycine and proline-rich region in fibrillin-2 [35]. Consequently, some homology to collagen is given, which lends credibility to the elastin–collagen interaction as seen in the above described LSM recordings; however, other proteins such as fibulin-5 are also central to elastic fiber formation [36]. The list of important proteins continues and their non-existence in our system may be a likely explanation for the lack of formation of distinct elastic fibers. This consideration hardens our claim of a lateral, or at least, homogenous incorporation of elastin into collagen fibers, as elastin monomers simply do not experience guidance and as such are subjected to following collagen fibrillogenesis.

The image in Figure 5 contains the clustered collagen and elastin which was already discussed in the section, "plate reader", where the high elastin concentration sample displayed an absorption peak prior to the main maximum. These clusters are said to contribute to clouding as the early binding of elastin to collagen might form these clusters which, by chance, are not polymerized into the final network. As we can see the clusters also in the "collagen only" channel (Figure 5a), they must have at least contained some collagen, but as the clusters also appeared after the elastin addition, they must have also contained elastin.

An interesting accordance is seen when the 3D pore size of the networks is compared (Figure 6). The addition of elastin lowered the median pore size by only ~4%. Additionally, the interquartile range was smaller after the elastin addition (0.46 μm for the collagen and 0.32 μm for the collagen–elastin). A likely explanation for this effect is an increase in the fiber diameter because of a lateral fiber alignment between the elastin and collagen chains. The resulting thicker fibers would automatically lead to a decreased pore size when the network architecture remains unchanged, which was shown earlier. Indeed, Figure 6 reveals an increase in the fiber thickness by ~10% which is, again, similar to the percentage changes for the pore size. In fact, other imaginable polymerization types, i.e., a branched fiber alignment, should significantly reduce the median fiber thickness. Consider also the network illustration shown in Figure 7. If the addition of elastin to the network would connect random points along the turquois fibers, the pore size would be halved or at least significantly reduced. This effect must lead to a significant decrease in the median pore size which, presently however, was not observed.

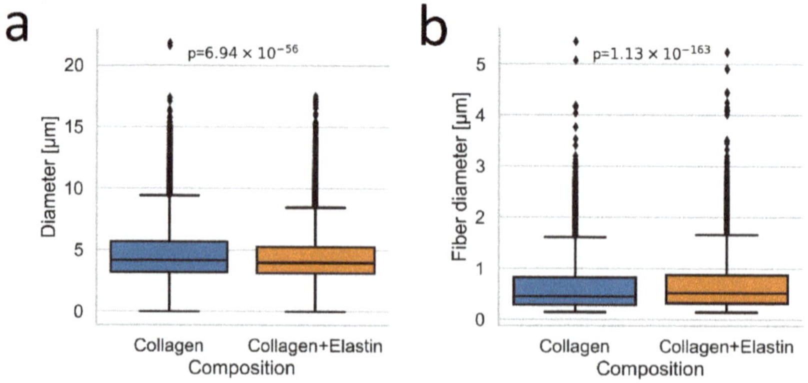

Figure 6. (**a**) Pore diameter of a 2 mg/mL collagen gel and a 33.3 w% elastin collagen gel. (**b**) Fiber diameter of the same gel. Ten positions for each condition were used and 100 planes were summed each prior to analysis. Significance was tested with the Mann–Whitney U test.

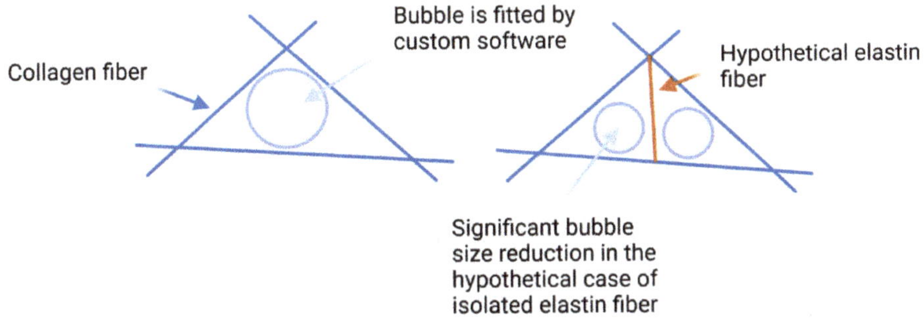

Figure 7. Drawn intersection of collagen fibers (blue lines), which enclose a pore (circle); however, a hypothetical elastin fiber (red line) will divide the pore in two much smaller pores. As we did not see a significant decrease in the pore size, the only other mechanism must be lateral polymerization.

4. Discussion

The most remarkable finding within our experiments was that the addition of elastin to a collagen solution at pH 7.5 does neither induce significant changes within the polymerization process nor structural changes within the network later. Initially, we discussed two extremes of a collagen–elastin interaction, namely, perpendicular and lateral polymerization. While we expected a mixed state between these two extremes prior to the experiments, it became quickly clear that the experiments favored the lateral state over the perpendicular and mixed state (Figure 8). This was also initially proposed as we had observed a Euler buckling-like behavior of the hybrid gels under heating in earlier experiments [13]. This phase transition-like behavior can only manifest if elastin conveys a compressive force on collagen fibers in the axial direction. In the opposite case of a perpendicular connection, one would expect a linear decline in the volume with the temperature, as the collagen network would gradually follow the contractive force conveyed by elastin. An homogenous incorporation by a lateral fiber alignment is also likely from another perspective. The persistence length l_p of a polymer describes the length over which bending fluctuations are correlated, where a larger value means that the respective polymer is rather inflexible. The literature reports values for collagen of l_p = 10 nm to 20 nm and for elastin of l_p = 0.3 nm to 0.6 nm [37–39]; therefore, elastin monomers are around 30 times more flexible than collagen monomers. Together with only 1/4th of collagen's mass, it is easily imaginable that elastin monomers attach quickly and in a highly adaptive manner to collagen monomers.

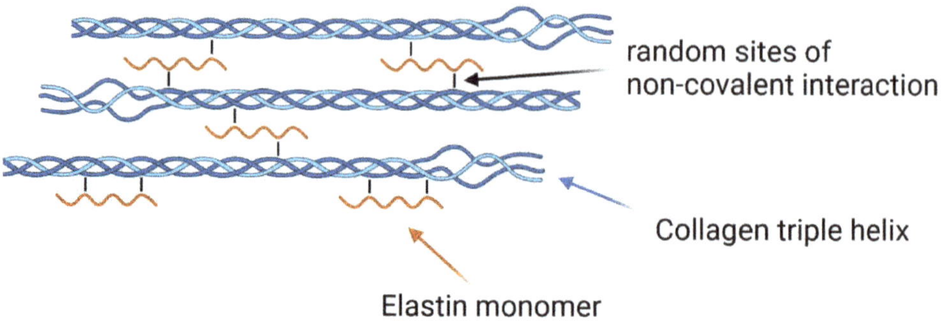

Figure 8. Proposed incorporation of elastin into a collagen fibril (hydrophilic and hydrophobic segments of elastin are not shown). The random coil ends of collagen should signify the suggested interference of elastin with collagen's secondary structure during polymerization. Elastin monomers are thought to bind to collagen through local H-bonds, van-der-Waals bonds and ionic bonds, although the latter is less likely due to the low zeta potential.

5. Conclusions

We showed insights into the polymerization features of elastin–collagen hydrogels. Especially, it was shown that elastin and collagen chains interact in a lateral fashion. This was directly demonstrated with the LSM recordings of collagen and collagen–elastin gels where the collagen was separately stained over the collagen–elastin and further indirectly, as the addition of elastin did not change the structural metrics pore size, fiber thickness or 2D anisotropy. Although we did not quantify changes in the axial and lateral polymerization rate, a visual inspection of the Videos S1–S4 highlights no changes in this polymerization metric after the elastin addition, i.e., the axial fiber growth still starts earlier than for lateral growth; however, the plate reader experiments revealed an elastin concentration-dependent acceleration of the polymerization rate and no signs of elastin coacervation in the presence of collagen. This is a strong sign that a lateral elastin–collagen association precedes the temperature-induced loss of the hydration shell in both, leading to homogenous elastin–collagen hybrid fibers. Further, the zeta potential experiments confirmed a similarly low potential for elastin and collagen, confirming optimal conditions for aggregation. Taken together, the addition of bovine neck ligament elastin to type I collagen solutions accelerated the polymerization rate, although no significant structural changes of the resulting gels could be observed. To generalize our findings, we showed elastin's propensity to bind to other bio polymers such as collagen in a lateral manner and our findings can help to guide the preparation of other elastin-based bio materials with or without actuatoric application.

Supplementary Materials: The following supporting information can be downloaded at: https://www.mdpi.com/article/10.3390/polym14204434/s1, Figure S1. Left: 10 stacked images of a 2 mg/mL collagen type I gel. Right: cluster detection as indicated by black dots. These were ignored for the fiber thickness estimation as shown in Figure S2; Figure S2. Left: exemplary 2 mg/mL collagen gel. Right: fiber thickness estimation. Detection was similarly undertaken for the elastin–collagen after cluster removal as shown in Figure S1; Figure S3. Top: extended polymerization curves. Mean polymerization curves at 37 °C for a 2 mg/mL collagen solution as well as two collagen elastin solutions containing 20 w% (0.6 mg/mL elastin) and 33.3 w% (1.2 mg/mL elastin), respectively. Bottom: derivative of the mean of the top curves. Six wells were recorded per sample and the color-coded curves in the top figure denote one standard deviation; Figure S4. Measured zeta potential values over time. Three samples were analyzed per condition. Video S1: Collagen; Video S2: 20 Elastin + Collagen; Video S3: 33.3 Elastin + Collagen; Video S4: Comparison.

Author Contributions: Conceptualization, N.W.; methodology, N.W., T.F., A.H.; software, N.W., T.F., A.H.; validation, N.W., T.F.; formal analysis, N.W., T.F., A.H.; investigation, N.W., T.F., A.H.; resources, N.W., T.F., A.H.; data curation, N.W., T.F., A.H.; writing—original draft preparation, N.W.; writing—review and editing, T.F., A.H., S.G.M..; visualization, N.W., T.F.; supervision, S.G.M.; project administration, S.G.M.; funding acquisition, S.G.M. All authors have read and agreed to the published version of the manuscript.

Funding: The work was financially supported by the Deutsche Forschungsgemeinschaft (DFG–Project MA 2432/6-3) as well as the Saxonian Ministry for Higher Education, Research and the Arts (SMWK) (100331694 (MUDIPlex)) is gratefully acknowledged. The LSM employed in these studies was funded by INST 268/357-1 FUGG (project number 323490432).

Institutional Review Board Statement: Not applicable.

Data Availability Statement: Not applicable.

Acknowledgments: We gratefully acknowledge Jan Griebel (IOM) and Nadja Schönherr (IOM) for the zeta potential measurements and discussion as well as Christian Elsner (IOM) for the plate reader and antibody experiments. This project was partially performed within the Leipzig Graduate School of Natural Sciences–Building with Molecules and Nano-objects (BuildMoNa).

Conflicts of Interest: The authors declare no conflict of interest.

References

1. Gandhi, A.; Paul, A.; Sen, S.O.; Sen, K.K. Studies on thermoresponsive polymers: Phase behaviour, drug delivery and biomedical applications. *Asian J. Pharm. Sci.* **2015**, *10*, 99–107. [CrossRef]
2. Klouda, L. Thermoresponsive hydrogels in biomedical applications. *Eur. J. Pharm. Biopharm.* **2015**, *97*, 338–349. [CrossRef]
3. Varghese, J.M.; Ismail, Y.A.; Lee, C.K.; Shin, K.M.; Shin, M.K.; Kim, S.I.; So, I.; Kim, S.J. Thermoresponsive hydrogels based on poly(N-isopropylacrylamide)/chondroitin sulfate. *Sens. Actuators B Chem.* **2008**, *135*, 336–341. [CrossRef]
4. Mills, C.E.; Ding, E.; Olsen, B. Protein Purification by Ethanol-Induced Phase Transitions of the Elastin-like Polypeptide (ELP). *Ind. Eng. Chem. Res.* **2019**, *58*, 11698–11709. [CrossRef]
5. Wang, E.; Desai, M.S.; Lee, S.-W. Light-Controlled Graphene-Elastin Composite Hydrogel Actuators. *Nano Lett.* **2013**, *13*, 2826–2830. [CrossRef]
6. MacEwan, S.R.; Chilkoti, A. Elastin-like polypeptides: Biomedical applications of tunable biopolymers. *Biopolymers* **2010**, *94*, 60–77. [CrossRef]
7. Bessa, P.C.; Machado, R.; Nürnberger, S.; Dopler, D.; Banerjee, A.; Cunha, A.M.; Rodríguez-Cabello, J.C.; Redl, H.; van Griensven, M.; Reis, R.L.; et al. Thermoresponsive self-assembled elastin-based nanoparticles for delivery of BMPs. *J. Control. Release* **2010**, *142*, 312–318. [CrossRef]
8. Costa, R.R.; Custódio, C.A.; Arias, F.J.; Rodríguez-Cabello, J.C.; Mano, J.F. Layer-by-Layer Assembly of Chitosan and Recombinant Biopolymers into Biomimetic Coatings with Multiple Stimuli-Responsive Properties. *Small* **2011**, *7*, 2640–2649. [CrossRef]
9. Kantardjiev, A.; Ivanov, P.M. Entropy Rules: Molecular Dynamics Simulations of Model Oligomers for Thermoresponsive Polymers. *Entropy* **2020**, *22*, 1187. [CrossRef]
10. Hou, L.; Wu, P. LCST transition of PNIPAM-b-PVCL in water: Cooperative aggregation of two distinct thermally responsive segments. *Soft Matter* **2014**, *10*, 3578. [CrossRef]
11. Cooperstein, M.A.; Canavan, H.E. Assessment of cytotoxicity of (N-isopropyl acrylamide) and Poly(N-isopropyl acrylamide)-coated surfaces. *Biointerphases* **2013**, *8*, 19. [CrossRef] [PubMed]
12. Mithieux, S.M.; Weiss, A.S. Elastin. In *Advances in Protein Chemistry*; Academic Press: Cambridge, MA, USA, 2005; pp. 437–461. [CrossRef]
13. Wilharm, N.; Fischer, T.; Ott, F.; Konieczny, R.; Zink, M.; Beck-Sickinger, A.G.; Mayr, S.G. Energetic electron assisted synthesis of highly tunable temperature-responsive collagen/elastin gels for cyclic actuation: Macroscopic switching and molecular origins. *Sci. Rep.* **2019**, *9*, 12363. [CrossRef] [PubMed]
14. Usha, R.; Ramasami, T. Structure and conformation of intramolecularly cross-linked collagen. *Colloids Surf. B Biointerfaces* **2005**, *41*, 21–24. [CrossRef]
15. Daamen, W.; Veerkamp, J.; Vanhest, J.; Vankuppevelt, T. Elastin as a biomaterial for tissue engineering. *Biomaterials* **2007**, *28*, 4378–4398. [CrossRef] [PubMed]
16. Gaar, J.; Naffa, R.; Brimble, M. Enzymatic and non-enzymatic crosslinks found in collagen and elastin and their chemical synthesis. *Org. Chem. Front.* **2020**, *7*, 2789–2814. [CrossRef]
17. Fischer, T.; Hayn, A.; Mierke, C.T. Effect of Nuclear Stiffness on Cell Mechanics and Migration of Human Breast Cancer Cells. *Front. Cell Dev. Biol.* **2020**, *8*, 393. [CrossRef]
18. Fischer, T.; Hayn, A.; Mierke, C.T. Fast and reliable advanced two-step pore-size analysis of biomimetic 3D extracellular matrix scaffolds. *Sci. Rep.* **2019**, *9*, 8352. [CrossRef]

19. Hayn, A.; Fischer, T.; Mierke, C.T. Inhomogeneities in 3D Collagen Matrices Impact Matrix Mechanics and Cancer Cell Migration. *Front. Cell Dev. Biol.* **2020**, *8*, 593879. [CrossRef]
20. Na, G.C.; Phillips, L.J.; Freire, E.I. In vitro collagen fibril assembly: Thermodynamic studies. *Biochemistry* **1989**, *28*, 7153–7161. [CrossRef]
21. Streeter, I.; de Leeuw, N.H. A molecular dynamics study of the interprotein interactions in collagen fibrils. *Soft Matter* **2011**, *7*, 3373–3382. [CrossRef]
22. Znidarsic, W.J.; Chen, I.-W.; Shastri, V.P. ζ-potential characterization of collagen and bovine serum albumin modified silica nanoparticles: A comparative study. *J. Mater. Sci.* **2009**, *44*, 1374–1380. [CrossRef]
23. Panduranga Rao, K. Recent developments of collagen-based materials for medical applications and drug delivery systems. *J. Biomater. Sci. Polym. Ed.* **1996**, *7*, 623–645. [CrossRef] [PubMed]
24. Partridge, S.M.; Davis, H.F.; Adair, G.S. The chemistry of connective tissues. 2. Soluble proteins derived from partial hydrolysis of elastin. *Biochem. J.* **1955**, *61*, 11–21. [CrossRef] [PubMed]
25. Vrhovski, B.; Jensen, S.; Weiss, A.S. Coacervation Characteristics of Recombinant Human Tropoelastin. *Eur. J. Biochem.* **1997**, *250*, 92–98. [CrossRef] [PubMed]
26. Vazquez-Portalatin, N.; Alfonso-Garcia, A.; Liu, J.C.; Marcu, L.; Panitch, A. Physical, Biomechanical, and Optical Characterization of Collagen and Elastin Blend Hydrogels. *Ann. Biomed. Eng.* **2020**, *48*, 2924–2935. [CrossRef]
27. Paderi, J.E.; Sistiabudi, R.; Ivanisevic, A.; Panitch, A. Collagen-Binding Peptidoglycans: A Biomimetic Approach to Modulate Collagen Fibrillogenesis for Tissue Engineering Applications. *Tissue Eng. Part A* **2009**, *15*, 2991–2999. [CrossRef]
28. Cisneros, D.A.; Hung, C.; Franz, C.M.; Muller, D.J. Observing growth steps of collagen self-assembly by time-lapse high-resolution atomic force microscopy. *J. Struct. Biol.* **2006**, *154*, 232–245. [CrossRef]
29. Kumar, A.; Dixit, C.K. 3-Methods for characterization of nanoparticles. In *Advances in Nanomedicine for the Delivery of Therapeutic Nucleic Acids*; Nimesh, S., Chandra, R., Gupta, N., Eds.; Woodhead Publishing: Buckingham, UK, 2017; pp. 43–58. [CrossRef]
30. Kadler, K.E.; Hojima, Y.; Prockop, D.J. Assembly of collagen fibrils de novo by cleavage of the type I pC-collagen with procollagen C-proteinase. Assay of critical concentration demonstrates that collagen self-assembly is a classical example of an entropy-driven process. *J. Biol. Chem.* **1987**, *262*, 15696–15701. [CrossRef]
31. Ravikumar, K.M.; Hwang, W. Role of Hydration Force in the Self-Assembly of Collagens and Amyloid Steric Zipper Filaments. *J. Am. Chem. Soc.* **2011**, *133*, 11766–11773. [CrossRef]
32. Ahmed, A.; Joshi, I.M.; Mansouri, M.; Ahamed, N.N.N.; Hsu, M.-C.; Gaborski, T.R.; Abhyankar, V.V. Engineering fiber anisotropy within natural collagen hydrogels. *Am. J. Physiol. Cell Physiol.* **2021**, *320*, C1112–C1124. [CrossRef]
33. Mostaço-Guidolin, L.B.; Smith, M.S.D.; Hewko, M.; Schattka, B.; Sowa, M.G.; Major, A.; Ko, A.C.-T. Fractal dimension and directional analysis of elastic and collagen fiber arrangement in unsectioned arterial tissues affected by atherosclerosis and aging. *J. Appl. Physiol.* **2019**, *126*, 638–646. [CrossRef]
34. Gibson, M.A. Microfibril-Associated Glycoprotein-1 (MAGP-1) and Other Non-fibrillin Macromolecules Which May Possess a Functional Association with the 10 nm Microfibrils. *Landes Biosci.* **2013**. Available online: https://www.ncbi.nlm.nih.gov/books/NBK6448/ (accessed on 17 June 2021).
35. Trask, T.M.; Trask, B.C.; Ritty, T.M.; Abrams, W.R.; Rosenbloom, J.; Mecham, R.P. Interaction of tropoelastin with the amino-terminal domains of fibrillin-1 and fibrillin-2 suggests a role for the fibrillins in elastic fiber assembly. *J. Biol. Chem.* **2000**, *275*, 24400–24406. [CrossRef] [PubMed]
36. Yanagisawa, H.; Davis, E.C.; Starcher, B.C.; Ouchi, T.; Yanagisawa, M.; Richardson, J.A.; Olson, E.N. Fibulin-5 is an elastin-binding protein essential for elastic fibre development in vivo. *Nature* **2002**, *415*, 168–171. [CrossRef] [PubMed]
37. Chang, S.-W.; Buehler, M.J. Molecular biomechanics of collagen molecules. *Mater. Today* **2014**, *17*, 70–76. [CrossRef]
38. Tarakanova, A.; Chang, S.-W.; Buehler, M.J. Computational Materials Science of Bionanomaterials: Structure, Mechanical Properties and Applications of Elastin and Collagen Proteins. In *Handbook of Nanomaterials Properties*; Bhushan, B., Luo, D., Schricker, S.R., Sigmund, W., Zauscher, S., Eds.; Springer: Berlin/Heidelberg, Germany, 2014; pp. 941–962. [CrossRef]
39. Fluegel, S.; Fischer, K.; McDaniel, J.R.; Chilkoti, A.; Schmidt, M. Chain Stiffness of Elastin-Like Polypeptides. *Biomacromolecules* **2010**, *11*, 3216–3218. [CrossRef]

polymers

MDPI

Article

5-Fluorouracil-Loaded Folic-Acid-Fabricated Chitosan Nanoparticles for Site-Targeted Drug Delivery Cargo

Shafi Ullah [1,†], Abul Kalam Azad [2,*,†], Asif Nawaz [1], Kifayat Ullah Shah [1], Muhammad Iqbal [1], Ghadeer M. Albadrani [3], Fakhria A. Al-Joufi [4], Amany A. Sayed [5] and Mohamed M. Abdel-Daim [6,7,*]

1 Advanced Drug Delivery Lab, Gomal Center of Pharmaceutical Sciences, Faculty of Pharmacy, Gomal University, Dera Ismail Khan 29050, Pakistan; shafikustian@gmail.com (S.U.); asifnawaz676@gmail.com (A.N.); kifayatrph@gmail.com (K.U.S.); iqbalmiani@gmail.com (M.I.)
2 Pharmaceutical Technology Unit, Faculty of Pharmacy, AIMST University, Bedong 08100, Malaysia
3 Department of Biology, College of Science, Princess Nourah bint Abdulrahman University, Riad 11671, Saudi Arabia; gmalbadrani@pnu.edu.sa
4 Department of Pharmacology, College of Pharmacy, Jouf University, Sakaka 72341, Saudi Arabia; faaljoufi@ju.edu.sa
5 Zoology Department, Faculty of Science, Cairo University, Giza 12613, Egypt; amanyasayed@sci.cu.edu.eg
6 Department of Pharmaceutical Sciences, Pharmacy Program, Batterjee Medical College, Jeddah 21442, Saudi Arabia
7 Pharmacology Department, Faculty of Veterinary Medicine, Suez Canal University, Ismailia 41522, Egypt
* Correspondence: azad@aimst.edu.my or aphdukm@gmail.com (A.K.A.); abdeldaim.m@vet.suez.edu.eg (M.M.A.-D.)
† These authors contributed equally to this work.

Citation: Ullah, S.; Azad, A.K.; Nawaz, A.; Shah, K.U.; Iqbal, M.; Albadrani, G.M.; Al-Joufi, F.A.; Sayed, A.A.; Abdel-Daim, M.M. 5-Fluorouracil-Loaded Folic-Acid-Fabricated Chitosan Nanoparticles for Site-Targeted Drug Delivery Cargo. *Polymers* **2022**, *14*, 2010. https://doi.org/10.3390/polym14102010

Academic Editors: Ariana Hudita and Bianca Gălăţeanu

Received: 18 March 2022
Accepted: 9 May 2022
Published: 13 May 2022

Abstract: Nanoparticles play a vital role in cancer treatment to deliver or direct the drug to the malignant cell, avoiding the attacking of normal cells. The aim of the study is to formulate folic-acid-modified chitosan nanoparticles for colon cancer. Chitosan was successfully conjugated with folic acid to produce a folic acid–chitosan conjugate. The folate-modified chitosan was loaded with 5-FU using the ionic gelation method. The prepared nanoparticles were characterized for size, zeta potential, surface morphology, drug contents, entrapment efficiency, loading efficiency, and in vitro release study. The cytotoxicity study of the formulated nanoparticles was also investigated. The conjugation of folic acid with chitosan was confirmed by FTIR and NMR spectroscopy. The obtained nanoparticles were monodispersed nanoparticles with a suitable average size and a positive surface charge. The size and zeta potential and PDI of the CS-5FU-NPs were 208 ± 15, 26 ± 2, and +20 ± 2, respectively, and those of the FA-CS-5FU-NPs were 235 ± 12 and +20 ± 2, respectively, which are in the acceptable ranges. The drug contents' % yield and the %EE of folate-decorated NPs were 53 ± 1.8% and 59 ± 2%, respectively. The in vitro release of the FA-CS-5FU-NPs and CS-5FU-NPs was in the range of 10.08 ± 0.45 to 96.57 ± 0.09% and 6 ± 0.31 to 91.44 ± 0.21, respectively. The cytotoxicity of the nanoparticles was enhanced in the presence of folic acid. The presence of folic acid in nanoparticles shows much higher cytotoxicity as compared to simple chitosan nanoparticles. The folate-modified nanoparticles provide a potential way to enhance the targeting of tumor cells.

Keywords: colon cancer; targeted delivery; folate-conjugated nanoparticles; cytotoxicity

1. Introduction

Colon cancer (CC) is one of the leading causes of mortality and morbidity in the world. Nine percent of all cancer cases are colon cancer. Throughout the world, CC is the third most common cancer type and the fourth most common cause of death [1]. CC is the most common type of cancer among all cancer types in western countries. Cases of colorectal cancer have increased remarkably in the past fifty years, and such cases have become the second highest in females and the third highest in males. The primary means of treatment for CC is surgery, with chemotherapy and/or radiotherapy being indicated depending on the nature and severity of the disease. Chemotherapy using different anticancer agents can be used as an adjuvant treatment on the second number after surgery, as a neo-adjuvant treatment before surgery, or as a main treatment to abate tumor size and growth as well as metastasis risk [2]. The 5-fluorouracil (5-FU) drug is an analog of the pyrimidines, and therefore uses the same metabolic routes as uracil and thymine. It is classified as an antimetabolite drug, interfering with nucleoside metabolism in RNA and DNA, and is used for the treatment of various tumors, such as those found in breast adenocarcinoma, the gastrointestinal tract, the ovary, the head, and the neck [3]. Despite its effectiveness, 5-FU presents some drawbacks. After oral administration, 5-FU bioavailability is highly variable due to its inconsistent absorption in the gastrointestinal tract and first-pass metabolism through the liver. Thus, the 5-FU half-life is extremely short (6–20 min), and frequent and high doses are required to maintain adequate plasma concentrations [4]. An alternative to overcoming these drawbacks and improving drug bioavailability, promoting controlled drug release, and choosing for more cell selectivity is the application of polymeric nanoparticles as 5-FU carriers. The oral delivery of drugs is of tremendous interest for patients seeking safe and controlled drug delivery. Compared to injections, the oral administration of anticancer drugs via oral route is cost-effective, reducing the hospitalization duration of the patient, as well as improving the patient's quality of life. Examples of some drugs that are used for cancer treatment are as follows: 5-flourouracil (5-FU), hexacarbonyl-(5-FU), and N^4 pentoxylcarbonyl-5-deoxy-5flourocytidine (capecitabine) [5].

Chitosan is a semi-crystalline, linear polysaccharide that is composed of (1-4)-2-acetamido-2-deoxy-β-D-glucan (*N*-acetyl D-glucosamine) and (1-4)-2-amino-2-deoxy-β-D-glucan (D-glucosamine) units. Chitosan is not extensively present in the environment in its original form but can be derived easily from chitin (a natural polymer) by removing its acetyl group. The ratio of D-glucosamine to the sum of D-glucosamine and N-acetyl D-glucosamine gives the degree of deacetylation (DD) of chitosan. DD indicates the number of amino (NH_2) groups along the chains [6]. Chitosan provides a valuable tool for the current system of novel drug delivery owing to its intrinsic biological and physicochemical properties. The characteristics of chitosan nanoparticles (NPs), such as their small size, better stability, inexpensiveness, easy manufacturing process, lower toxicity, and versatile method of administration, made them favorable drug and gene delivery carriers. Chitosan can be easily chemically modified due to the presence of its active functional groups such as amine (NH_2) and hydroxyl (OH) groups. Due to pH changes and electrostatic interactions throughout the gastrointestinal tract (GIT) that are vital for maintaining the NP's stability, the permanent positive charge of chitosan favors their mucoadhesion property in the intestinal mucosa layer. This characteristic has been used to develop enhanced drug delivery systems that could help in CC treatment [7].

The surface morphology of nanoparticles (NPs) can be modified with the conjugation of targeting ligands, such as folic acid (FA), antibodies, integrins, transferrin, and polysaccharides, to improve receptor affinity and internalization by target tissues. Many tumor cell surfaces overexpress folate receptors (FRs), which are less often expressed in normal and healthy cells. This feature makes tumor cells an excellent target for tumor-targeting drug delivery [8]. Folic acid (FA) has emerged as an optimal targeting ligand for the selective delivery of attached imaging and therapeutic agents to cancer cells and inflammation sites. The use of FA as a target ligand has arisen primarily from its following features: (1) its easy conjugation to both therapeutic and diagnostic agents; (2) its great affinity for the folate receptors (FRs); and (3) the distribution of folate receptors (FRs) in limited numbers in normal tissues. Folic acid as a targeting ligand has been investigated by many scientists [9]. In one study, it was demonstrated that the folic-acid-modified chitosan NPs were excellent vectors for the colon-specific delivery of 5-aminolevulinic acid (5-ALA) for fluorescent endoscopic detection [10]. The FA decoration upheld the establishment of a genuine affinity for FRs+ cancer cells even when co-cultured closely with higher numbers of healthy cells [11]. In one study, FA was evaluated in vitro, in which it was conjugated with carboxymethyl chitosan, and its nanoparticles were loaded with doxorubicin for targeted drug delivery. These FA-modified NPs manifested FA feasibility as an excellent targeted delivery carrier [12]. In the present study, folic-acid–chitosan-conjugated nanoparticles for oral delivery were prepared and evaluated for in vitro release and cytotoxicity studies.

2. Materials and Methods

2.1. Materials

Chitosan (deacetylation degree—83% and mol wt—310,000–375,000), 5-flourouracil and folic acid were obtained from Sigma-Aldrich (lot# A263299) (Sigma-Aldrich, Inc. St. Louis, MO, USA). TPP (85%), potassium dihydrogen phosphate, calcium chloride, 1-ethyl-3-(3 dimethylaminopropyl) carbodiimide (EDC), and sodium hydroxide were obtained from Sigma Chemicals (Merck Pte. Ltd. 2 Science Park Drive, Singapore). Acetic acid, hydrochloric acid, ethanol, and DMSO were obtained from Merck (Merck KGaA, Darmstadt, Germany).

2.2. Conjugation of Folic Acid (FA) with Chitosan (CS)

The conjugation process of folic acid (FA) to chitosan (CS) is described as follows. FA and 1-ethyl-3-(3-dimethylaminopropyl) carbodiimide (EDC) solution in anhydrous dimethylsulfoxide (DMSO) (20 mL), with 1:1 molar ratio, was made and stirred at room temperature until the EDC and FA were mixed well. The solution was then slowly added to 0.5% (w/v) CS in an aqueous solution of 0.1 M of acetic acid with a pH of 4.7, and then stirred at 25 °C in the dark area for 16 h to let the FA conjugate onto the CS molecules. Then, 1 M of NaOH was added to adjust the pH of the solution to 9.0. The solution was centrifuged at 2500 rpm to settle down the FA–CS conjugate. The sediment was first dialyzed against a phosphate buffer with a pH of 7.4 for 3 days, and then against water for 4 days. Finally, the FA–CS conjugate was collected as a sponge by freeze-drying and kept for further study [13].

2.2.1. Fourier Transform Infrared Spectroscopy

The Fourier transform infrared spectroscopy (FTIR) was performed using an ATR FTIR spectrometer (L1600300, PerkinElmer, Beaconsfield, UK). The FTIR spectra of chitosan, folic acid, and its conjugate (FA–CS) were obtained. The recording range of the spectrum was 600–4000 cm^{-1} at 32 scans per minute with a resolution of 4 cm^{-1} in absorbance mode. After recording, the spectra were baseline, corrected, and normalized using Spectra software to identify the characteristic peaks and differences [14].

2.2.2. H-NMR

For NMR spectroscopic analyses, solutions of CS, FA, and their conjugates were prepared in 1.97 mL of $CDCl_3$ and kept at room temperature until their complete dissolution. Acetic acid was used as a co-solvent for the solubility of chitosan in $CDCl_3$. ^{1}H-NMR spectra were obtained using a Bruker AV-500 MHz NMR spectrometer. Bruker–Topspin software (version 4.1.1) was used for the analysis of NMR spectra.

2.2.3. Determination of Folate (FA) Content

The FA–CS conjugates were accurately weighed and then dissolved in 50 mL of 0.2 molar sodium bicarbonate buffer solution with a pH of 10 at 25 °C with magnetic stirring. The solution was centrifuged at 3500 rpm for 10 min (Laboratory Centrifuge, YJ03-0434000, Shanghai, China). The supernatant was tested for determining folate (FA) using a UV–visible spectrophotometry technique with a wavelength of 365 nm. The folate content was calculated as the percentage of FA in a unit weight of conjugate. For each experiment, at least three duplicates were carried out and the results were averaged.

2.3. Preparation of Nanoparticles (NPs)

The FA–CS NPs were synthesized by ionic cross-linking with tripolyphosphate (TPP) using the method described by Salar and Kumar, 2016, with slight modifications. The FA–CS conjugate solution (0.2%, w/v, pH 2.5) was prepared using 1% v/v acetic acid at room temperature. The TPP (0.2%, w/v) solution in distilled water was prepared. For the synthesis of the 5-FU-loaded nanoparticles, an aqueous solution of 5-FU (500 mg/10 mL) was prepared separately. A solution of 5-FU was added drop-wise into the FA–chitosan conjugate solution. The TPP solution was added into the conjugate solution drop-wise in a 1:3 ratio. The solution was allowed to stir for 1 h on a magnetic stirrer at room temperature. The nanoparticle's suspension was centrifuged at 5000 rpm for 10 min for separating the nanoparticles from the solution, and then freeze-dried for 24 h to obtain the final product of the NPs. NPs without an FA conjugation were also prepared in the same manner.

2.4. Characterization

2.4.1. NP Size and Zeta Potential

Photon correlation spectroscopy was used to determine the particle size and zeta potential of FA-CS-5FU-NPs and CS-5FU-NPs at 25 °C in quartz cell and zeta potential cell with a detect angle of 90°, respectively, using a Malvern Zetasizer Nano ZS 90 (Malvern Instruments Ltd., Malvern, UK). In 5.0 mL of deionized water, one mg of NPs was added, and vortex stirring (Velp Scientifica, Usmate Velate, Italy) was used to fortify the mixture [15].

2.4.2. Nanoparticle Morphology

The surface morphology of NPs was examined using the scanning electron microscopy (SEM) technique (JSM6360LA, JEOL, Tokyo, Japan). The NPs were fixed with carbon tape onto studs and directly examined under the SEM. Images of the NPs were captured at a 20,000× magnification level [15].

2.4.3. Percentage Yield, Drug Entrapment Efficiency (%EE), and Drug-Loading Efficiency

Precisely weighed 15 mg of FA-CS-5FU-NPs and CS-5FU-NPs was dispersed in 15 mL of distilled water under magnetic stirring at 200 rpm for 2 h in two separate beakers followed by centrifugation (Laboratory Centrifuge, YJ03-0434000, Shanghai, China) at 5000 rpm for 45 min. The supernatant of both formulations was isolated and analyzed for free 5-FU using the UV spectroscopy technique. The percentage (%) yield was calculated using the formula given in Equation (1) [16]:

$$\%\ yield = \text{Mass of NPs obtained}/\ \text{total weight of drug and polymer} \times 100 \quad (1)$$

Both drug entrapment efficiency (%EE) and drug-loading efficiency were determined indirectly using free drug concentration. After centrifugation, the obtained sediments of the formulations were dissolved in ethanol, aliquot filtered, and analyzed at 265 nm in UV spectroscopy for drug entrapment efficiency using Equation (2).

$$EE\ (\%) = \frac{\text{5FUtotal} - \text{5FU Free}}{\text{5FUtotal}} \times 100 \quad (2)$$

The total drug load collected from the supernatant and sediment was used to calculate the drug-loading efficiency using Equation (3). Triplicates were conducted and the results were averaged [16].

$$Drug\ loading\ efficiency = \frac{\text{5FUtotal} - \text{5FUfree}}{\text{Weight of 5FU loaded nanoparticles taken}} \times 100 \quad (3)$$

2.5. In Vitro Release of Nanoparticles

The release rate for the designed formulations was studied for up to 2 h in 900 mL of release media such as simulated gastric fluid (solution of 0.2 MHCL and 0.2 MKCl, pH 1.2) and simulated intestinal fluid (solution of 0.2 M potassium dihydrogen phosphate and 0.1 M sodium hydroxide, pH 6.5) for up to 24 h using a dissolution tester (basket method type 1) at 37.5 ± 0.5 °C. The stirring speed was set at 100 rpm. Then, 15 mg of FA-CS-5FU-NPs and CS-5FU-NPs was placed in two separate baskets and run the apparatus. At predetermined time intervals (0.5, 1, 1.5, 2, 4, 8, 12, 16, 20, and 24 h), a 5 mL sample was withdrawn and replaced with a fresh dissolution medium. All the samples were analyzed using a UV–visible spectrophotometer at a wavelength of 265 nm. The cumulative percentage of the drug released was calculated [17].

2.6. Cytotoxicity Studies

For the cytotoxicity study, human colon carcinoma cell lines (Caco2) were used. Cells were cultured in Eagle's minimum essential medium supplemented with 2 mM of glutamine, 20% fetal bovine serum (FBS), 1.5 g/L of sodium bicarbonate, 1 mM of sodium pyruvate, and 0.1 mM of nonessential amino acid. Cells were equilibrated with 5% CO_2. Growing temperature was set at 37 °C in an incubator allowing humidified air to pass through. Cytotoxicity studies of the 5-FU solution, CS-5-FU-NPs, and FA-CS-5-FU-NPs were performed on the Caco-2 using the MTT (3-(4,5-dimethylthiazol-2-yl)-2,5-diphenyltetrazolium bromide) assay. First, 5×10^3 cells were seeded in 96-well plates and incubated for 24 h without drug/formulations. The cells were then treated with the 5-FU solution, CS-5FU-NPS, and FA-CS-5FU-NPS, and were then incubated for 24 h. The cells in the absence the 5-FU solution, CS-5FU-NPS, and FA-CS-5FU-NPSNPs were considered as the control group. The MTT solution was added to assess the cytotoxicity of drug and nanoparticles. Cells were incubated for 4 h in MTT solution followed by the addition of DMSO to dissolve formazan and quantified spectrophotometrically using a microplate reader (Thermo Varioscan Multiplate Reader).

2.7. Data Analysis and Statistics

The obtained data were statistically analyzed using ANOVA (one-way analysis of variation) and the student's *t*-test (IBM® SPSS® Statistics version 19, Armonk, NY, USA), and the Statistical Package Minitab® version 20 (Minitab, LLC, State College, PA, USA). Data with values of $p < 0.05$ were considered statistically significant. All the tested data were described as triplicate ($n = 3$) and mean ± standard deviation (S.D.).

3. Results and Discussion

3.1. Synthesis of FA–CS Conjugate

The synthesis of the FA–CS conjugates was carried out by means of carbodiimide chemistry using the water-soluble 1-ethyl-3-(3-dimethylaminopropyl) carbodiimide (EDC) (Figure 1). The EDC is a "zero-length" crosslinked chemical. It is used in the formation of conjugate via amide linkage without leaving a spacer molecule [18]. The EDC reacted with the COO^- of the FA and 5-FU to form an intermediate of active ester. The intermediate reacted further with the primary amine (NH_2) group of the CS, giving rise to an amide (N–H) bond, with an isourea by-product that was removed easily by filtration or dialysis [18]. The FTIR and ^{1}H-NMR spectra (Figures 2 and 3) successfully confirmed the conjugation of folic acid onto chitosan molecules.

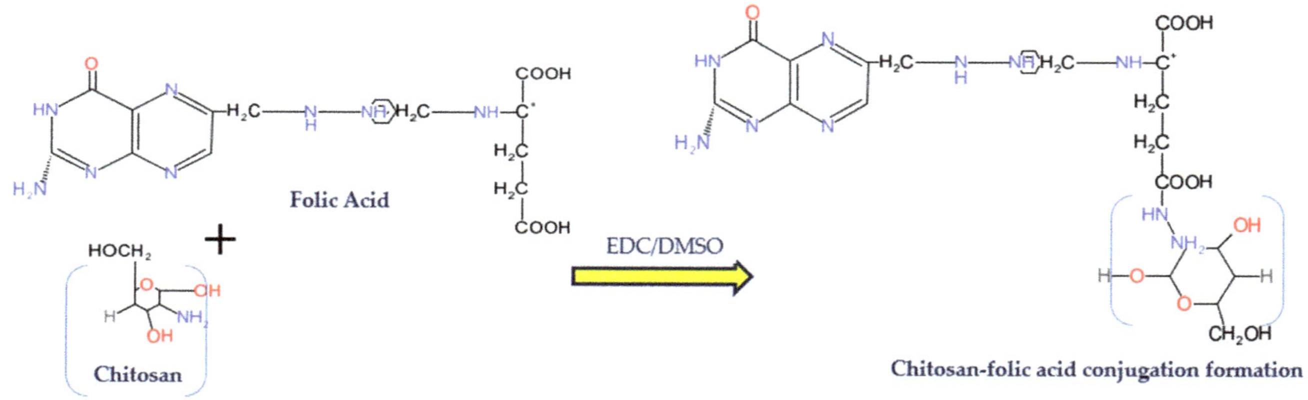

Figure 1. A scheme illustrating the reaction of chitosan with folic acid.

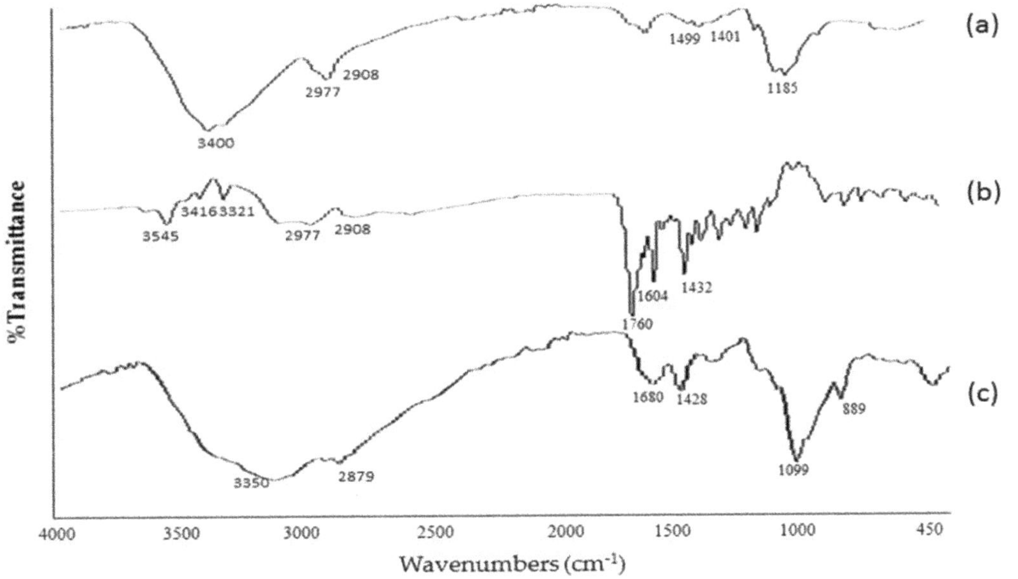

Figure 2. FTIR spectra of (**a**) pure chitosan, (**b**) pure FA, and (**c**) FA–CS conjugate.

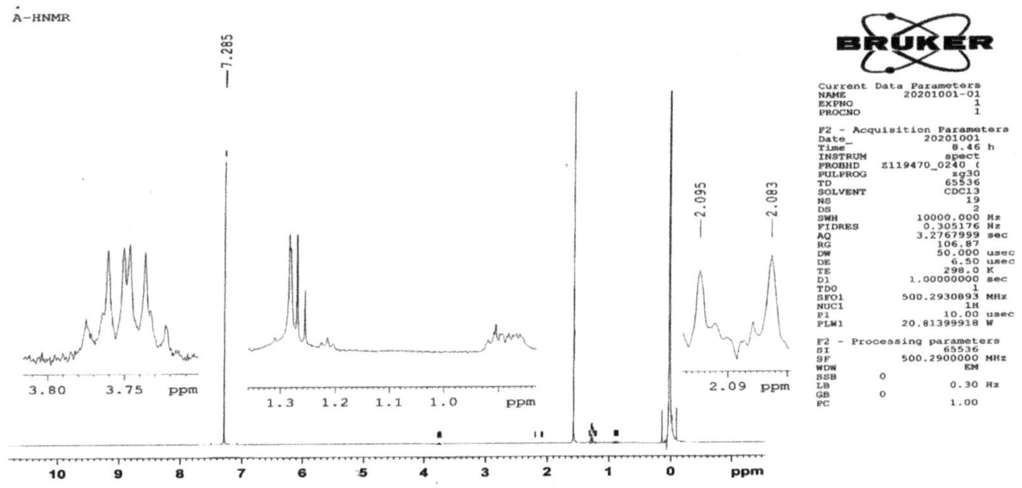

Figure 3. *Cont.*

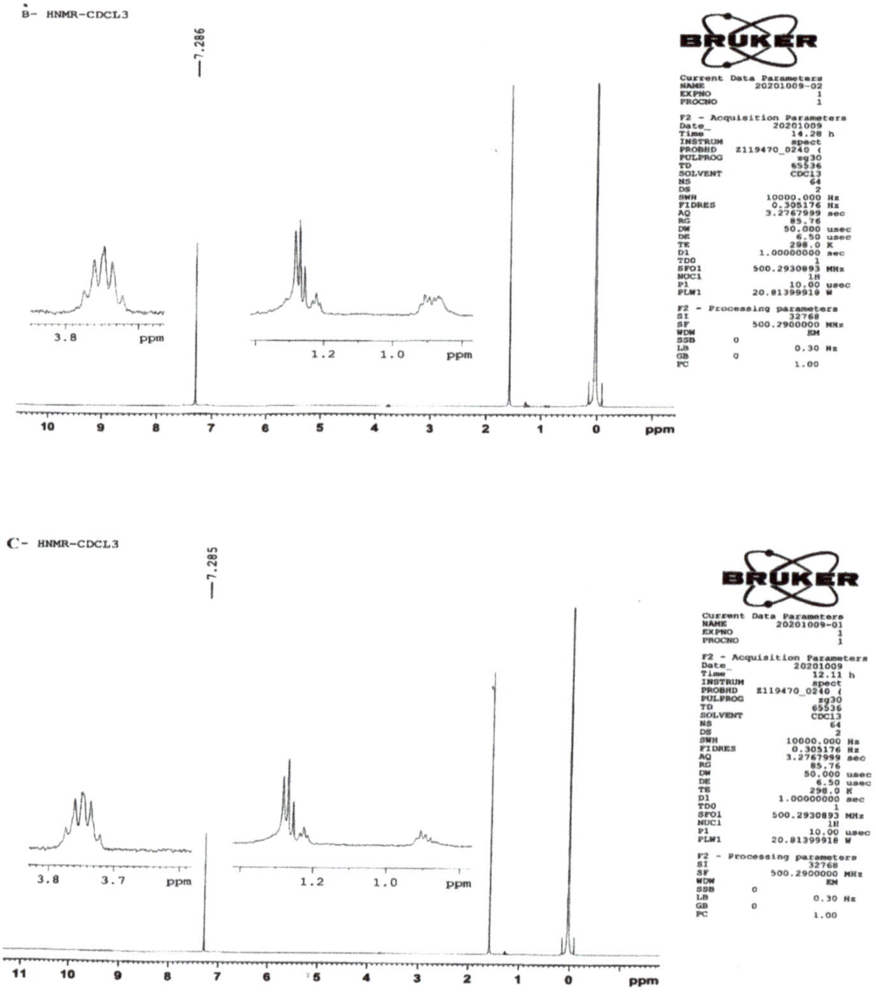

Figure 3. ^{1}H-NMR spectra of (**A**) pure chitosan, (**B**) pure FA, and (**C**) FA–CS conjugate.

3.1.1. FTIR Studies

In an FTIR of chitosan (Figure 2), a strong band at the region of 3400 cm^{-1} represents NH functional groups (primary amine). The absorption band at around 2977 cm^{-1} can be attributed to CH symmetric stretching. Bands at 1401 cm^{-1} indicate a methyl group (CH_3). Symmetrical bending in the range of 1260–800 cm^{-1} belong to the glycosidic ring; in particular, the band at 1156 cm^{-1} corresponds to the glycosidic linkage. Similarly, an FTIR of pure folic acid showed the IR spectrum at 3100–3500 cm^{-1} which can be attributed to the OH carboxylic of glutamic acid moiety and the NH group of the pterin ring stretching. Absorption at 1760 cm^{-1} represents C=O carboxylic acid in pure FA. Similarly, the absorption band at 1432 cm^{-1} represents the phenyl and the pterin ring. The band at 3321 cm^{-1} pure folic acid is absent/overlapped in the conjugate formulation, indicating the coupling of folate with a chitosan polymer [19]. The primary amine of the chitosan reacted with the carboxylic acid group of folic acid, forming an amide bond. The amide bond formation between the chitosan and folic acid was evidenced by a shift of the FTIR wavenumber of folic acid from 1760 to 1680 cm^{-1}. The assignment of FTIR peaks was correlated with earlier studies [20–22]. The peaks at 2.07 ppm attributed to the acetamino group CH3, and the CH peak appeared at 3.50–3.95 ppm, corresponding to carbons 3, 4, 5, and 6 of the glucosamine rings of CS.

3.1.2. H-NMR Study

FA has two active –COOH groups at its end point. Among these, γ-COOH is more sensitive to the reaction, owing to its high reactivity [23]. The final product of FA-CS was synthesized by the reaction between the activated FA ester and the primary amine NH_2 groups of CS through the formation of an amide bond under homogeneous conditions. The peaks at 2.08 ppm attributed to the hydrogen atom of the methyl group (CH_3) of the acetamino groups of chitosan, as well as CH peaks at 3.77–3.8 ppm, can be explained by hydrogen bonded to carbons 3, 4, 5, and 6 of the glucosamine rings of CS [24]. The CS conjugation was confirmed by the peculiar signals at 2.5 ppm, which attributed to the aromatic protons of the FA, and characteristic peaks at 2.84 ppm corresponded to the FA proton from the H22 [25]. This was ascribed to the development of amide linkage after the folic acid–chitosan conjugation. Ji et al. previously reported similar results at 2.45 ppm in relation to the FA proton from the H10 and H22 [24], respectively, which is in line with the current study.

3.1.3. Folate (FA) Content

Folate content was found to be 5% of the total weight of the FA-CS-5FU-NPs formulation. Folic acid is commonly engaged as a ligand for targeting cancer cells, as its receptors are over-expressed on the surface of several human cancer cells. Integrating folic acid into chitosan-based drug delivery inventions directs the systems with a well-organized targeting ability [26].

3.2. Characterization of Nanoparticles

3.2.1. Size, Zeta Potential, and Surface Morphology

The size of the CS-5-FU NPs was found to be 208 ± 14.65 nm, while the FA-CS-5-FU NPs was 235 ± 11.5 nm, as shown in Table 1 and Figure 4A,B. The poly dispersity index (PDI) was found to be 0.2 and 0.19 for of the FA-CS-5-FU NPs and the CS-5-FU NPs. This small size of NPs is important since such NPs in anticancer drugs can easily escape the leaky tumor vasculature and accumulate within the tumor region to exert cytotoxic effects on proliferating cells [27]. The zeta potential of the FA-CS-5-FU NPs was found to be +20 ± 2. The FA-CS-5-FU NPs show an insignificant decrease in zeta potential as compared to the CS-5-FU NPs (Table 1; $p > 0.05$). Other studies have also found a decrease in the value of zeta potential after folate conjugation. [28–30]. The obtained zeta potential value of +26 ± 2 mV in the instance of folic-acid-modified chitosan NPs indicates that folic acid binds to chitosan quite strongly. The free positive NH_2 groups of chitosan molecules may account for the positive value. This positive zeta potential is helpful in crossing the negatively charged membrane of cancer cells. The value of the zeta potential (ZP) indicates the repulsive interactions between suspended particles and can therefore be used to forecast the stability of colloidal aqueous dispersions. The prepared nanoparticles were spherical in shape and smooth in surface, as shown in Figure 4B.

Table 1. Physicochemical characterization of folic-acid-modified 5-FU-loaded chitosan NPs. Data were presented as triplicate (n = 3) and mean ± SD.

Formulation Code	Size (nm)	Zeta Potential (mV)	PDI	Drug Content (%)	Percent Yield	%EE	%LE
CS-5-FU NPs	208 ± 15.00	+26 ± 2.00	0.19 ± 0.01	55 ± 1.00	90 ± 4.24	61 ± 2.00	43 ± 3.00
FA-CS-5-FU NPs	235 ± 12.00	+20 ± 2.00	0.25 ± 0.01	53 ± 1.00	80.8 ± 3.19	59 ± 2.00	39 ± 2.00

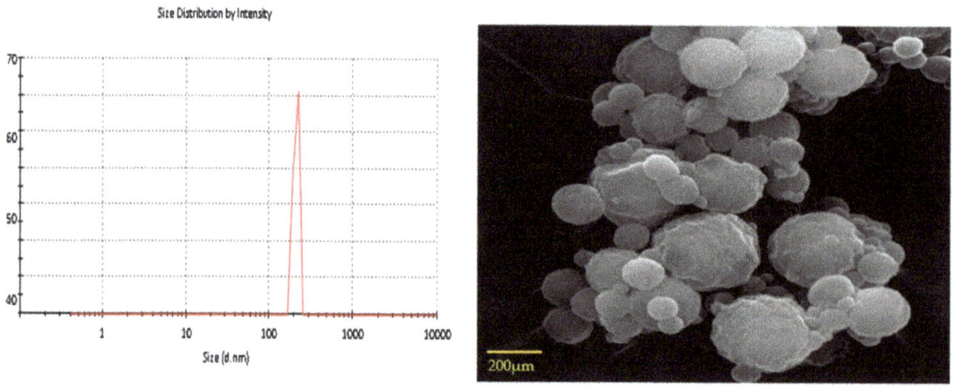

(**A**) (**B**)

Figure 4. (**A**) Size distribution of nanoparticles, (**B**) surface morphology of folic-acid-modified 5-FU-loaded chitosan NPs.

3.2.2. Drug Content, Encapsulation Efficiency, and Drug-Loading Efficiency

TPP was used as a cross linker in folate-modified chitosan nanoparticles loaded with 5-FU. The drug content and %EE of the drug were estimated based on the amount of the drug in the supernatant and the sedimented pellets of dispersed nanoparticles after centrifugation. The FA-CS-5-FU NPs demonstrated 5-FU content of 53 ± 0.14% and EE of 59 ± 0.23%. The drug-loading efficiency was 43 ± 3% and 39 ± 2% for CS-5-FU NPs and FA-CS-5-FU NPs, respectively. A decrease in the loading efficiency of NPs with FA conjugation occurred because the folic acid changed a number of amino groups on the chitosan molecules, lowering their positive charges and thereby attracting drug molecules [13]. Consequently, it emerged that the amount of folic acid conjugations in the mixture had a significant effect on the loading efficiency (LE) (Table 1; $p < 0.05$).

3.3. *In Vitro Release*

In vitro drug release was evaluated at a pH of 1.2 and 6.5 to measure the 5-FU release from the FA-CS-5FU-NPs and the CS-5FU-NPs using a USP dissolution apparatus 1. Such conditions were set to simulate the acidic gastric and physiological environment of the intestine. The percentage of drug released from the FA-CS-5FU-NPs and the CS-5FU-NPs was in the range of 10.08 ± 0.45% to 96.57 ± 0.09% and 6 ± 0.31% to 91.44 ± 0.21%, respectively. In artificial gastric liquid, 17.02 ± 0.12% and 14.5 ± 0.41% of 5-FU were released from the FA-CS-5FU-NPs and CS-5FU-NPs, respectively, in the first 2 h. The difference in the release pattern of these two formulations was insignificant, as shown in Figure 5 ($p > 0.05$). The initial release of 5-FU at an acidic pH was followed by a sustained release of up to 24 h. The initial release may be due to weakly bound drugs on the surface of nanoparticles [20].

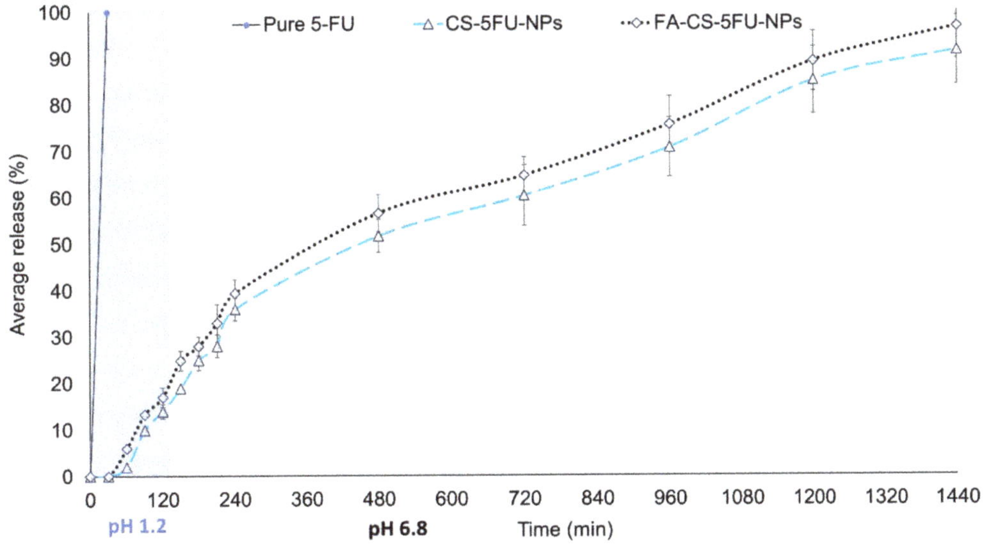

Figure 5. In vitro release study of pure 5-FU, CS-5FU-NPs, and FA-CS-5FU-NPs.

Drug release at a pH of 6.5 within the first 2 h from the FA-CS-5FU-NPs and CS-5FU-NPs was 39.37 ± 3% and 36 ± 2.45%, and the accumulative release in 24 h (1440 min) was 96.57 ± 7% and 91.44 ± 7.45%, respectively. These in vitro values indicate that the FA-decorated nanoparticles can be used as a 5-FU delivery vector with a typical controlled release process. The remarkably high release rate of 5-FU from the folic-acid-conjugated nanoparticles, more interestingly at a pH of 6.5, may be due to the increased acidity of the respective release media caused by the presence of folic acid on the targeted nanoparticles. The improved hydrophilicity of the FA–CS nanoparticles due to the addition of folate was linked to the increased release rate [31]. Taken together, the acidic environments of tumor cells are likely to elicit the release of 5-FU from the developed delivery vehicles, and the sustained drug release profile from the vehicles over time can reduce dosing regimens [32].

3.4. Cytotoxicity Studies

Cytotoxicity studies of free drug and NPs were performed on caco-2 cell lines. The percentage of cell death was determined and shown in Figure 5. The IC50 value of free 5-FU was found to be 4.21 µg/mL. This value was reduced to 3.43 µg/mL (CS-5-FU-NPs) and 2.67 µg/mL (FA-CS-5-FU-NPs) when 5-FU was incorporated into nanoparticles, showing significantly more cytotoxicity than the free drug. Up to 9% of cell death was induced by the free drug (5-FU solution). The percentage of cell death increased when the CS-5-FU-NPs and FA-CS-5-FU-NPs were applied. This increase might be due to the combined effect of drug and hydrophilicity of the FA–CS nanoparticles due to the addition of folate, which began to increase the release rate of 5-FU from NPs. However, a significant effect ($p < 0.05$) on the percentage of cell death was produced when the FA–CS-conjugated NPs were applied (Figure 6). This was the resultant effect of the combination of the drug and folic acid conjugation with chitosan. The FA receptors are more expressed on cancer cells; therefore, the introduction of folic acid on NPs makes them more targeted and cytotoxic in action.

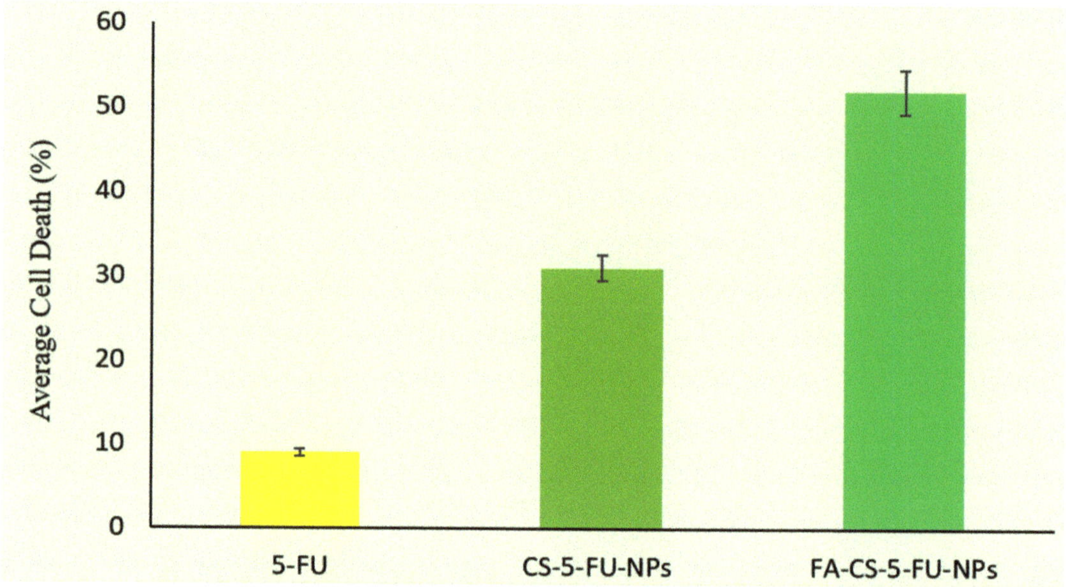

Figure 6. Cytotoxicity study shows the % cell death of 5-FU, CS-5-FU-NPs, and FA-CS-5-FU-NPs.

4. Conclusions

Nanoparticles were successfully prepared using the ionic gelation method. The size and zeta potential and PDI of the CS-5FU-NPs were 208 $\pm$ 15, 26 $\pm$ 2, -20 ± 2, respectively, and those of the FA-CS-5FU-NPs were 235 $\pm$ 12, +20 $\pm$ 2 and 0.25, respectively, which are within acceptable ranges. FTIR and ^{1}H-NMR studies confirmed the conjugation of folic acid with the nanoparticles. The drug contents' % yield and the %EE of folate-decorated NPs were 53 $\pm$ 1, 80.8 and 59 $\pm$ 2%, respectively. The in vitro release of FA-CS-5FU-NPs and CS-5FU-NPs was in the range of 10.08 $\pm$ 0.45 to 96.57 $\pm$ 0.09% and 6 $\pm$ 0.31 to 91.44 $\pm$ 0.21, respectively. The percentage of cell death increased in the presence of folic acid, as compared to the free drug and chitosan nanoparticles due to the overexpression of folate receptors on the cancer cells. The results of all these parameters indicate that folate-modified chitosan 5-FU nanoparticles can be used successfully for the delivery of 5-FU with enhanced cytotoxicity and targeted delivery to the tumors.

Author Contributions: Conceptualization, A.K.A. and M.M.A.-D.; Data curation, A.K.A. and A.N.; Formal analysis, M.I.; Funding acquisition, A.K.A. and M.M.A.-D.; Investigation, S.U. and M.I.; Methodology, S.U., A.N. and K.U.S.; Validation, A.A.S.; Visualization, A.K.A.; Writing—original draft, S.U.; Writing—review & editing, A.K.A., A.N., K.U.S., M.I., G.M.A., F.A.A.-J., A.A.S. and M.M.A.-D. All authors have read and agreed to the published version of the manuscript.

Funding: This research was supported by Princess Nourah bint Abdulrahman University Researchers Supporting Project number (PNURSP2022R30), Princess Nourah bint Abdulrahman University, Riyadh, Saudi Arabia; and Faculty of Pharmacy, AIMST University, Kedah, Malaysia.

Institutional Review Board Statement: The animal study protocol was approved by the Institutional Review Board of office of Research, Innovation, and Commercialization (ORIC, 1600/ORIC/2019-ag-394).

Informed Consent Statement: Not applicable.

Data Availability Statement: Not applicable.

Acknowledgments: This research was supported by Princess Nourah bint Abdulrahman University Researchers Supporting Project number (PNURSP2022R30), Princess Nourah bint Abdulrahman University, Riyadh, Saudi Arabia.; and Faculty of Pharmacy, AIMST University, Kedah, Malaysia.

Conflicts of Interest: The authors declare no conflict of interest.

References

1. Haggar, F.A.; Boushey, R.P. Colorectal cancer epidemiology: Incidence, mortality, survival, and risk factors. *Clin. Colon Rectal Surg.* **2009**, *22*, 191–197. [CrossRef] [PubMed]
2. Kumar, C.S.; Thangam, R.; Mary, S.A.; Kannan, P.R.; Arun, G.; Madhan, B. Targeted delivery and apoptosis induction of trans-resveratrol-ferulic acid loaded chitosan coated folic acid conjugate solid lipid nanoparticles in colon cancer cells. *Carbohydr. Polym.* **2020**, *231*, 115682. [CrossRef] [PubMed]
3. Croisier, F.; Jereme, C. Chitosan-based biomaterials for tissue engineering. *Eur. Polym. J.* **2013**, *49*, 780–792. [CrossRef]
4. Akhlaq, M.; Azad, A.K.; Ullah, I.; Nawaz, A.; Safdar, M.; Bhattacharya, T.; Uddin, A.B.M.H.; Abbas, S.A.; Mathews, A.; Kundu, S.K.; et al. Methotrexate-Loaded Gelatin and Polyvinyl Alcohol (Gel/PVA) Hydrogel as a pH-Sensitive Matrix. *Polymers* **2021**, *13*, 2300. [CrossRef] [PubMed]
5. Handali, S.; Moghimipour, E.; Rezaei, M.; Ramezani, Z.; Kouchak, M.; Amini, M.; Dorkoosh, F.A. A novel 5-Fluorouracil targeted delivery to colon cancer using folic acid conjugated liposomes. *Biomed. Pharmacother.* **2018**, *118*, 1259–1273. [CrossRef]
6. Low, P.S.; Henne, W.A.; Doorneweerd, D.D. Discovery and Development of Folic-Acid-Based Receptor Targeting for Imaging and Therapy of Cancer and Inflammatory Diseases. *Acc. Chem. Res.* **2008**, *41*, 120–129. [CrossRef]
7. Zhang, M.; Kim, Y.K.; Cui, P.; Zhang, J.; Qiao, J.; He, Y.; Jiang, H. Folate-conjugated polyspermine for lung cancer–targeted gene therapy. *Acta Pharm. Sin. B* **2016**, *6*, 336–343. [CrossRef]
8. Akhlaq, M.; Azad, A.K.; Fuloria, S.; Meenakshi, D.U.; Raza, S.; Safdar, M.; Nawaz, A.; Subramaniyan, V.; Sekar, M.; Sathasivam, K.V.; et al. Fabrication of Tizanidine Loaded Patches Using Flaxseed Oil and Coriander Oil as a Penetration Enhancer for Transdermal Delivery. *Polymers* **2021**, *13*, 4217. [CrossRef]
9. Chanphai, P.; Thomas, T.J.; Tajmir-Riahi, H.A. Design of functionalized folic acid–chitosan nanoparticles for delivery of tetracycline, doxorubicin, and tamoxifen. *J. Biomol. Struct. Dyn.* **2018**, 1–7. [CrossRef]
10. Yang, S.J.; Lin, F.H.; Tsai, H.M.; Lin, C.F.; Chin, H.C.; Wong, J.M.; Shieh, M.J. Alginate-folic acid-modified chitosan nanoparticles for photodynamic detection of intestinal neoplasms. *Biomaterials* **2011**, *32*, 2174–2182. [CrossRef]
11. Gaspar, V.M.; Costa, E.C.; Queiroz, J.A.; Pichon, C.; Sousa, F.; Correia, I.J. Folate-targeted multifunctional amino acid-chitosan nanoparticles for improved cancer therapy. *Pharm. Res.* **2015**, *32*, 562–577. [CrossRef] [PubMed]
12. Sahu, S.K.; Mallick, S.K.; Santra, S.; Maiti, T.K.; Ghosh, S.K.; Pramanik, P. In vitro evaluation of folic acid modified carboxymethyl chitosan nanoparticles loaded with doxorubicin for targeted delivery. *J. Mater. Sci. Mater. Med.* **2010**, *2*, 1587–1597. [CrossRef] [PubMed]
13. Yang, S.J.; Lin, F.H.; Tsai, K.C.; Wei, M.F.; Tsai, H.M.; Wong, J.M.; Shieh, M.J. Folic acid-conjugated chitosan nanoparticles enhanced protoporphyrin IX acolon cancerumulation in colorectal cancer cells. *Bioconj. Chem.* **2010**, *2*, 679–689. [CrossRef] [PubMed]
14. Tan, Y.L.; Liu, C.G. Preparation and characterization of self-assemblied nanoparticles based on folic acid modified carboxymethyl chitosan. *J. Mat. Sci. Mat. Med.* **2011**, *22*, 1213–1220. [CrossRef]
15. Malviya, R.; Sundram, S.; Fuloria, S.; Subramaniyan, V.; Sathasivam, K.V.; Azad, A.K.; Sekar, M.; Kumar, D.H.; Chakravarthi, S.; Porwal, O.; et al. Evaluation and Characterization of Tamarind Gum Polysaccharide: The Biopolymer. *Polymers* **2021**, *13*, 3023. [CrossRef]
16. Khan, T.A.; Azad, A.K.; Fuloria, S.; Nawaz, A.; Subramaniyan, V.; Akhlaq, M.; Fuloria, N.K. Chitosan-Coated 5-Fluorouracil Incorporated Emulsions as Transdermal Drug Delivery Matrices. *Polymers* **2021**, *13*, 3345. [CrossRef]
17. Khan, M.A.; Azad, A.K.; Safdar, M.; Nawaz, A.; Akhlaq, M.; Paul, P.; Hossain, M.K.; Rahman, M.H.; Baty, R.S.; El-kott, A.F.; et al. Synthesis and Characterization of Acrylamide/Acrylic Acid Co-Polymers and Glutaraldehyde Crosslinked pH-Sensitive Hydrogels. *Gels* **2022**, *8*, 47. [CrossRef]
18. Bandara, S.; Carnegie, C.A.; Johnson, C.; Akindoju, F.; Williams, E.; Swaby, J.M.; Carson, L.E. Synthesis and characterization of Zinc/Chitosan-Folic acid complex. *Heliyon* **2018**, *4*, e00737. [CrossRef]
19. Musalli, A.H.; Talukdar, P.D.; Roy, P.; Kumar, P.; Wong, T.W. Folate-induced nanostructural changes of oligochitosan nanoparticles and their fate of cellular internalization by melanoma. *Carbohydr. Polym.* **2020**, *244*, 116488. [CrossRef]
20. Akinyelu, J.; Singh, M.J.A.N. Folate-tagged chitosan-functionalized gold nanoparticles for enhanced delivery of 5-fluorouracil to cancer cells. *Appl. Nanosci.* **2019**, *9*, 7–17. [CrossRef]
21. Shah, M.K.A.; Azad, A.K.; Nawaz, A.; Ullah, S.; Latif, M.S.; Rahman, H.; Alsharif, K.F.; Alzahrani, K.J.; El-Kott, A.F.; Albrakati, A.; et al. Formulation Development, Characterization and Antifungal Evaluation of Chitosan NPs for Topical Delivery of Voriconazole In Vitro and Ex Vivo. *Polymers* **2022**, *14*, 135. [CrossRef] [PubMed]
22. Mauricio-Sánchez, R.A.; Salazar, R.; Luna-Bárcenas, J.G.; Mendoza-Galván, A.J.V.S. FT-IR spectroscopy studies on the spontaneous neutralization of chitosan acetate films by moisture conditioning. *Vib. Spectrosc.* **2018**, *94*, 1–6. [CrossRef]

23. Latif, M.S.; Azad, A.K.; Nawaz, A.; Rashid, S.A.; Rahman, M.H.; Al Omar, S.Y.; Bungau, S.G.; Aleya, L.; Abdel-Daim, M.M. Ethyl Cellulose and Hydroxypropyl Methyl Cellulose Blended Methotrexate-Loaded Transdermal Patches: In Vitro and Ex Vivo. *Polymers* **2021**, *13*, 3455. [CrossRef] [PubMed]
24. Ji, J.; Wu, D.; Liu, L.; Chen, J.; Xu, Y. Preparation, characterization, and in vitro release of folic acid-conjugated chitosan nanoparticles loaded with methotrexate for targeted delivery. *Polym. Bull.* **2012**, *68*, 1707–1720. [CrossRef]
25. Wan, A.; Sun, Y.; Li, H. Characterization of folate-graft-chitosan as a scaffold for nitric oxide release. *Int. J. Biol. Macromol.* **2008**, *43*, 415–421. [CrossRef] [PubMed]
26. John, A.; Jaganathan, S.K.; Ayyar, M.; Krishnasamy, N.P.; Rajasekar, R.; Supriyanto, E. Folic acid decorated chitosan nanoparticles and its derivatives for the delivery of drugs and genes to cancer cells. *Curr. Sci.* **2017**, 1530–1542. [CrossRef]
27. Danhier, F.; Feron, O.; Préat, V. To exploit the tumor microenvironment: Passive and active tumor targeting of nanocarriers for anti-cancer drug delivery. *J. Control. Release* **2010**, *148*, 135–146. [CrossRef]
28. Ince, I.; Yildirim, Y.; Guler, G.; Medine, E.I.; Ballıca, G.; Kusdemir, B.C.; Goker, E. Synthesis and characterization of folic acid-chitosan nanoparticles loaded with thymoquinone to target ovarian cancer cells. *J. Radioanal. Nucl. Chem.* **2020**, *324*, 71–85. [CrossRef]
29. Luong, D.; Kesharwani, P.; Alsaab, H.O.; Sau, S.; Padhye, S.; Sarkar, F.H.; Iyer, A.K. Folic acid conjugated polymeric micelles loaded with a curcumin difluorinated analog for targeting cervical and ovarian cancers. *Colloids Surf. B Biointerfaces* **2017**, *157*, 490–502. [CrossRef]
30. Zhang, H.; Li, X.; Gao, F.; Liu, L.; Zhou, Z.; Zhang, Q. Preparation of folate-modified pullulan acetate nanoparticles for tumor-targeted drug delivery. *Drug. Deliv.* **2009**, *17*, 48–57. [CrossRef]
31. Ramezani Farani, M.; Azarian, M.; Heydari Sheikh Hossein, H.; Abdolvahabi, Z.; Mohammadi Abgarmi, Z.; Moradi, A.; Rabiee, N. Folic acid-adorned curcumin-loaded iron oxide nanoparticles for cervical cancer. *ACS Appl. Bio Mater.* **2022**, *5*, 1305–1318. [CrossRef] [PubMed]
32. Mattos, A.C.D.; Altmeyer, C.T.; Tania, T.; Khalil, N.M.; Mainardes, R.M. Polymeric nanoparticles for oral delivery of 5-fluorouracil: Formulation optimization, cytotoxicity assay and pre-clinical pharmacokinetics study. *Eur. J. Pharm. Sci.* **2016**, *8*, 83–91. [CrossRef] [PubMed]

polymers

MDPI

Article

pH-Responsive PVA/BC-*f*-GO Dressing Materials for Burn and Chronic Wound Healing with Curcumin Release Kinetics

Wafa Shamsan Al-Arjan [1], Muhammad Umar Aslam Khan [2,3,*], Hayfa Habes Almutairi [1], Shadia Mohammed Alharbi [1] and Saiful Izwan Abd Razak [4,5]

1 Department of Chemistry, College of Science, King Faisal University, Al-Ahsa 31982, Saudi Arabia; walarjan@kfu.edu.sa (W.S.A.-A.); halmutairi@kfu.edu.sa (H.H.A.); smalharbi@kfu.edu.sa (S.M.A.)
2 Biomedical Research Center, Qatar University, Doha 2713, Qatar
3 Department of Mechanical and Industrial Engineering, Qatar University, Doha 2713, Qatar
4 BioInspired Device and Tissue Engineering Research Group, School of Biomedical Engineering and Health Sciences, Universiti Teknologi Malaysia, Johor Bahru 81300, Johor, Malaysia; saifulizwan@utm.my
5 Centre of Advanced Composite Materials, Faculty of Engineering, Universiti Teknologi Malaysia, Johor Bahru 81300, Johor, Malaysia
* Correspondence: umar007khan@gmail.com

Abstract: Polymeric materials have been essential biomaterials to develop hydrogels as wound dressings for sustained drug delivery and chronic wound healing. The microenvironment for wound healing is created by biocompatibility, bioactivity, and physicochemical behavior. Moreover, a bacterial infection often causes the healing process. The bacterial cellulose (BC) was functionalized using graphene oxide (GO) by hydrothermal method to have bacterial cellulose-functionalized-Graphene oxide (BC-*f*-GO). A simple blending method was used to crosslink BC-*f*-GO with polyvinyl alcohol (PVA) by tetraethyl orthosilicate (TEOS) as a crosslinker. The structural, morphological, wetting, and mechanical tests were conducted using Fourier-transform infrared spectroscopy (FTIR), Scanning electron microscope (SEM), water contact angle, and a Universal testing machine (UTM). The release of Silver-sulphadiazine and drug release kinetics were studied at various pH levels and using different mathematical models (zero-order, first-order, Higuchi, Hixson, Korsmeyer–Peppas, and Baker–Lonsdale). The antibacterial properties were conducted against Gram-positive and Gram-negative severe infection-causing pathogens. These composite hydrogels presented potential anticancer activities against the U87 cell line by an increased GO amount. The result findings show that these composite hydrogels have physical-mechanical and inherent antimicrobial properties and controlled drug release, making them an ideal approach for skin wound healing. As a result, these hydrogels were discovered to be an ideal biomaterial for skin wound healing.

Keywords: antibacterial; biopolymers; curcumin; drug delivery; hydrogels; pH-responsive; skin wound healing

Citation: Al-Arjan, W.S.; Khan, M.U.A.; Almutairi, H.H.; Alharbi, S.M.; Razak, S.I.A. pH-Responsive PVA/BC-*f*-GO Dressing Materials for Burn and Chronic Wound Healing with Curcumin Release Kinetics. *Polymers* **2022**, *14*, 1949. https://doi.org/10.3390/polym14101949

Academic Editor: Ariana Hudita

Received: 16 April 2022
Accepted: 6 May 2022
Published: 11 May 2022

1. Introduction

A wound is a type of injury caused by surgery, trauma, infection, diabetes, or other factors. Chronic wounds on the skin have become a life-threatening issue in recent years. According to reports, approximately 10 million people experience burning worldwide, leading to death if not handled properly [1]. Skin wounds of this nature necessitate immediate attention and treatment. Dehydration is caused by burning, which can lead to fatal complications. Choosing a suitable wound dressing is critical in caring for and the treatment of burn and chronic wounds. The material choice is crucial in facilitating quick wound healing and retaining fluidic contents with minimal dehydration [2,3]. An ideal wound dressing should have enough mechanical stability to support skin injuries, flexibility, water retention, antibacterial properties, and tear resistance, and be easily removable. It should also absorb a large amount of wound exudate from the wound surface, as the exudate volume in burning wounds is usually relatively high [4,5]. It should protect against pathogens that cause severe disease by preventing their growth in the wound environment and surface. It should be biodegradable and easily removable to avoid skin peeling after application. Long-term drug delivery and growth factors should be easily swellable with controlled biodegradation. Various advanced wound dressings are commercially available for the care and treatment of skin-burning wounds. Ointment, cream, gels, natural oils, mists, gauzes, and bandages are among the wound dressings available [6,7]. Most chronic wound healing products are effective in wound healing, but their role in skin burns has yet to be determined.

Mechanical strength, swelling, water retention, and controlled therapeutic agent release are all problems with biomaterials. In addition, some of them are poor blood coagulators, while others lack antibacterial properties [8–10]. The non-biodegradability of these wound dressings is a significant issue, and most of the materials used in these dressings are also expensive. Taking all of this into account, developing an advanced dressing material with all of the necessary wound-healing and tissue engineering properties [11,12]. Hydrogel-based wound healing materials with all of the above properties can treat chronic and burn wounds. Hydrogels are typically made up of natural or synthetic polymers. However, they lack mechanical properties, biodegradation, and a low swelling ratio, limiting their clinical use as chronic wound dressings [13,14].

It is essential to develop composite hydrogels with enhanced functionalities to address these limitations. Bacterial nanocellulose is a biopolymer and polysaccharide that can retain enough biofluid. Water absorption, swelling, biodegradation, and long-term drug delivery properties make it a good candidate for biomedical applications. Cellulose is the most abundantly available natural polymer and is extensively used in several biomedical applications [15,16]. Graphene oxide is a potential material well-known due to its multifunctional behavior, and it has become popular with researchers due to its potential and versatility, including large surface area, π-π stacking, etc. The most widely credited material for developing next-generation biomaterials is graphene oxide (GO). On the basal plane and at the edge, GO has both hydrophobic and hydrophilic parts, with oxygen-containing functional groups like hydroxyl, epoxy, carbonyl, and carboxyl groups [17]. In addition, GO contains several oxygen-containing functional groups in biomedical applications and essential properties. Furthermore, GO-based materials have multifunctional properties [18].

We have reported the fabrication of composite hydrogels for burn and chronic wound healing applications. The bacterial nanocellulose and different graphene oxide amounts were functionalized via the hydrothermal method. Then, BC-*f*-GO was crosslinked with PVA using TEOS as a crosslinker to fabricate composite hydrogel. According to the best of our knowledge, these formulations have never been reported. The curcumin has been loaded into the composite hydrogel to determine its release behavior and kinetics. FTIR, SEM, water contact angle, and UTM determined the structural, morphological, wetting, and mechanical properties. The swelling was analyzed at different pH in PBS and aqueous media. The in vitro biodegradation and drug release analysis was conducted in PBS media. Antimicrobial were performed against Gram-positive and Gram-negative infection-causing pathogens to observe their antibacterial performance in burn and chronic wound healing.

2. Materials and Methods

2.1. Materials

A well-reported method was used to synthesize bacterial nanocellulose by a method by Saiful et al. [19], polyvinyl alcohol (PVA), Graphene oxide (CAS No. 763713-1G) tetraethyl orthosilicate (CAS No. 78-10-4), potassium persulfate (initiator), hydrochloric acid (CAS No. 7647-01-0), phosphate buffer saline (MDL No. 806552-1L) solution, and Sigma-Aldrich, Petaling Jaya, Selangor, Malaysia, supplied the ethanol. These chemicals were analytically graded and used as received.

2.2. Hydrogel Fabrication

The GO-functionalized-bacterial cellulose (GO-*f*-BC) was synthesized by the hydrothermal approach, in which bacterial cellulose (0.7 g) and various amounts of GO (0.01, 0.02, 0.03, and 0.04 mg) were placed in a 50 mL stainless-steel autoclave for 30 min. The stainless-steel autoclave was placed in the oven at 50 °C overnight to have BC-*f*-GO, and the stainless-steel autoclave was removed after the overnight reaction to obtain BC-*f*-GO, which was then added in 25 mL double deionized water (DDW) and stirred at 55 °C for 1 h to have a homogenized mixture. PVA (0.3 g) was added in 10 mL DDW and dissolved at 80 °C. These were mixed and stirred at 55 °C for 1 h to have a homogeneous suspension. The crosslinker (TEOS (240 μL)) was added to 5 mL ethanol, added into the whorl of the mixture solution, and allowed to stir at 55 °C for another 2 h. Finally, 0.2 g of Potassium persulfate as initiator was dissolved into 5 mL of DDW, added to the solution as an initiator, and stirred at 55 °C for 3 h for successful crosslinking. After 3 h, we had a light yellowish color of the fabrication composite hydrogels and shifted the composite hydrogels into Petri plates. These Petri plates were oven-dried at 45 °C, and different codes (BSG-1, BSG-2, BSG-3, and BSG-4) were assigned to these composite hydrogels after a different GO amount (0.01, 0.02, 0.03, and 0.04 mg). The proposed chemical schematic of the composite hydrogel has shown in Scheme 1.

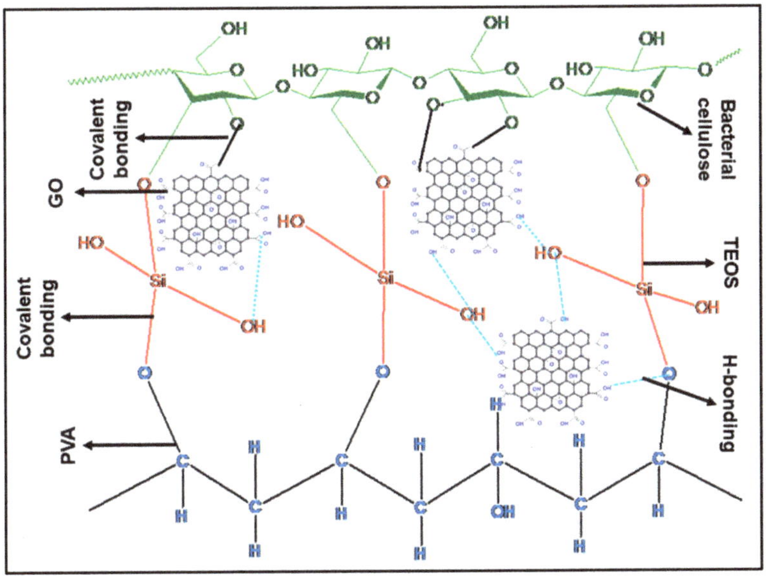

Scheme 1. The proposed chemical interaction of the bacterial cellulose, polyvinyl alcohol, GO, and crosslinked via TEOS.

3. Characterizations

3.1. FT-IR

The structural and functional group information was analyzed by Fourier transform infrared (FT-IR) Nicolet 5700, Waltham, MA, USA, spectrophotometer. FT-IR analysis was carried out at attenuated total reflectance (ART) mode with wavenumber (4000 to 400 cm^{-1}) and 120 scans per sample.

3.2. SEM

The surface morphologies of well-dried hydrogels were examined (JSM-6701S, JEOL, Peabody, MA, USA). First, double-sided carbon tape was used to secure the hydrogels to the aluminum stubs, which were then gold-sputtered. Then, micrographs were taken to examine their morphologies.

3.3. Water Contact Angle

The wetting behavior of composite hydrogels was studied using a water contact angle system (JY-82, Dingsheng, Chengde, China) to determine their hydrophilicity and hydrophobicity. We recorded wetting at zero and ten seconds to investigate wetting behavior over time.

3.4. Swelling

Swell tests were performed in PBS and aqueous media at various pH levels to determine their pH-responsive behavior. The sliced hydrogels were weighed (50 mg), and the initial weight (W_i) was taken. These hydrogels were submerged in PBS and aqueous media-containing beakers at room temperature. Hydrogels that had swelled were removed from the corresponding media. After 12 h, the extra surface media was carefully removed and weighed (W_f) as the final weight and swelling (%) were calculated using Equation (1).

$$Swelling\ (\%) = \frac{W_f - W_i}{W_i} \times 100 \quad (1)$$

whereas: W_f = Hydrogel final weight, W_i = Hydrogel initial weight.

3.5. Biodegradation

The hydrogels were tested in vitro in PBS buffer solution (pH 7.4, at 37 °C with 5% CO_2) to see how they degraded. The squarely sliced hydrogels were carefully weighed (45 mg) and placed in PBS buffer solution to determine biodegradation behavior. Equation (2) was used to calculate the percentage of biodegradation.

$$Weight\ loss\ (\%) = \frac{W_i - W_t}{W_i} \times 100 \quad (2)$$

whereas: W_t = Hydrogel weight at "t", W_i = Hydrogel initial weight.

3.6. Curcumin Loading and Franz Diffusion Release

The dermatologic antibiotic curcumin prevents infection in partial-thickness and full-thickness chronic and burns wounds. However, it is still commonly used to treat third-degree burns. Curcumin (10 mg) was dissolved in ethanol (5 mL) and dropped into the polymeric mixture dropwise (BC 0.7 g, PVA 0.3 g, and GO 0.4 mg). The heterogeneous mixture was stirred for two hours at 55 °C, then crosslinked with TEOS (240 μL), and stirred for three hours at the same temperature. In vitro drug release was measured using the Franz diffusion transdermal method at three different pH levels (6.4, 7.4, and 8.4), as Saiful et al. [15] reported in PBS buffer solution at 37 °C. In addition, the calibration analysis was conducted to evaluate the release of Silver-sulfadiazine from hydrogel at different pH levels.

3.7. Drug Release Kinetics

We used mathematical models to study drug release mechanisms (3–8) (zero order, first order, Higuchi, Hixson–Crowell, Korsmeyer–Peppas, and Baker–Lonsdale). In addition, data from Franz diffusion at various pH levels were used to evaluate drug release kinetics.

$$\text{Zero-order} \qquad M_t = M_o + K_o t \quad (3)$$

$$\text{First order} \qquad logC = logC_o - \frac{kt}{2.303} \quad (4)$$

$$\text{Higuchi model} \qquad ft = Q = K_H \times t^{1/2} \quad (5)$$

$$\text{Hixson Crowell model} \qquad {W_o}^{1/3} - {W_t}^{1/3} = kt \quad (6)$$

Korsmeyer-Peppas model $$ln\frac{M_t}{M_o} = n\ lnt + lnK \quad (7)$$

Baker-Lonsdale model $$F_t = \frac{3}{2}\left[1 - \left(1 - \frac{M_t}{M_o}\right)^{\frac{2}{3}}\right]\frac{M_t}{M_o} = K(t)^{0.5} \quad (8)$$

M_t = amount of drug release at "t", and K_H, K, and K_o are constants.

3.8. In Vitro Analysis

3.8.1. Antimicrobial Activities

The disc diffusion method has been used to evaluate the antibacterial activities of the composite hydrogels, which have been assessed against Gram-positive (*Staphylococcus aureus* (*S. aureus*)) and Gram-negative (*Escherichia coli* (*E. coli*) and *Pseudomonas aeruginosa* (*P. aeruginosa*)) bacterial strains. The bacterial strains were provided by ATCC, USA. The molten agar was poured on the Petri dishes and allowed to cool at ambient temperature and bacterial media was applied via cotton swab. Then, the hydrogels (70 μL) were placed on bacterial cultured Petri dishes using a micropipette. The Petri dishes were incubated at 37 °C for 24 h, and the diameter of zone inhibition was measured in millimeters (mm).

3.8.2. Anticancer Analysis

The cytotoxicity of the materials was investigated against the U87 cell line by MTT assay to determine cell viability and cell morphology under in vitro conditions. The purpose of this study is to determine the anticancer behavior of composite hydrogels and drug-loaded composite hydrogels [20].

3.9. Statistical Analysis

SPSS Statistics 21 (IBM SPSS Statistics 21, SPSS Inc., New York, NY, USA) analyzed the triplicate data and presented it as mean ± standard error. Figures with Y-error bars represent the standard error. $p < 0.05$, n = 3.

4. Results and Discussions

4.1. FTIR Analysis

The FT-IT spectral profile of the hydrogel can, as shown in Figure 1, determine the structural and functional analysis of the material present in hydrogels. The vibration band from 1110 to 1000 cm^{-1} is attributed to the asymmetric stretching of –Si–O–C and –Si–O–Si due to TEOS and confirmed the successful crosslinking of bacterial cellulose and polyvinyl alcohol. These polymers and GO were also crosslinked due to hydrogen bonding (H–bonding) that presents the absence peak at 1759 cm^{-1} and broadband at 3600–3200 cm^{-1}. The increased broadband valley is due to increased intra and inter-hydrogen bonding [21]. The characteristic peaks of BC are hydroxyl, COO^-, and pyranose rings. The absorption peak of the saccharine structure and pyranose ring is confirmed at 1060 cm^{-1} and 876 cm^{-1}, respectively. The stretching peak at 2950 cm^{-1} is due to alkyl –CH of BC. The stretching peaks at 1643 and 1469 cm^{-1} are attributed to functional groups of C=O and C–C and confirm the presence of GO and it is associated with the polymeric matrix via H-bonding. Hence, the FT-IR spectral confirms successful crosslinking of the polymers and GO interaction with the polymeric matrix that has been determined via available functional groups and interaction.

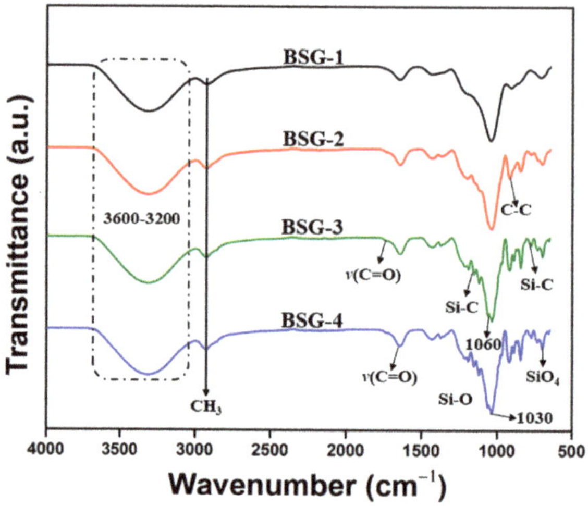

Figure 1. FTIR spectra of composite hydrogels to determine the structural, and functional groups and their physicochemical interactions.

4.2. SEM Morphology

The surface of hydrogel is a fundamental phenomenon for drug release and interaction with the body's biological system. Therefore, SEM analysis was performed to investigate the surface properties of the hydrogel materials, as shown in Figure 2. The increasing amount of GO causes more particulate-like (GO-flakes) morphology than smooth surface morphology. These GO-flakes impart their unique role in the hydrogels' morphology by increasing surface roughness and closing the packing of the hydrogel. Such surface morphology helps burn and chronic wounds by providing them with important hydration [22,23]. However, it was also observed that an increasing GO amount also causes cracking on drying. Hence, it is essential to introduce the only optimized amount of GO to have the desired surface morphology with structural integrity.

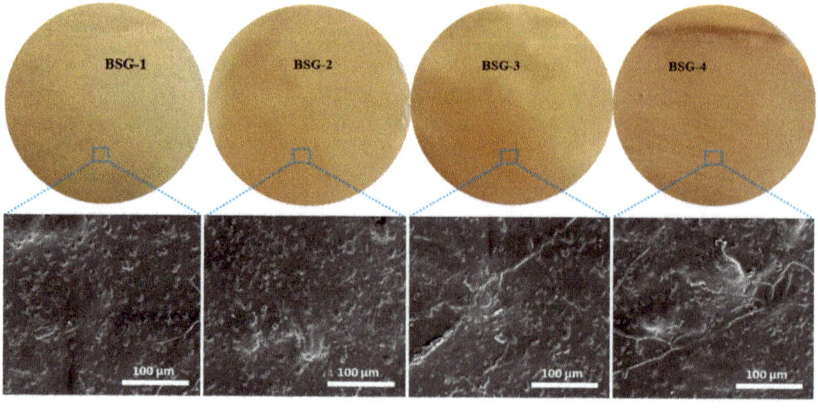

Figure 2. The surface morphology of the hydrogels via SEM.

4.3. Water Contact Angle

The ability of a liquid to contact the solid surface is known as wetting. When the liquid and solid surfaces are brought together, they result from intermolecular interactions. A force balance between adhesive and cohesive forces determines the degree of wetting.

Wettability is essential in the bonding or adherence of two different materials. Wettability describes the material hydrophilicity and hydrophobicity caused by the surface forces that control wetting. A liquid drop spreads across the surface due to adhesive forces between the liquid and the solid. Due to cohesive forces within the liquid, the drop balls up and avoid contact with the surface [24]. We have observed the wetting behavior of the hydrogels at different times (zero and ten seconds) and with an increased amount of GO. As time and GO amount increase, the hydrogel's wetting behavior shifts from hydrophobicity to hydrophilicity. BSG-4 and BSG-1 had water contact angles of 103.8° and 65.4° at zero seconds, respectively. However, after ten seconds, the water contact angles for BSG-4 and BSG-1 were 63.40° and 96.10°, respectively. The wetting of BSG-2 = 75.50° and BSG-3 = 83.40° was observed at zero seconds as shown in Figure 3. The wetting analysis clearly shows that the hydrophilicity to hydrophobicity ratio has shifted. That is due to the increasing amount of GO, as it contains several oxygen-based functional groups that caused hydrogen bonding and other weak interactions [25]. A more hydrophilic nature for BSG-4 was observed with increased hydrophilicity by an increasing GO amount and prolonged contact time. Hence, BSG-4 will be helpful to absorb more wound exudate and keep the burn and chronic wound hydrated.

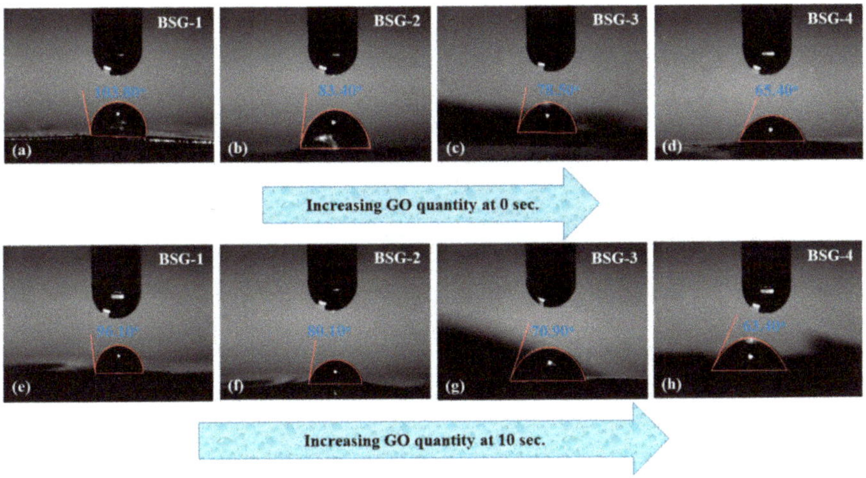

Figure 3. The wetting analysis of the hydrogel was conducted via water contact angle to determine the hydrophilicity and hydrophobicity nature of the hydrogels. The wetting behavior was investigated at different time intervals and 4(**a–d**) wetting behavior at 0 s while 4(**e–h**) wetting behavior of hydrogels at 10 s.

4.4. Swelling

Swelling is a significant feature of hydrogels for burn and chronic wound healing and drug delivery. The impacts of pH on the swelling behavior of hydrogels were studied at various pH levels, as shown in Figure 4a,b. All hydrogels showed minimal swelling in acidic and basic pH, with amazing swelling at pH 7. It reveals that hydrogels were extremely sensitive to changes in pH. The hydrophilic groups on the bacterial cellulose, PVA, and GO structures caused swelling due to H-bonding. The water penetrates the void space inside the hydrogels in neutral media. Functional groups in BC, such as carboxyl groups, were protonated to produce more electrostatic repulsion forces [26,27]. The available oxygen-based functional facilitate hydrogen bonding. However, in acidic media, the –OH functional groups have protonated those decreases in swelling, dissociating at higher pH. The hydrogels also have different swelling due to a variable amount of GO. It may act as a crosslinker as it has such a closely packed hydrogel structure due to the enhanced weak intra-hydrogel bonding and electrostatic interactions [28,29]. BSG-1 has maximum, and BSG-4 has minimum swelling in aqueous and PBS media. At pH 7, electrostatic repulsion causes hydrogen bonding and void spaces in hydrogels, resulting in maximum swelling. As a result, the tested hydrogels cause the hydrogel to swell dramatically.

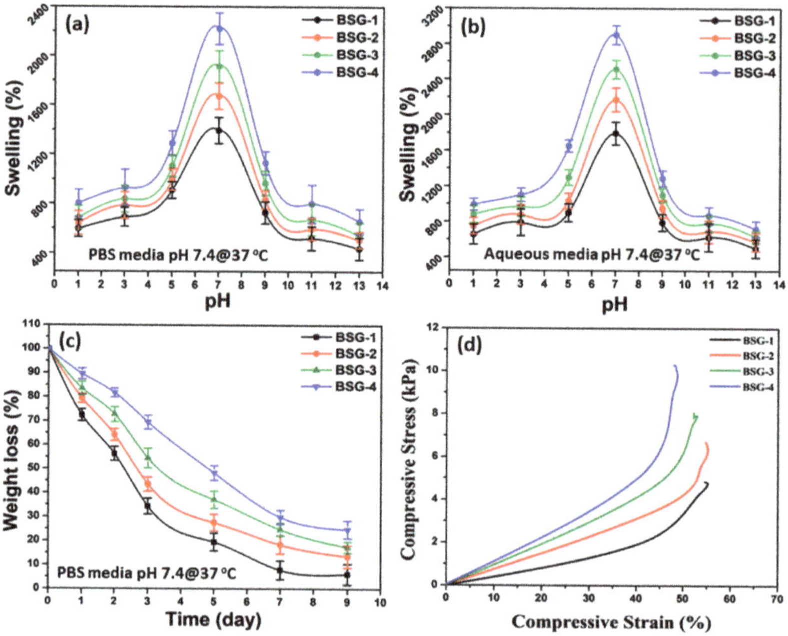

Figure 4. The physicochemical analysis of the hydrogels was conducted to study to analyze the behavior of the hydrogel with biofluid during application. Swelling in PBS (**a**) and aqueous (**b**) media, (**c**) biodegradation in PBS buffer solution (**d**), and stress–strain curve of hydrogels.

4.5. Biodegradation

Biodegradation is a characteristic of drug delivery hydrogels that are likely related to drug release. In vitro degradation of hydrogels was carried out under in vitro conditions in PBS buffer solution. The hydrogels exhibited different biodegradation due to different GO amounts, as shown in Figure 4c. The hydrogel BSG-1 has maximum weight loss, and BSG-4 has the most minor. It may be due to the increasing amount of GO that acts as a crosslinker to hold the polymeric chain tightly. Biodegrading also occurs due to alkyl linkage and glycosidic bonds in BC, facilitating drug release [30]. The different biodegradation also confirms the successful hydrogel fabrication, and the difference in degradation ratio was due to varying amounts of GO. The increased GO amount provides a closely packed structure with increased crosslinking density. The available alkyl linkage and glycosidic bonds of BC breakdown cause biodegradation [31]. Hence, controlled swelling facilitates controlled drug release and other therapeutic agents essential for wound healing.

4.6. Mechanical Testing

The mechanical behavior of hydrogels was studied using a stress–strain curve. The mechanical properties of hydrogels provide structural integrity and substantial strength to resist swelling and degradation. The mechanical properties of hydrogels can be controlled by optimizing crosslinker and filler (GO) amounts. It is worth mentioning that increasing the GO amount improved the tensile strength and elastic modulus from BSG-1 to BSG-4, as shown in Figure 4d. Hence, the mechanical strength confirms that our hydrogels have been successfully fabricated with sufficient strength to resist swelling after absorbing wound exudate.

4.7. In Vitro Drug Release

Stimulated release, degradation-controlled release, solvent-controlled release, and diffusion-controlled release are the four mechanisms of drugs released from the polymer network of hydrogels. Solvent-controlled mechanisms include osmotic pressure-controlled and swelling-controlled mechanisms. The swelling of hydrogels is an essential factor in their structural design. It is strongly associated with the controlled release of drugs from hydrogels. The extracellular matrix of hydrogel expands on swelling and makes the drug available on the surface via diffusion pathways. We have taken BSG-4 to load curcumin, which is a natural antibacterial and anticancer drug, due to its optimum physicochemical characterizations. In vitro drug release of curcumin-BSG-4 loaded was determined via the Franz diffusion method under different pH-level (6.4, 7.4, and 8.4) in PBS media, and the drug release profile has been shown in Figure 5. The hydrogels are pH-responsive and offer other drug release mechanisms under different pH levels. The drug release mechanism was found in 7.4 > 6.4 > 8.4 pH, and it is strongly related to swelling and biodegradation of hydrogel, as shown in Figure 6a–c. The pore size of hydrogels increases during swelling due to the diffusion process and the drug is released [32]. Hence, maximum drug release was found at 7.4 pH, and the residual drug could not be determined as the hydrogel films broke into tiny pieces. The drug release can be prolonged by increasing the crosslinker amount. The increasing crosslinker amount can control swelling and degradation, essential factors for the drug delivery system [33]. The drug release at different pH is strongly linked to the hydrogels' swelling behavior that facilitates drug release on swelling under various other mechanisms. From the drug release profile, a smooth drug release was observed $\approx$ 91.38% at pH 7.4 after 33 h as seen in Figure 6b. The drug contains hydrogel broken down into pieces and it is not easy to determine drug release from small pieces. The controlled drug release time can be increased or decreased purely depending on crosslinking factors, which optimizes the swelling and degradation of the hydrogels that are necessary parameters for controlled drug release. Hence, these hydrogels could be desired dressing materials with pH-sensitive behavior that could be helpful in controlled drug delivery for burn and chronic wound healing.

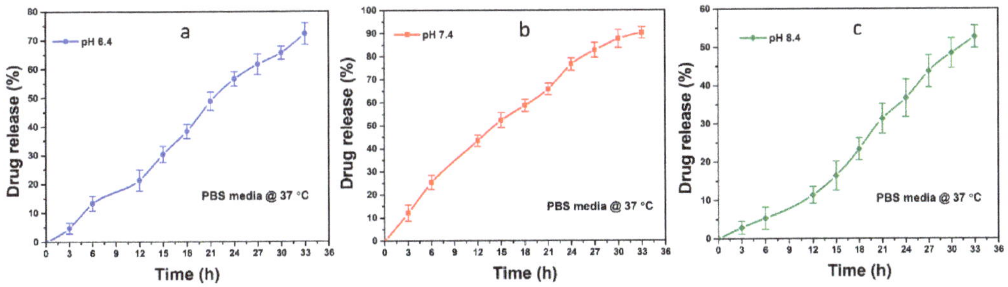

Figure 5. The drug release of curcumin was studied at 37 °C under different pH levels to determine the release analysis behavior: (**a**) Curcumin release at pH 6.4; (**b**) Curcumin release at pH 7.4 and; (**c**) Curcumin release at pH 6.4.

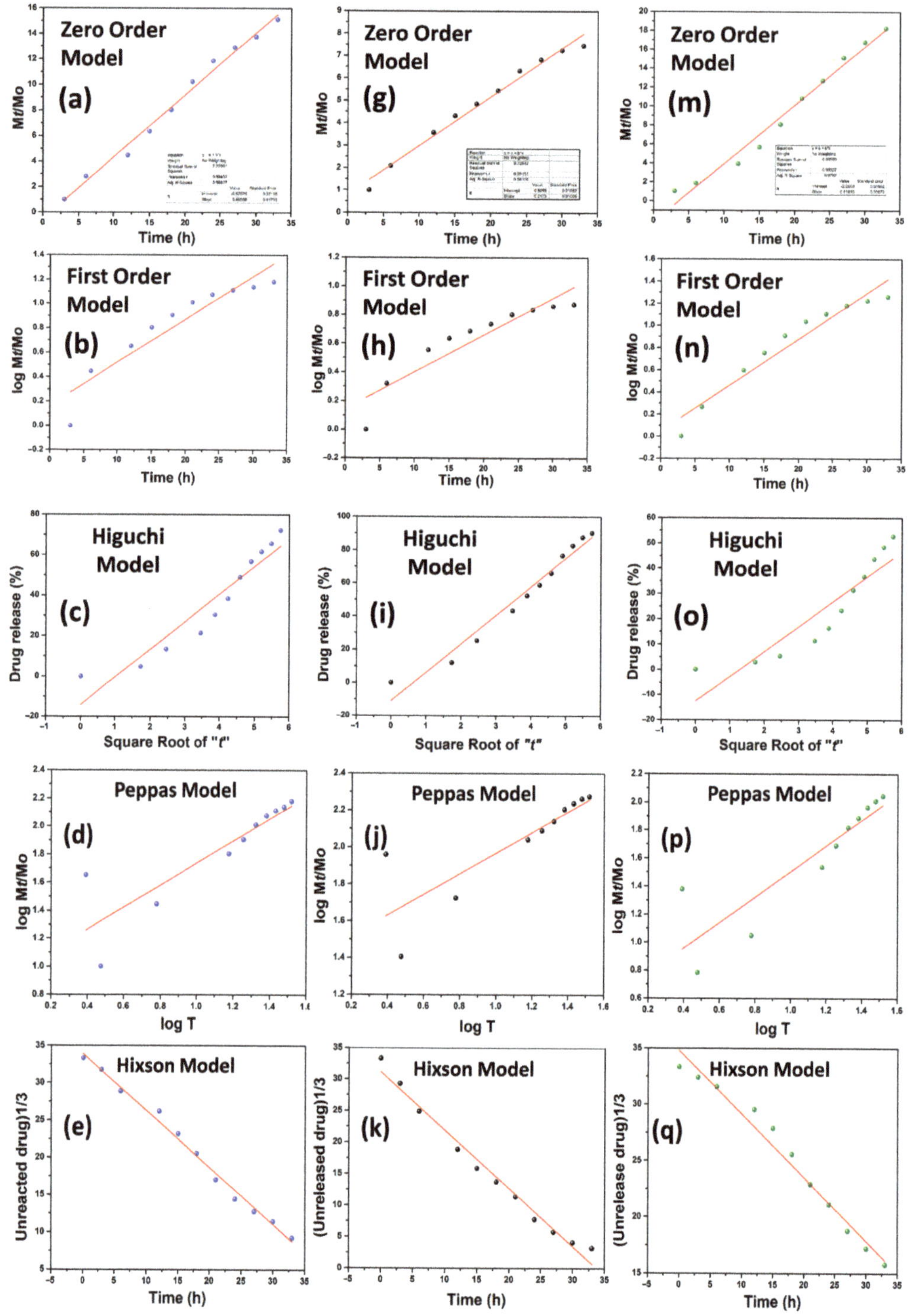

Figure 6. *Cont.*

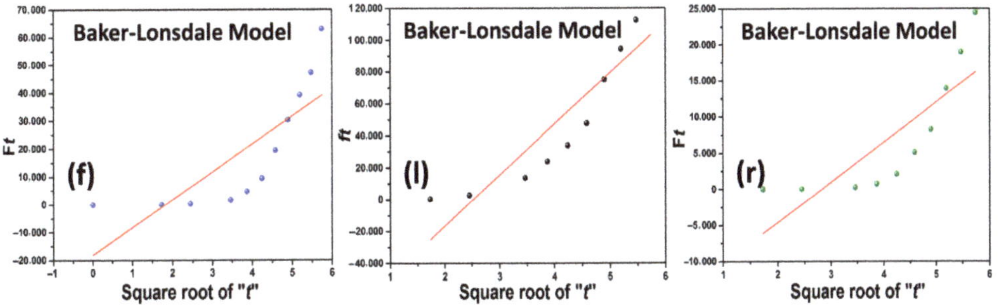

Figure 6. The hydrogel sample BSG-3 was taken to on a random basis to determine curcumin release kinetics at different pH levels (6.4 (**a–f**), 7.4 (**g–l**), and 8.4 (**m–r**)) against different mathematical models (Zero order, First order, Higuchi, Korsmeyer-Peppas, Hixson & Bakers-Lonsdale).

4.8. Drug Release Kinetics

Drug release provides a deep understanding of the mass transport mechanisms involved. A therapeutic system must provide a specific drug release profile to mathematically calculate the resulting drug release kinetics. Various mathematical models were used to develop simple and complex drug delivery systems and predict overall release behavior. They make it possible to measure an extensive range of vital parameters and apply a model fitting to the experimental release data of curcumin from BSG-4 as presented in Figure 6. Understanding how to use these equations is crucial to grasping the various factors influencing drug release velocity. Their dissolution behaviors affect the efficacy of a patient's therapeutic regimen [34,35]. These mathematical models have different values of the regression coefficient (R^2) at different pH levels. All the different parameters have been summarized in Table 1, these values provide drug release behavior through the polymeric matrix. Among all the factors, R^2 is the most prominent factor that describes the drug release behavior from different therapeutic systems. The release of curcumin follows various mechanisms at different pH levels that can be described based on the maximum of R^2. We have calculated the value of R^2 against all fitting models. It was observed that at pH 6.4, curcumin follows the Hixson model with a maximum R^2 value (0.99054). The Hixson model describes the release of the curcumin dissolution approach. The release of curcumin was observed to follow zero-order at 7.4 pH by R^2 value (098106). The zero-order describes the quick release of therapeutic agents that is very important to control the attack of the pathogens. However, the curcumin release behavior was observed at pH 8.4 to follow zero-order due to the maximum R^2 value (0.9782). The release of curcumin from the composite hydrogel at pH 8.4 is found to be as effective as it is at pH 7.4. However, the mathematical fitting models have been summarized in Table 1.

Table 1. The drug release kinetics summarizes against different models to determine the best-fitted model at different pH levels (6.4, 7.4, and 8.4).

Drug Release at Different pH-Levels	Models	Intercept/Standard Error	Slop/Standard Error	Regression Coefficient (R^2)
Drug release kinetics at pH 6.4	Zero order	−0.52979/0.37126	0.48568/0.01755	0.98837
	First order	0.16701/0.09621	0.03517/0.00455	0.86721
	Higuchi	−14.19939/6.18472	13.67722/1.49206	0.89251
	Korsmeyer-Peppas	0.94986/0.17899	0.78764/0.15055	0.74554
	Hixson	−0.7623/0.2354	33.8974/0.47476	0.99054
	Bakers-Lonsdale	−18,183.23/10,589.06	9974.14/2554.60	0.58753

Table 1. *Cont.*

Drug Release at Different pH-Levels	Models	Intercept/Standard Error	Slop/Standard Error	Regression Coefficient (R^2)
Drug release kinetics at pH 7.4	Zero order	0.8095/0.21262	0.2173/0.01005	098106
	First order	0.14378/0.07845	0.02573/0.00371	0.83965
	Higuchi	−11.10565/4.33	17.22369/1.04596	0.96431
	Korsmeyer-Peppas	1.4071/0.15032	0.56002/0.12644	0.67412
	Hixson	31.14956/0.84	−0.92297/0.04151	0.98014
	Bakers-Lonsdale	−80,632.11/21,021.06	32,004.02/4835.30	0.82628
Drug release kinetics at pH 8.4	Zero order	−2.2668/0.65	0.61835/0.03	0.9782
	First order	0.04723/0.08	0.04155/0.003	0.93277
	Higuchi	−12.57362/5.93	9.85459/1.43	0.82256
	Korsmeyer-Peppas	0.59247/0.19	0.91234/0.16	0.76697
	Hixson	34.79252/0.59	−0.56507/0.03	0.97366
	Bakers-Lonsdale	−15,727.59/5918.95	5546.23/1361.49	0.63407

4.9. Antimicrobial Activities

Chronic and burn wounds, and many surgical procedures are all susceptible to bacterial infection. It can lead to severe disease, muscle tissue death, septicemia, and even death. Antimicrobial agents that are effective in treating bacteria are desperately needed. The prepared hydrogels were evaluated against severe wound infection-causing microbes (*S. aureus*, *E. coli*, and *P. aeruginosa*) via the disc diffusion method. The prepared hydrogels were applied to the bacterial lawns in Petri dishes, and these plates were kept in incubation for 12 h. Antibacterial activities were determined via measuring zone inhibition, as shown in Figure 7. These hydrogels have different zone inhibitions against treated bacterial strains. The hydrogel BSG-1 performed minimum, and BSG-4 exhibited maximum antibacterial activity. The maximum antibacterial activities of BSG-4 may be due to the maximum GO amount that ruptures the bacterial membrane with sharp edges of GO [9,36]. On the other hand, the bacterial cell membrane is composed of lipopolysaccharides and phospholipids. The polymeric part of the hydrogel may interact with the bacterial membrane to surround and penetrate several available functional groups to hinder bacterial growth [37,38]. The covalently crosslinked hydrogels via TEOS have inter/intra-hydrogen bonding and weak interactions. BSG-4 exhibited maximum antimicrobial activities, which can be polymeric, and GO components of hydrogels that may communicate with DNA to stop bacterial replication. Hence, the prepared hydrogel could be a potential biomaterial with promising antibacterial activities, which is helpful to treat burns and chronic wounds.

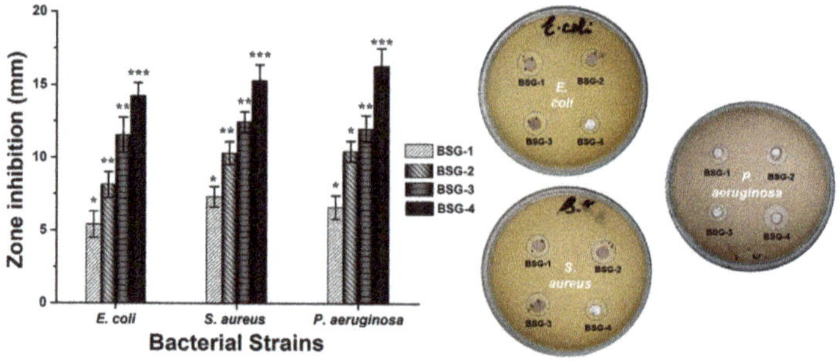

Figure 7. The antibacterial activities of composite hydrogels against different severe skin infections causing Gram-positive and Gram-negative pathogens. * $p < 0.05$, ** $p < 0.01$ and *** $p < 0.001$ and n = 3.

4.10. Anticancer Activities

Figure 8 presents the anticancer activity via cell viability after different time intervals (24, 48, and 72 h) and cell morphology of all samples of composite hydrogels and curcumin-loaded composite hydrogels against U87 cell lines. After 72 h of contact time with composite hydrogels and curcumin-BSG-4 composite hydrogels, nearly 79.56 to 87.58 percent of U87 cells were determined to be nonviable, respectively, as exhibited in Figure 8a. Curcumin combined with PVA/BC-*f*-GO was discovered to have improved anticancer properties. The BSG-4 composite hydrogel also had the highest cell nonviability among the composite hydrogels, which could be owing to the increased GO content [20,39]. The curcumin-loaded-BSG-4 had the strongest antitumor effect. This could be due to the interaction of the composite hydrogel with the cellular membrane, allowing curcumin to conduct anticancer functions. Cell clustering may also be found for BSG-1 and BSG-2, as shown by the red circle, even after 72 h, whereas some cell adherence can be seen for both samples as presented in Figure 8(b–e). However, the red arrows indicated cell detachment, which might occur as a result of cell death, as well as the cell repute. Curcumin was delivered via the drug-loaded BSG-3 composite hydrogel membrane, resulting in increased anticancer activity [36,40]. As a result of the synergetic impact of the release of curcumin-loaded-BSG-3 composite hydrogels after 72 h, the anticancer activities are at their peak. As a result, the curcumin-loaded-BSG-3 composite hydrogel could be used as a biomaterial for wound care and treatment in cancer patients.

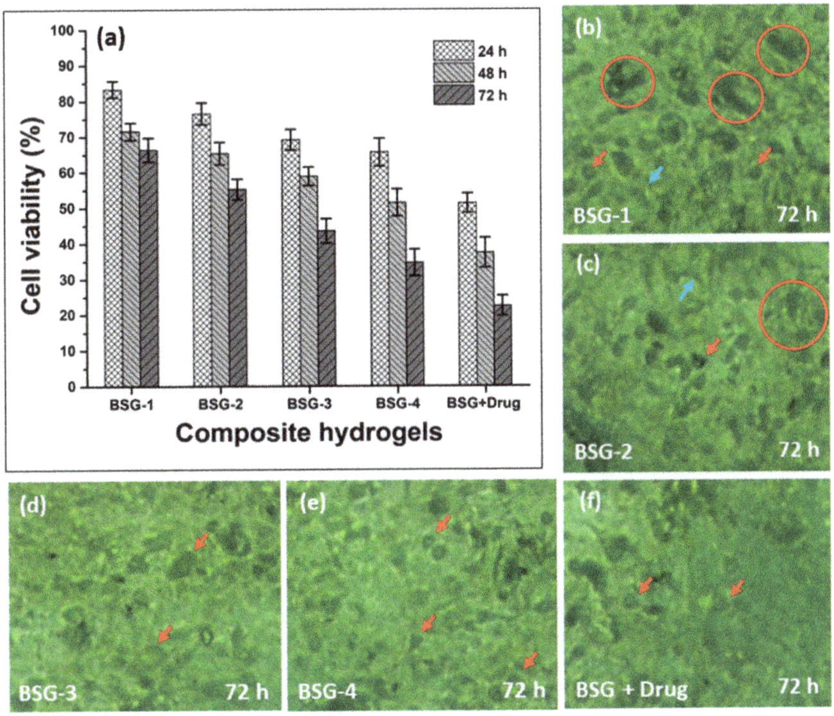

Figure 8. The anticancer activities of composite hydrogels against U87 cell line, (**a**) cell viability in terms of anticancer activities, and (**b–f**) cell morphology with different indications.

5. Conclusions

The pH-sensitive composite hydrogels were prepared by blending bacterial nanocellulose with PVA and GO and crosslinked with optimized TEOS. The controlled drug delivery was determined at various pH levels, and the drug release mechanism was investigated using different mathematical models. In hydrogels, FTIR reveals all of the functional groups of bacterial nanocellulose, GO, and TEOS and well-defined crosslinking and hydrogen bonding. GO-flakes were observed via SEM and increasing GO-flakes were observed due to the increasing GO amount. BSG-4 was more stable than other hydrogels. Due to complex polymerization formation, the biodegradation rate of hydrogels was slowed by increasing the GO content. In addition, increasing the GO amount increased the mechanical and hydrophilicity properties. Swelling in the buffer and non-buffer solutions demonstrates that the prepared hydrogels are pH-sensitive. Therefore, hydrogels are a good fit for controlled drug release due to their responsive behavior, and they could be used to release curcumin in a controlled manner. Furthermore, these newly composite hydrogels have exhibited substantial potential, but BSG-4 and curcumin-loaded-BSG-4 have potential antibacterial and anticancer activities. Hence, these composite hydrogels could be promising biomaterials for chronic wound healing applications.

Author Contributions: Conceptualization, M.U.A.K.; Data curation, M.U.A.K. and S.I.A.R.; Formal analysis, M.U.A.K. and W.S.A.-A.; Funding acquisition, M.U.A.K. and S.I.A.R.; Investigation, S.I.A.R. and S.M.A.; Methodology, M.U.A.K., S.I.A.R. and H.H.A.; Project administration, S.I.A.R., H.H.A. and S.M.A.; Resources, M.U.A.K., S.M.A. and S.I.A.R.; Software, M.U.A.K.; Supervision, M.U.A.K., S.I.A.R. and W.S.A.-A.; Validation, M.U.A.K. and S.I.A.R.; Visualization, M.U.A.K. and W.S.A.-A.; Writing—original draft, M.U.A.K.; Writing—review & editing, M.U.A.K. All authors have read and agreed to the published version of the manuscript.

Funding: This work is supported through the Annual Funding track by the Deanship of Scientific Research, Vice Presidency for Graduate Studies and Scientific Research, King Faisal University, Saudi Arabia (Project No. AN000593).

Institutional Review Board Statement: Not applicable.

Informed Consent Statement: Not applicable.

Data Availability Statement: The data is available in the article.

Acknowledgments: We highly acknowledge the countless efforts and expertized of Muhammad Umar Aslam Khan for the completion and publication of the project.

Conflicts of Interest: All authors declared no conflict of interest.

References

1. Yao, Z.; Niu, J.; Cheng, B. Prevalence of chronic skin wounds and their risk factors in an inpatient hospital setting in northern China. *Adv. Ski. Wound Care* **2020**, *33*, 1–10. [CrossRef] [PubMed]
2. Negut, I.; Dorcioman, G.; Grumezescu, V. Scaffolds for Wound Healing Applications. *Polymers* **2020**, *12*, 2010. [CrossRef] [PubMed]
3. Aslam Khan, M.U.; Abd Razak, S.I.; Al Arjan, W.S.; Nazir, S.; Sahaya Anand, T.J.; Mehboob, H.; Amin, R. Recent Advances in Biopolymeric Composite Materials for Tissue Engineering and Regenerative Medicines: A Review. *Molecules* **2021**, *26*, 619. [CrossRef] [PubMed]
4. Khan, M.U.A.; Iqbal, I.; Ansari, M.N.M.; Razak, S.I.A.; Raza, M.A.; Sajjad, A.; Jabeen, F.; Riduan Mohamad, M.; Jusoh, N. Development of Antibacterial, Degradable and pH-Responsive Chitosan/Guar Gum/Polyvinyl Alcohol Blended Hydrogels for Wound Dressing. *Molecules* **2021**, *26*, 5937. [CrossRef]
5. Khan, M.U.A.; Razaq, S.I.A.; Mehboob, H.; Rehman, S.; Al-Arjan, W.S.; Amin, R. Antibacterial and hemocompatible pH-responsive hydrogel for skin wound healing application: In vitro drug release. *Polymers* **2021**, *13*, 3703. [CrossRef]
6. Kim, H.S.; Sun, X.; Lee, J.-H.; Kim, H.-W.; Fu, X.; Leong, K.W. Advanced drug delivery systems and artificial skin grafts for skin wound healing. *Adv. Drug Deliv. Rev.* **2019**, *146*, 209–239. [CrossRef]
7. Shafiei, M.; Ansari, M.N.M.; Razak, S.I.A.; Khan, M.U.A. A Comprehensive Review on the Applications of Exosomes and Liposomes in Regenerative Medicine and Tissue Engineering. *Polymers* **2021**, *13*, 2529. [CrossRef]
8. Panduranga Rao, K. Recent developments of collagen-based materials for medical applications and drug delivery systems. *J. Biomater. Sci. Polym. Ed.* **1996**, *7*, 623–645. [CrossRef]
9. Aslam Khan, M.U.; Al-Arjan, W.S.; Binkadem, M.S.; Mehboob, H.; Haider, A.; Raza, M.A.; Abd Razak, S.I.; Hasan, A.; Amin, R. Development of Biopolymeric Hybrid Scaffold-Based on AAc/GO/nHAp/TiO_2 Nanocomposite for Bone Tissue Engineering: In-Vitro Analysis. *Nanomaterials* **2021**, *11*, 1319. [CrossRef]

10. Yang, P.; Zhu, F.; Zhang, Z.; Cheng, Y.; Wang, Z.; Li, Y. Stimuli-responsive polydopamine-based smart materials. *Chem. Soc. Rev.* **2021**, *50*, 8319–8343. [CrossRef]
11. Abruzzo, A.; Cappadone, C.; Sallustio, V.; Picone, G.; Rossi, M.; Nicoletta, F.P.; Luppi, B.; Bigucci, F.; Cerchiara, T. Development of Spanish Broom and Flax Dressings with Glycyrrhetinic Acid-Loaded Films for Wound Healing: Characterization and Evaluation of Biological Properties. *Pharmaceutics* **2021**, *13*, 1192. [CrossRef] [PubMed]
12. Aslam Khan, M.U.; Haider, A.; Abd Razak, S.I.; Abdul Kadir, M.R.; Haider, S.; Shah, S.A.; Hasan, A.; Khan, R.; Khan, S.U.D.; Shakir, I. Arabinoxylan/graphene-oxide/nHAp-NPs/PVA bionano composite scaffolds for fractured bone healing. *J. Tissue Eng. Regen. Med.* **2021**, *15*, 322–335. [CrossRef] [PubMed]
13. Huang, W.; Wang, Y.; Huang, Z.; Wang, X.; Chen, L.; Zhang, Y.; Zhang, L. On-demand dissolvable self-healing hydrogel based on carboxymethyl chitosan and cellulose nanocrystal for deep partial thickness burn wound healing. *ACS Appl. Mater. Interfaces* **2018**, *10*, 41076–41088. [CrossRef]
14. Yang, L.; Wang, C.; Li, L.; Zhu, F.; Ren, X.; Huang, Q.; Cheng, Y.; Li, Y. Bioinspired integration of naturally occurring molecules towards universal and smart antibacterial coatings. *Adv. Funct. Mater.* **2022**, *32*, 2108749. [CrossRef]
15. Jiji, S.; Udhayakumar, S.; Rose, C.; Muralidharan, C.; Kadirvelu, K. Thymol enriched bacterial cellulose hydrogel as effective material for third degree burn wound repair. *Int. J. Biol. Macromol.* **2019**, *122*, 452–460. [CrossRef] [PubMed]
16. Loh, E.Y.X.; Mohamad, N.; Fauzi, M.B.; Ng, M.H.; Ng, S.F.; Amin, M.C.I.M. Development of a bacterial cellulose-based hydrogel cell carrier containing keratinocytes and fibroblasts for full-thickness wound healing. *Sci. Rep.* **2018**, *8*, 1–12. [CrossRef] [PubMed]
17. Zhou, W.; Zhuang, W.; Ge, L.; Wang, Z.; Wu, J.; Niu, H.; Liu, D.; Zhu, C.; Chen, Y.; Ying, H. Surface functionalization of graphene oxide by amino acids for Thermomyces lanuginosus lipase adsorption. *J. Colloid Interface Sci.* **2019**, *546*, 211–220. [CrossRef]
18. Singh, D.P.; Herrera, C.E.; Singh, B.; Singh, S.; Singh, R.K.; Kumar, R. Graphene oxide: An efficient material and recent approach for biotechnological and biomedical applications. *Mater. Sci. Eng. C* **2018**, *86*, 173–197. [CrossRef]
19. Abba, M.; Ibrahim, Z.; Chong, C.S.; Zawawi, N.A.; Kadir, M.R.A.; Yusof, A.H.M.; Abd Razak, S.I. Transdermal delivery of crocin using bacterial nanocellulose membrane. *Fibers Polym.* **2019**, *20*, 2025–2031. [CrossRef]
20. Nazir, S.; Khan, M.U.A.; Al-Arjan, W.S.; Abd Razak, S.I.; Javed, A.; Kadir, M.R.A. Nanocomposite hydrogels for melanoma skin cancer care and treatment: In-vitro drug delivery, drug release kinetics and anti-cancer activities. *Arab. J. Chem.* **2021**, *14*, 103120. [CrossRef]
21. Khan, M.U.A.; Razak, S.I.A.; Ansari, M.N.M.; Zulkifli, R.M.; Ahmad Zawawi, N.; Arshad, M. Development of Biodegradable Bio-Based Composite for Bone Tissue Engineering: Synthesis, Characterization and In Vitro Biocompatible Evaluation. *Polymers* **2021**, *13*, 3611. [CrossRef] [PubMed]
22. Caló, E.; Khutoryanskiy, V.V. Biomedical applications of hydrogels: A review of patents and commercial products. *Eur. Polym. J.* **2015**, *65*, 252–267. [CrossRef]
23. Karahaliloglu, Z.; Kilicay, E.; Denkbas, E.B. Antibacterial chitosan/silk sericin 3D porous scaffolds as a wound dressing material. *Artif. Cells Nanomed. Biotechnol.* **2017**, *45*, 1172–1185. [CrossRef]
24. Huang, Y.; Hao, M.; Nian, X.; Qiao, H.; Zhang, X.; Zhang, X.; Song, G.; Guo, J.; Pang, X.; Zhang, H. Strontium and copper co-substituted hydroxyapatite-based coatings with improved antibacterial activity and cytocompatibility fabricated by electrodeposition. *Ceram. Int.* **2016**, *42*, 11876–11888. [CrossRef]
25. Tan, G.; Wang, Y.; Li, J.; Zhang, S. Synthesis and characterization of injectable photocrosslinking poly (ethylene glycol) diacrylate based hydrogels. *Polym. Bull.* **2008**, *61*, 91–98. [CrossRef]
26. Jeddi, M.K.; Mahkam, M. Magnetic nano carboxymethyl cellulose-alginate/chitosan hydrogel beads as biodegradable devices for controlled drug delivery. *Int. J. Biol. Macromol.* **2019**, *135*, 829–838. [CrossRef]
27. Khan, M.U.A.; Raza, M.A.; Razak, S.I.A.; Abdul Kadir, M.R.; Haider, A.; Shah, S.A.; Mohd Yusof, A.H.; Haider, S.; Shakir, I.; Aftab, S. Novel functional antimicrobial and biocompatible arabinoxylan/guar gum hydrogel for skin wound dressing applications. *J. Tissue Eng. Regen. Med.* **2020**, *14*, 1488–1501. [CrossRef]
28. Khan, M.U.A.; Yaqoob, Z.; Ansari, M.N.M.; Razak, S.I.A.; Raza, M.A.; Sajjad, A.; Haider, S.; Busra, F.M. Chitosan/Poly Vinyl Alcohol/Graphene Oxide Based pH-Responsive Composite Hydrogel Films: Drug Release, Anti-Microbial and Cell Viability Studies. *Polymers* **2021**, *13*, 3124. [CrossRef]
29. Khan, M.U.A.; Haider, S.; Raza, M.A.; Shah, S.A.; Abd Razak, S.I.; Kadir, M.R.A.; Subhan, F.; Haider, A. Smart and pH-sensitive rGO/Arabinoxylan/chitosan composite for wound dressing: In-vitro drug delivery, antibacterial activity, and biological activities. *Int. J. Biol. Macromol.* **2021**, *192*, 820–831. [CrossRef]
30. Hong, Y.; Song, H.; Gong, Y.; Mao, Z.; Gao, C.; Shen, J. Covalently crosslinked chitosan hydrogel: Properties of in vitro degradation and chondrocyte encapsulation. *Acta Biomater.* **2007**, *3*, 23–31. [CrossRef]
31. McBath, R.A.; Shipp, D.A. Swelling and degradation of hydrogels synthesized with degradable poly (β-amino ester) crosslinkers. *Polym. Chem.* **2010**, *1*, 860–865. [CrossRef]
32. Son, G.-H.; Lee, B.-J.; Cho, C.-W. Mechanisms of drug release from advanced drug formulations such as polymeric-based drug-delivery systems and lipid nanoparticles. *J. Pharm. Investig.* **2017**, *47*, 287–296. [CrossRef]
33. Mirzaei, B.E.; Ramazani SA, A.; Shafiee, M.; Danaei, M. Studies on glutaraldehyde crosslinked chitosan hydrogel properties for drug delivery systems. *Int. J. Polym. Mater. Polym. Biomater.* **2013**, *62*, 605–611. [CrossRef]
34. Bruschi, M.L. Mathematical models of drug release. In *Strategies to Modify the Drug Release from Pharmaceutical Systems*; Woodhead Publishing: Cambridge, UK, 2015; p. 63.

35. Khan, M.U.A.; Abd Razak, S.I.; Haider, S.; Mannan, H.A.; Hussain, J.; Hasan, A. Sodium alginate-f-GO composite hydrogels for tissue regeneration and antitumor applications. *Int. J. Biol. Macromol.* **2022**, *208*, 475–485. [CrossRef] [PubMed]
36. Zamri, M.F.M.A.; Bahru, R.; Amin, R.; Khan, M.U.A.; Abd Razak, S.I.; Hassan, S.A.; Kadir, M.R.A.; Nayan, N.H.M. Waste to health: A review of waste derived materials for tissue engineering. *J. Clean. Prod.* **2021**, *290*, 125792. [CrossRef]
37. Huang, J.P.; Mojib, N.; Goli, R.R.; Watkins, S.; Waites, K.B.; Ravindra, R.; Andersen, D.T.; Bej, A.K. Antimicrobial activity of PVP from an Antarctic bacterium, Janthinobacterium sp. Ant5-2, on multi-drug and methicillin resistant Staphylococcus aureus. *Nat. Prod. Bioprospect.* **2012**, *2*, 104–110. [CrossRef]
38. Aslam Khan, M.U.; Mehboob, H.; Abd Razak, S.I.; Yahya, M.Y.; Mohd Yusof, A.H.; Ramlee, M.H.; Sahaya Anand, T.J.; Hassan, R.; Aziz, A.; Amin, R. Development of polymeric nanocomposite (xyloglucan-co-methacrylic acid/hydroxyapatite/sio2) scaffold for bone tissue engineering applications—in-vitro antibacterial, cytotoxicity and cell culture evaluation. *Polymers* **2020**, *12*, 1238. [CrossRef]
39. Iqbal, M.S.; Khan, M.U.; Akbar, J.; Shad, M.A.; Masih, R.; Chaudhary, M.T. Isoconversional thermal and pyrolytic GC–MS analysis of street samples of hashish. *J. Anal. Appl. Pyrolysis* **2016**, *122*, 175–182. [CrossRef]
40. Mirzaie, Z.; Reisi-Vanani, A.; Barati, M. Polyvinyl alcohol-sodium alginate blend, composited with 3D-graphene oxide as a controlled release system for curcumin. *J. Drug Deliv. Sci. Technol.* **2019**, *50*, 380–387. [CrossRef]

polymers

MDPI

Article

Tailoring of Geranium Oil-Based Nanoemulsion Loaded with Pravastatin as a Nanoplatform for Wound Healing

Waleed Y. Rizg [1,2], Khaled M. Hosny [1,2,*], Bayan A. Eshmawi [1], Abdulmohsin J. Alamoudi [3], Awaji Y. Safhi [4], Samar S. A. Murshid [5], Fahad Y. Sabei [4] and Adel Al Fatease [6]

1 Department of Pharmaceutics, Faculty of Pharmacy, King Abdulaziz University, Jeddah 21589, Saudi Arabia; wrizq@kau.edu.sa (W.Y.R.); beshmawi@kau.edu.sa (B.A.E.)
2 Center of Excellence for Drug Research and Pharmaceutical Industries, King Abdulaziz University, Jeddah 21589, Saudi Arabia
3 Department of Pharmacology and Toxicology, Faculty of Pharmacy, King Abdulaziz University, Jeddah 21589, Saudi Arabia; ajmalamoudi@kau.edu.sa
4 Department of Pharmaceutics, Faculty of Pharmacy, Jazan University, Jazan 82817, Saudi Arabia; asafhi@jazanu.edu.sa (A.Y.S.); fsabei@jazanu.edu.sa (F.Y.S.)
5 Department of Natural Products and Alternative Medicine, Faculty of Pharmacy, King Abdulaziz University, Jeddah 21589, Saudi Arabia; samurshid@kau.edu.sa
6 Department of Pharmaceutics, College of Pharmacy, King Khalid University, Abha 62529, Saudi Arabia; afatease@kku.edu.sa
* Correspondence: kmhomar@kau.edu.sa; Tel.: +966-561-682-377

Abstract: The healing of a burn wound is a complex process that includes the re-formation of injured tissues and the control of infection to minimize discomfort, scarring, and inconvenience. The current investigation's objective was to develop and optimize a geranium oil–based self-nanoemulsifying drug delivery system loaded with pravastatin (Gr-PV-NE). The geranium oil and pravastatin were both used due to their valuable anti-inflammatory and antibacterial activities. The Box–Behnken design was chosen for the development and optimization of the Gr-PV-NE. The fabricated formulations were assessed for their droplet size and their effects on the burn wound diameter in experimental animals. Further, the optimal formulation was examined for its wound healing properties, antimicrobial activities, and ex-vivo permeation characteristics. The produced nanoemulsion had a droplet size of 61 to 138 nm. The experimental design affirmed the important synergistic influence of the geranium oil and pravastatin for the healing of burn wounds; it showed enhanced wound closure and improved anti-inflammatory and antimicrobial actions. The optimal formulation led to a 4-fold decrease in the mean burn wound diameter, a 3.81-fold lowering of the interleukin-6 serum level compared to negative control, a 4-fold increase in the inhibition zone against *Staphylococcus aureus* compared to NE with Gr oil, and a 7.6-fold increase in the skin permeation of pravastatin compared to PV dispersion. Therefore, the devised nanoemulsions containing the combination of geranium oil and pravastatin could be considered a fruitful paradigm for the treatment of severe burn wounds.

Keywords: nanotechnology; burn wound; ex-vivo permeation; essential oil; statins; Box–Behnken design

Citation: Rizg, W.Y.; Hosny, K.M.; Eshmawi, B.A.; Alamoudi, A.J.; Safhi, A.Y.; Murshid, S.S.A.; Sabei, F.Y.; Al Fatease, A. Tailoring of Geranium Oil-Based Nanoemulsion Loaded with Pravastatin as a Nanoplatform for Wound Healing. *Polymers* **2022**, *14*, 1912. https://10.3390/polym14091912

Academic Editors: Ariana Hudita and José Antonio Gabaldón Hernández

Received: 26 March 2022
Accepted: 29 April 2022
Published: 7 May 2022

1. Introduction

Skin, the largest organ of the human body, is an important protective organ that can safeguard the body against various dangers such as pathogens, toxins, ultraviolet radiation, and mechanical damage [1]. Skin has three main layers, namely, the epidermis, dermis, and hypodermis, each of which has a unique composition and function that contributes to the overall protection against external environmental factors [2]. Critical burns and wounds are amongst the most prevalent life-threatening dangers. The healing process of open skin includes the restoration of the skin's normal protective layer following trauma caused by accident or an intentional surgical procedure. This healing process has several integrated and coordinated successive phases, namely, hemostasis, inflammation, proliferation, revascularization, and remodeling [3,4]. A person's age, sex, level of stress, and pre-existing medical conditions all affect the time needed for complete wound healing [5]. Wounds associated with diabetes are a significant type that result from the stress diabetes causes in the body. The wound healing process is impaired in diabetic patients due to neuropathy, hypoxia, decreased immunity, fibroblastic dysfunction, and impaired angiogenesis [6,7]. Treatment strategies to accelerate the wound healing process were assessed, and many of them showed positive results [8–10].

Statin drugs are among the important cholesterol-lowering agents used currently [11]. Besides their powerful cholesterol-lowering activity, they have desirable effects on wound healing in many animal models [12,13]. Treatment of different wound animal models with statins led to an increase in vascular endothelial growth factor (VEGF), an important inducer of angiogenesis and promoter of wound healing [14,15]. Statins also improved epithelialization and enhanced the mechanical strength of the skin, both of which improved the healing of wounds [16]. Pravastatin (PV) is one of the statins; it was patented in 1980 and approved for clinical use in 1989. Like other statins, PV exerts its hypolipidemic action by inhibiting the enzyme HMG-CoA reductase, which is expressed in the liver and plays a pivotal role in cholesterol output [17]. In a recent clinical trial involving patients who received radiation for head and neck cancer, PV was administered in a dosage of 40 mg/day for 12 months. The results showed that PV was an efficient antifibrotic agent, and this finding supports the theory that PV could reverse radio-induced fibrosis and enhance the wound-healing process [18].

Geranium (Gr) oil is an essential oil extracted from the plant *Pelargonium graveolens* that was cultivated originally in South Africa and then distributed in Asia and the Middle East [19]. Gr oil is known traditionally for its calming effects on emotional distresses such as frustration, anger, and anxiety and its ability to lower high blood pressure [20]. Gr oil possesses immune-boosting properties and immune-modulating properties against natural killer cells [21]. Recent studies showed anticancer activity of this essential oil that resulted from two of its major components, citronellol and trans-geraniol [22,23]. The aerial parts of *P. graveolens* have been used for wound healing because they possess antimicrobial activity against *Staphylococcus aureus, Candida glabrata, Bacillus subtilis, Enterococcus faecalis, Candida krusei, Mycobacterium tuberculosis,* and *Mycobacterium intracellulare* [24]. Rose geranium oil has its own market in the cosmetic, perfume, and aromatherapy industries. Its anti-inflammatory and palliative effects on the skin make this essential oil a good candidate for many topical formulations [25,26].

Self-nanoemulsifying drug delivery systems (SNEDDS) are anhydrous mixtures of a low-solubility drug, an oil, a surfactant, and a cosurfactant with a droplet size of less than 100 nm [27]. These nanoemulsions have been used extensively in recent times in dermal applications owing to their nanosized droplets, which enhance the solubility and transdermal permeation of the incorporated drug [28–30]. Different low-solubility drugs from different drug classes, such as antiviral drugs, immunomodulators, nonsteroidal anti-inflammatory drugs, and lipid-lowering agents, have been incorporated in nanoemulsion formulations [31–33]. Besides the ease of making nanoemulsion formulations, they readily gain acceptance from authorities because they are considered to be drugs that are generally recognized as safe during manufacturing [34].

A design of experiment (DOE) approach was used in the formulation of the varying dosage forms in this study to choose the optimal formulation and test its biological activity. The DOE has been used extensively in nanotechnology formulations to help researchers study the interactions of drug excipients, solve problems during formulation, study the process parameters and how they affect each other, and reduce the number of experiments to reach the optimal formulation [35].

To our knowledge, no studies have been done to investigate the combined effect of statins and Gr oil in treating burn wounds. Therefore, the objective of this study was to introduce a transdermal drug delivery system containing a Gr oil–based nanoemulsion loaded with PV (Gr-PV-NE). The intended choice of a transdermal drug delivery system was made based on its localized action and the droplet size of the nanoemulsion, which could promote skin membrane permeation and avoid presystemic metabolism and efflux mechanisms [36,37].

2. Materials and Methods

2.1. Materials

PV was acquired as a generous gift from the Saudi Arabian Japanese (SAJA) Pharmaceutical Company Limited (Jeddah, Saudi Arabia). Gr oil was purchased from the Beutysway Commercial Foundation (Jeddah, Saudi Arabia). Tween 80 and Span 80 were purchased from Sigma-Aldrich Co. (St. Louis, MO, USA). Transcutol was a gift from Gattefosse (Saint-Priest, France). High-performance liquid chromatography–grade methanol and acetonitrile were obtained from Merck (Darmstadt, Germany). All other reagents and chemicals were of analytical grade.

2.2. Methods

2.2.1. Experimental Design and Optimization of Self-Nanoemulsion Formulations

With reference to this study, the Box–Behnken design (BBD) was pursued to scrutinize the influence of independent variables on dependent ones. Twenty-three formulations were made with the chosen design using Design-Expert Version 13 Software (Stat-Ease, Inc., Minneapolis, MN, USA). The selected statistical design produced various relationships between the independent variables, and they are summarized in Table 1. The three explored factors were the amount of Gr oil in milligrams (A), amount of PV in milligrams (B), and hydrophilic-lipophilic balance (HLB) of the surfactant mixture (Tween80/Span80) (C) in the prepared nanoemulsion. The estimated responses were the globule size of the prepared NEs (Y_1) and the mean burn wound diameter (Y_2). Preliminary studies were followed to select the factors' levels.

Table 1. Independent variables and their levels, along with dependent variables and their constraints, in the BBD of the nanoemulsion formulations.

Independent Variables	Levels		
	−1	**0**	**1**
A = Geranium oil amount (mg)	100	200	300
B = Pravastatin amount (mg)	10	20	40
C = HLB of surfactant mixture	11	12	13
Dependent Variables		Goal	
Y_1 = Globule size (nm)		Minimize	
Y_2 = Mean burned wound diameter (mm)		Minimize	

2.2.2. Gr-PV-NEs Preparation

The production procedure had two steps. The first step was the fabrication of the plain SNEDDS, in which a Gr oil concentration of 10%, 20%, or 30% (according to the design) was blended with 60% of the surfactant mixture with an HLB of 11, 12, or 13 (according to the design) and then brought to 100% by the Transcutol cosurfactant. In the second step, PV was mixed with the plain NEs with the aid of sonication in concentrations of 10, 20, or 40 mg/g of the prepared self-nanoemulsion according to the design, as shown in Table 2.

Table 2. BBD and responses of Gr-PV-NEs.

	A	B	C	Y_1	Y_2	
Run	Geranium Oil Amount (mg)	Pravastatin Amount (mg)	HLB	Droplet Size (nm)	Mean Burned Wound Diameter (mm)	Polydispersity Index
1	200	10	13	110 ± 2.0	7.5 ± 0.30	0.09 ± 0.02
2	300	20	11	128 ± 4.5	4.5 ± 0.18	0.11 ± 0.03
3	100	20	11	99 ± 7.0	8.0 ± 0.90	0.15 ± 0.03
4	100	10	12	62 ± 1.5	9.0 ± 1.10	0.21 ± 0.04
5	300	20	13	129 ± 3.2	4.5 ± 0.62	0.13 ± 0.02
6	100	40	12	61 ± 1.5	5.0 ± 0.51	0.32 ± 0.04
7	200	20	12	81 ± 5.0	6.5 ± 0.25	0.28 ± 0.05
8	200	20	12	80 ± 4.5	6.0 ± 0.18	0.19 ± 0.04
9	300	40	12	89 ± 3.5	3.0 ± 0.09	0.22 ± 0.05
10	200	10	11	109 ± 4.0	7.5 ± 0.64	0.38 ± 0.06
11	200	40	13	108 ± 8.0	5.0 ± 0.33	0.40 ± 0.05
12	200	40	11	110 ± 2.5	4.5 ± 0.21	0.10 ± 0.02
13	200	20	12	79 ± 1.9	6.0 ± 0.30	0.35 ± 0.02
14	100	20	13	99 ± 4.0	7.5 ± 0.42	0.30 ± 0.04
15	300	10	12	87 ± 3.5	5.0 ± 0.17	0.29 ± 0.05
16	100	10	11	95 ± 2.9	10 ± 1.20	0.26 ± 0.05
17	300	40	13	135 ± 5.5	3.5 ± 0.09	0.18 ± 0.04
18	100	10	13	101 ± 2.1	9.5 ± 0.50	0.32 ± 0.03
19	100	40	11	97 ± 6.1	5.5 ± 0.11	0.27 ± 0.05
20	300	20	12	90 ± 1.5	4.0 ± 0.30	0.19 ± 0.06
21	300	40	11	138 ± 3.1	2.5 ± 0.30	0.11 ± 0.03
22	100	20	12	65 ± 2.0	7.0 ± 1.90	0.25 ± 0.05
23	200	10	12	77 ± 1.8	7.0 ± 0.21	0.39 ± 0.06

2.2.3. Determination of Globule Size

The droplet size and polydispersity index (PDI) of each of the fabricated Gr-PV-NEs was examined by diluting 100 µL of each formulation with 900 µL of double distilled water in a volumetric flask. Next, the diluted samples were vigorously mixed, and 100 µL of the dispersed sample was withdrawn to determine its droplet size; this was done with a Zeta track particle size analyzer (Microtrac, Inc., Montgomeryville, PA, USA) [38]. Assessment of the sample was carried out in triplicate, and the results were presented as the mean ± standard deviation (SD). PDI is a good tool for evaluating foemulation homogeneity.

2.2.4. Assessment of Wound Healing

Animal Handling and Care

The guidelines of the Animal Ethics Committee, Beni Suef Center for Laboratory Animals, Beni Suef, Egypt, were adopted in animal handling and care. Investigators abided by the guidelines articulated in the Declaration of Helsinki and its Guiding Principles in the Care and Use of Animals (NIH Publication No. 85-23, 1985 revision) and adopted the ethical approval of the protocol before experimentation (Approval No. 26/4-21).

The experimental rats were obtained and kept in laboratory cages with free access to food and water. Animals' suffering was minimized according to the guidelines. The animals were kept for a minimum of 14 days prior to the investigation under standard conditions of temperature (25 ± 1 °C) and relative humidity (55 ± 5%) with a 12 h light and 12 h dark cycle.

Sixty-nine rats were used in the study; they were divided into 23 groups, with 3 rats per group. Each group was treated with one of the formulations developed by the experimental design. An intraperitoneal dose of thiopental of 15 mg/kg was administered simultaneously with an intramuscular dose of 25 mg/kg of ketamine. The hair on the rat's back was shaved, and the skin was sterilized with an alcohol swab. Following that, burn wounds were induced on the back skin using 1.5 cm skin biopsy heated bunches. Then, the investigated formulation for each group was applied once daily for a period of 14 days [39].

Measurement of Burn Wound Diameter

After the application of each formulation once a day for 14 days, a caliper was used to measure the average diameter of the induced wound lesions in each group to evaluate the second response for each formulation. Measurements were done in triplicate, and the findings were introduced as the mean ± SD.

2.3. *Optimization of the Gr-PV-NEs*

The desirability function was the basis of the Design-Expert software for obtaining the optimal NE formulation. The main objective of the optimization process was to find a formulation with the smallest droplet size and mean burn wound diameter. The software chose the solution with the desirability value closest to 1. To guarantee the model's validity and adequacy, the elected formulation was fabricated, depicted, and eventually compared with the response values expected by the software.

2.4. *Characterization of the Optimized Formulation*

2.4.1. Determination of Entrapemnt Efficiency

Percentage drug entrapment efficiency (EE%) was determined for the optimum Gr-PV-NE formulation (containing 40 mg PV) using an indirect method. Sample was centrifuged at 15,000 rpm for 30 min using cooling centrifuge (SIGMA 3–30K, Steinheim Germany). After centrifuge, 1 mL of supernatant transparent layer was diluted with 10 mL distilled water. The PV amount in the sample was determined by a reported high-performance liquid chromatography (HPLC) method. In short, the withdrawn samples were diluted with the mobile phase, which composed of 10 mM ammonium acetate, methanol, and triethylamine in a ratio of 40:60:0.17 *v/v/v*. Ten microliters of the prepared samples were injected with a flow rate of 1.0 mL min^{-1}. The PV detection wavelength and elution time were set to be 239 nm and 2.15 min, respectively. Results were taken in triplicate and the average was taken into consideration and EE% was calculated using the following equation.

$$EE\% = [TD - FD]/TD \times 100 \quad (1)$$

where TD is the total added drug amount, FD is the free unentrapped drug

2.4.2. Zeta Potential Determination

Zeta potentiap (ZP) of optimum Gr-PV-NEs (containing 40 mg PV) was examined via diluting 100 µL of the formulation with 900 µL of double distilled water in a volumetric flask. The diluted samples were thoroughly mixed, and 100 µL of the mixed samples were employed to determine its ZP using a Zeta track analyzer (Microtrac, Inc., Montgomeryville, PA, USA) [38]. Assessment of the sample was carried out in triplicates, and the results were presented as the mean ± standard deviation (SD).

2.4.3. Measurement of Burn Wound Diameter and Interlukin-6 Level

As in previously described evaluations of different Gr-PV-NE formulations, the test in this study was repeated on 15 rats that were divided into 5 groups. The first group (Group A) was treated with the optimum formulation proposed by the design and followed to determine the outcomes of the first stage of the animal test. The second group (Group B) was treated with the optimum formulation that contained no PV. The third group (Group C) was treated with the optimum formulation prepared utilizing oleic acid in lieu of Gr oil. The fourth group (Group D) was treated with Gr-PV mixture. The fifth group (Group E) was treated with normal saline and served as a negative control. The test continued for 14 days. The formulations were applied once a day and the required parameters were evaluated. All tested formulations contained PV amount equal to 40 mg/g of the tested formulation.

For the assessment of interleukin-6 (IL-6), a previously described quantitative sandwich enzyme-linked immunoassay technique (R&D Systems, Inc., Minneapolis, MN, USA) was employed. A microplate was covered with a particular monoclonal antibody that was specified for rat IL-6. Consequently, the animals' IL-6 that was present in the tested sample was engaged by the antibody. Following the elimination of any unbound interleukin, an enzyme-linked polyclonal antibody particular for rat IL-6 was appended, and the original blue color was transformed into a yellow one. Finally, the IL-6 level in the tested samples was evaluated by measuring the intensity of the produced color [36].

2.4.4. Ex-Vivo Permeation Study

The test was performed according to a previously described procedure [40]. A freshly excised section of abdominal rat skin was obtained from a male Wistar rat and utilized to determine the permeation profile of PV from various formulations (i.e., Gr-PV-NE optimum formulation, PV-NE prepared with oleic acid instead of Gr oil, Gr-NE prepared without PV, physical mixture of PV and Gr oil, PV aqueous dispersion, and plain NE prepared without PV or Gr oil). All formulations except for the plain nanoemulsion contained PV amount equal to 40 mg/g of the tested formulation. The institutional animal ethics committee approved the experimental protocol. In brief, the abdominal skin hair was guardedly clipped and removed without causing any skin damage. Then, rats were slaughtered, and abdominal skin was separated from the subsidiary connective tissues and cleaned with a Ringer solution before experimentation. Automated Franz diffusion cells (MicroettePlus; Hanson Research, Chatsworth, CA, USA) having a permeation area of 1.76 cm^2 were employed for determining the ex-vivo skin permeation of the PV from the previously stated formulations. The skin (2.5 $\times$ 2.5 cm^2) was inserted between the donor and receptor chambers of the cells with the stratum corneum layer oriented toward the donor compartment. The receptor milieu was phosphate buffer saline (PBS, pH 7.4) adjusted at 37 $\pm$ 0.5 °C in a sufficient volume to meet the sink condition for ex-vivo permeation studies. One milliliter of the various tested formulations was applied to the stratum corneum layer of the skin through the donor chamber and covered with Parafilm to diminish evaporation. At time intervals of 1, 2, 4, 6, 8, 10, 12, and 24 h, 0.5 mL aliquots were extracted from the receptor compartment. Immediately, a fresh receptor milieu previously heated to 37 $\pm$ 0.5 °C replaced the withdrawn samples. The PV amount in the gathered samples was determined by a reported high-performance liquid chromatography (HPLC) method previously described in Section 2.4.1.

2.4.5. Antibacterial Activity Evaluation

The antimicrobial action of the previously stated formulations (Section 2.4.2) was examined applying a disc diffusion method [41]. The well-known Gram-positive bacterium *S. aureus*, ordinarily present in infected burns, was utilized as a test bacterium. Following the procedure endorsed by the Clinical and Laboratory Standards Institute, an *S. aureus* suspension was made to the 0.5 McFarland turbidity standard and mounted on Mueller–Hinton agar plates. Filter-paper discs (with a diameter of 10 mm) were immersed in the tested samples and then put in the center of the agar plate and incubated for 24 h at 37 °C. When inhibitory concentrations were reached, a clear so-called inhibition zone containing no colonies could be seen around the discs. Lastly, the inhibition zones' diameters were measured for each examined formulation.

2.4.6. Statistical Analysis

Compiled IL-6 serum level data were tested using the one-way ANOVA followed by the post-hoc Tukey honest significant difference (HSD) test for multiple comparisons, and the level of significance was set at a p-value of less than 0.05 using SPSS software (version 22, Chicago, IL, USA). The obtained data were tested for normality using the Kolmogorov–Smirnov (K-S) test.

3. Results and Discussion

3.1. *Assessment of Gr-PV-NE Droplet Size*

Regarding the topical administration of active agents, the physicochemical specifications of nanoemulsions are fundamental factors that must be determined in the fabrication process [42].

The droplet size is an essential key parameter and can identify an emulsion as being a microemulsion or a nanoemulsion. The present investigation revealed that the developed formulations had a droplet size of between 61 ± 1.5 and 138 ± 3.1 nm (see Table 2), with an acceptable polydispersity index ranging from 0.09 to 0.40, confirming the acceptable homogeneity and favorable size distribution of the developed formulations.

A quadratic model of polynomial analysis was employed to scrutinize the collected data on droplet size. The chosen design showed the employed model's competence in exploring the influence of the amount of Gr oil (A), amount of PV (B), and HLB of the surfactant mixture (C) on the developed emulsions' droplet size. The advocated mathematical model had an adjusted R^2 value of 0.9729 and predicted R^2 value of 0.9445, which were closely related, as shown in Table 3. The ANOVA analysis of the data provided the following equation.

$$\text{Globule size} = +78.26 + 14.87A + 1.29B + 0.1909C + 3.04AB - 0.1560AC - 1.51BC + 0.5087A^2 - 2.68B^2 + 35.64C^2 \quad (2)$$

As perceived, the amount of Gr oil (factor A) had a significantly agonistic action on the droplet size at a *p*-value of less than 0.0001; thus, any increase in the amount of Gr oil would eventually increase the droplet size. Increasing the amount of Gr oil might have provided more space in which the PV could be housed within the nanoemulsion, giving droplets a larger diameter. Further, the increase in the amount of Gr oil would be accompanied with a corresponding decrease in the amount of cosurfactant used and its capacity to downsize a nanoemulsion's droplets, leading to the formation of larger droplets, as similar findings had previously reported [36]. On the other hand, the amount of PV and HLB value of the surfactant mixture exerted an insignificant effect on the droplet size.

Based on the above equation and Figure 1, which show the effect of the studied factors on the droplet size, an interesting result was detected: There was a significant impact exerted by the quadratic term of factor C (i.e., C^2) on the droplet size of the developed formulations. As shown in Figure 1A, such an observation implies that the peripheral levels of factor C had a positive effect on the droplet size, whereas the middle level exerted a negative effect, yielding droplets with a smaller diameter. Such phenomena could be due to the fact that each lipid material must have a particular HLB value to lower the nanoemulsion droplet size, as reported by other researchers [43,44]. Therefore, it might be deduced that the best HLB value for the Gr oil used in the formulations should be the middle value among those used in the proposed design.

Table 3. Regression analysis results for Y_1 and Y_2 responses.

Dependent Variables	R^2	Adjusted R^2	Predicted R^2	F-Value	*p*-Value	Adequate Precision
Y_1	0.9840	0.9729	0.9445	88.79	0.0001	30.3760
Y_2	0.9892	0.9817	0.9642	132.17	0.0001	38.5081

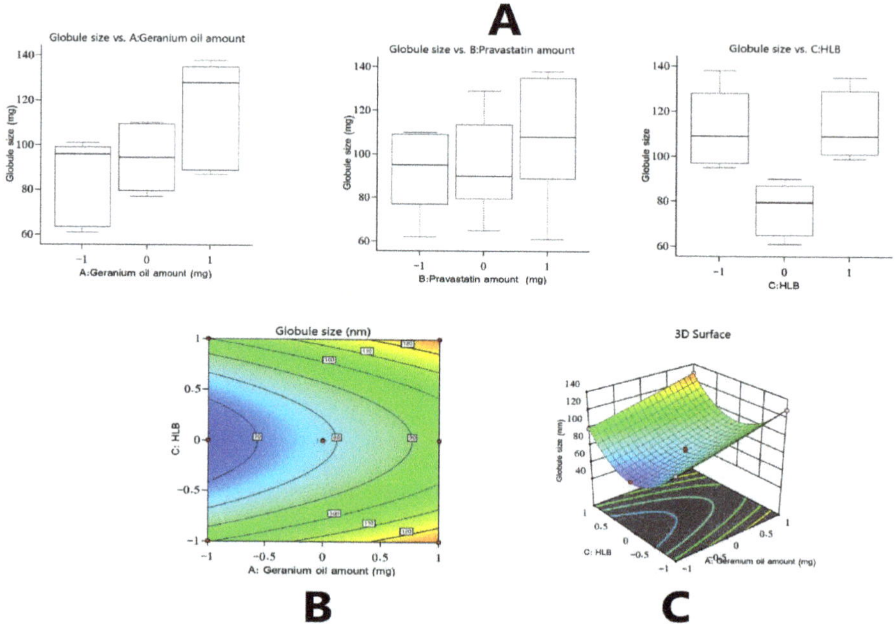

Figure 1. Main effect diagram (**A**), contour plot (**B**), and three-dimensional (3D) surface plot (**C**) showing the effects of different independent variables on the droplet size (Y_1) of different Gr-PV-NE formulations.

3.2. Assessment of Wound Healing

The healing of open wounds is a complicated dynamic chain of events that encompasses some sequential and overlapping processes, including hemostasis, inflammation, epithelialization, cell proliferation, revascularization, and collagen development [45]. In the present investigation, the healing impact of the fabricated formulations on burn wounds was determined by measuring the wound diameter. The diameters fluctuated between 2.5 ± 0.30 and 10 ± 1.2 mm (see Table 2) and followed a quadratic model of polynomial analysis. The chosen experimental design employed the model's adequacy to observe the impact of independent variables on the diameter of burn wounds (Y_2). The statistical model had an adjusted R^2 value of 0.9817, which was in line with an expected R^2 value of 0.9642, as shown in Table 3. The ANOVA analysis of the collected data yielded the following equation.

$$\text{Mean burn wound diameter} = +6.03 - 1.64A - 1.51B + 0.0377C + 0.5102AB + 0.1786AC + 0.1548BC - 0.3194A^2 - 0.2792B^2 + 0.41108C^2 \quad (3)$$

It was noticed that there was an inverse relationship between the amounts of Gr oil and PV and response Y_2 values at a p-value of less than 0.0001. The potential of the PV to decrease the burn wound size might be related to its ability, as a member of the statin class of drugs, to promote the output of the VEGF at the injury site; VEGF is a key element for developing new blood vessels [46]. Further, PV is also thought to inhibit mevalonate and farnesyl pyrophosphate production, leading to enhanced epithelialization and renovation of tissues of wounded skin [47]. Additionally, the wound healing action of Gr oil (factor A) might be due to its strong antibacterial activity against the Gram-negative bacterial strains that are mostly responsible for wound infections and that are resistant to treatments, and similar findings were reported in the literature [45]. The influence of the studied factors on the mean burn wound diameter is shown in Figure 2.

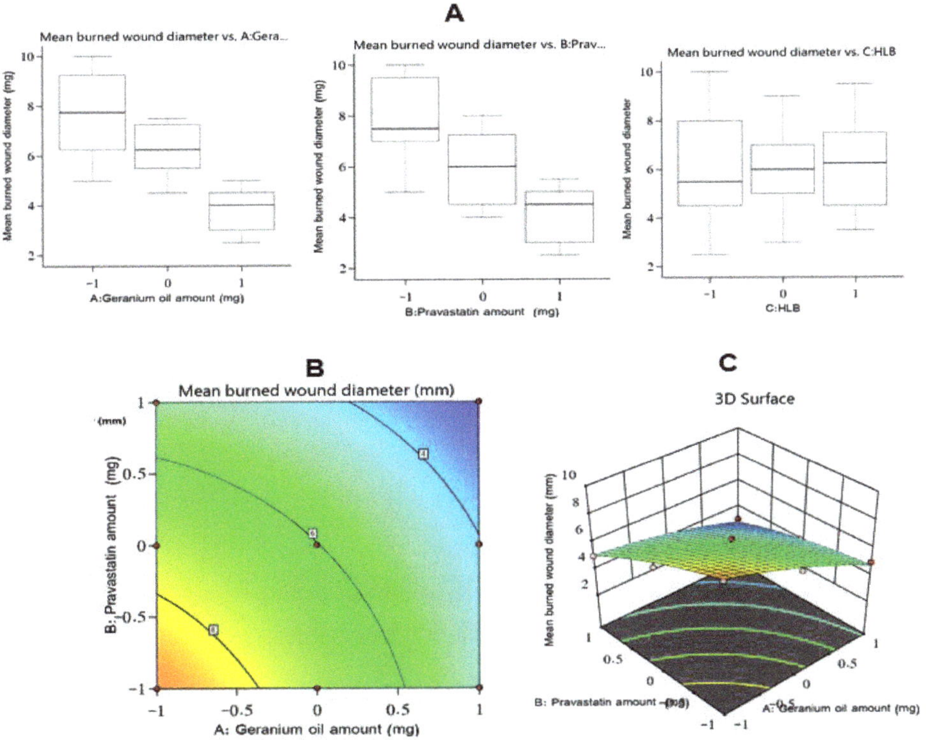

Figure 2. Main effect diagram (**A**), contour plot (**B**), and 3D surface plot (**C**) showing the effects of different independent variables on the mean burn wound diameter (Y_2) obtained after the application of different Gr-PV-NE formulations.

3.3. Optimization and Evaluation of Nanoemulsion Formulations

Following the completion of the tests described, a nanoemulsion formulation with the most appropriate specifications (i.e., the optimum formulation) was defined. Varying combinations of independent variables were suggested by the experimental design. The optimum formulation had 275 mg of Gr oil, 40 mg of PV, and a surfactant mixture with an HLB of 12 with a desirability value of 0.784. The fabricated optimal Gr-PV-NE had a droplet size of 95 ± 2.4 nm and a mean burn wound diameter of 3 ± 0.3 mm. Such results were in close accordance with the predicted values of the same responses, which were 90 nm for the droplet size and 3.18 mm for the mean burn wound diameter. Figure 3 clarifies the desirability ramp and bar chart for different levels of the studied factors and predicted dependent variables of the optimal formulation. Optimum formulation acquired an EE% of 91.3 ± 2.6% and ZP value of −17.3 ± 1.2 mV indcating the acceptable drug loading and stability of the developed optimal formulation.

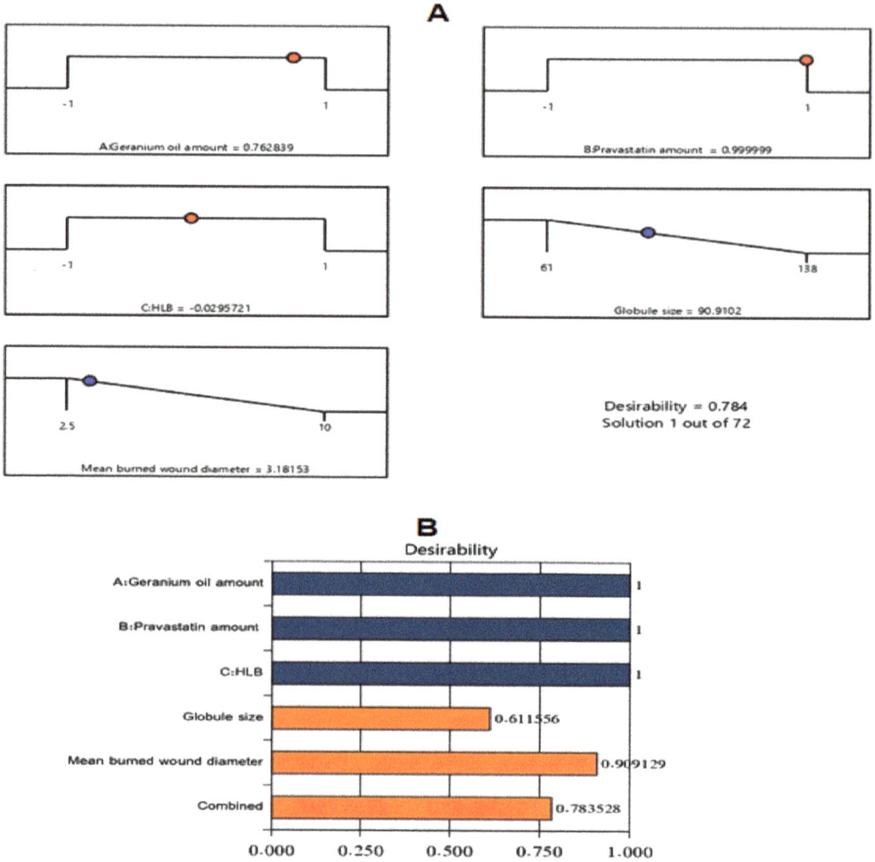

Figure 3. Desirability ramp and bar chart for optimization. (**A**) The desirability ramp shows the levels of independent variables and predicted values for the responses of the optimum formulation. (**B**) The bar chart shows the desirability values for the combined responses.

3.3.1. Wound Healing Action Assessment

Burn Wound Diameter Evaluation

As can be observed in Figure 4, the optimum Gr-PV-NE (formulation A) had the lowest mean burn wound diameter of 3 ± 0.5 mm, while the group treated with normal saline (formulation E) had the largest wound diameter, 12 ± 1.5 mm, compared with the other tested formulations. It was also noticed that the nanoemulsion containing no PV (formulation B) had a wound diameter (i.e., 6 ± 0.75 mm) greater than that of formulation A but very close to that of formulation C (i.e., 5.5 ± 0.5 mm), which was a nanoemulsion containing oleic acid instead of Gr oil. Such outcomes affirm the synergistic wound healing activity of PV and Gr oil. On the other hand, the Gr-PV mixture (formulation D) had a mean burn wound diameter of 8 ± 1 mm, which was greater than that of the optimum nanoemulsion formulation, indicating the predominant wound healing activity of the nanosized formulation compared with the mixture of Gr oil and PV. Figure 5 also illustrated the wound healing results of rat skin after 14 days of treatment The obtained rsults were found to be significant at asignificance level of 0.05 (p-value < 0.05)

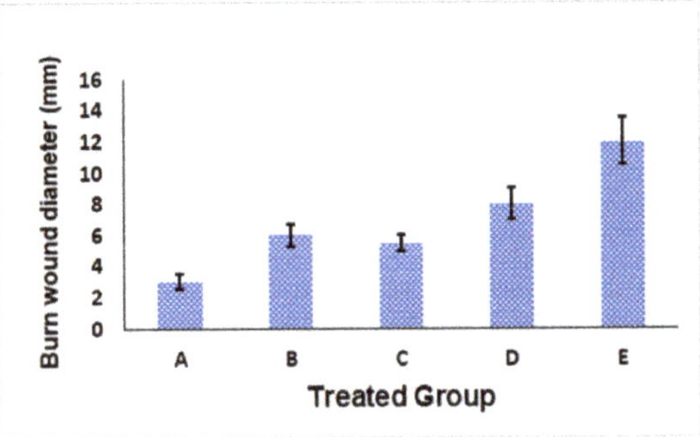

Figure 4. Mean burn wound diameter for different formulations: optimum Gr-PV-NE (**A**), Gr-NE (**B**), PV-NE (**C**), PV-Gr mixture (**D**), and normal saline (**E**).

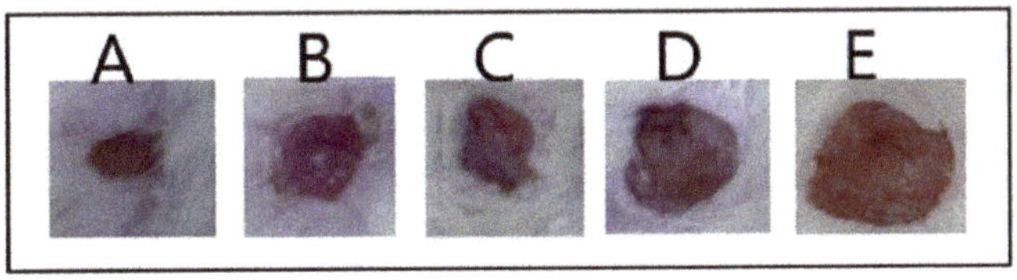

Figure 5. Mean burn wound diameter for different formulations: optimum Gr-PV-NE (**A**), Gr-NE (**B**), PV-NE (**C**), PV-Gr mixture (**D**), and normal saline (**E**).

IL-6 Level Evaluation

IL-6 is an abundant cell protein that helps in modulating immune system responses. The IL-6 level is usually raised by triggers such as injuries, inflammatory conditions, microbial infections, disturbances of the immune system, and malignant tumors; accordingly, IL-6 could be an advantageous marker for detecting inflammation and immune system activation [48].

As could be seen in Figure 6, formulation A (i.e., Gr-PV-NE) had the lowest IL-6 serum level of 944 $\pm$ 100 U/mL, whereas formulation E (i.e., normal saline) had the highest level of IL-6 (3600 $\pm$ 450 U/mL); this indicated the superiority of formulation A in counteracting inflammation compared with the other tested formulations. The superior anti-inflammatory activity of the optimum nanoemulsion formulation could be due to its content of PV and Gr oil [49]. It is well known that statins can depress the output of pro-inflammatory cytokines [50] via inhibiting HMG-CoA reductase, which could invigorate the mevalonate pathway. As a result, PV might minimize the occurrence of the isoprenylation and geranylgeranylation of proteins, particularly Ras protein prenylation. The suppression of Ras diminishes the efficiency of transcription factor nuclear factor kappa B, which plays a pivotal role in many inflammatory reactions [51].

The anti-inflammatory action of Gr oil might be connected to the prohibition of some intracellular signaling pathways encompassing several inflammatory mediators' actions. Abe et al. [52] revealed the ability of Gr oil to depress the adherence response of human neutrophils in vitro and diminish the induced neutrophil mobilization in the peritoneal cavity following the intraperitoneal administration of the oil. Several studies have explored the components that might be responsible for such bioactivity [53]. It was found that the major components of the oil, namely, geraniol, citronellol, and linalool, were proven to have anti-inflammatory influences [54,55].

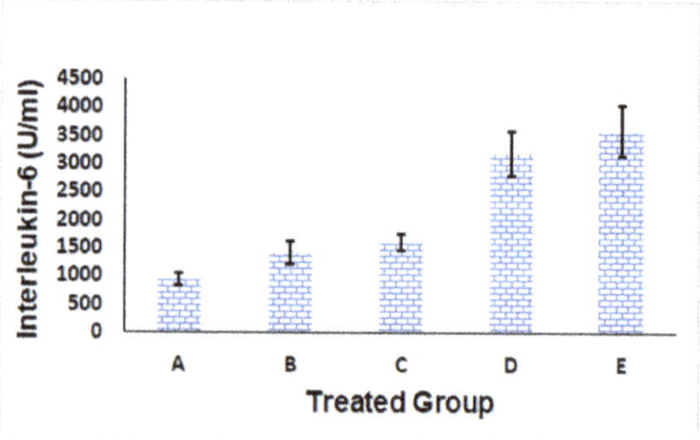

Figure 6. IL-6 levels for different formulations: optimum Gr-PV-NE (**A**), Gr-NE (**B**), PV-NE (**C**), PV-Gr mixture (**D**), and normal saline (**E**).

Formulations B and C had Il-6 values of 1415 ± 200 U/mL and 1611 ± 150 U/mL, respectively. Although these values indicated a higher effect of formulation C, which contained PV and oleic acid instead of Gr oil in the nanoemulsion, compared with formulation B, which contained only Gr oil in the nanoemulsion, such difference was found to be insignificant using Tukey post test. Moreover, formulation D, which was composed of PV and Gr oil, had an IL-6 value of 3200 ± 400 U/mL, affirming the favorable anti-inflammatory behavior of the optimum nanoemulsion formulation (formulation A).

IL-6 serum levels of the different tested groups were examined for normal distribution characteristics using the K-S test of normality. The observed high *p*-values and low D-values for all of the formulations suggested that there was no considerable difference between the collected data and the data that were normally distributed.

The outcomes of the ANOVAs confirmed that all of the tested formulations (except for formulation D) had much higher IL-6 levels than formulation E, with a *p*-value of less than 0.01; therefore, the noticed variations between formulations could not happen by chance. The Tukey HSD post-hoc test revealed that the IL-6 serum levels of all of the formulations varied significantly from each other (p-value < 0.01), except that the comparisons between formulations C and B and D and E were found to be insignificant. Such findings are quite reasonable because formulation B and formulation C each contained one component (i.e., PV in the case of B and Gr oil in the case of C) that had an IL-6–lowering effect.

3.3.2. Ex-Vivo Permeation Study

Upon reviewing the ex-vivo permeation results presented in Table 4 and Figure 7, the following information was detected. First, the nanoemulsion formulation that contained PV and oleic acid instead of Gr oil promoted drug permeation across the skin by more than 5.8 when compared with the PV aqueous dispersion. More importantly, the optimized Gr-PV-NE encouraged skin permeation of PV by 7.6-fold and 2.7-fold in comparison with the PV aqueous dispersion and PV-Gr physical mixture, respectively. Such enhancement might be due to a synergistic effect of a nanoemulsion as a drug delivery system and the effect of Gr oil (which contains components like citronellol and geraniol, which acted as penetration enhancers) on the permeation by PV; similar results were found in the literature [44]. Nanosized emulsions are known to offer a large surface area for drug permeation, in addition to their surfactant and cosurfactant contents, which are thought to fluidize the stratum corneum layer, which is the main barrier to the permeation of drugs through the skin [56]. The obtained rsults were found to be significant at a significance level of 0.05 (p-value < 0.05).

Table 4. Inhibition zones against *S. aureus* and percentage PV permeated for varying formulations.

Run	A: Geranium Oil Amount	B: Pravastatin Amount	C: HLB Value	Pravastatin Permeated %	Inhibition Zone against *S. aureus* (mm)
Optimum formulation (Gr-PV-NE)	275 mg	40	12	84% ± 3.1	20.0 ± 1.8
Gr-NE	275 mg	0	12	0	19.0 ± 1.4
Oleic acid-PV-NE	0	40	12	64% ± 2.4	6.00 ± 0.8
Gr-PV mixture	275 mg	40	0	31% ± 1.2	12.0 ± 1.1
PV aqueous dispersion	0	40	0	11% ± 0.6	4.50 ± 0.3
Plain NE	0	0	12	0	5.00 ± 0.5

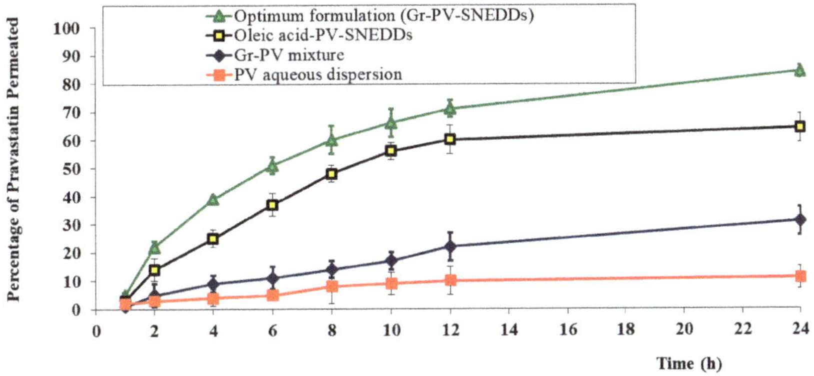

Figure 7. Ex-vivo permeation profiles of PV from different tested formulations.

3.3.3. Antibacterial Activity Assessment

Table 4 shows the collected antibacterial activity data, which revealed the following. The optimum Gr-PV-NE had the largest growth inhibition zone against *S. aureus* (20 ± 1.8 mm) compared with the other tested formulations. The antimicrobial activity of the optimized formulation appeared to be mainly due to its Gr oil content, which is known to have significant antibacterial activity against *S. aureus* due to its components of citronellol, geraniol, linalool, isomenthone, nerol, and citronellyl formate [57]. Such components exert their antimicrobial effects by interacting with the bacterial cell membrane, increasing its fluidity and leading to the leakage of cell components [57]. Such results were further confirmed by comparing the growth inhibition zones for formulations that contained Gr oil with those that contained no Gr oil. Results indicated that the formulations that contained Gr oil enhanced the antimicrobial activity by four times compared with those that did not contain Gr oil. The NE containing Gr oil had 1.6 times the antimicrobial activity as the physical mixture; in the N, the Gr oil present in the nanosized dispersion enhanced the permeation of the formulation across the microbial membrane and potentiated its action. The obtained rsults were found to be significant at asignificance level of 0.05 (p-value < 0.05).

4. Conclusions

The present investigation utilized the BBD for the characterization and optimization of a Gr oil–based nanoemulsion loaded with PV for the transdermal management of burn wounds. The fabricated nanoemulsions had a reasonable droplet size of 61 to 138 nm. The experimental design confirmed the substantial synergistic effect of the Gr oil and PV for burn wound healing. Such a blend increased the management of the wound healing and the anti-inflammatory character and antibacterial effects of each of the constituents. The optimal formulation had up to a 4-fold decrease in the mean burn wound diameter, a 3.81-fold lowering of the IL-6 serum level, a 4-fold increase in the inhibition zone against *S. aureus*, and a 7.6-fold increase in PV permeation upon comparison with different formulations. It was conclusively seen that the obtained nanoemulsions that contained a combination of Gr oil and PV could be considered a promising paradigm for handling burn wounds.

Author Contributions: Conceptualization, W.Y.R. and K.M.H.; methodology, S.S.A.M., B.A.E., A.A.F., A.J.A. and A.Y.S.; software, F.Y.S., W.Y.R. and S.S.A.M.; validation, K.M.H., A.J.A.; formal analysis, A.Y.S. and B.A.E.; investigation, F.Y.S.; resources, A.A.F. and W.Y.R.; data curation, A.Y.S.; writing—original draft preparation, F.Y.S. and A.A.F.; writing—review and editing, K.M.H.; visualization, W.Y.R.; supervision, W.Y.R., A.J.A. and K.M.H.; project administration, W.Y.R. and S.S.A.M.; funding acquisition, W.Y.R. All authors have read and agreed to the published version of the manuscript.

Funding: The Deanship of Scientific Research (DSR) at King Abdulaziz University, Jeddah, Saudi Arabia, funded this project, under grant no. (G-098-166-1442). The authors therefore acknowledge with thanks DSR technical and financial support.

Institutional Review Board Statement: Not applicable.

Informed Consent Statement: Not applicable.

Data Availability Statement: All data available are reported in the article.

Acknowledgments: The Deanship of Scientific Research (DSR) at King Abdulaziz University, Jeddah, Saudi Arabia, funded this project, under grant no. (G-098-166-1442). The authors therefore acknowledge with thanks DSR technical and financial support.

Conflicts of Interest: The authors declare no conflict of interest.

References

1. Montagna, W. *The Structure and Function of Skin*; Elsevier: Amsterdam, The Netherlands, 2012.
2. Yousef, H.; Alhajj, M.; Sharma, S. *Anatomy, Skin (Integument), Epidermis, in StatPearls*; StatPearls Publishing LLC: Treasure Island, FL, USA, 2022.
3. Velnar, T.; Bailey, T.; Smrkolj, V. The Wound Healing Process: An Overview of the Cellular and Molecular Mechanisms. *J. Int. Med. Res.* **2009**, *37*, 1528–1542. [CrossRef] [PubMed]
4. Cañedo-Dorantes, L.; Cañedo-Ayala, M. Skin Acute Wound Healing: A Comprehensive Review. *Int. J. Inflamm.* **2019**, *2019*, 3706315. [CrossRef] [PubMed]
5. Guo, S.; Dipietro, L.A. Factors affecting wound healing. *J. Dent. Res.* **2010**, *89*, 219–229. [CrossRef] [PubMed]
6. Brem, H.; Tomic-Canic, M. Cellular and molecular basis of wound healing in diabetes. *J. Clin. Investig.* **2007**, *117*, 1219–1222. [CrossRef]
7. Vincent, A.M.; Russell, J.W.; Low, P.; Feldman, E.L. Oxidative stress in the pathogenesis of diabetic neuropathy. *Endocr. Rev.* **2004**, *25*, 612–628. [CrossRef]
8. Schneider, K.L.; Yahia, N. Effectiveness of Arginine Supplementation on Wound Healing in Older Adults in Acute and Chronic Settings: A Systematic Review. *Adv. Skin Wound Care* **2019**, *32*, 457–462. [CrossRef]
9. Garros Ide, C.; Campos, A.C.; Tâmbara, E.M.; Tenório, S.B.; Torres, O.J.; Agulham, M.A.; Araújo, A.C.; Santis-Isolan, P.M.; Oliveira, R.M.; Arruda, E.C. Extract from Passiflora edulis on the healing of open wounds in rats: Morphometric and histological study. *Acta Cir. Bras.* **2006**, *21*, 55–65.
10. Givol, O.; Kornhaber, R.; Visentin, D.; Cleary, M.; Haik, J.; Harats, M. A systematic review of Calendula officinalis extract for wound healing. *Wound Repair Regen.* **2019**, *27*, 548–561. [CrossRef]
11. Grundy, S.M. HMG-CoA reductase inhibitors for treatment of hypercholesterolemia. *N. Engl. J. Med.* **1988**, *319*, 24–33.
12. Toker, S.; Gulcan, E.; Cayc, M.K.; Olgun, E.G.; Erbilen, E.; Ozay, Y. Topical atorvastatin in the treatment of diabetic wounds. *Am. J. Med. Sci.* **2009**, *338*, 201–204. [CrossRef]
13. Ko, J.H.; Kim, P.S.; Zhao, Y.; Hong, S.J.; Mustoe, T.A. HMG-CoA Reductase Inhibitors (Statins) Reduce Hypertrophic Scar Formation in a Rabbit Ear Wounding Model. *Plast. Reconstr. Surg.* **2012**, *129*, 252e–261e. [CrossRef] [PubMed]
14. Bitto, A.; Minutoli, L.; Altavilla, D.; Polito, F.; Fiumara, T.; Marini, H.; Galeano, M.; Calò, M.; Lo Cascio, P.; Bonaiuto, M.; et al. Simvastatin enhances VEGF production and ameliorates impaired wound healing in experimental diabetes. *Pharmacol. Res.* **2008**, *57*, 159–169. [CrossRef]
15. Hoeben, A.; Landuyt, B.; Highley, M.S.; Wildiers, H.; Van Oosterom, A.T.; De Bruijn, E.A. Vascular endothelial growth factor and angiogenesis. *Pharmacol. Rev.* **2004**, *56*, 549–580. [CrossRef] [PubMed]

16. Karadeniz Cakmak, G.; Irkorucu, O.; Ucan, B.H.; Emre, A.U.; Bahadir, B.; Demirtas, C.; Tascilar, O.; Karakaya, K.; Acikgoz, S.; Kertis, G.; et al. Simvastatin improves wound strength after intestinal anastomosis in the rat. *J. Gastrointest. Surg.* **2009**, *13*, 1707–1716. [CrossRef] [PubMed]
17. Fischer, J.; Ganellin, C.R. *Analogue-Based Drug Discovery*; John Wiley & Sons: Hoboken, NJ, USA, 2006; p. 472; ISBN 9783527607495.
18. Bourgier, C.; Auperin, A.; Rivera, S.; Boisselier, P.; Petit, B.; Lang, P.; Lassau, N.; Taourel, P.; Tetreau, R.; Azria, D.; et al. Pravastatin Reverses Established Radiation-Induced Cutaneous and Subcutaneous Fibrosis in Patients with Head and Neck Cancer: Results of the Biology-Driven Phase 2 Clinical Trial Pravacur. *Int. J. Radiat. Oncol. Biol. Phys.* **2019**, *104*, 365–373. [CrossRef] [PubMed]
19. Ayad, H.S.; Reda, F.; Abdalla, M.A. Effect of putrescine and zinc on vegetative growth, photosynthetic pigments, lipid peroxidation and essential oil content of geranium (*Pelargonium graveolens* L.). *World J. Agric. Sci.* **2010**, *6*, 601–608.
20. Lis-Balchin, M. A chemotaxonomic study of the *Pelargonium* (Geraniaceae) species and their modern cultivars. *J. Hortic. Sci.* **1997**, *72*, 791–795. [CrossRef]
21. Standen, M.D.; Connellan, P.A.; Leach, D.N. Natural killer cell activity and lymphocyte activation: Investigating the effects of a selection of essential oils and components in vitro. *Int. J. Aromather.* **2006**, *16*, 133–139. [CrossRef]
22. Zhuang, S.R.; Chen, S.L.; Tsai, J.H.; Huang, C.C.; Wu, T.C.; Liu, W.S.; Tseng, H.C.; Lee, H.S.; Huang, M.C.; Shane, G.T.; et al. Effect of citronellol and the Chinese medical herb complex on cellular immunity of cancer patients receiving chemotherapy/radiotherapy. *Phytother. Res.* **2009**, *23*, 785–790. [CrossRef]
23. Burke, Y.D.; Stark, M.J.; Roach, S.L.; Sen, S.E.; Crowell, P.L. Inhibition of pancreatic cancer growth by the dietary isoprenoids farnesol and geraniol. *Lipids* **1997**, *32*, 151–156. [CrossRef]
24. Asgarpanah, J.; Ramezanloo, F. An overview on phytopharmacology of *Pelargonium graveolens* L. *Indian J. Tradit. Knowl.* **2015**, *14*, 558–563.
25. Lohani, A.; Verma, A.; Hema, G.; Pathak, K. Topical Delivery of Geranium/Calendula Essential Oil-Entrapped Ethanolic Lipid Vesicular Cream to Combat Skin Aging. *BioMed. Res. Int.* **2021**, *2021*, 4593759. [CrossRef] [PubMed]
26. Boukhatem, M.N.; Kameli, A.; Ferhat, M.A.; Saidi, F.; Mekarnia, M. Rose geranium essential oil as a source of new and safe anti-inflammatory drugs. *Libyan J. Med.* **2013**, *8*, 22520. [CrossRef]
27. Singh, B.; Bandopadhyay, S.; Kapil, R.; Singh, R.; Katare, O. Self-emulsifying drug delivery systems (SEDDS): Formulation development, characterization, and applications. *Crit. Rev. Ther. Drug Carrier Syst.* **2009**, *26*, 427–521. [CrossRef] [PubMed]
28. Portugal, I.; Jain, S.; Severino, P.; Priefer, R. Micro-and Nano-Based Transdermal Delivery Systems of Photosensitizing Drugs for the Treatment of Cutaneous Malignancies. *Pharmaceuticals* **2021**, *14*, 772. [CrossRef]
29. Mossa, A.H.; Afia, S.I.; Mohafrash, S.M.; Abou-Awad, B.A. Rosemary essential oil nanoemulsion, formulation, characterization and acaricidal activity against the two-spotted spider mite *Tetranychus urticae* Koch (Acari: Tetranychidae). *J. Plant Prot. Res.* **2019**, *59*, 102–112.
30. Hosny, K.M. Development of Saquinavir Mesylate Nanoemulsion-Loaded Transdermal Films: Two-Step Optimization of Permeation Parameters, Characterization, and Ex Vivo and In Vivo Evaluation. *Int. J. Nanomed.* **2019**, *14*, 8589–8601. [CrossRef] [PubMed]
31. Chae, G.S.; Lee, J.S.; Kim, S.H.; Seo, K.S.; Kim, M.S.; Lee, H.B.; Khang, G. Enhancement of the stability of BCNU using self-emulsifying drug delivery systems (SEDDS) and in vitro antitumor activity of self-emulsified BCNU-loaded PLGA wafer. *Int. J. Pharm.* **2004**, *301*, 6–14. [CrossRef] [PubMed]
32. Odeberg, J.M.; Kaufmann, P.; Kroon, K.G.; Höglund, P. Lipid drug delivery and rational formulation design for lipophilic drugs with low oral bioavailability, applied to cyclosporine. *Eur. J. Pharm. Sci.* **2003**, *20*, 375–382. [CrossRef]
33. Priya, K.; Bhikshapathi, D.V.R.N.; Ramesh, B. Design and Characterization of Self-Nanoemulsifying Drug Delivery System of Lovastatin. *Int. J. Pharm. Sci. Nanotechnol.* **2018**, *11*, 4042–4052. [CrossRef]
34. Harwansh, R.K.; Deshmukh, R.; Rahman, M.A. Nanoemulsion: Promising nanocarrier system for delivery of herbal bioactives. *J. Drug Deliv. Sci. Technol.* **2019**, *51*, 224–233. [CrossRef]
35. N Politis, S.; Colombo, P.; Colombo, G.; M Rekkas, D. Design of experiments (DoE) in pharmaceutical development. *Drug. Dev. Ind. Pharm.* **2017**, *43*, 889–901. [CrossRef] [PubMed]
36. Hosny, K.M.; Alhakamy, N.A.; Sindi, A.M.; Khallaf, R.A. Coconut Oil Nanoemulsion Loaded with a Statin Hypolipidemic Drug for Management of Burns: Formulation and In Vivo Evaluation. *Pharmaceutics* **2020**, *12*, 1061. [CrossRef] [PubMed]
37. Gadekar, R.; Saurabh, M.K.; Saurabh, A. Study of formulation, characterisation and wound healing potential of transdermal patches of Curcumin. *Asian J. Pharm. Clin. Res.* **2012**, *5*, 225–230.
38. Hussein, R.M.; Kandeil, M.A.; Mohammed, N.A.; Khallaf, R.A. Evaluation of the hepatoprotective effect of curcumin-loaded solid lipid nanoparticles against paracetamol overdose toxicity: Role of inducible nitric oxide synthase. *J. Liposome Res.* **2022**, *8*, 1–11. [CrossRef]
39. Salem, H.F.; Nafady, M.M.; Ewees, M.G.E.; Hassan, H.; Khallaf, R.A. Rosuvastatin calcium-based novel nanocubic vesicles capped with silver nanoparticles-loaded hydrogel for wound healing management: Optimization employing Box-Behnken design: In vitro and in vivo assessment. *J. Liposome Res.* **2022**, *32*, 45–61. [CrossRef]
40. Salem, H.F.; El-Menshawe, S.F.; Khallaf, R.A.; Rabea, Y.K. A novel transdermal nanoethosomal gel of lercanidipine HCl for treatment of hypertension: Optimization using Box-Benkhen design, in vitro and in vivo characterization. *Drug Deliv. Transl. Res.* **2020**, *10*, 227–240. [CrossRef]

41. Hosny, K.M.; Khallaf, R.A.; Asfour, H.Z.; Rizg, W.Y.; Alhakamy, N.A.; Sindi, A.M.; Alkhalidi, H.M.; Abualsunun, W.A.; Bakhaidar, R.B.; Almehmady, A.M.; et al. Development and Optimization of Cinnamon Oil Nanoemulgel for Enhancement of Solubility and Evaluation of Antibacterial, Antifungal and Analgesic Effects against Oral Microbiota. *Pharmaceutics* **2021**, *13*, 1008. [CrossRef]
42. Ustündağ Okur, N.; Apaydın, S.; Karabay Yavaşoğlu, N.Ü.; Yavaşoğlu, A.; Karasulu, H.Y. Evaluation of skin permeation and anti-inflammatory and analgesic effects of new naproxen microemulsion formulations. *Int. J. Pharm.* **2011**, *416*, 136–144. [CrossRef]
43. Orafidiya, L.O.; Oladimeji, F.A. Determination of the required HLB values of some essential oils. *Int. J. Pharm.* **2002**, *237*, 241–249. [CrossRef]
44. Hosny, K.; Asfour, H.; Rizg, W.; Alhakamy, N.A.; Sindi, A.; Alkhalidi, H.; Abualsunun, W.; Bakhaidar, R.; Almehmady, A.M.; Akeel, S.; et al. Formulation, Optimization, and Evaluation of Oregano Oil Nanoemulsions for the Treatment of Infections Due to Oral Microbiota. *Int. J. Nanomed.* **2021**, *16*, 5465–5478. [CrossRef] [PubMed]
45. Sienkiewicz, M.; Poznańska-Kurowska, K.; Kaszuba, A.; Kowalczyk, E. The antibacterial activity of geranium oil against Gram-negative bacteria isolated from difficult-to-heal wounds. *Burns* **2014**, *40*, 1046–1051. [CrossRef] [PubMed]
46. Rezvanian, M.; Amin, M.C.I.M.; Ng, S.F. Development and physicochemical characterization of alginate composite film loaded with simvastatin as a potential wound dressing. *Carbohydr. Polym.* **2016**, *137*, 295–304. [CrossRef] [PubMed]
47. Farsaei, S.; Khalili, H.; Farboud, E.S. Potential role of statins on wound healing: Review of the literature. *Int. Wound J.* **2012**, *9*, 238–247. [PubMed]
48. Tanaka, T.; Narazaki, M.; Kishimoto, T. IL-6 in Inflammation, Immunity, and Disease. *Cold Spring Harb. Perspect. Biol.* **2014**, *6*, a016295. [CrossRef] [PubMed]
49. Kolodziejczyk, A.M.; Targosz-Korecka, M.; Szymonski, M. Nanomechanical testing of drug activities at the cellular level: Case study for endothelium-targeted drugs. *Pharm. Rep.* **2017**, *69*, 1165–1172. [CrossRef] [PubMed]
50. Saggini, A.; Anogeianaki, A.; Maccauro, G.; Teté, S.; Salini, V.; Caraffa, A.; Conti, F.; Fulcheri, M.; Galzio, R.; Shaik-Dasthagirisaheb, Y.B. Cholesterol, cytokines and diseases. *Int. J. Imm. Pharm.* **2011**, *24*, 567–581. [CrossRef]
51. Lahera, V.; Goicoechea, M.; de Vinuesa, S.G.; Miana, M.; de las Heras, N.; Cachofeiro, V.; Luño, J. Endothelial dysfunction, oxidative stress and inflammation in atherosclerosis: Beneficial effects of statins. *Curr. Med. Chem.* **2007**, *14*, 243–248. [CrossRef]
52. Abe, S.; Maruyama, N.; Hayama, K.; Inouye, S.; Oshima, H.; Yamaguchi, H. Suppression of neutrophil recruitment in mice by geranium essential oil. *Mediat. Inflamm.* **2004**, *13*, 21–24. [CrossRef]
53. Su, Y.W.; Chao, S.H.; Lee, M.H.; Ou, T.Y.; Tsai, Y.C. Inhibitory effects of citronellol andgeraniol on nitric oxide and prostaglandin E2 production in macrophages. *Plant. Med.* **2010**, *76*, 1666–1671. [CrossRef]
54. Pérez, G.S.; Zavala, S.M.; Arias, G.L.; Ramos, L.M. Anti-inflammatory activity of some essential oils. *J. Essent. Oil Res.* **2011**, *23*, 38–44. [CrossRef]
55. Chen, W.; Viljoen, A.M. Geraniol—A review of a commercially important fragrance material. *S. Afr. J. Bot.* **2010**, *76*, 643–651. [CrossRef]
56. Rizg, W.Y.; Hosny, K.M.; Elgebaly, S.S.; Alamoudi, A.J.; Felimban, R.I.; Tayeb, H.H.; Alharbi, M.; Bukhary, H.A.; Abualsunun, W.A.; Almehmady, A.M.; et al. Preparation and Optimization of Garlic Oil/Apple Cider Vinegar Nanoemulsion Loaded with Minoxidil to Treat Alopecia. *Pharmaceutics* **2021**, *13*, 2150. [CrossRef] [PubMed]
57. Bigos, M.; Wasiela, M.; Kalemba, D.; Sienkiewicz, M. Antimicrobial activity of geranium oil against clinical strains of Staphylococcus aureus. *Molecules* **2012**, *17*, 10276–10291. [CrossRef] [PubMed]

polymers

MDPI

Article

Biopolymer Material from Human Spongiosa for Regenerative Medicine Application

Ilya L. Tsiklin, Evgeniy I. Pugachev, Alexandr V. Kolsanov, Elena V. Timchenko *, Violetta V. Boltovskaya, Pavel E. Timchenko and Larisa T. Volova

Biotechnology Center "Biotech", Samara State Medical University, 443079 Samara, Russia; tsiklin.i@yandex.ru (I.L.T.); e.i.pugachev@samsmu.ru (E.I.P.); avkolsanov@mail.ru (A.V.K.); violetta.boltovskaya@yandex.ru (V.V.B.); timpavel@mail.ru (P.E.T.); volovalt@yandex.ru (L.T.V.)
* Correspondence: laser-optics.timchenko@mail.ru; Tel.: +79-277-111-387

Abstract: Natural biopolymers demonstrate significant bone and connective tissue-engineering application efficiency. However, the quality of the biopolymer directly depends on microstructure and biochemical properties. This study aims to investigate the biocompatibility and microstructural properties of demineralized human spongiosa Lyoplast® (Samara, Russian Federation). The graft's microstructural and biochemical properties were analyzed by scanning electron microscopy (SEM), micro-computed tomography, Raman spectroscopy, and proteomic analysis. Furthermore, the cell adhesion property of the graft was evaluated using cell cultures and fluorescence microscopy. Microstructural analysis revealed the hierarchical porous structure of the graft with complete removal of the cellular debris and bone marrow components. Moreover, the proteomic analysis confirmed the preservation of collagen and extracellular proteins, stimulating and inhibiting cell adhesion, proliferation, and differentiation. We revealed the adhesion of chondroblast cell cultures in vitro without any evidence of cytotoxicity. According to the study results, demineralized human spongiosa Lyoplast® can be effectively used as the bioactive scaffold for articular hyaline cartilage tissue engineering.

Keywords: biopolymers; demineralized human spongiosa; scanning electron microscopy; micro-computed tomography; Raman spectroscopy; proteomic analysis; chondroblasts; tissue engineering; scaffold; fluorophores

Citation: Tsiklin, I.L.; Pugachev, E.I.; Kolsanov, A.V.; Timchenko, E.V.; Boltovskaya, V.V.; Timchenko, P.E.; Volova, L.T. Biopolymer Material from Human Spongiosa for Regenerative Medicine Application. *Polymers* **2022**, *14*, 941. https://doi.org/10.3390/polym14050941

Academic Editors: Ariana Hudita and Bianca Gălăţeanu

Received: 20 January 2022
Accepted: 22 February 2022
Published: 26 February 2022

1. Introduction

Current treatment options for degenerative bone and cartilage tissue pathology aim to enhance post-traumatic and post-operative defects regeneration using various biological or synthetic products.

Biodegradable scaffolds, including calcium phosphate, aerogels [1–6], autologous [7–9], allogeneic [10–12], or xenogeneic grafts [13–17] demonstrate significant efficiency as bone substitute materials. Ideally, biocompatible materials' resorption rate coincides with the formation rate of the new organotypic tissue. Allogeneic products incorporate identical structural and biological components and provide optimal conditions for genetically programed physiological regeneration in the human body [10,18]. Original technology of the human bone tissue products manufacturing developed at Samara Tissue Bank at Samara State Medical University has been successfully applied in bone tissue repair for more than twenty years. This technology provides thorough mechanical cleaning and complete removal of the antigenic components from human spongiosa while preserving its biological activity [18,19]. Microstructural and biochemical properties of the natural biopolymers play a crucial role in the regeneration process and directly depend on the manufacturing technology.

This study aims to investigate the microstructure and biocompatibility of the novel biopolymer material from demineralized human spongiosa.

2. Materials and Methods

2.1. Manufacturing and Characterizing Materials

The biopolymer Lyoplast® analyzed in this study is lyophilized demineralized human spongiosa manufactured at the Samara tissue bank at the "BioTech" Biotechnology Center, Samara State Medical University (RF patent No. 2366173 of 15.05.2008; certificate of conformity ISO 13485:2016, reg. No. RU CMS-RU.PT02.00115; certificate ISO 9001:2015, reg. No. TIC 15 100 159171) (Figure 1).

Figure 1. Samples of demineralized lyophilized human spongiosa Lyoplast®.

Experimental samples of Lyoplast® material underwent compulsory low-frequency ultrasonic treatment using ultrasonic bath "Sapphire" TTC (RMD), (Sapphire LTD, Moscow, Russia with a frequency of 24–40 kHz.

Lyophilization of the material (vacuum drying by sublimation) was performed using a sublimation unit ALPHA2-4LSC (Martin Christ Gefriertrocknungsanlagen GmbH, Osterode am Harz, Germany).

Demineralization of human spongiosa was carried out in a weak HCl solution. Hermetically packaged lyophilized product was then sterilized with gamma rays using a certified GU-200 M (NIIP Joint-Stock Company, Moscow, Russia).

The residual content of lipids in the biomaterial was estimated using a spectrophotometer (SF-56 "Lomo-Spektr", St. Petersburg, Russia). Finally, the humidity of the product was determined using a thermogravimetric infrared moisture meter (Sartorius-MA-150, Malente, Germany).

The study was carried out using physical, chemical, biological, and cultural methods.

2.2. Scanning Electron Microscopy (SEM)

The samples were examined using a JEOLJSM-6390 A Analysis Station SEM (Tokyo, Japan). Bioimplant samples were washed and fixed with a 2.5% aqueous solution of glutaric aldehyde. After that, they were spiked with ethanol of increasing concentration, followed by drying at room temperature for 24 h. Immediately before the study, the biomaterial was sprayed with gold or carbon to improve the surface electrical conductivity required for SEM.

2.3. Micro-Computed Tomography (Micro-CT)

Micro-CT scanning of the Lyoplast® lyophilized allogeneic spongiosa samples was performed in Laboratory of Microanalysis in Skolkovo Technopark (Moscow, Russia) using high-resolution 3D X-ray microscope VersaXRM-500 (Xradia, Inc. Pleasanton, CA, USA) with voltage range 30–160 kV, maximum power 10 W, 360° rotation, and maximum spatial resolution < 0.7–1 μm (True Spatial Resolution™). At the first stage, the scanning of the sample was performed using a resolution of 8.6 μm/pixel at a voltage of 80 kV with a set of

1081 projections and 0.5 s exposition ti0me. Next, a region of interest (ROI), including bone trabeculae [20] was selected and scanned with a resolution of 1.1 µm/pixel at a voltage of 80 kV with 1441 projections and 0.5 s exposition time. The obtained data were reconstructed with the Filtered Back Projection method using the XRM Reconstructor software. Computed microtomography data were saved in TXRM and DICOM formats, and 3D models of the sample structure were saved in TXM and TIFF formats.

2.4. Raman Spectroscopy (Raman Spectroscopy)

This research was performed at the Department of Laser and Biotechnical Systems of Samara National Research University. Spectral characteristics of lyophilized, demineralized human spongiosa Lyoplast® were studied using an experimental setup that included a high-resolution digital spectrometer Shamrock SR-303I (Oxford Instruments PLC, Abingdon, UK) with a built-in cooling chamber AndorDV420A-OE (Oxford Instruments PLC, Abingdon, UK) and an RPB785 fiber-optic probe combined with a laser module LuxxMasterLML-785.0RB-04(Laser Components Germany GmbH, Olching, Germany), all under the control of a PC workstation. This spectrograph provided 0.15 nm wavelength image resolution with low intrinsic noise. To exclude the autofluorescence contribution in the Raman spectra, we used a method for subtracting the fluorescence component of the polynomial approximation with additional filtration of random noise effects. In this work, the Raman spectra were analyzed in 350–2200 cm^{-1}. The laser power of 400 mW was applied for 30 s exposure time, without evident degradation of the samples. Raman spectra were registered using an optical probe, which was placed above the object at a distance of 7 mm. We used the method of spectral contour fitting and deconvolution of the Gaussian function in the software environment MagicPlotPro 2.7.2. Thus, we conducted a non-linear regression analysis of Raman spectra to decompose the signal into spectral lines [21–24].

2.5. Mass Spectroscopy (Proteomic Analysis)

Samples of lyophilized, demineralized spongiosa were subjected to heat treatment (100 °C, 5 min) in a medium containing 2% SDS (sodium dodecyl sulfate) and 5% mercaptoethanol, centrifuged, and separated in a 10% polyacrylamide gel in the presence of SDS. After electrophoresis, the gel was stained with Coomassie G250 and cut into fragments according to the visually detectable fractions. For qualitative identification of tightly bound proteins, gel fragment samples were washed with ammonium hydrogen carbonate and acetonitrile solution (1:1) at 50 °C, dehydrated in 100% acetonitrile, and then treated with trypsin solution in 50 mM ammonium hydrogen carbonate. The reaction was stopped with the addition of 0.1% trifluoroacetic acid, and the released peptides were extracted from the gel by placement in an ultrasonic bath and separated by high-performance liquid chromatography (Dionex Ultimate 3000, Country) using an AcclaimPepMap C18 analytical column (2 µm, 100 Å, 75 µm × 15 cm) (Thermo Scientific). Mass spectra of the eluant were obtained on a maXis Impact mass spectrometer (Bruker, Germany) equipped with a CaptiveSpray (Bruker, Germany) ion source. The mass spectra were processed using DataAnalysis 4.1 software to obtain the mass lists, following a preset script to analyze continuous chromatograms. Proteins were identified from the mass lists using the Mascot 2.4.0 program (Matrix Science). Protein representation in the sample was evaluated using Multi Quant 3.0.2 software from the MRM transition peak area.

2.6. Obtaining a Line of Human Juvenile Chondroblasts

We used a culture of juvenile chondroblasts obtained from cartilage fragments of the interphalangeal joints of removed extra toes in healthy children diagnosed with polydactyly. Collection of biological material for cell culture was performed after parents/legal representatives had signed voluntary informed consent and with approval from the Bioethics Committee at Samara State Medical University. Tissue donors were somatically healthy and negative for HPV, HIV, HBsAg, and HCV infections. Material preparation was performed in the Biotech Department "BioTech" cell culture laboratory at Samara State Medical University. This laboratory is equipped with a suite of class B "clean rooms", with the possibility of upgrading to class A areas following ISO 5 standard. Cartilage tissue fragments were washed three times with sterile Hanks' solution, mechanically crushed, and then subjected to enzymatic treatment with 0.1% collagenase solution Biolot LLC (Saint Petersburg, Russia) for two hours in a shaker incubator BioSan (Riga, Latvia). The enzyme was inactivated by adding sterile 0.02% Versene solution (Biolot LLC, Russia). The material was transferred to centrifuge tubes and was washed in complete growth medium 199 containing 10% fetal calf serum (Biolot LLC, Russia) and spun at 1500 rpm on a low-temperature centrifuge Eppendorf 572 R (Eppendorf SE, Hamburg, Germany) at 4 °C for 20 min. The pellet was transferred into sterile plastic culture dishes (TRR Techno Plastic Products AG, Trasadingen, Switzerland) with an area of 25 cm^2. A fresh portion of 10% 199 complete growth medium was added to the culture vial, then placed in a CO_2 incubator. The condition of cells in the culture vial was monitored daily by observation under an inverted microscope Olympus CKX-41 (Olympus Corporation, Tokyo, Japan). The culture medium was changed every three days until the culture had attained 80% confluence, at which point the cells were transferred into new culture vials. The obtained chondroblast lines were analyzed for viability at the second passage using fluorescence microscopy.

2.7. Creation of a Cell-Tissue Graft

The Lyoplast® tissue engineering product is a 3D carrier of demineralized human spongiosa seeded with juvenile chondroblasts. The chondroblasts were seeded on the scaffold. Cell culture was removed from the bottom of the culture plate at four passages in a traditional manner and seeded at a density of 5 × 104 cells per 27 mm^3 (3 mm × 3 mm × 3 mm) block of the medium. The tissue-engineered constructs were placed in 7 mL vials with a complete growth medium (2 samples per vial) and were cultured in an incubator at 5% CO_2 and 20% O_2 at 37 °C. Samples of demineralized spongiosa Lyoplast® of similar size but without chondroblasts served as a control group.

2.8. Fluorescence Microscopy

Fluorescence microscopy was performed using the Leica DMIL LED fluorescence module (Germany). For this purpose, a Live/Dead® Viability/Cytotoxicity AssayKitfluorophore kit (Thermo Fisher Scientific Inc, Waltham, MA, USA) was used. Staining was performed according to the manufacturer's protocol. The kit contains Calcein-AM and ethidium homodimer-1 solutions and is designed for simultaneous fluorescent staining of live and dead cells. Calcein-AM, the acetoxymethyl ester of calcein, is highly lipophilic and permeable to cell membranes. Although calcein-AM is not a fluorescent molecule, calcein produced by esterase reaction with calcein-AM in viable cells emits strong green fluorescence (490 nm excitation, 515 nm emission), while calcein-AM stains only viable cells. On the other hand, the nucleus staining dye ethidium homodimer-1 cannot pass through cell membranes, but it penetrates the membrane of dead cells, reaches the nucleus, and intercalates with the double helix of the cell DNA, where it shows red fluorescence (535 nm excitation, 617 nm emission). Thus, living cells glow green, while dead cells glow red.

2.9. MTT Assay for the Study of Demineralized Human Spongiosa Cytotoxicity

A culture of juvenile human chondroblasts of the seventh passage was used to study the cytotoxicity of the human spongiosa. A culture of human chondroblasts was seeded on a 24-well plate (SPL LIFE SCIENCES, Gyeonggi-do, Korea) at a dose of 4 × 10^4 cells/well and incubated in a CO_2 incubator Sanyo MSO-18AC (SANYO Electric Co., Ltd., Osaka, Japan) at 37 °C and 5% CO_2. Control №1—medium only; control №2—medium with spongiosa. Probes №1 and №2 are cells only and cells with spongiosa, respectively. After the monolayer reached 80% confluency, the nutrient medium 199 Biolo t LLC, (Saint Petersburg, Russia) with a serum content of 10% (Biolo T, Russia) was changed in the wells, and Lyoplast® spongiosa samples Lyocell (Samara, Russia) were placed in cubes 5 mm × 5 mm × 5 mm in size. Permeable plate inserts SPL LIFE SCIENCES, Gyeonggi-do,

Korea were used to hold the material above the monolayer to exclude mechanical damage to the monolayer. After 48 h of cultivation with the material, cell viability was determined using 3-(4,5-dimethylthiazol-2-yl)-2, 5-diphenyltetrazolium bromide (MTT) (Sigma-Aldrich, Merck KGaA, Darmstadt, Germany), which allows evaluating the ability of live cells to convert soluble tetrazolium salt into an insoluble purple precipitate of formazan due to the action of cellular dehydrogenases. After removing the medium, each well was incubated with 0.3 mg/mL MTT in a growth medium at 37 °C for 3.5 h. At the end of the incubation period, the medium was removed by pipetting; intracellular formazan was dissolved in 200 μL of DMSO; and the optical density was measured at 550 nm on a Tecan Infinite M200 Pro microplate reader (Tecan Group Ltd., Männedorf, Switzerland). The percentage of viable cells was determined based on the calculated optical density, taking as 100% the values in the wells with cells without material.

3. Results and Discussion

3.1. Materials Characterization

Low-frequency ultrasound exposure, such as cavitation and microwaves, ensured complete destruction and removal of all cellular and bone marrow components, including bone cells, stroma, myeloid cells, and lipids from the spongiosa. Quality control of the scaffold delipidization was performed using physical and biochemical methods listed below. After lyophilization in a sublimation unit, the humidity of the biomaterial did not exceed 5%.

Exposure of the material to hydrochloric acid resulted in the final destruction of bone cells and demineralization of the scaffold.

The shelf life of lyophilized demineralized spongiosa Lyoplast® is three years after gamma-ray sterilization. Storage and transportation of the finished product does not require any special temperature conditions.

3.2. Scanning Electron Microscopy

SEM examination of human spongiosa that underwent all stages of the Lyoplast® bioimplant production process, including ultrasound treatment of the biomaterial, demineralization, lyophilization, and γ-sterilization, showed complete preservation of the original trabecular bone architectonics. According to the results of the SEM, the completed biopolymer product is a porous three-dimensional (3D) matrix with a hierarchical structure of pores of various calibers (300–800 μm), free of cellular and bone marrow components. Trabeculae of lamellar bone are visualized on the image; their contours are precise and interconnected to form fine pores. Similarly, these fine pores do not contain cellular and bone marrow components. The 3D photos demonstrate a self-similar hierarchical architecture of the human spongiosa organization (Figure 2).

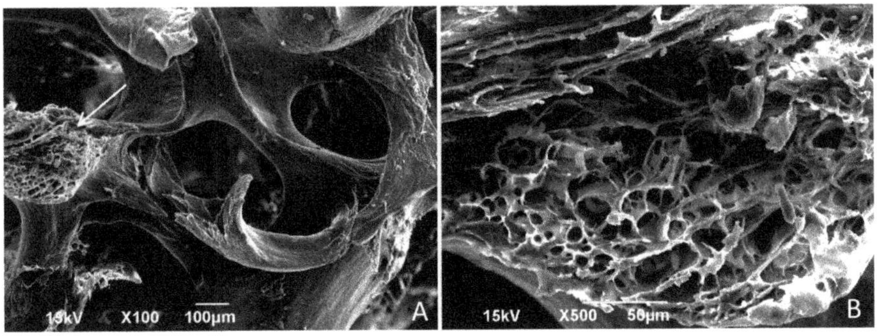

Figure 2. Architectonics of demineralized lyophilized human spongiosa (SEM). Trabecular architecture: (**A**) cross section of trabeculae (marked by an arrow), ×100; internal structure of the trabeculae (**B**) ×500.

3.3. Micro-Computed Tomography

Non-destructive microstructural analysis of the Lyoplast® human spongiosa bioimplant using computed microtomography confirmed the preservation of spongy bone's trabecular architectonics and porous microstructure with a pore caliber of 300 to 800 μm. Three-dimensional reconstruction of bone tissue samples using a resolution of 8.6 μm/pixel and a voltage of 80 kV made it possible to visualize the surface of the pores, free from cellular debris and fatty components of the bone marrow (Figure 3).

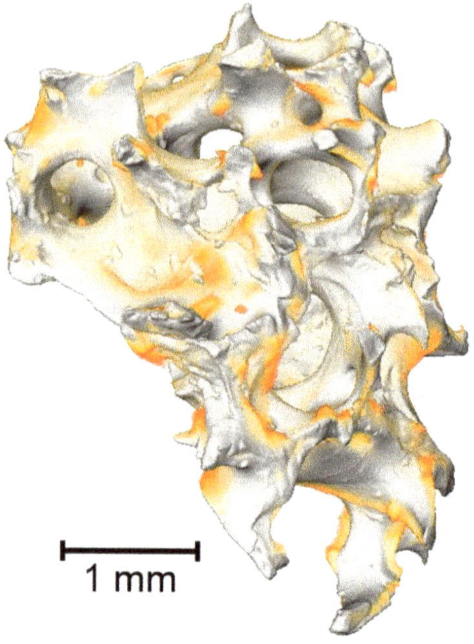

Figure 3. Architectonics of demineralized lyophilized human spongiosa (Micro-CT). Trabecular architecture: (80 kV, 8.6 μm/pixel; 1081 projections; 0.5 s exposition time).

Using a resolution of 1.1 μm/pixel and a voltage of 80 kV confirmed the hierarchical bone tissue architectonics and allowed us to visualize osteocyte lacunae in the intrinsic structure of bone trabeculae. The average caliber of osteocyte lacunae was 10–30 μm. The investigation also confirmed the absence of a cellular component in the lacunae. In addition, three-dimensional image inversion allowed a detailed analysis of the number and condition of osteocyte lacunae. (Figure 4). An essential advantage of micro-computed tomography compared to scanning electron microscopy (SEM) is the combination of high-throughput fashion and ultra-high resolution, making it possible to simultaneously visualize a significant number of osteocyte lacunae while maintaining high image quality.

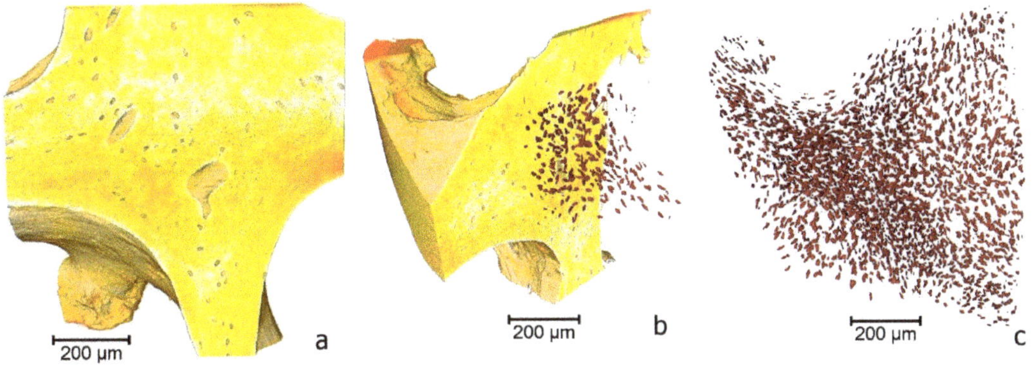

Figure 4. Architectonics of demineralized lyophilized human spongiosa (Micro-CT). Trabecular architecture: (80 kV, 8.6 µm/pixel; 1081 projections; 0.5 s exposition time) (**a**)—visualization of the osteocyte lacunae; (**b**,**c**)—image inversion and osteocyte lacunae detection).

3.4. Raman Spectroscopy

Raman spectroscopy allowed us to obtain detailed information about spectral contour decomposition of demineralized spongiosa samples using a Gaussian function as a trial (Figure 5). The mean value of the coefficient of correlation between the recovered and input spectrum (R2) in the region of 750–2050 cm^{-1} was 0.99, indicating near-perfect agreement.

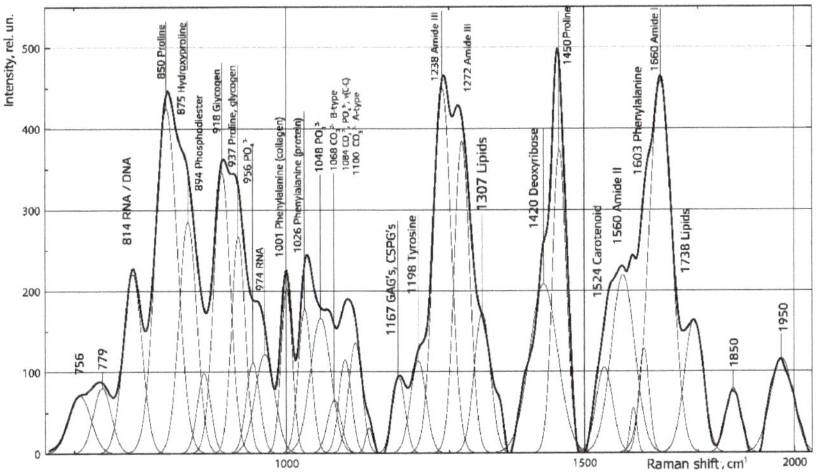

Figure 5. Spectral contour decomposition of demineralized spongiosa samples. The solid line-original Raman spectrum; the dashed lines-the Raman lines after separation.

We have established no mineral components in the demineralized spongiosa, as indicated by the disappearance of the Raman line at 956 cm^{-1}, corresponding to $PO43-(\nu 1)$. As can be seen in Figure 3, the demineralized spongiosa lacks fat, as indicated by the absence of an intense Raman line at 1307 cm^{-1} (lipids). At the same time, the preservation of the organic matrix is observed, as indicated by the presence of Raman lines at 850 cm^{-1} (proline), 1238 cm^{-1}–1272 cm^{-1} (Amide III), 1450 cm^{-1} (proline), 1167 cm^{-1} (glycosaminoglycans), and 1660 11 cm^{-1} (Amide I). Collagen is the main protein component of bone tissue, and it forms the fibrillar framework of the bone matrix. The amino acid sequence of collagen is notably rich in proline, about half of which is hydroxylated during collagen breakdown to form hydroxyproline.

3.5. Mass Spectroscopy

Proteomic analysis of the demineralized human spongiosa demonstrated the presence of collagenous and extracellular bone matrix proteins. The ability of these proteins to stimulate and inhibit cell adhesion, proliferation, and differentiation is noteworthy. We identified five main types of collagen (I, IV, VI, XII, XIV), fibronectin, vitronectin, osteopontin, matrix Gla-protein, along with TGF-β mimecan, decorin, and other proteins in the human spongiosa organic matrix. A list of these proteins, their localization, and their mass are presented in Table 1.

Table 1. List of organic components identified in the analysis of demineralized human spongiosa Lyoplast®.

Proteins (Polypeptide Chains)	Localization	Mass, kDa
Matrix Gla-protein	Bone	12,353
Secreted phosphoprotein 24	Bone	24,338
Transforming growth factor beta-1 (TGF-β1)	Bone	25,000
Mimecan	Bone	33,922
SPARC (Osteonectin)	Bone	34,632
Bone sialoprotein 2	Bone	35,148
Osteopontin	Bone	35,423
Lumican	Extracellular matrix (ECM)	38,429
Decorin	ECM	39,747
Chondroadherin	Cartilage matrix protein	40,476
Biglycan	ECM	41,654
Fibromodulin	ECM	43,179
Prolargin	ECM	43,810
Osteomodulin	Bone	49,492
Vitronectin	Plasma, ECM	75,000
Collagen (I) alpha chain	ECM	108–168
Collagen (IV) alpha chain	ECM	108–168
Collagen (VI) alpha chain	ECM	108–168
Collagen (XII) alpha chain	ECM	108–168
Collagen (XIV) alpha chain	ECM	108–168
Tenascin	ECM	240,853
Fibronectin	Plasma, ECM	262,625

We assign properties and functions of the identified proteins according to Baghy et al. [25–28]. In summary, collagen acts as a cell-binding protein that performs an adhesive function by integrating collagen bundles, a major component of the extracellular matrix. Bone, basal membrane, and soft tissue collagens are found in human spongiosa. Fibronectins bind cell surfaces and compounds, including collagen, fibrin, heparin, DNA, and actin. Fibronectins are involved in cell adhesion, cell motility, opsonization, wound healing, and maintenance of cell shape. They also participate in the regulation of type I collagen deposition by osteoblasts. Vitronectin is a glycoprotein of the hemopexin family, which is abundant in serum and the extracellular matrix of bone tissue. It is involved in fibrinolysis, mediates cell adhesion and migration, and binds glycosaminoglycans, collagen, and plasminogen. These three organic substances (collagen, fibronectin, vitronectin) are widely used in biotechnology to create the cytoadhesive surface of the culture plate.

Mimecan, or osteoglycin, is a small proteoglycan rich in leucine, important for collagen fibrillogenesis. Decorin, lumican, and biglican are small proteoglycans of the extracellular matrix that bind to fibronectin, inhibit cell adhesion, attach to tumor growth factor, and reduce tumor cells' mitogenic activity. They play a regulatory role in connective tissue development and repair processes. TGF-β1 is a secreted protein that performs many cellular functions, including control of cell growth, cell proliferation, cell differentiation, and apoptosis. Osteopontin is involved in cell proliferation, migration, and adhesion, including bone marrow mesenchymal stem cells, hematopoietic stem cells, osteoclasts, and osteoblasts. Osteonectin is a bone tissue glycoprotein that binds calcium. It is released by osteoblasts during bone formation, initiating mineralization and promoting the formation of mineral crystals. Osteonectin also has an affinity for collagen. Finally, bone sialoprotein is a critical component with a high sialic acid content of the extracellular bone matrix, constituting approximately 8% of all non-collagenous proteins. It has the function of forming the hydroxyapatite core during bone mineralization. Matrix Gla-protein associates with the organic matrix of bone and cartilage and acts as an inhibitor of bone formation, thereby playing an essential role in bone mineralization but acting as a mineralization inhibitor in cartilage and vessels. Finally, the extracellular matrix protein tenascin is involved in the control of migrating neurons and axons during neuronal development, synaptic plasticity, and regeneration. Its role in osteoblasts differentiation is unknown. This comprehensive analysis of protein constituents complements the results of Raman spectroscopy and provides a broader understanding of the biochemical composition of the organic matter of human spongiosa Lyoplast®.

3.6. Obtaining a Line of Human Juvenile Chondroblasts and Creating a Cell-Tissue Graft

Observations of the native culture showed that cells had excellent adhesive properties.

On the first day in culture, most chondroblasts from the suspension were deposited from the medium to the bottom of the plate, where they attached and spread out. They took on an elongated shape with a well-defined boundary, connecting through 3–5 processes. The cytoplasm contained many vacuoles in the peripheral zone. The nucleus was oval shaped, usually located in the center of the cell, and contained 1–3 nuclei. The number of cells gradually increased during cultivation, forming a uniform monolayer. When the culture reached 80% confluence, the chondroblasts adhered tightly to each other, and there were practically no areas free of cells. At this stage, cell transplantation was performed using the standard method. A qualitative culture assessment was carried out using cultural and morphological methods in the fourth passage. Our earlier preclinical animal studies, using combined cell-tissue grafts based on rabbit allogenic demineralized spongiosa and rabbit rib cartilage cell culture, showed the healing of the animal joint's simulated bone and cartilage defects. Finally, organotypic hyaline cartilage tissue was revealed [19], and it is promising for translational use of Lyoplast® human spongiosa. The unique structure and composition of the demineralized human spongiosa allowed us to use it as an effective bioscaffold for creating the cell-tissue graft.

The findings reported by Doran and his study group members in 2021 [29] also demonstrated the efficiency of using cartilaginous differon cells for articular cartilage tissue restoration. Obtained results demonstrate the prospect for further use of demineralized human spongiosa for articular hyaline cartilage defects repair. Obtaining such cell-tissue grafts is relatively simple and does not require complex bioreactor systems.

Cell cultures stained with hematoxylin and eosin showed the geometric pattern typical of cartilage differon cells. The cells were aligned, forming concentric and ellipsoidal figures resembling osteons and insertion plates of compact bone tissue. The presence of polygonal cells was noted. The cytolemma of chondroblasts in culture was smooth, and the weakly oxyphilic cytoplasm contained vacuoles. Nuclei of regular round shape with a smooth envelope were located mainly in the center. Chromatin in the form of fine granularity was located diffusely in the nuclei. The large processes were shortened, and the intercellular substance was visualized as translucent layers (Figure 6a). The fluorescence microscopy with a Live/Dead® fluorophore kit revealed 93% viable cells (Figure 6b).

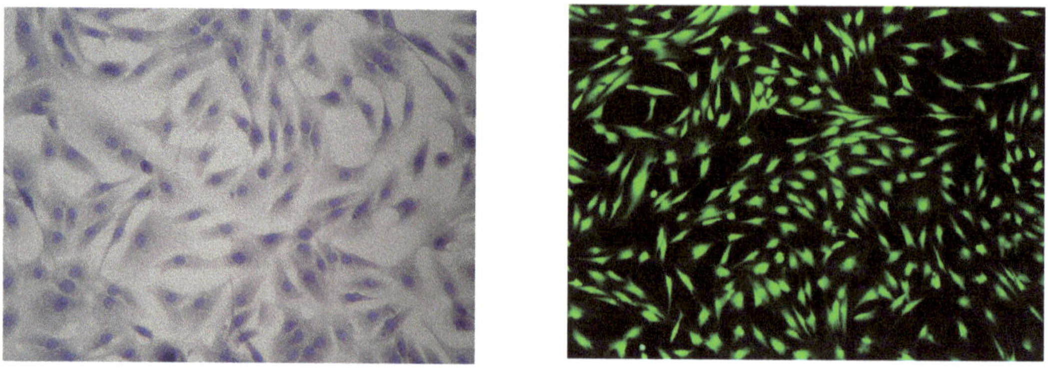

(**a**) (**b**)

Figure 6. Human juvenile chondroblast culture: (**a**) hematoxylin and eosin staining. Light microscopy, ×100; (**b**) live chondroblasts exhibiting bright green glow. Fluorescence microscopy. Staining with Live/Dead® fluorophores, ×100.

Upon examining living and damaged cells populated on a 3D carrier made of Lyoplast® human spongiosa using the Live/Dead® kit, we detected viable cells with bright green coloration and damaged or dead cells with a bright red nucleus (Figure 7a). On closer examination, the cells on the surface of the trabeculae were evenly distributed throughout the scaffold depth, and the shape was polygonal (elongated, triangular, rounded). We visualized the outgrowths with which the chondroblasts were connected, creating a uniform monolayer, which was indicative of good cell adhesion to this material. When the cell-tissue graft was examined using SEM, we saw adsorption of proteins on the trabeculae surface, where cells of elongated shape were connected by outgrowths (Figure 7b).

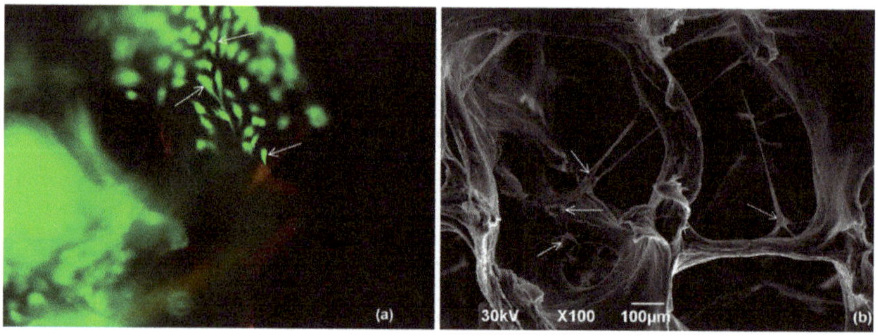

Figure 7. Demineralized spongiosa using the Lyoplast® technology with populated chondroblasts: (**a**) populated viable chondroblasts on the surface of the biocarrier (arrows indicate attached cells). Live/Dead® fluorophore staining. Fluorescence microscopy, ×100; (**b**) cells inhabiting the surface of trabeculae are marked by arrows. SEM, ×100.

3.7. MTT Assay of the Demineralized Human Spongiosa MTT Test Results

MTT Assay of the Demineralized Human Spongiosa MTT Test Results Are Presented in Table 2 and Figure 8.

Table 2. MTT test. Indicators of optical density in experimental and control wells.

	Controls		Probes	
	Medium	**Medium + Spong**	**Cells**	**Cells + Spong**
OD, units	0.0716	0.0782	0.491	0.512
STDV	0.007163155	0.000260349	0.023076494	0.002760717

Using MTT test (48 h treatment), it was found that for the experimental group (Cells + Spong and controls (Cells only)), the optical density was 0.491 ± 0.023 and 0.512 ± 0.003, correspondingly. Statistical analyses revealed that no significant statistical differences in cell viability ($p = 0.513$) were observed between control cells (Cells only) and cells grown in the presence of the investigated biomaterial (Cells + Spong).

Thus, the investigated biomaterial—demineralized human spongiosa Lyoplast®—is not cytotoxic.

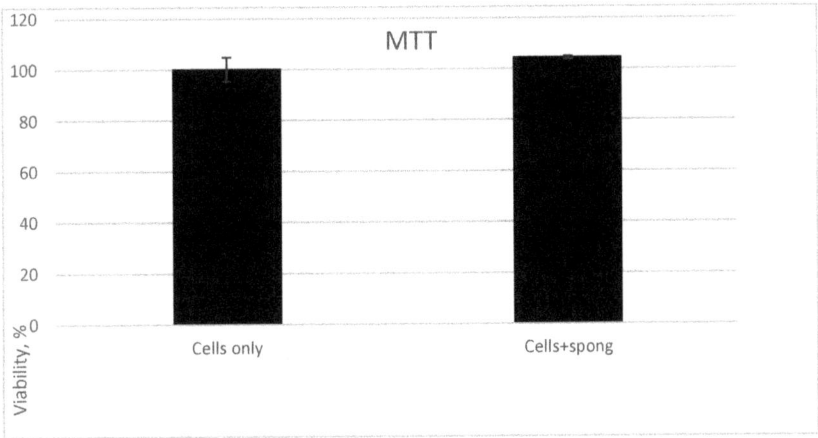

Figure 8. MTT test. No difference in cell viability between control (Cells only) and probe (Cells + Spong) was detected by MTT test.

4. Conclusions

Microstructural analysis of the demineralized human spongiosa convincingly demonstrates its preserved hierarchical porous structure. Pores of various configurations and calibers are entirely free of cellular debris and bone marrow components. Furthermore, the proteomic analysis confirmed the complete preservation of collagen and other extracellular matrix proteins. These proteins play a crucial role in inhibiting and stimulating cell adhesion, proliferation, and differentiation. We revealed the adhesion of chondroblast cell cultures in vitro without any evidence of cytotoxicity. Revealed features of the demineralized human spongiosa create optimal conditions for hyaline articular cartilage and subchondral bone regeneration. Thus, according to the study results, demineralized human spongiosa Lyoplast® can be effectively used as the bioactive scaffold for articular hyaline cartilage tissue engineering. For the future perspective, the identified microstructural and biochemical features of the demineralized human spongiosa can be considered in 3D bioprinting and creating tissue-engineering constructs of the cartilage tissue.

5. Patents

Patent RF № 2170016 from 17.02.1999 "Method of saturation of bone spongy tissue grafts with medication" Volova L.T., Kirilenko A.G., Uvarovsky B.B.

Patent RF № 2156139 from 15.03.1999 "Method of sterilization of lyophilized bone transplants" Volova L.T., Kirilenko A.G., Uvarovsky B.B.

Patent RF № 99108699 from 21.04.1999 "Method of bone marrow removal from cancellous bone grafts" Volova L.T., Kirilenko A.G.

Patent RF № 2366173 of 15.05.2008. "Method of manufacturing large-block lyophilized bone implants" Volova L.T.

Author Contributions: Production of demineralized lyophilized spongiosis Lyoplast®, interpreting the received experimental results, I.L.T.; Conception, methodology, research design, analysis of results, conclusions, original writing, L.T.V.; project management, review preparation, editing, revision, A.V.K.; staging and conducting in vitro studies, drafting, visualization, E.I.P.; construction of experiments methodology using Raman spectroscopy and processing of the obtained experimental data, E.V.T.; conducting research using fluorescence microscopy, interpretation of the results obtained, visualization, V.V.B.; carrying out the experimental study using the Raman spectroscopy method, processing and interpreting the received results of the study, finalizing the article, P.E.T. All authors have read and agreed to the published version of the manuscript.

Funding: This research received no external funding.

Institutional Review Board Statement: The study was conducted according to the guidelines of the Declaration of Helsinki. The protocol was approved by the Ethics Committee (extract 20.01.2021no.215 of minutes of the meeting of the Committee on Bioethics of Samara State Medical University).

Informed Consent Statement: Not applicable.

Data Availability Statement: Data available on request due to restrictions eg privacy and ethical.

Conflicts of Interest: The authors declare no conflict of interest.

References

1. Eremin, I.I.; Bozo, I.Y.; Volozhin, G.A.; Deev, R.V.; Rozhkov, S.I.; Eremin, P.S.; Komlev, V.S.; Zorin, V.L.; Pulin, A.A.; Timashkov, D.A.; et al. Biological effects of tissue-engineered bone grafts from tri-calcium phosphate and multipotent mesenchymal stromal cells in orthotopic conditions in vivo. *Kremlin. Med. Clin. Bull.* **2015**, *4*, 144–150.
2. Reverchon, E.; Baldino, L.; Cardea, S.; De Marco, I. Biodegradable synthetic scaffolds for tendon regeneration. *Muscles Ligaments Tendons J.* **2012**, *2*, 181–186.
3. Baldino, L.; Cardea, S.; Scognamiglio, M.; Reverchon, E. A new tool to produce alginate-based aerogels for medical applications, by supercritical gel drying. *J. Supercrit. Fluids* **2019**, *146*, 152–158. [CrossRef]
4. Lermontov, S.A.; Sipyagina, N.A.; Malkova, A.N.; Baranchikov, A.E.; Erov, K.E.; Petukhov, D.I.; Ivanov, V.K. Elastic aerogels based on methyltrimethoxysilane: The influence of supercritical fluid on structure-sensitive properties. *J. Inorg. Chem.* **2015**, *60*, 549–553.
5. Osorio, D.A.; Lee, B.E.J.; Kwiecien, J.M.; Wang, X.; Shahid, I.; Hurley, A.L.; Cranston, E.D.; Grandfield, K. Cross-linked cellulose nanocrystal aerogels as viable bone tissue scaffolds. *Acta Biomater.* **2019**, *87*, 152–165. [CrossRef] [PubMed]
6. Khalil, A.; Adnan, A.S.; Yahya, E.B.; Olaiya, N.G.; Safrida, S.; Hossain, M.S.; Balakrishnan, V.; Gopakumar, D.A.; Abdullah, C.K.; Oyekanmi, A.A.; et al. A Review on plant cellulose nanofibre-based aerogels for biomedical applications. *Polymers* **2020**, *12*, 1759. [CrossRef] [PubMed]
7. Schmidt, A.H. Autologous bone graft: Is it still the gold standard? *Injury* **2021**, *52*, 18–22. [CrossRef]
8. Hovius, S.E.; de Jong, T. Bone grafts for scaphoid nonunion: An overview. *Hand Surg.* **2015**, *20*, 222–227. [CrossRef]
9. Falk, S.S.I.; Mittlmeier, T. Autologe spongiosa- und trikortikale spanentnahme aus dem hinteren beckenkamm [Harvesting cancellous bone or composite corticocancellous bone grafts from the posterior iliac crest]. *Oper. Orthop. Traumatol.* **2021**, *33*, 341–357. [CrossRef]
10. Nather, A.; Yusof, N.; Hilmy, N. *Allograft Procurement, Processing and Transplantation: A Comprehensive Guide for Tissue Banks;* World Scientific: Singapore, 2010; 565p.
11. Chaushu, L.; Chaushu, G.; Kolerman, R.; Vered, M.; Naishlos, S.; Nissan, J. Anterior atrophic mandible restoration using cancellous bone block allograft. *Clin. Implant. Dent. Relat. Res.* **2019**, *21*, 903–909. [CrossRef]
12. Jiang, X.Q. Biomaterials for bone defect repair and bone regeneration. *Chin. J. Stomatol.* **2017**, *52*, 600–604. [CrossRef]
13. Jordana, F.; Visage, L.C.; Weiss, P. Substituts osseux [Bone substitutes]. *Med. Sci.* **2017**, *33*, 60–65. [CrossRef]
14. Temmerman, A.; Cortellini, S.; Van Dessel, J.; De Greef, A.; Jacobs, R.; Dhondt, R.; Teughels, W.; Quirynen, M. Bovine-derived xenograft in combination with autogenous bone chips versus xenograft alone for the augmentation of bony dehiscences around oral implants: A randomized, controlled, split-mouth clinical trial. *J. Clin. Periodontol.* **2020**, *47*, 110–119. [CrossRef] [PubMed]
15. Mendoza-Azpur, G.; Fuente, A.; Chavez, E.; Valdivia, E.; Khouly, I. Horizontal ridge augmentation with guided bone regeneration using particulate xenogenic bone substitutes with or without autogenous block grafts: A randomized controlled trial. *Clin. Implant. Dent. Relat. Res.* **2019**, *4*, 521–530. [CrossRef] [PubMed]
16. Salamanca, E.; Hsu, C.C.; Huang, H.M.; Teng, N.C.; Lin, C.T.; Pan, Y.H.; Chang, W.J. Bone regeneration using a porcine bone substitute collagen composite in vitro and in vivo. *Sci. Rep.* **2018**, *8*, 984. [CrossRef] [PubMed]

17. Baldwin, P.; Li, D.J.; Auston, D.A.; Mir, H.S.; Yoon, R.S.; Koval, K.J. Autograft, allograft, and bone graft substitutes: Clinical evidence and indications for use in the setting of orthopaedic trauma surgery. *J. Orthop. Trauma.* **2019**, *33*, 203–213. [CrossRef]
18. Volova, L.T.; Rossinskaya, V.V.; Milyakova, M.N.; Boltovskaya, V.V.; Nefedova, I.F.; Kulagina, L.N.; Pugachev, E.I. Study of the influence of spaceflight factors on chondroblast culture in 3D model. *Aerosp. Environ. Med.* **2016**, *50*, 11–17. [CrossRef]
19. Volova, L.T.; Kotelnikov, G.P.; Lartsev, Y.V.; Dolgushkin, D.A.; Boltovskaya, V.V.; Terteryan, M.A. Features of regenerative processes in plasty of bone-cartilage defects by combined transplants based on autologous and allogenic cell cultures of rib cartilage. *Morphology* **2014**, *4*, 23–31.
20. Goff, E.; Buccino, F.; Bregoli, C.; McKinley, J.P.; Aeppli, B.; Recker, R.R.; Shane, E.; Cohen, A.; Kuhn, G.; Müller, R. Large-scale quantification of human osteocyte lacunar morphological biomarkers as assessed by ultra-high-resolution desktop micro-computed tomography. *Bone* **2021**, *152*, 116094. [CrossRef]
21. Timchenko, E.; Timchenko, P.; Volova, L.; Frolov, O.; Zibin, M.; Bazhutova, I. Raman spectroscopy of changes in the tissues of teeth with periodontitis. *Diagnostics* **2020**, *10*, 876. [CrossRef]
22. Butler, H.J.; Ashton, L.; Bird, B.; Cinque, G.; Curtis, K.; Dorney, J.; Esmonde-White, K.; Fullwood, N.J.; Gardner, B.; Martin-Hirsch, P.L.; et al. Using Raman spectroscopy to characterize biological materials. *Nat. Protoc.* **2016**, *11*, 664–687. [CrossRef] [PubMed]
23. Shah, F.A. Towards refining Raman spectroscopy-based assessment of bone composition. *Sci. Rep.* **2020**, *10*, 16662. [CrossRef] [PubMed]
24. Ilin, Y.; Kraft, M.L. Secondary ion mass spectrometry and Raman spectroscopy for tissue engineering applications. *Curr. Opin. Biotechnol.* **2015**, *31*, 108–116. [CrossRef] [PubMed]
25. Database UniProt Knowledge Base (UniProtKB). Available online: http://www.uniprot.org (accessed on 1 November 2019).
26. Baghy, K.; Reszegi, A.; Tátrai, P.; Kovalszky, I. Decorin in the tumor microenvironment. *Adv. Exp. Med. Biol.* **2020**, *1272*, 17–38. [CrossRef]
27. Paganini, C.; Costantini, R.; Superti-Furga, A.; Rossi, A. Bone and connective tissue disorders caused by defects in glycosaminoglycan biosynthesis: A panoramic view. *FEBS J.* **2019**, *286*, 3008–3032. [CrossRef]
28. Li, C.S.; Tian, H.; Zou, M.; Zhao, K.W.; Li, Y.; Lao, L.; Brochmann, E.J.; Duarte, M.E.; Daubs, M.D.; Zhou, Y.H.; et al. Secreted phosphoprotein 24 kD (Spp24) inhibits growth of human pancreatic cancer cells caused by BMP-2. *Biochem. Biophys. Res. Commun.* **2015**, *466*, 167–172. [CrossRef]
29. Futrega, K.; Music, E.; Robey, P.G.; Gronthos, S.; Crawford, R.; Saifzadeh, S.; Klein, T.J.; Doran, M.R. Characterisation of ovine bone marrow-derived stromal cells (oBMSC) and evaluation of chondrogenically induced micro-pellets for cartilage tissue repair in vivo. *Stem Cell Res. Ther.* **2021**, *12*, 1–19. [CrossRef]

polymers

Article

Antibacterial Electrospun Polycaprolactone Nanofibers Reinforced by Halloysite Nanotubes for Tissue Engineering

Viera Khunová [1,*], Mária Kováčová [2], Petra Olejniková [3], František Ondreáš [4,5], Zdenko Špitalský [2], Kajal Ghosal [6] and Dušan Berkeš [7]

1 Department of Plastics and Rubber, Faculty of Chemical and Food Technology, Slovak University of Technology, Radlinského 9, 81237 Bratislava, Slovakia
2 Polymer Institute, Slovak Academy of Sciences, Dúbravská Cesta 9, 84541 Bratislava, Slovakia; m.kovacova@savba.sk (M.K.); zdeno.spitalsky@savba.sk (Z.Š.)
3 Department of Biochemistry and Microbiology, Slovak University of Technology, Radlinského 9, 81237 Bratislava, Slovakia; petra.olejnikova@stub.sk
4 Central European Institute of Technology, Brno University of Technology, Purkynova 656/123, 61200 Brno, Czech Republic; frantisek.ondreas@ceitec.vutbr.cz
5 CONTIPRO a.s., Dolní Dobrouč 401, 56102 Dolní Dobrouč, Czech Republic
6 Department of Pharmaceutical Technology, Jadavpur University, Kolkata 700032, India; kajal.ghosal@gmail.com
7 Department of Organic Chemistry, Slovak University of Technology, Radlinského 9, 81237 Bratislava, Slovakia; dusan.berkes@stuba.sk
* Correspondence: viera.khunova@stuba.sk

Abstract: Due to its slow degradation rate, polycaprolactone (PCL) is frequently used in biomedical applications. This study deals with the development of antibacterial nanofibers based on PCL and halloysite nanotubes (HNTs). Thanks to a combination with HNTs, the prepared nanofibers can be used as low-cost nanocontainers for the encapsulation of a wide variety of substances, including drugs, enzymes, and DNA. In our work, HNTs were used as a nanocarrier for erythromycin (ERY) as a model antibacterial active compound with a wide range of antibacterial activity. Nanofibers based on PCL and HNT/ERY were prepared by electrospinning. The antibacterial activity was evaluated as a sterile zone of inhibition around the PCL nanofibers containing 7.0 wt.% HNT/ERY. The morphology was observed with SEM and TEM. The efficiency of HNT/ERY loading was evaluated with thermogravimetric analysis. It was found that the nanofibers exhibited outstanding antibacterial properties and inhibited both Gram- (*Escherichia coli*) and Gram+ (*Staphylococcus aureus*) bacteria. Moreover, a significant enhancement of mechanical properties was achieved. The potential uses of antibacterial, environmentally friendly, nontoxic, biodegradable PCL/HNT/ERY nanofiber materials are mainly in tissue engineering, wound healing, the prevention of bacterial infections, and other biomedical applications.

Keywords: biocompatible; antibacterial; halloysite; erythromycin; polycaprolactone; nanofibers; electrospinning; tissue engineering

Citation: Khunová, V.; Kováčová, M.; Olejniková, P.; Ondreáš, F.; Špitalský, Z.; Ghosal, K.; Berkeš, D. Antibacterial Electrospun Polycaprolactone Nanofibers Reinforced by Halloysite Nanotubes for Tissue Engineering. *Polymers* **2022**, *14*, 746. https://doi.org/10.3390/polym14040746

Academic Editors: Ariana Hudita and Bianca Gălăţeanu

Received: 20 January 2022
Accepted: 8 February 2022
Published: 15 February 2022

1. Introduction

Halloysite nanotubes (HNTs) are natural, nontoxic, biocompatible, eco-friendly, and low-cost materials recognized by the Environmental Protection Agency as nanomaterials (EPA 4). Presently, HNTs play a significant role in drug-carrier systems suitable for different biomedical applications, e.g., tissue engineering [1,2].

HNTs, naturally occurring in 1:1 layered aluminosilicate clay, consist of aluminum and silicon oxide layers rolled into tubes. The layers are rolled into tubes because of differences in the sizes of silicon and aluminum ions [3]. The typical length of the nanotubes is about 1–2 μm. Their outer and inner diameters range from 50 to 100 nm and from 10 to 50 nm, respectively [4,5]. For biomedical applications, the most significant advantages of HNTs compared to other tubular silicates are that they present a unique combination of structure, natural availability, rich functionality, good biocompatibility, and cytotoxicity [2,3,6–9].

One of the most remarkable features of HNTs is their different surface chemistries at the inner and outer sides of the tubes: silica sheets make up the external surfaces of the tubes and aluminum oxide makes up the inner (lumen) surface chemistry. Furthermore, alumina has a positive charge up to pH 8.5, and silica has a negative charge at pH values above 1.5 [10,11]. Due to differently charged outer and inner sides, it is possible to utilize HNTs as multifunctional nanocontainers for the selective modification of the outer and inner sides of nanotubes [7,11–16].

Halloysite's inner diameter fits well to macromolecules and proteins [6]. In this regard, drugs of smaller molecular size are typically vacuum-loaded within the inner lumen of the nanotube, and drugs with larger molecular size can attach to the outer surface of the halloysite [17].

HNTs have also been successfully used as low-cost nanocontainers for several antibiotics such aminoglycoside gentamicin [18] and β-lactam antibiotic amoxicillin [19]. In addition, a wide range of applications of vancomycin-loaded halloysite nanotubes have been presented in alginate-based wound dressing [20] and silk fibroin hydrogel applicable for bone tissue engineering [21].

Electrospinning is a technology for the fabrication of continuous nanofibers with a simple setup [22]. Electrospun nanofibers can be prepared from natural or synthetic polymers or their blends. In the past decade, significant progress has been achieved in researching advanced electrospun nanofibers for biomedical applications or one-dimensional nanofibers made of intrinsically conducting polymers [23]. Recent advances in the electrospinning of functional scaffolds for tissue engineering and nanofiber scaffolds were summarized in the work of Hanumantharao et al. [24]. Simultaneously with the improvement of electrospun process technology, a lot of effort is being made to study new types of antimicrobial nanoparticles or make the known ones much more effective against microbial effects [25–28].

At present, there have been many examples where a biodegradable polymer matrix was combined with antimicrobial nanofillers (e.g., metals and/or metal oxides) for the preparation of polymer nanocomposites for biomedical applications [29–31]. An extensively used synthetic, biodegradable, semi-crystalline polymer used for biomedical applications is polycaprolactone (PCL) [32]. PCL is well known for its versatile use, biocompatibility, chemical stability, thermal stability, slow biodegradation (around 24 months), tissue compatibility, and easy processing [31,33]. PCL is often used in biomedicine as an FDA-approved material in the form of nanofibers that have evolved as controllable drug delivery systems [34,35]. Furthermore, due to its slow degradation rate, PCL is a preferred polymer mainly used as a long-term drug delivery carrier. Another advantage of PCL is its compatibility with a wide range of drugs, which provides a homogenous distribution of predominantly lipophilic drugs in the carrier matrix due to its hydrophobic nature [36]. Moreover, in our earlier work, it was confirmed that except for an important and strong reinforcing effect, all studied PCL/Gel nanofibers with an HNT content from 0.5 to 9.0 wt.% were non-toxic and had no effect on cell behavior [35]. Thus, through a combination of low-strength PCL with drug-loaded halloysite, it is possible prepare reinforced biodegradable polymer nanocomposites with regular drug release.

In this work, electrospun nanofibers based on PCL and drug-loaded HNTs were studied. HNTs were used as a nanocarrier for erythromycin (ERY) as a model antibacterial active compound with wide range of antibacterial activity on both Gram-positive and Gram-negative bacteria. ERY—a macrolide antibiotic used to treat a number of bacterial infections, e.g., respiratory tract infections and pelvic inflammatory disease—was used for HNT loading. In addition, erythromycin is clinically used in dermatology as a very effective topical antibiotic drug in treating bacterial skin disease.

ERY is soluble in methanol but has minimal solubility in an aqueous solution and presents acid instability, which limit its broader application. Therefore, drug-carrier systems based on HNTs and ERY have great potential to overcome these weaknesses. Moreover, cheap ERY was chosen as a model substrate for another type of pH-sensitive macrolide antibiotics intensively studied by our group [37].

2. Materials and Methods

2.1. Materials

PCL CAPA® 6800 (M_w = 80,000) and Erythromycin E6376 were obtained from Sigma-Aldrich Saint Louis, MO 63103, USA and ULTRA HalloPure, respectively, produced by I-Minerals Inc., Vancouver, BC, Canada. The ULTRA HalloPure comprised purified HNTs with 93.5% of halloysite, 6.1% of kaolinite, and 0.4% of quartz from Dragon Mine in Utah of USA, produced by Applied Minerals Inc., New York, NY, USA. The dimensions are presented in Table 1. Methanol and distilled water were purchased from Lachner s.r.o., Neratovice, Czech Republic.

Table 1. Structural parameters of ULTRA HalloPure.

Length (µm)	Inner Diameter (nm)	Outer Diameter (µm)	Aspect Ratio
1.0–2.0	15–20	0.10–0.20	Typically 15

2.2. Loading of a Drug in Halloysite Nanotubes

The loading of active agents to halloysite is based on the diffusion of molecules from an external solution into the inner part of HNTs due to the concentration gradient. The evaporation of the solvent under the vacuum elevates the concentration of the active agents in the solution and enhances the diffusion rate. Therefore, fast-drying solvents with low viscosities such as acetone or ethanol are preferable for organic substances [38]. To eliminate potential water on the outer part of HNTs before the loading procedure, HNTs were dried in the oven for 2 h at 150 °C. Since ERY is soluble in methanol, after the drying, it was dissolved in methanol. HNTs were dispersed in the ERY solution and sonicated for 20 min. The rate of halloysite to erythromycin was 60:40. After the sonification, halloysite loaded with ERY was dried in a vacuum and washed with distilled water using a 0.2 µm membrane filter. Residual unloaded ERY was removed with methanol (4 h).

HNTs were dried in the oven for 2 h at 150 °C. We dissolved 400 mg of the drug in 25 mL of methanol. We dispersed 600 mg of dried halloysite in a drug solution (ratio 40:60 ERY:HNT), which was then sonicated for 20 min. Then, this dispersion was dried in a vacuum. After vacuum drying, it was washed with distilled water using a 0.2 µm membrane filter. Residual unloaded ERY was removed with methanol (4 h).

2.3. Optimization of Electrospinning Process Parameters

The electrospinning process for PCL composites (Figure 1) was optimized and previously described by our group [29]. Briefly, a solution of 10% w/v PCL was prepared in a mixture of chloroform and methanol in a ratio of 4:1 and continuously stirred for 2 h. Then, the mixture was kept unstirred for another 15 min to remove any bubbles present in the solution. Prepared HNTs loaded with ERY were added to the PCL solution and stirred for another 1 h. Electrospun fibers were fabricated using a Spellman high voltage power source (Spellman High Voltage Electronics Corporation, New York, NY, USA) and syringe pump (New Era Pump Systems, Inc., New York, NY, USA). The electrospinning apparatus consisted of a 10 mL syringe that was integrated with a grounded electrode, and the needle diameter was 0.41 mm. The distance between tee collector and source was kept to 13 cm. One thin aluminum sheet was fixed over the static collector. The feeding rate for the electrospinning solution was set to 1 mL/h at a voltage of 25 kV. The temperature for the location varied between 21 and 29 °C, and the humidity varied from 73% to 93%.

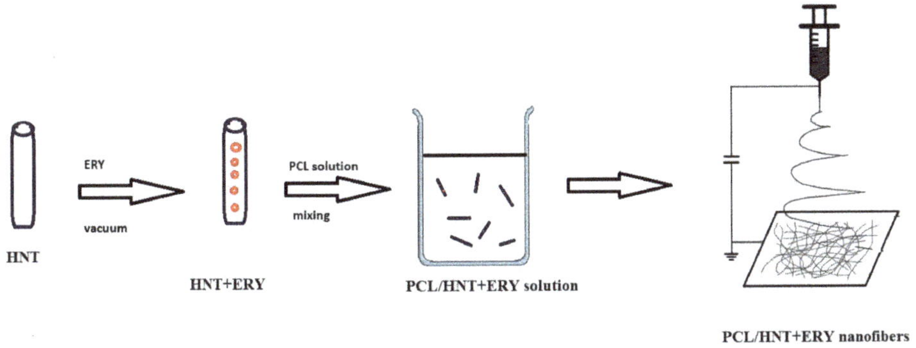

Figure 1. Scheme of electrospun nanofiber preparation.

2.4. Scanning Electron Microscopy

For the needs of scanning electron microscopy (SEM), the sample's surface was covered in gold with a Sample Preparation System Quorum Technologies Q150R S/E/ES sputter coater evaporator (Quorum Technologies, Laughton, England, and the micrographs were obtained with an FIB Microscope Quanta 3D 200i (FEI Company, Tokyo, Japan) in a secondary electron mode at different magnitudes. The surface morphologies of pure and modified fibers were obtained.

2.5. Transmission Electron Microscopy

The structure of HNTs loaded with ERY was characterized by transmission electron microscopy (Jeol TEM 1200EX, JEOL Ltd., Tokyo, Japan) at an accelerating voltage of 100 kV. The sample was dispersed on a copper grid with carbon support film.

2.6. Antibacterial Activity

The antibacterial activity of HNT/ERY before and after removing residual ERY from HNTs was assayed. The antibacterial activity of PCL/HNT/ERY nanofibers was assessed with the diffusion method by placing the HNT, HNT:ERY, 1 cm^2 of prepared PCL fiber, PCL/HNT fiber, and PCL/HNT and ERY fiber on the inoculated (10^6 cells/mL) MHA (Mueller Hinton Agar) growth media. For inoculation, the model bacteria *S. aureus* CCM 3953 (Czech Collection of Microorganisms) and *E. coli* CCM 3988 were used. The antibacterial activity of prepared nanofibers was evaluated as the occurrence of a sterile zone of inhibition around the modified HNTs with ERY and modified fibers of PCL/HNT and ERY, respectively, after cultivation for 24 h at 37 °C. With the aim to wash out the residual ERY from the surface of the prepared materials, modified HNTs with ERY and the PCL/HNT and ERY fibers were washed two times in methanol and two times in water. The antibacterial activity was evaluated again as described above. To evaluate the sustained release of ERY from the PCL/HNT fiber, serial cultivation was examined as follows. The fresh prepared PCL/HNT and ERY fibers were first placed on the inoculated growth media, and the antibacterial effect was evaluated after 24 h of cultivation at 37 °C. Then, these same fibers from the grown bacterial cultures were replaced with fresh ones that were again cultivated and considered. This procedure was repeated for 5 days.

2.7. Mechanical Property Measurement

Mechanical properties were tested in uniaxial tension at a crosshead speed of 5 $mm \cdot min^{-1}$ and ambient temperature of 22 °C using a Zwick Roell Z010 (Zwick-Roell, Ulm, Germany) equipped with a 10 kN load cell. Six specimens of rectangle shape (approximately 30 × 5 mm^2) were tested for each type of material, and the averages and standard deviations were determined. The thickness of the specimens was measured with a micrometer.

2.8. Thermogravimetric Analysis

Thermal stability was determined by thermogravimetric analysis (TGA) using a Q1500 D instrument (from MON Budapest, with TA Universal Analysis software). Each sample (100 mg) was heated from 30 to 600 °C at a heating rate of 10 °C/min in the presence of air with a flow of 50 mL/min. The corresponding weight loss was recorded as a function of temperature.

3. Results and Discussion

3.1. Microscopic HNT/ERY' Structure Analysis

As our task was the preparation of antimicrobial nanofibers for biomedical applications, the main criterion for the selection of an appropriate type of halloysite was its purity. For this reason, we selected ULTRA HalloPureTM tubular halloysite. This particular HNT type has a very high purity (93.5%), with only trace levels of feldspar. The second reason for our study was ULTRA HalloPure's geometry. From the point of view reinforcing effect, which is essential for nanofibers based on low-strength PCL, it is necessary to use HNTs with long aspect ratios. Moreover, long-aspect-ratio HNTs provide more prolonged antibacterial effects due to the release of an active compound. The geometry of the used HNT type (length, inner and outer diameter, and aspect ratio) is presented in Table 1.

The micromorphology and nanomorphology of HNTs both modified and loaded with ERY were studied with transmission electron microscopy. The structure of HNTs before ERY loading is shown in Figure 2. TEM images of HNTs loaded with ERY are shown in Figure 3.

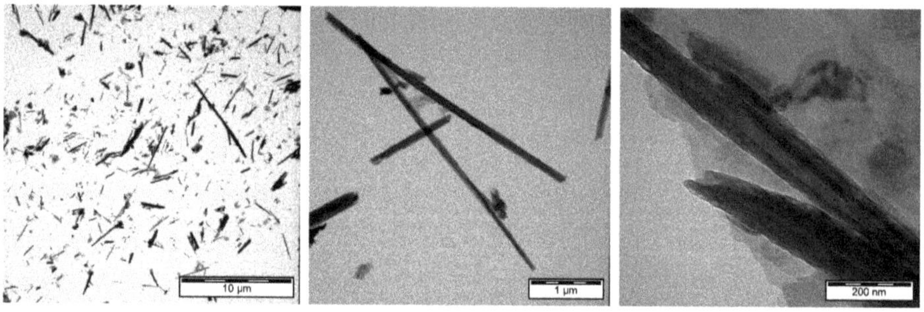

Figure 2. TEM of unmodified HNTs before ERY loading.

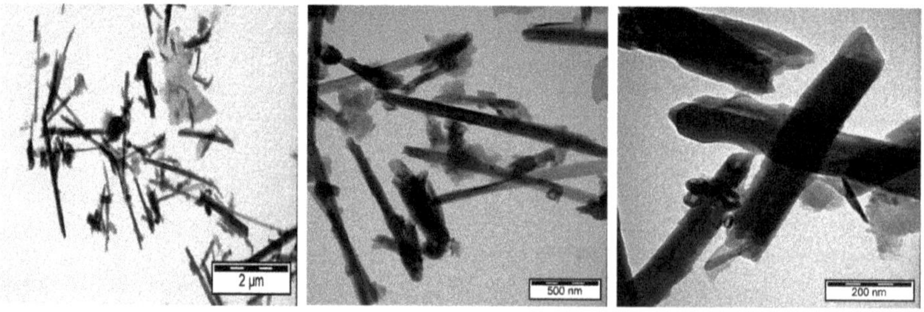

Figure 3. TEM of HNTs loaded with ERY.

3.2. Loading Efficiency of HNTs

A thermogravimetric study of HNTs and HNT/ERY was performed to determine loading efficiency, i.e., the amount of ERY loaded into HNTs. As mentioned earlier, the loading of active agents to HNTs is based on the diffusion of molecules from external solutions into the inner part of HNTs due to the concentration gradient. As shown in Figure 3, the unmodified HNTs were stable up to 500 °C, at which point de-hydroxylation [6,39] occurred with a disruption of the tube-wall multilayer packing. At 600 °C, the weight loss of the unmodified HNTs was 12.5 wt.% (blue curve in Figure 4). Pure erythromycin showed steep decomposition around 300 °C followed by a slow degradation process up to 600 °C. (red curve in Figure 4). The efficiency of the used loading method was evaluated via a comparison TGA of pristine HNTs with HNTs loaded with ERY (HNT/ERY) washed with water (green curve in Figure 4) and methanol (orange curve in Figure 4).

The sample of HNT/ERY washed in water (HNT/ERY/WAT) showed the main decomposition around 300 °C followed by gradual degradation process and further degradation step around 500 °C, reflecting the behavior of precursors.

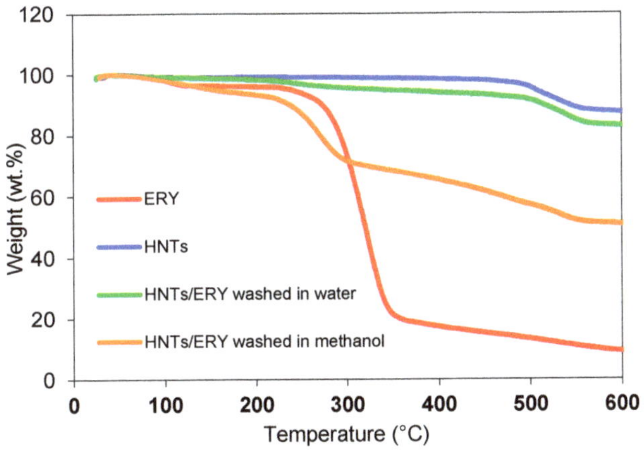

Figure 4. TG curves of ERY, HNTs, HNT/ERY washed in water, and HNT/ERY washed in methanol.

The weight loss of HNT/ERY/WAT—37.9 wt.% at 500 °C—corresponded to the rate of erythromycin loaded on both the inner and outside surface of HNTs (HNT:ERY at 60:40) and the rate of ERY loaded exclusively onto the inner surface of HNTs, as evaluated on the sample from which residual ERY was removed with methanol: 4.6 and 11.5 vol.%, respectively. In summary, we obtained extremely highly-loaded HNTs suitable for antibacterial applications in biomedicine.

3.3. Morphological Observation of PCL/HNT/ERY Nanofibers

Our previous study described the morphology of electrospun PCL antibacterial composites with different fillers—hydrophobic quantum dots working on the principle of photodynamic therapy—for wound healing in tissue engineering in detail [29]. A strongly porous material with interconnected structures was obtained via the electrospinning method. This 3D structure is very suitable for use in tissue engineering. Despite this fact, we created relatively homogeneous nanofibers with an average thickness of 3–4 μm, and the same effect was not observed in the case of HNT-loaded samples. As shown in Figure 5, the fibers contained a lot of beads in their structure due to the very high probability of the electric conductivity of the electrospinning solution to change, which is one of the key factors of the electrospinning process. Hydrophobic quantum dots are semiconductors, and HNTs or ERY-loaded HNTs are insulators. Therefore, the final electric conductivity of the solution was significantly lower and the viscosity of solution was reduced. All those parameters affected the final structure of the fibers. These beads had only a minor effect on antibacterial and mechanical properties because these were determined by releasing ERY from HNTs, as shown later.

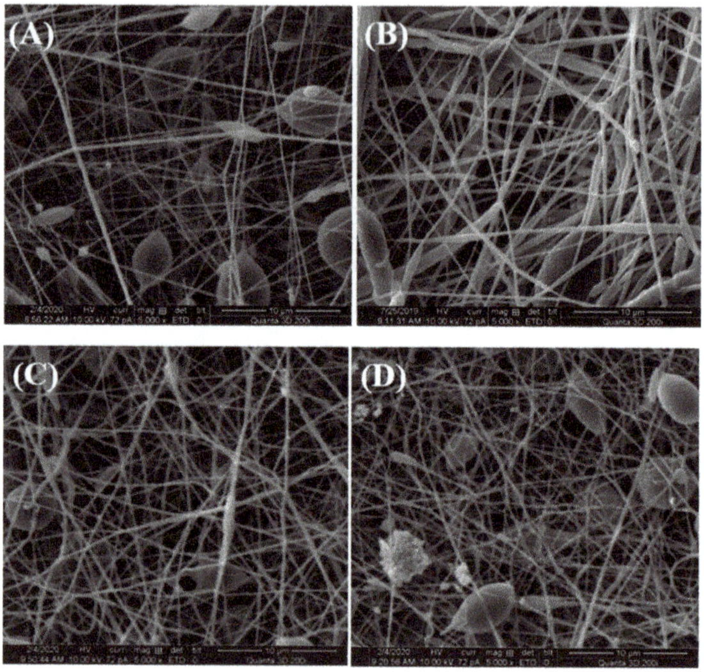

Figure 5. SEM images at high magnification (5000×) in the secondary electron mode of electrospun (**A**) pure PCL, (**B**) PCL and 6 wt.% HNT nanofibers, (**C**) PCL and 6 wt.% HNT/ERY (80:20), and (**D**) PCL and 6 wt.% HNT/ERY 60:40.

3.4. Antibacterial Properties of PCL/Halloysite/ERY

The antibacterial activity of HNT/ERY before and after the removal of residual ERY from HNT surfaces was assayed (Figure 6), and the obtained antibacterial properties of PCL/HNT and ERY are shown in Figure 7. Because PCL and HNT/PCL were not antibacterial active, as is clear from Figure 7, no inhibition zone was formed; the HNT/ERY and PCL/HNT/ERY nanofibers revealed significant antibacterial activity.

Zones of inhibition were observed around the HNT and ERY fibers and the = PCL/HNT and ERY fibers. After the successful incorporation of ERY into the carrier HNTs (proving that it was also a part of the prepared PCL/HNT fibers), the washing of the fibers in methanol followed by washing in water was conducted to ensure that the whole residual ERY of the surface was removed. Next, the antibacterial assay was repeated with washed materials. As is clearly shown in Figures 6 and 7, even though the zones of inhibition of washed samples were smaller, antibacterial activity was recorded again, which means that the antibiotic was incorporated into the HNT nanostructures. Accordingly, HNTs are an appropriate carrier for ERY.

Finally, the antibacterial activity was assayed when the PCL/HNT and ERY fibers were repeatedly (three times) replaced on fresh inoculated growth media. The results are shown in Table 2. These results support the idea of sustained release from the prepared fiber because the antibacterial effect was still observed after the second replacement (72 h) for *E. coli* and after the first replacement (48 h) for *S. aureus* (Table 2).

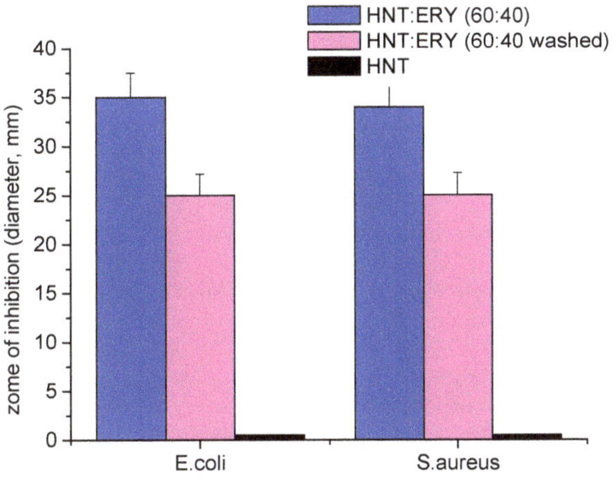

Figure 6. Comparison of the antibacterial activity of HNT/ERY before and after removing ERY from the outside surface of HNTs assayed with the disk diffusion method on *E. coli* and *S. aureus*. The antibacterial activity was recorded as the inhibition zone diameter ± SE (standard error).

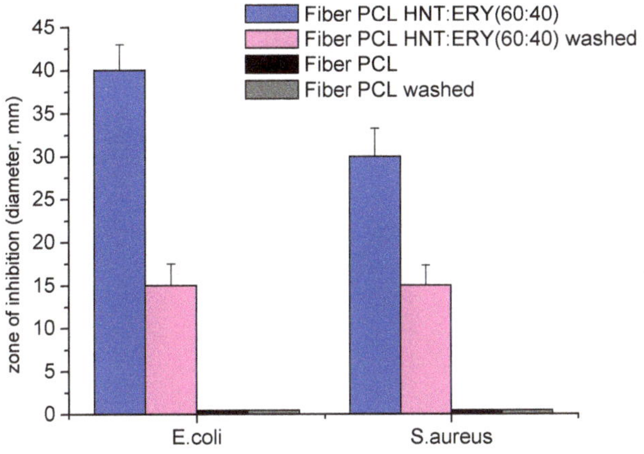

Figure 7. Comparison of the antibacterial activity of washed and unwashed PCL/HNT fibers with ERY assayed with the disk diffusion method on *E. coli* and *S. aureus*. The antibacterial activity was recorded as the inhibition zone diameter ± SE (standard error).

Table 2. Antibacterial activity of PCL/HNT and ERY fiber after 3x replacement.

Model Bacteria	Zone of Inhibition (mm)		
	Mode of Application	**Fiber PCL/HNT**	**Fiber PCL/HNT and ERY**
E. coli	Placed on the growth media	0	40
	Replaced for the 1st time	0	35
	Replaced for the 2nd time	**0**	**21**
	Replaced for the 3rd time	0	0
S. aureus	Placed on the growth media	0	30
	Replaced for the 1st time	**0**	**30**
	Replaced for the 2nd time	0	0
	Replaced for the 3rd time	0	0

3.5. Mechanical Properties of PCL/HNT/ERY Nanofibers

As it is clearly evident, the mechanical properties of investigated PCL nanofiber system were significantly influenced by the presence of both unmodified and HNT-loaded ERY (Figure 8, Table 3). The introduction of HNT particles into nanofibers significantly increased the Young's modulus and tensile strength. Furthermore, elongation at break remained unchanged for this system. The observed enhancement correlated well with the results obtained on PCL [40]. Stress transfer, volume replacement, and segmental immobilization [41–43] reinforcing mechanisms were responsible for this enhancement of mechanical performance. The loading of ERY into HNTs decreased Young's modulus to the value of the PCL while elongation at break was significantly increased. The observed trend re-affirmed the effect of plasticizers—small compatible molecules that are able to decrease the glass transition temperature of polymers—and suggested the good compatibility of ERY with the PCL/HNT nanofibers. More importantly, tensile strength also decreased upon ERY loading. However, the decrease was so slight that the values of the PCL/HNT and ERY systems were still more than 100% higher than those of the pure PCL system. A similar tensile strength enhancement (about 100%) was observed for PCL/gelatin/HNT microfiber system loaded with metronidazole [44].

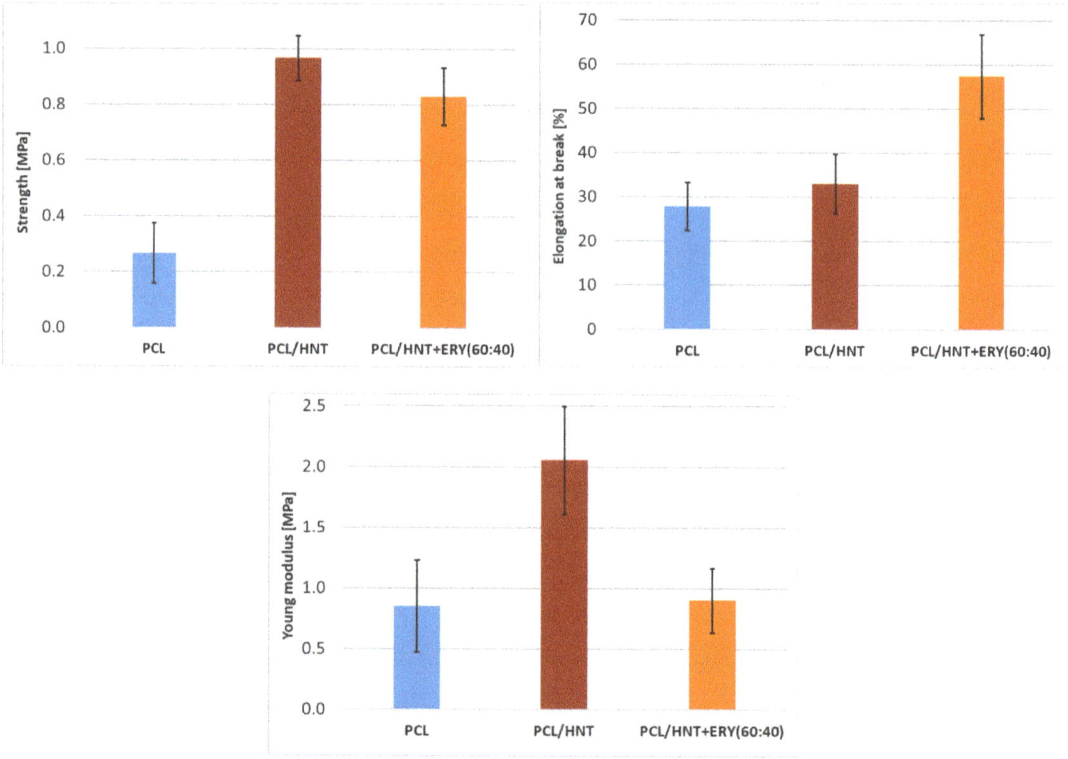

Figure 8. Tensile strength, elongation at break, and Young's modulus of PCL, PCL/HNT, and PCL/HNT and ERY nanofibers.

Table 3. Young's modulus, tensile strength, and elongation at break of PCL, PCL/HNT, and PCL/HNT and ERY samples.

Sample	Young's Modulus (MPa)	Strength (MPa)	Elongation at Break (%)
PCL	0.9 ± 0.4	0.3 ± 0.1	27.9 ± 5.4
PCL/HNT	2.1 ± 0.4	1.0 ± 0.1	32.9 ± 6.8
PCL/HNT and ERY(80:20)	0.7 ± 0.4	0.76 ± 0.1	56.8 ± 9.2
PCL/HNT and ERY(60:40)	0.9 ± 0.3	0.83 ± 0.1	57.4 ± 9.5

4. Conclusions

In this work, antibacterial, biodegradable, environmentally friendly, and nontoxic nanofibers with good biocompatibility and cost-effective production were prepared via the electrospinning of PCL and HNTs loaded with ERY. The PCL/HNT/ERY nanofibers exhibited outstanding antibacterial properties and resulted in the inhibition of both Gram-negative (*Escherichia coli*) and Gram-positive (*Staphylococcus aureus*) bacteria. We observed the gradual release of ERY out of the nanofibers, that was detected as repetitive antimicrobial activity after the material replacement on fresh prepared inoculated growth media. In addition to the antibacterial properties, the significant enhancement of mechanical properties was achieved via the incorporation of both unmodified and ERY-loaded HNTs, thus improving suitability of the system for medical applications. To summarize, PCL/HNT/ERY nanofibers with strong antibacterial effects have great potential as a progressive new biomedical material in both tissue engineering and a number of other biomedical applications.

Author Contributions: Conceptualization, V.K.; methodology, V.K.; investigation, V.K., M.K., K.G., F.O., P.O. and D.B.; writing—original draft preparation, V.K.; writing—review and editing, V.K., M.K, and Z.Š.; supervision, V.K.; funding acquisition, V.K and Z.Š. All authors have read and agreed to the published version of the manuscript.

Funding: The authors would like to thank the Slovak Grant Agency for financial assistance, project VEGA 1/0486/19 and VEGA 1/0411/22. Frantisek Ondreas greatly acknowledges the funding under the project LTAUSA19059 Inter-Excellence Grant from MEYS CR.

Institutional Review Board Statement: Not applicable.

Informed Consent Statement: Not applicable.

Data Availability Statement: The data presented in this study are available on request from the corresponding author.

Acknowledgments: The authors also would like to thank I-Minerals Inc., Vancouver, BC, Canada, for providing of ULTRA HalloPure™ halloysite sample.

Conflicts of Interest: The authors declare no conflict of interest.

References

1. Santos, A.C.; Pereira, I.; Reis, S.; Veiga, F.; Saleh, M.; Lvov, Y. Biomedical potential of clay nanotube formulations and their toxicity assessment. *Expert Opin. Drug Deliv.* **2019**, *16*, 1169–1182. [CrossRef] [PubMed]
2. Vergaro, V.; Abdullayev, E.; Lvov, Y.M.; Zeitoun, A.; Cingolani, R.; Rinaldi, R.; Leporatti, S. Cytocompatibility and uptake of halloysite clay nanotubes. *Biomacromolecules* **2010**, *11*, 820–826. [CrossRef]
3. Atyaksheva, L.F.; Kasyanov, I.A. Halloysite, Natural Aluminosilicate Nanotubes: Structural Features and Adsorption Properties (A Review). *Pet. Chem.* **2021**, *61*, 932–950. [CrossRef]
4. Papoulis, D.; Komarneni, S.; Panagiotaras, D. Geochemistry of halloysite-7Å formation from plagioclase in trachyandesite rocks from Limnos Island, Greece. *Clay Miner.* **2014**, *49*, 75–89. [CrossRef]
5. Hanif, M.; Jabbar, F.; Sharif, S.; Abbas, G.; Farooq, A.; Aziz, M. Halloysite nanotubes as a new drug-delivery system: A review. *Clay Miner.* **2016**, *51*, 469–477. [CrossRef]
6. Joussein, E.; Petit, S.; Churchman, J.; Theng, B.; Righi, D.; Delvaux, B. Halloysite clay minerals—A review. *Clay Miner.* **2005**, *40*, 383–426. [CrossRef]
7. Leporatti, S. Halloysite clay nanotubes as nano-bazookas for drug delivery. *Polym. Int.* **2017**, *66*, 1111–1118. [CrossRef]

8. Jaurand, M.C. Chapter 20—An Overview on the Safety of Tubular Clay Minerals. In *Developments in Clay Science*; Yuan, P., Thill, A., Bergaya, F., Eds.; Elsevier: Amsterdam, The Netherlands, 2016; Volume 7, pp. 485–508. [CrossRef]
9. Stavitskaya, A.; Batasheva, S.; Vinokurov, V.; Fakhrullina, G.; Sangarov, V.; Lvov, Y.; Fakhrullin, R. Antimicrobial applications of clay nanotube-based composites. *Nanomaterials* **2019**, *9*, 708. [CrossRef]
10. Veerabadran, N.G.; Price, R.R.; Lvov, Y.M. Clay Nanotubes for Encapsulation and Sustained Release of Drugs. *Nano* **2007**, *2*, 115–120. [CrossRef]
11. Abdullayev, E.; Lvov, Y. Halloysite for Controllable Loading and Release. *Dev. Clay Sci.* **2016**, *7*, 554–605. [CrossRef]
12. Lazzara, G.; Riela, S.; Fakhrullin, R.F. Clay-based drug-delivery systems: What does the future hold? *Ther. Deliv.* **2017**, *8*, 633–646. [CrossRef]
13. Yendluri, R.; Lvov, Y.; de Villiers, M.M.; Vinokurov, V.; Naumenko, E.; Tarasova, E.; Fakhrullin, R. Paclitaxel Encapsulated in Halloysite Clay Nanotubes for Intestinal and Intracellular Delivery. *J. Pharm. Sci.* **2017**, *106*, 3131–3139. [CrossRef] [PubMed]
14. Massaro, M.; Cavallaro, G.; Colletti, C.G.; D'Azzo, G.; Guernelli, S.; Lazzara, G.; Pieraccini, S.; Riela, S. Halloysite nanotubes for efficient loading, stabilization and controlled release of insulin. *J. Colloid Interface Sci.* **2018**, *524*, 156–164. [CrossRef] [PubMed]
15. Tan, D.; Yuan, P.; Annabi-Bergaya, F.; Liu, D.; Wang, L.; Liu, H.; He, H. Loading and in vitro release of ibuprofen in tubular halloysite. *Appl. Clay Sci.* **2014**, *96*, 50–55. [CrossRef]
16. Nazir, M.S.; Mohamad Kassim, M.H.; Mohapatra, L.; Gilani, M.A.; Raza, M.R.; Majeed, K. Characteristic Properties of Nanoclays and Characterization of Nanoparticulates and Nanocomposites. In *Nanoclay Reinforced Polymer Composites*; Springer: Berlin/Heidelberg, Germany, 2016; pp. 35–55. [CrossRef]
17. Santos, A.C.; Ferreira, C.; Veiga, F.; Ribeiro, A.J.; Panchal, A.; Lvov, Y.; Agarwal, A. Halloysite clay nanotubes for life sciences applications: From drug encapsulation to bioscaffold. *Adv. Colloid Interface Sci.* **2018**, *257*, 58–70. [CrossRef] [PubMed]
18. Luo, Y.; Humayun, A.; Mills, D.K. Surface modification of 3D printed PLA/halloysite composite scaffolds with antibacterial and osteogenic capabilities. *Appl. Sci.* **2020**, *10*, 3971. [CrossRef]
19. Wei, W.; Minullina, R.; Abdullayev, E.; Fakhrullin, R.; Mills, D.; Lvov, Y. Enhanced efficiency of antiseptics with sustained release from clay nanotubes. *RSC Adv.* **2014**, *4*, 488–494. [CrossRef]
20. Kurczewska, J.; Pecyna, P.; Ratajczak, M.; Gajęcka, M.; Schroeder, G. Halloysite nanotubes as carriers of vancomycin in alginate-based wound dressing. *Saudi Pharm. J.* **2017**, *25*, 911–920. [CrossRef]
21. Avani, F.; Damoogh, S.; Mottaghitalab, F.; Karkhaneh, A.; Farokhi, M. Vancomycin loaded halloysite nanotubes embedded in silk fibroin hydrogel applicable for bone tissue engineering. *Int. J. Polym. Mater. Polym. Biomater.* **2020**, *69*, 32–43. [CrossRef]
22. Hu, X.; Liu, S.; Zhou, G.; Huang, Y.; Xie, Z.; Jing, X. Electrospinning of polymeric nanofibers for drug delivery applications. *J. Control. Release* **2014**, *185*, 12–21. [CrossRef] [PubMed]
23. Pierini, F.; Lanzi, M.; Lesci, I.G.; Roveri, N. Comparison between inorganic geomimetic chrysotile and multiwalled carbon nanotubes in the preparation of one-dimensional conducting polymer nanocomposites. *Fibers Polym.* **2015**, *16*, 426–433. [CrossRef]
24. Hanumantharao, S.N.; Rao, S. Multi-functional electrospun nanofibers from polymer blends for scaffold tissue engineering. *Fibers* **2019**, *7*, 66. [CrossRef]
25. Gharpure, S.; Akash, A.; Ankamwar, B. A Review on Antimicrobial Properties of Metal Nanoparticles. *J. Nanosci. Nanotechnol.* **2019**, *20*, 3303–3339. [CrossRef]
26. Kováčová, M.; Špitalská, E.; Markovic, Z.; Špitálský, Z. Carbon Quantum Dots as Antibacterial Photosensitizers and Their Polymer Nanocomposite Applications. *Part. Part. Syst. Charact.* **2020**, *37*, 1900348. [CrossRef]
27. Shanmuganathan, R.; LewisOscar, F.; Shanmugam, S.; Thajuddin, N.; Alharbi, S.A.; Alharbi, N.S.; Brindhadevi, K.; Pugazhendhi, A. Core/shell nanoparticles: Synthesis, investigation of antimicrobial potential and photocatalytic degradation of Rhodamine B. *J. Photochem. Photobiol. B Biol.* **2020**, *202*, 111729. [CrossRef]
28. Jee, S.C.; Kim, M.; Shinde, S.K.; Ghodake, G.S.; Sung, J.S.; Kadam, A.A. Assembling ZnO and Fe3O4 nanostructures on halloysite nanotubes for anti-bacterial assessments. *Appl. Surf. Sci.* **2020**, *509*, 145358. [CrossRef]
29. Ghosal, K.; Kováčová, M.; Humpolíček, P.; Vajďák, J.; Bodík, M.; Špitalský, Z. Antibacterial photodynamic activity of hydrophobic carbon quantum dots and polycaprolactone based nanocomposite processed via both electrospinning and solvent casting method. *Photodiagnosis Photodyn. Ther.* **2021**, *35*, 102455. [CrossRef]
30. Kraśniewska, K.; Galus, S.; Gniewosz, M. Biopolymers-based materials containing silver nanoparticles as active packaging for food applications–A review. *Int. J. Mol. Sci.* **2020**, *21*, 698. [CrossRef]
31. Mochane, M.J.; Motsoeneng, T.S.; Sadiku, E.R.; Mokhena, T.C.; Sefadi, J.S. Morphology and properties of electrospun PCL and its composites for medical applications: A mini review. *Appl. Sci.* **2019**, *9*, 2205. [CrossRef]
32. Malikmammadov, E.; Tanir, T.E.; Kiziltay, A.; Hasirci, V.; Hasirci, N. PCL and PCL-based materials in biomedical applications. *J. Biomater. Sci. Polym. Ed.* **2018**, *29*, 863–893. [CrossRef] [PubMed]
33. Karuppuswamy, P.; Reddy Venugopal, J.; Navaneethan, B.; Luwang Laiva, A.; Ramakrishna, S. Polycaprolactone nanofibers for the controlled release of tetracycline hydrochloride. *Mater. Lett.* **2015**, *141*, 180–186. [CrossRef]
34. Jiang, S.; Chen, Y.; Duan, G.; Mei, C.; Greiner, A.; Agarwal, S. Electrospun nanofiber reinforced composites: A review. *Polym. Chem.* **2018**, *9*, 2685–2720. [CrossRef]
35. Pavliňáková, V.; Fohlerová, Z.; Pavliňák, D.; Khunová, V.; Vojtová, L. Effect of halloysite nanotube structure on physical, chemical, structural and biological properties of elastic polycaprolactone/gelatin nanofibers for wound healing applications. *Mater. Sci. Eng. C* **2018**, *91*, 94–102. [CrossRef]

36. Williamson, M.R.; Chang, H.I.; Coombes, A.G.A. Gravity spun polycaprolactone fibres: Controlling release of a hydrophilic macromolecule (ovalbumin) and a lipophilic drug (progesterone). *Biomaterials* **2004**, *25*, 5053–5060. [CrossRef] [PubMed]
37. Ferko, B.; Zeman, M.; Formica, M.; Veselý, S.; Doháňošová, J.; Moncol, J.; Olejníková, P.; Berkeš, D.; Jakubec, P.; Dixon, D.J.; et al. Total Synthesis of Berkeleylactone A. *J. Org. Chem.* **2019**, *84*, 7159–7165. [CrossRef] [PubMed]
38. Abdullayev, E.; Lvov, Y. Halloysite clay nanotubes for controlled release of protective agents. *J. Nanosci. Nanotechnol.* **2011**, *11*, 10007–10026. [CrossRef]
39. Duce, C.; Vecchio Ciprioti, S.; Ghezzi, L.; Ierardi, V.; Tinè, M.R. Thermal behavior study of pristine and modified halloysite nanotubes: A modern kinetic study. *J. Therm. Anal. Calorim.* **2015**, *121*, 1011–1019. [CrossRef]
40. Nitya, G.; Nair, G.T.; Mony, U.; Chennazhi, K.P.; Nair, S.V. In vitro evaluation of electrospun PCL/nanoclay composite scaffold for bone tissue engineering. *J. Mater. Sci. Mater. Med.* **2012**, *23*, 1749–1761. [CrossRef]
41. Ondreas, F.; Lepcio, P.; Zboncak, M.; Zarybnicka, K.; Govaert, L.E.; Jancar, J. Effect of Nanoparticle Organization on Molecular Mobility and Mechanical Properties of Polymer Nanocomposites. *Macromolecules* **2019**, *52*, 6250–6259. [CrossRef]
42. Zboncak, M.; Ondreas, F.; Uhlir, V.; Lepcio, P.; Michalicka, J.; Jancar, J. Translation of segment scale stiffening into macroscale reinforcement in polymer nanocomposites. *Polym. Eng. Sci.* **2020**, *60*, 587–596. [CrossRef]
43. Jancar, J.; Ondreas, F.; Lepcio, P.; Zboncak, M.; Zarybnicka, K. Mechanical properties of glassy polymers with controlled NP spatial organization. *Polym. Test.* **2020**, *90*, 106640. [CrossRef]
44. Xue, J.; Niu, Y.; Gong, M.; Shi, R.; Chen, D.; Zhang, L.; Lvov, Y. Electrospun microfiber membranes embedded with drug-loaded clay nanotubes for sustained antimicrobial protection. *ACS Nano* **2015**, *9*, 1600–1612. [CrossRef] [PubMed]

polymers

MDPI

Article

Naproxen-Loaded Poly(2-hydroxyalkyl methacrylates): Preparation and Drug Release Dynamics

Abeer Aljubailah [1], Saad M. S. Alqahtani [1,*], Tahani Saad Al-Garni [1], Waseem Sharaf Saeed [1], Abdelhabib Semlali [2] and Taieb Aouak [1,*]

1 Chemistry Department, College of Science, King Saud University, Riyadh 11451, Saudi Arabia; akaljubailah@imamu.edu.sa (A.A.); tahanis@ksu.edu.sa (T.S.A.-G.); wsaeed@ksu.edu.sa (W.S.S.)

2 Groupe de Recherche en Écologie Buccale, Faculté de Médecin Dentaire, Université Laval, Quebec City, QC G1V 0A6, Canada; abdelhabib.semlali@greb.ulaval.ca

* Correspondence: salqahtani2@ksu.edu.sa (S.M.S.A.); taouak@ksu.edu.sa (T.A.)

Abstract: Poly(2-hydroxyethylmethacrylate)/Naproxen (NPX/pHEMA) and poly (2-hydroxypropyl methacrylate)/Naproxen (NPX/pHPMA) composites with different NPX content were prepared in situ by free radical photopolymerization route. The resulted hybrid materials were characterized by Fourier transform infrared spectroscopy (FTIR), differential scanning calorimetry (DSC), scanning Electron microscopy (SEM), and X-ray diffraction (XRD). These composites have been studied as drug carrier systems, in which a comparison of the in vitro release dynamic of NPX between the two drug carrier systems has been conducted. Different factors affecting the performance of the release dynamic of this drug, such as the amount of Naproxen incorporated in the drug carrier system, the pH of the medium and the degree of swelling, have been investigated. The results of the swelling study of pHEMA and pHPMA in different media pHs revealed that the diffusion of water molecules through both polymer samples obeys the Fickian model. The "in vitro" study of the release dynamic of Naproxen from NPX/pHEMA and NPX/pHPMA drug carrier systems revealed that the higher percentage of NPX released was obtained from each polymer carrier in neutral pH medium, and the diffusion of NPX trough these polymer matrices also obeys the Fickian model. It was also found that the less the mass percent of NPX in the composites, the better its release will be. The comparison between the two drug carrier systems revealed that the pHEMA leads to the best performance in the release dynamic of NPX. Regarding Naproxen solubility in water, the results deducted from the "in vitro" study of NPX/pHEMA10 and NPX/pHPMA10 drug carrier systems revealed a very significant improvement in the solubility of NPX in media pH1 (2.33 times, 1.43 times) and 7 (3.32 times, 2.60 times), respectively, compared to those obtained by direct dissolution of Naproxen powder.

Keywords: poly(hydroxyalkylmethacrylate)/Naproxen; drug release; solubility enhancement; drug-polymer miscibility; cell adhesion; toxicity; performance comparison

Citation: Aljubailah, A.; Alqahtani, S.M.S.; Al-Garni, T.S.; Saeed, W.S.; Semlali, A.; Aouak, T. Naproxen-Loaded Poly(2-hydroxyalkyl methacrylates): Preparation and Drug Release Dynamics. *Polymers* **2022**, *14*, 450. https://doi.org/10.3390/polym14030450

Academic Editors: Ariana Hudita and Bianca Gălățeanu

Received: 30 December 2021
Accepted: 20 January 2022
Published: 23 January 2022

1. Introduction

Medications are introduced into the human body through various drug delivery routes. For example, administered orally (by mouth), intravenously, intramuscularly, or breathed into the lungs (inhaled). The oral route remains the most popular way of drug administration [1]. In fact, it is the most preferred route by patients due to the ease of self-administration, pain avoidance, and cost-effectiveness. Some other advantages are that the oral ingestion route provides, for the patients, the least amount of sterility constraints, the minimal possibility to introduce systemic infection as a complication of treatment, the versatility to accommodate various types of drugs, and, most importantly, high patient compliance. Hence, it is the most employed route of drug delivery [2,3]. However, an orally administered drug must reach its target site at a concentration sufficient to induce the desired therapeutic effect. The term "bioavailability" refers to the fractional extent to which an administered dose of drug reaches its site of action or bloodstream from which the drug has access to finally reach its site of action [4,5]. Although the definition applies to any route of administration, practically-speaking the term is usually used for the oral route [6]. A severe drawback of oral ingestion of drugs is the limited absorption of some drugs due to their physical characteristics (e.g., poor aqueous solubility and low membrane permeability) [4]. According to the biopharmaceutical industry, all drugs must meet certain minimal requirements to achieve clinical effectiveness. More than 40% of newly discovered chemical entities entering the drug development pipeline fail to reach therapeutic range due to their poor water solubility, which in turn influences the absorption of the drug from the gastrointestinal tract, thereby leading to low bioavailability [7]. Among the severe drawbacks of oral drug administration is that, in some cases, a significant portion of the drug is destroyed in the stomach (very acidic pH) before reaching the intestines (neutral pH) where it will be absorbed, and, therefore, additional amounts of drug are required to reach the therapeutic threshold, not to mention the side effects that can cause fragments resulting from the degradation of these drugs. Other medications cause direct irritation to the gastric mucosa due to the inhibition of prostaglandins and prostacyclins and thus causes ulceration, epigastric distress, and/or hemorrhage [8]. In order to minimize these inconveniences, several authors have run towards the encapsulation of these drugs in the form of intelligent systems labeled as "drug-carrier" acting according to the environment where they are found. Sustained release of aspirin formulation would reduce the undesired side effects, reduce frequency of administration, and improve patient compliance [9]. Zheng et al. [10] investigated the ibuprofen/montmorillonite intercalation composites as the drug release system, and the in vitro results revealed that the release of ibuprofen from this system was affected by the pH value of the dispersion. The release rate in simulated intestinal fluid (pH = 7.4) was noticeably higher than that in simulated gastric fluid (pH = 1.2).

Polymers have played a key role in the advancement of drug delivery technology. The "in vitro" release of ibuprofen was also investigated by Carreras et al. [11]. These authors encapsulated this drug by poly(ε-caprolactone) using the solvent casting method. The results obtained revealed that the system has low homogeneity in particle size distribution, with a particle size average of 846.9 nm, and, therefore, they are microspheres. Mangindaan et al. [12] developed a controlled release system composed of surface modified porous polycaprolactone (PCL) membranes combined with a layer of tetraorthosilicate (TEOS)–chitosan sol–gel. The drugs chosen in this investigation were silver-sulfadiazine (AgSD) and ketoprofen, which were impregnated in the TEOS–chitosan sol–gel, and the results obtained revealed that the release of AgSD on O_2 plasma-treated porous PCL membranes was prolonged when compared with the pristine sample. On the contrary, the release rate of ketoprofen revealed no significant difference on pristine and plasma-treated PCL membranes. Diclofenac sodium(DS) combined with an electrospinning nano-and-nanofiber (DS-NNEM) mesh of polycaprolactone (PCL)-chitosan (CH) prepared by the electrospinning technique was the subject of a controlled release study of soluble DS in water [13]. Because of the very slow degradation of nanofiber mats, these authors suggested that DS is released either by diffusion or by permeation through DS-NNEMs structure. In summary, the results suggest that NNEMs technology can potentially serve as a biomimetic platform for loading and the sustained release of biologically active therapeutic compounds and other drugs for prolonged periods.

Naproxen (NPX), also known by its trade name "Proxen" (Scheme 1), belongs to the family of aryl propanoic acids such as ibuprofen, ketoprofen and diclofenac. This medication is a potent nonsteroidal anti-inflammatory drug (NSAIDs) that is used to treat acute pain, inflammation, as well as pain related to arthritis and rheumatic diseases [14,15]. However, the pharmaceutical applications of Naproxen is hampered by its poor water solubility [16]. Naproxen is a weak acid drug (pKa 4.2) that belongs to BCS class II drugs. It is a highly lipophilic drug (log P 3.18) with an aqueous solubility of 0.0159 mg·mL^{-1} at 25 °C [15,17].

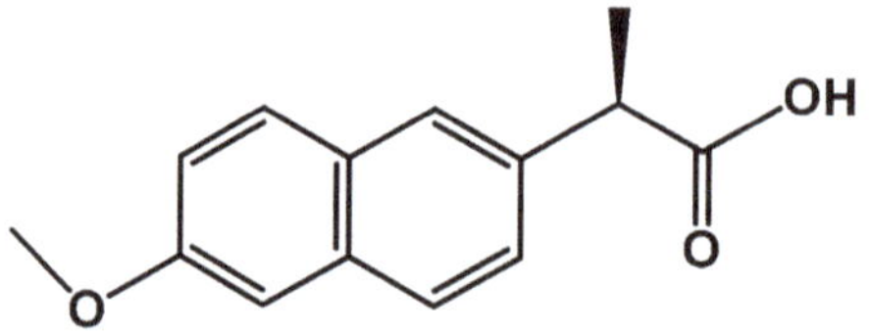

Scheme 1. Chemical structure of Naproxen.

Among the polymers that have attracted the attention of many researchers in the biomedical field, poly(2-hydroxyethyl methacrylate)(pHEMA) was found as a suitable compatible biomaterial [18,19] and a good candidate for drug delivery [20–26] and bone implantation [27–30]. This material can be prepared by bulk polymerization with low water content or by suspension polymerization to form microbeads [31,32]. pHEMA is usually reported to be biocompatible but less biodegradable [33]. However, for oral administration of this polymer, it must not be biodegradable.

Although pHEMA has been extensively studied in the biomedical field, its analog with additional methyl group, poly (2-hydroxypropyl methacrylate)(pHPMA) is very little known in this field. Therefore, it will be curious to know the reasons why this polymer has not been able to place itself among polymers selected as potential candidates as a carrier in drug delivery or as scaffolds used in the biomedical field.

In order to have an idea on the performance of pHPMA in the drug release domain, a comparative study on the effect of the 2-hydroxyalkyl methacrylate substituent on the release dynamics of Naproxen from the Naproxen/poly (2-hydroxyalkyl methacrylate) drug carrier system was carried out. To reach this goal, two series of composites involving Naproxen combined with pHEMA and Naproxen combined with pHPMA as polymer composites were prepared with different compositions by solvent casting route. The distribution of NPX particles in the resulted systems were studied by FTIR, DSC, XRD, and SEM methods, while the cell viability and the cell adhesions were examined by MTT test and LDH essay. A comparative study of the efficiency of these two drug carrier systems was carried out on the release dynamic of Naproxen by varying different parameters that affect the release performance of NPX, such as the percentage of medication incorporated in the polymer matrix and the pH media. The improvement in the solubility of Naproxen in the different pH media was also deduced from the release process.

2. Materials and Methods

2.1. Chemicals

HEMA (purity, ≥99%), HPMA (purity, ≥99%), and AIBN (purity, 98%) were provided by Sigma Aldrich (Taufkirchen, Germany). Proxen tablets manufactured by GRUNENTHAL were purchased from Riyadh Pharma (Saudi Arabia). Monomers were purified from hydroquinone (inhibitor) by distillation under reduced pressure. AIBN was purified three times by dissolution and recrystallization in ethanol. Human oral cancer cell line Ca9-22 cells were obtained from the laboratory of Dr. Abdelhabib Semlali (GREB–laval University, Quebec City, QC, Canada). RPMI-1640 medium was purchased from ThermoFisher (Burlington, ON, Canada), and the fetal bovine serum (FBS, Gibco) and 1% penicillin/streptomycin solution were from Sigma Aldrich (St. Louis, MO, USA).

2.2. Naproxen Extraction

Commercialized Proxen 500 mg tablets were ground into a fine powder using an electric grinder. The Proxen powder obtained is added to a 3M HCl solution and stirred for 24 h then left to stand for 1 h. Naproxen (NPX) or (S)-2-(6-methoxynaphthalen-2-yl) propionic acid, which is very poorly soluble in water precipitates, the additives dissolve, and the precipitant is then recovered by filtration. The extracted NPX powder is washed several times with water and then dried under a vacuum at 25 °C to constant weight. In order to remove the residual matter from the organic phase, the dry precipitate obtained is dissolved in chloroform and then transferred to a separating funnel containing an equivalent amount of distilled water. The whole is then stirred until the complete dissolution of Naproxen. The two phases are finally separated by settling. This process was repeated three times to ensure the purification of the product. Pure Naproxen is then extracted from the isolated organic phase by evaporating chloroform at room temperature (25 °C) using a rotary. The melting point of the pure NPX white crystals obtained, measured by DSC analysis, indicates 166 °C, which agree with the literature [34].

2.3. Preparation of NPX/pHEMA and NPX/pHPMA

NPX/pHEMA and NPX/PHPMA composites containing 2, 5, 7, and 10 wt% of NPX were prepared in situ by free radical polymerization at 25 °C in the presence of NPX using camphorquinone as a photoinitiator. Using known amounts distillated under reduced pressure of HEMA and HPMA monomers, camphorquinone and NPX were weighed with precision and placed in a Teflon pan traversed with a stream of nitrogen U. These mixtures are irradiated throughout the reaction time by means of UV light coming from a UV lamp with a wavelength of 380 nm and a power of 13.3 MW. A solid film deposed in the Teflon pan is obtained indicating the completion of the polymerization reaction. To remove all traces of residual monomer encrusted in the film obtained, the Teflon pan and polymer film set are placed in a vacuum oven maintained at 40 °C until constant mass. The aggregated NPX particles deposited or glued to the film surface are removed by washing three times with distillated water. Two series of NPX/pHEMA and NPX/pHPMA mixtures containing 2, 5, 7, and 10 wt% of NPX content are prepared by this same method, and the preparation conditions are summarized in Table 1.

Table 1. Preparation conditions of NPX/pHEMA and NPX/pHPMA composites.

Drug-Carrier System	HEMA (g)	HPMA (g)	NPX (g)	NPX (wt%)	Camphorquinone (g)
NPX/pHEMA2	5.00	-	0.102	2.0	1.0
NPX/pHEMA5	5.00	-	0.263	5.0	1.0
NPX/pHEMA7	5.00	-	0.376	7.0	1.0
NPX/pHEMA10	5.00	-	0.555	10.0	1.0
NPX/pHPMA2	-	5.00	0.102	2.0	1.0
NPX/pHPMA5	-	5.00	0.263	5.0	1.0
NPX/pHPMA7	-	5.00	0.376	7.0	1.0
NPX/pHPMA10	-	5.00	0.555	10.0	1.0

2.4. Characterization

The FTIR spectra of NPX powder, PHEMA homopolymer, NPX/PHEMA, and NPX/PHPMA composites films were performed in the wavenumber range 400–4000 cm^{-1} on a Nicolet 6700 FT-IR from the company Thermo Scientific. A 30,000–200 cm^{-1} diameter diamond-like Smart orbit crystal reflector, supplied by the same company, was used to accomplish this task. The DSC thermograms of drug, polymer, and their mixtures were performed on a Shimadzu DSC-60 (Japan) previously calibrated with indium. An amount of 8–10 mg of NPX powder or film samples were deposited in an aluminum pan and then closed, being placed in the DSC analysis cell. All samples were scanned from −40 to +240 °C under nitrogen gas atmosphere at a heating rate of 20 °C·min^{-1}. All the thermograms taken from the second scan run revealed no traces of polymers or drug degradation. The T_g value of pure constituents or their mixture was taken precisely as the median point on the thermogram indicating the variation in the heat capacity versus the temperature. The T_m value was taken exactly at the top of the endothermic peak. The surface morphology of NPX particles, polymers, and their composites were examined by scanning electronic microscope using a JEOL JSM-6360LV SEM (Tokyo, Japan) at an accelerating voltage of 10 kV. The surface and cross-sections of samples were sputter-coated with a thin layer of gold prior and was imaged at a magnification range of 300–3000 nm. The crystalline structures of NPX powder, polymers, and mixtures were examined by XRD analysis on an X-ray diffractometer (Rigaku D/max 2000) equipped with a Cu anode tube. The applied voltage was 40 kV and a generator current of 100 mA. All samples were examined at $2\theta = 5°–80°$ at a scanning rate of $1.0° \cdot min^{-1}$.

A U-2910 spectrophotometer manufactured by Hitachi Company was used to measure ultraviolet and visible light absorbance of NPX released. Absorbance was measured using quartz cuvettes with a side length of 1 cm. The wavelength corresponding to the maximum absorbance of NPX was 230 nm. The released NPX concentration was deduced from a linear calibration curve indicating the change in absorbance versus concentration.

2.5. Cell Culture and Proliferation Assessment

The human oral cancer cell line Ca9–22 cells were cultured at 37 °C and 5% CO_2 in RPMI-1640 medium (Thermo Fisher Scientific, Burlington, ON, Canada), supplemented with L-glutamine, 5% fetal bovine serum (FBS, Gibco) provided by the same company, and 1% penicillin/streptomycin solution (Sigma-Aldrich, Oakville, ON, Canada). Cell proliferation was bi-evaluated using two MTT and LDH tests as described by Semlali et al. 2021 [35,36] and Contant et al. 2021 [37]. For the MTT assay, 10^5 Ca9-22 cells per well containing the sample were seeded in 24-well plates for 24 h. After adhesion and growth of the cells, the culture medium is replaced by a new one containing a solution of MTT at 5 $mg \cdot ml^{-1}$ in PBS and left for 3 h at 37 °C in the dark. Then, the cells were lysed with HCl, 0.05 N, in 1 mL of isopropanol. Addition of 100 µL of analysis buffer to test wells from 96-well microplates was required to measure absorbance at 550 nm by an iMark reader (Bio-Rad). The percentage of viable proliferating cells was determined using Equation (1)

$$Cell.viability(\%) = \frac{OD_T - OD_B}{OD_C - OD_B} \times 100, \quad (1)$$

where OD_T, OD_B, and OD_C are the optic densities of the treated cell, blank, and control, cell respectively.

Adhesion positive control was the plate for the treated cultured tissues. The negative control for cell adhesion was the plate for untreated cultured tissue. LDH assay was realized by the LDH Cytotoxicity Detection Kit from Roche, which allows to directly quantify the cell death in culture based on the measurement of lactate dehydrogenase released into growth media. As described in our previous work [35,36], 10^5 cells per well were seeded in 24-well plates containing NPX/pHEMA and NPX/pHPMA with different compositions. After adhesion for 24 h, 50 µL of each supernatant was transferred in triplicate into a 96-well plate and supplemented with 50 µL reconstituted substrate mixture. Then, the plates were incubated for 30 min at room temperature in the dark until the yellow color developed, before reading at 490 nm with an xMark microplate absorbance spectrophotometer (Bio-Rad, Mississauga, ON, Canada). Triton X-100 (1%) was used as a positive control for LDH, and the negative one was obtained with untreated cells. LDH release activity was calculated using Equation (2)

$$LDH.activity(\%) = \frac{ABS_p - ABS_{nc}}{ABS_{pc} - ABS_{nc}} \times 100, \quad (2)$$

where ABS_p, ABS_{nc}, and ABS_{pc} are the absorbance of drug-carrier system, positive control, and negative control, respectively.

2.6. Swelling Properties

The swelling behavior of pHEMA and pHPMA hydrogels was studied on samples of thin films of dimensions 3 cm × 3 cm and thickness varying between 2.20 and 2.56 mm. Each film sample of determined mass (m_o) was placed in 50 mL of an aqueous solution at known pH (1 or 7) and maintained at 37 °C, then stirred (260 rpm) until the swelling equilibrium was reached. The mass of the medium absorbed at each time interval (m_t) is obtained by weighing the film after delicately wiping the droplets deposited on the two surfaces using tissue paper. The swelling degree of pHEMA and pHPMA film samples was determined from Equation (3).

$$S(wt\%) = \frac{m_t - m_o}{m_o} \times 100, \tag{3}$$

2.7. Density Measurements

Polymer density values were determined at 25 °C using a pycnometer, in which cyclohexane was used as a non-solvent and Equation (4) [38]:

$$\rho = \frac{m_p \times \rho_{chx}}{m_p + m_{pc} + m_T}, \tag{4}$$

where m_p is the mass of the polymer, m_{pc} is the mass of the pycnometer with cyclohexane, and m_T is the mass of the pycnometer with cyclohexane and polymer. ρ_{chx} is the density of cyclohexane (0.78 g·cm^{-3}). Each experiment was triplicated, and the density was taken from the average arithmetic values obtained.

2.8. In Vitro Release Dynamic of NPX

The "in vitro" release dynamic of NPX from the NPX/pHEMA and NPX/pHPMA drug carrier systems was investigated at body temperature (37 °C) in aqueous media of pH1, 3, 5, and 7. NPX released was monitored for 72 h, in which 0.5 mL of the solution was withdrawn after each time interval, then dosed by UV analysis. The accumulative drug release percent ADR (wt %) at time *t* was calculated at chosen time intervals using the following equation:

$$ADR(wt\%) = \frac{m_t \times 100}{m_o}, \tag{5}$$

where m_t and m_o are the total mass of NPX released at a certain time *t* and the initial mass of drug loaded in the polymer.

3. Results and Discussions

3.1. Characterization

3.1.1. FTIR Analysis

A comparison between the FTIR spectra of the NPX/pHEMA composites with those of their components shown in Figure 1 reveals, for the composite absorption bands, that they are practically similar to those of pure pHEMA. The NPX/pHEMA spectra exhibits a shift in the broad absorption band of the hydroxyl group vibrations from 3472 to 3463 cm^{-1} and a shift of the sharp absorption band attributed to the carbonyl group (C=O) vibration of pHEMA from 1727.62 cm^{-1} to 1724.32 cm^{-1}. It is well known that the position of the vibration peak of the carbonyl group suggests that most of the carboxylic acid groups are associated with the intermolecular hydrogen bonds formed between the HEMA derived moieties and the acid groups [39–41]. In addition, the wide absorption band in the spectral region 3200–3600 cm^{-1} corresponding to the vibrations of the OH group also confirms that hydrogen bonds form in the structure of poly(2-hydroxyethyl methacrylate-co-acrylic acid). The deconvolution in Lorentzian peaks of the hydroxyl absorption band between 2600 cm^{-1} and 4000 cm^{-1} (Figure 2) also reveals the appearance of a new band at 3274.54 cm^{-1}, attributed to the vibration of the hydrogen bond between hydroxyl group of pHEMA and carbonyl group of NPX. On the carbonyls side, the deconvolution of the absorption band between 1600 cm^{-1} and 2000 cm^{-1} (Figure 3) reveals another new absorption band at 1723.5 cm^{-1}, thus confirming this finding. Similar results are also observed for NPX/pHPMA composites in Figure 4. Indeed, the comparison between the FTIR spectra of the NPX/pHPMA composites with that of its pure polymer reveals a small shift in the absorption band of the carbonyl group of pHPMA toward the lower wave number (from 1724.12 cm^{-1}to 1727.34 cm^{-1}) and in the hydroxyl group (from 3400.07 cm^{-1}

to 3395.36 cm^{-1}). These facts are without doubt attributed to a dynamic caused by the hydrogen bond interactions leading to a miscibility of these two components.

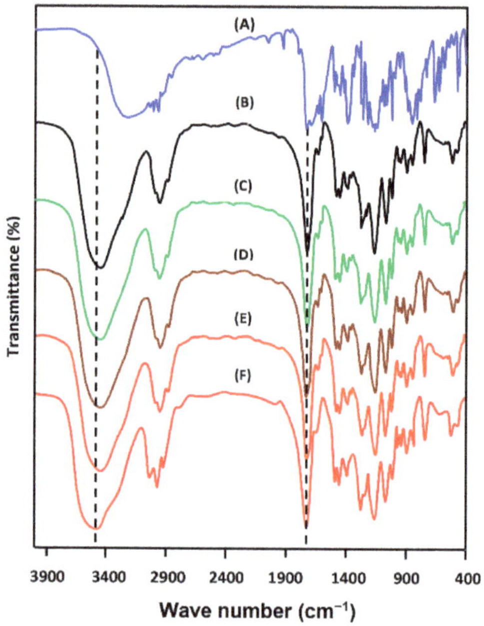

Figure 1. Comparative FTIR spectra of: (**A**) pure NPX; (**B**) NPX/pHEMA2; (**C**) NPX/pHEMA5; (**D**) NPX/pHEMA7; (**E**) NPX/pHEMA10; (**F**) virgin pHEMA.

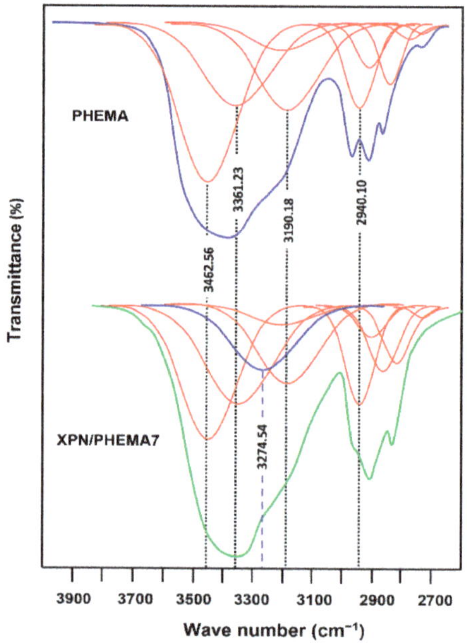

Figure 2. Deconvolution in Lorentzian of the FTIR spectra of pure pHEMA and NPX/pHEMA7 composite spectra between 4000 cm^{-1} and 2600 cm^{-1}.

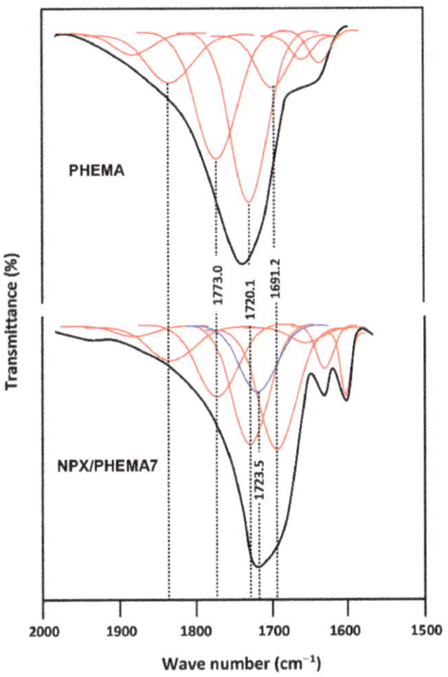

Figure 3. Deconvolution in Lorentzian curves of the FTIR spectra of pure pHEMA and NPX/pHEMA7 composite spectra between 2000 cm^{-1} and 1600 cm^{-1}.

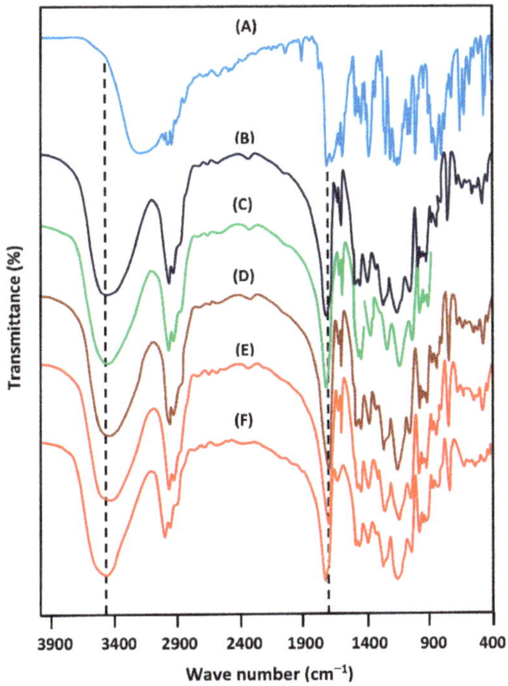

Figure 4. Comparative FTIR spectra of: (**A**) pure NPX; (**B**) NPX/pHPMA2; (**C**) NPX/pHPMA5; (**D**) NPX/pHPMA7; (**E**) NPX/pHPMA10; (**F**) virgin pHPMA.

3.1.2. XRD Analysis

The crystalline structure of NPX in the pHEMA and pHPMA polymer matrices was investigated by X-ray diffraction, and the results obtained are gathered with their pure components in Figures 5 and 6, respectively. As shown in Figure 5, the XRD pattern of pure NPX aggregated powder shows its highly crystalline structure and nature through the distinct peaks at 6.5°, 12.4°, 16.6°, 19°, 20°, 22.5°, 24°, and 28.6° 2θ, which are in good agreement with the literature [42,43]. The XRD spectra of pHEMA and pHPMA reveal an amorphous structure. The XRD motif of the NPX/pHEMA7 composite shows no new crystallinity signals or those characterizing the crystallinity of pure NPX; as for the pure polymer, this material exhibits a completely amorphous structure. Similar results are also obtained for the NPX/pHPMA systems as shown in Figure 6. This indicates that the NPX drug is uniformly distributed in its molecular level inside the polymer matrix.

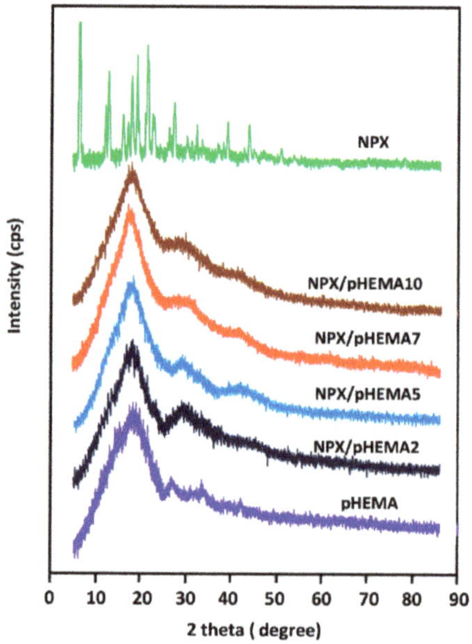

Figure 5. X-ray diffraction spectra of NPX, pHEMA, and NPX/pHEMA systems with different NPX contents.

3.1.3. DSC Analysis

The DSC thermograms of pure NPX, pHEMA, and NPX/pHEMA systems with different NPX contents are shown in Figure 7. The thermal curve of pure pHEMA shows a glass transition temperature (Tg) at 86 °C, which agrees with that of the literature [44], while the pure Naproxen shows, through its thermal plot, a sharp endothermic peak at 166 °C characterizing its melting temperature [34]. The NPX/pHEMA thermograms reveal a small shift in the Tg of pHEMA toward the low temperatures and a complete disappearance of the NPX melting peak. This reveals a uniform distribution of the NPX filler in the polymer matrix in its molecular level, in which Naproxen loses its crystallinity, thus confirming the results obtained by FTIR and XRD analysis. The decrease in the Tg value of the polymer in the mixture is probably due to the increase of the free volume between the polymer chains caused by the insertion of MPX molecules between them, thus promoting the chain sliding.

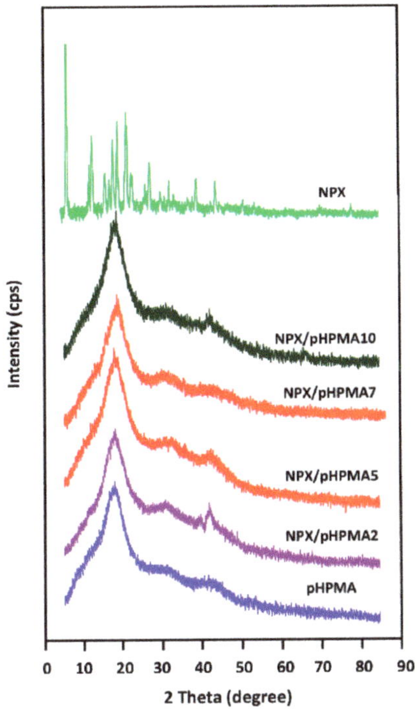

Figure 6. X-ray diffraction spectra of NPX, pHPMA, and NPX/pHPMA systems with different NPX contents.

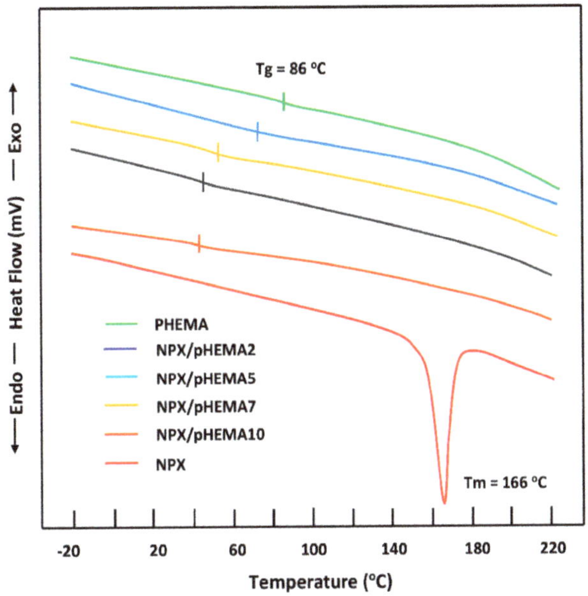

Figure 7. DSC thermograms of pure NPX, pHEMA, and NPX/pHEMA7 systems.

The thermal analysis of NPX/pHPMA systems by DSC technique led to the results of Figure 8. As it can be observed on the thermogram of pure pHPMA, a transition appears at 83 °C characterizing the glass transition of this polymer [45]. Concerning the NPX/pHEMA system, as for the system containing PHEMA carrier, a shift was shown in the Tg of the pHPMA from 82 °C to 54 °C as the NPX content in the mixture increased. A complete disappearance of the transition characterizing the fusion of the NPX is also observed on the thermograms of NPX/pHPMA mixtures, except that containing 10 wt% of NPX content, in which a weak transition at 131 °C attributed to the melting point of excess of NPX aggregates.

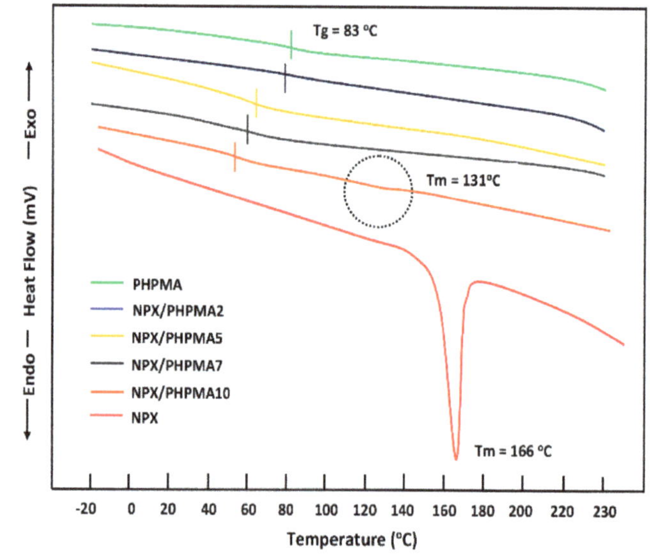

Figure 8. DSC thermograms of NPX, pHPMA, and NPX/pHPMA systems with different NPX contents.

3.2. Cells Adhesion and Toxicity

As shown in Figure 9 (in blue), the NPX/pHEMA drug-carrier system with different NPX contents presents, in general, a good adhesion compared to the negative (untreated tissue culture plate) and positive (tissue culture plate treated for cell adhesion) control used in this study. However, the NPX amount incorporated in the pHEMA in drug carrier system seemed to not significantly affect the cell adhesion when the Ca9-22 cells were treated with naproxen. These results were also confirmed by the LDH assay (Figure 10 in green). In addition, NPX/pHEMA drug carrier systems, as well as the pure pHEMA, induce low cytotoxicity compared to the negative and positive controls (2% Triton). Comparable results were also observed when the pHEMA was replaced by pHPMA in the drug carrier system, regardless of the range of the composition investigated (Figure 9 in blue and Figure 10 in green).

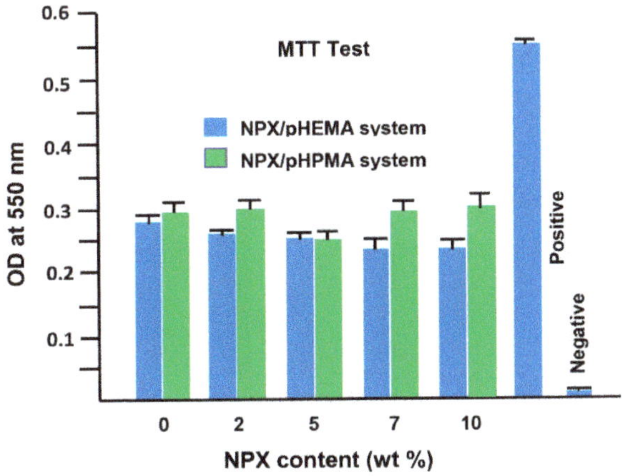

Figure 9. Effect of NPX content in NPX/pHEMA and NPX/pHPMA drug carrier systems on Ca9-22 cells adhesion

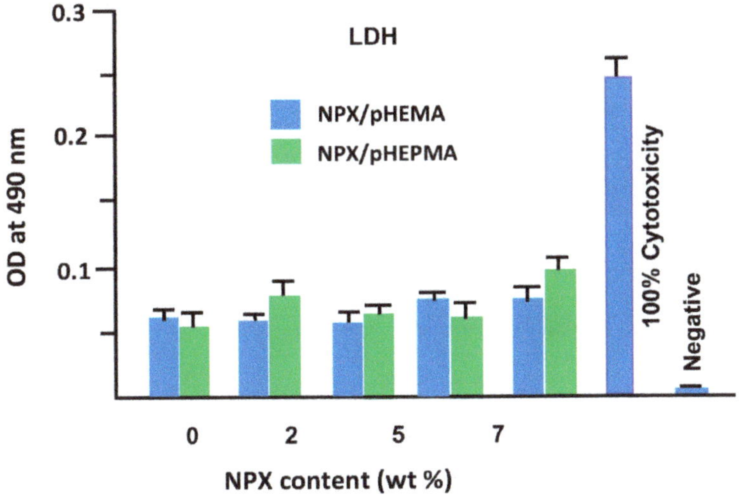

Figure 10. Effect of NPX content in NPX/pHEMA and NPX/pHPMA drug carrier systems on Ca9-22 cells cytotoxicity.

3.3. Swelling Behavior

Figure 11 shows the variation of the swelling degree of pHEMA and pHPMA film samples versus time, and Table 2 collects the deducted swelling degree at equilibrium.

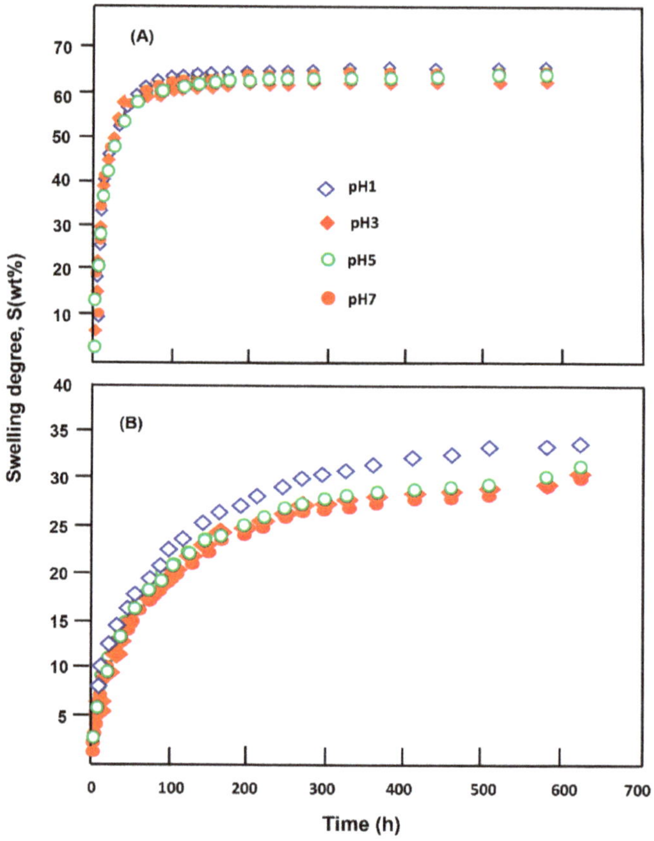

Figure 11. Variation of the swelling degree of (**A**) pHEMA, (**B**) pHPMA film samples versus time at 37 °C.

Table 2. Swelling capacity of pHEMA and pHPMA film samples in different pH of medium.

Film Sample	Swelling Degree at Equilibrium			
	pH			
	1	3	5	7
pHEMA	66.21	64.14	62.50	62.45
pHPMA	34.22	33.36	32.88	32.06

The comparison of the swelling degree values of these two polymer samples reveals that the absorption capacity of the pHEMA film is approximately double that of pHPMA, regardless of the pH of absorbed medium. This seems to be obvious and can be explained quite simply by the more hydrophilic character of the hydroxyethyl substitute belonging to the HEMA units, with regard to that of the hydroxypropyl of pHPMA, which contains an additional methyl. These results also reveal that, for both samples, the swelling capacity increased slightly when the pH medium decreased. This is probably due to the protonation of the oxygen of certain hydroxyl or carbonyl groups belonging to the monomeric units, thus increasing the hydrophilicity of the polymer. Generally, the swelling kinetics are used to investigate the diffusion of small molecules such as water through polymeric materials when immersed in a penetrating medium during a certain time. According to Comyn [46], the kinetics that govern the diffusion of small molecules through a polymer material are given by Equation (6)

$$\frac{m_t}{m_{max}} = 1 - \sum_{n=0}^{\infty} \frac{8}{(2n+1)^2\pi^2} \exp\left[\frac{-D(2n+1)^2\pi^2 t}{l^2}\right], \quad (6)$$

where m_t and m_{max} are the masses of the absorbed molecules during t time and at the maximum absorption (equilibrium), respectively. D and l are the diffusion coefficient with regard to the small molecules and the film thickness, respectively. For the short times of the initial stage of diffusion and when the m_t/m_{max} ratio is lower than 0.5, Equation (6) above takes the following expression:

$$\frac{m_t}{m_{max}} = 2 \times \left(\frac{D \times t}{\pi \times l^2}\right)^{1/2}, \quad (7)$$

in which, D can be deduced from the slope of the linear portion of the curve corresponding to the variation of m_t/m_{max} versus square root of time.

The fundamental equation of mass uptake by a polymer material is given by Equation (8) [47]:

$$\frac{m_t}{m_{max}} = k \times t^n, \quad (8)$$

where n exponent is the type of diffusion mechanism and k is the constant that depends on the diffusion coefficient and the film thickness. By analogy with Equation (7), k takes the following expression:

$$k = \frac{2}{l}\left(\frac{D}{\pi}\right)^n, \quad (9)$$

Equation (8) can be linearized by entering the logarithm of its two members as follows:

$$\ln\left(\frac{m_t}{m_{max}}\right) = \ln k + n \ln t, \quad (10)$$

The variation of $\ln(m_t/m_{max})$ versus ln$t$ for pHEMA and pHPMA materials is plotted in Figures 12 and 13, respectively. Straight lines were obtained, indicating that the diffusion of water molecules through these polymer materials obeys the Fick's model as long as their temperature in media (37 °C) is well above T_g (80 °C for pHEMA and 87 °C for pHPMA). This condition also indicates that the diffusion of water through the polymer film is purely and simply governed by a mechanical process and non-disturbed by a probable esterification reaction, which can occur in acidic media between the hydroxyl group contained in these polymers and the carboxylic group of Naproxen. The data of n, D, and k values deducted from these linear curves are gathered in Table 3. These results reveal, for both systems, an increase of the diffusion rate of water molecules when the pH of the medium increased. This property is highly valued in the field of drug delivery because this carrier is able to swell sufficiently and therefore delivers an appropriate amount of drug directly into the target organ (intestines, neutral pH medium).

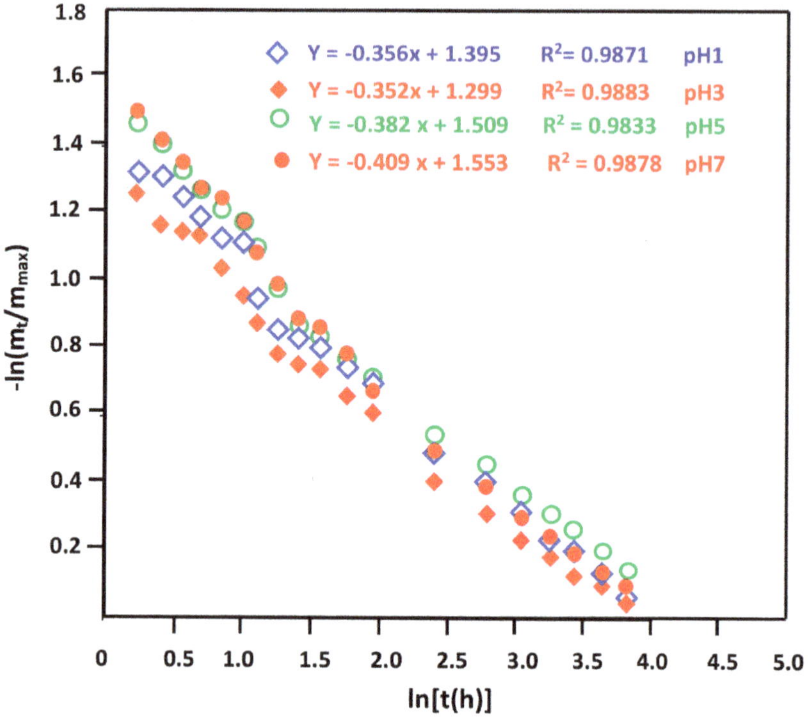

Figure 12. Variation of ln (m_t/m_{max}) versus ln (t) for the pHEMA material.

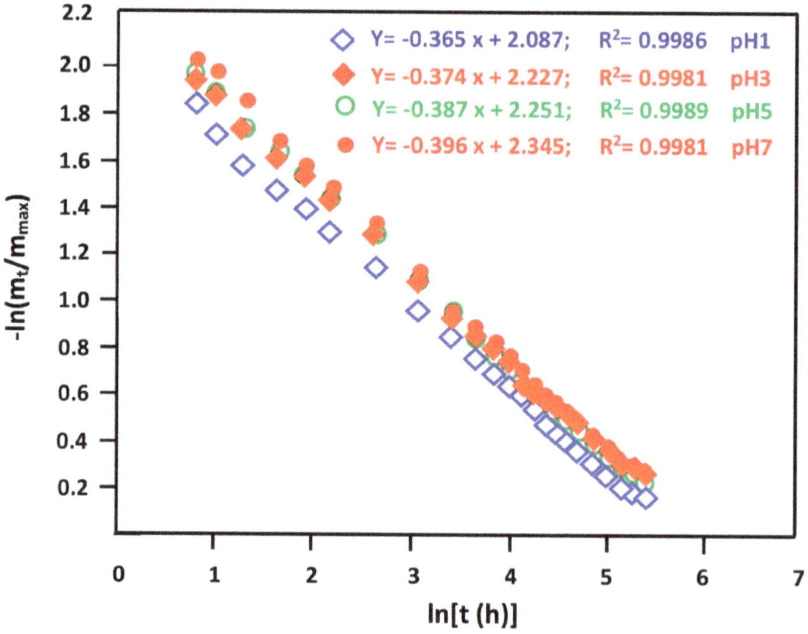

Figure 13. Variation of the ln (m_t/m_{max}) versus ln (t) for the pHPMA material.

As can be seen from these data, practically no change in the order of the water diffusion through each polymer material is observed, regardless the pH of medium, which is close to 0.40. The diffusion coefficient attributed to pHEMA material is higher than that of pHPMA, except those experimented in medium at neutral pH, which are both close to 0.80 $mm^2 \cdot h^{-1}$. This is probably due to the decrease of the affinity between pHPMA and water caused by the hydropropyl group of the substitute (less hydrophilic) compared to that of hydroxyethyl group of pHEMA (more hydrophilic). These data also reveal that, for the pHEMA material, the D value decreased when the pH of the medium increased. In this same pH order, this parameter increased for the pHPMA.

Table 3. Diffusion parameters of water at different pHs through pHEMA and pHPMA materials.

Polymer Sample	n				k				D ($mm^2 \cdot h^{-1}$)			
pH	1	3	5	7	1	3	5	7	1	3	5	7
pHEMA	0.36	0.35	0.38	0.41	4.03	3.67	4.52	4.73	1.16	1.24	1.06	0.80
pHPMA	0.37	0.37	0.39	0.40	8.06	9.27	9.50	10.43	0.54	0.44	0.51	0.79

3.4. Drug–Polymer Interactions

The drug–polymer Flory–Huggins interaction parameter denoted $\chi_{d,p}$ gives an important idea on the chemical affinity and the magnitude of the adhesion force between the drug and the polymer carrier through its sign and its absolute value, respectively. According to the Flory–Huggins theory [48], a negative value of $\chi_{d,p}$ indicates miscibility of a drug carrier system, and a positive value indicates its immiscibility. The $\chi_{d,p}$ values of the drug carrier systems involving NPX and pHEMA on the one hand and NPX and pHPMA on the other hand were estimated using the data of Table 4 and Equation (11) [49]:

$$\frac{\Delta H_f}{R}\left(\frac{1}{T_f} - \frac{1}{T}\right) = \ln v_d + \left(1 - \frac{1}{\lambda}\right) v_p + \chi_{d,p} v_p^2, \tag{11}$$

where ΔH_f and T_f are the enthalpy of fusion of Naproxen and the melting temperature of the pure drug. R is the gas constant, and T is the measured solubility temperature for a volume fraction v with subscripts d and p denoting drug and polymer, respectively. λ is the ratio of the molar volumes of the drug and polymer. $\chi_{d,p}$ is the drug–polymer Flory–Huggins interaction parameter. The results obtained are gathered for comparison in Table 5.

Table 4. Some characteristic data of pHEMA, pHEPMA, and Naproxen taken at 25 °C.

Compound	Density ($g \cdot mL^{-1}$)	Hansen Solubility Parameter ($KPa^{1/2}$)	Molar Volume ($mL \cdot mol^{-1}$)	ΔH_f ($kJ \cdot mol^{-1}$)	T_f (°C)
pHEMA	1.25		103.8	-	-
pHPMA	1.33		108.2	-	-
Naproxen	1.27 [c]	21.62 [c]	157.3[c]	31.50 [a,b]	154.6 [a,b]

[a] Ref. [50]; [b] Ref. [34]; [c] Ref. [51].

Table 5. Comparative values of the Flory–Huggins parameters of the NPX/pHEMA and NPX/pHPMA drug carrier systems determined at 25 °C using Equation (9).

Drug-Carrier System	$\chi_{d,p}$	Drug-Carrier System	$\chi_{d,p}$	$\Delta(\chi_{d,p})$
NPX/pHEMA2	−0.27	NPX/pHPMA2	−0.30	0.03
NPX/pHEMA5	−1.28	NPX/pHPMA5	−1.33	0.05
NPX/pHEMA7	−1.72	NPX/pHPMA7	−1.77	0.05
NPX/pHEMA10	−2.26	NPX/pHPMA10	−2.32	0.06

As it can be seen from these results, the values of $\chi_{d,p}$ are all negative regardless of the drug carrier system and its composition. According to the Flory–Huggins theory, a negative value of $\chi_{d,p}$ indicates the miscibility of the drug carrier system. These data also reveal that the values of $\chi_{d,p}$ increase with the NPX loading incorporated into the polymer. This means an increase in the affinity of NPX molecules with regard to the polymer when the drug loaded in the drug carrier system increased. This seems to be obvious because the density of hydrogen bonds between the hydroxyl groups of pHEMA or pHPMA and the carbonyl of the carboxyl group of NPX increases with the drug load in the drug carrier system. This leads to an increase of the attraction forces between these two components.

The comparison of the values $\chi_{d,p}$ of the NPX/pHPMA system with those of the NPX/pHEMA system indicates a slight increase in the interactions between NPX and pHPMA compared to those between NPX and pHEMA, whatever the composition of the mixture studied. These results also reveal that the more the amount of NPX increases in the drug carrier system, the greater the absolute value of the difference between the Flory–Huggins interaction parameters of the NPX/pHEMA and NPX/pHPMA drug systems, $\Delta\left(\chi_{d,p}\right)$ also increases.

3.5. *In Vitro Release Dynamic of NPX*

3.5.1. Release Kinetics of NPX

The release dynamic of NPX from NPX/pHEMA and NPX/pHPMA drug carrier systems with different compositions are shown in Figures 14 and 15, respectively. As it can be seen from the curve profiles obtained for both systems, the maximum percentage of NPX released is reached with the drug carrier systems containing 2 wt% of NPX content. The comparison between the NPX release capacities for these two different systems during 72 h of the release process reveals that the NPX/pHEMA shows the best performance. Indeed, for the NPX/pHEMA2, a maximum of 42 wt% of NPX was released in neutral pH medium during this period and about 31 wt% in acidic media (pH 1 and 3), while for the NPX/pHPMA2, only 10.5% wt% of NPX was released in neutral pH medium and 7.4% in acidic media (pH 1 and 3) during the same period. This represents a reduction in NPX release dynamics of about a quarter. The decrease in NPX release dynamics observed by replacing pHEMA by pHPMA appears to be due to a dramatic decrease in the hydrophilicity of the carrier polymer caused by the additional methyl group in the substituent. The decrease in hydrophilicity when passing from the pHEMA to pHPMA carrier reduces the degree of swelling as revealed in Table 2. This limits the water amount absorbed by the polymer carrier necessary for the dissolution of a significant part of NPX incorporated in this material. It was also observed from these same curves, for both systems, the behavior of NPX release versus time is characterized by two pseudo stable zones of the release dynamic. The first zone, which is rapid and short, is observed during about the first 4–7 h of the release process, depending on the nature of the drug-carrier used and the pH of the medium. The second zone, which is long and slow, is observed during the 65 h of the release process. The first step is mainly attributed to the leaching of a fraction of NPX particles deposited on the surface or slightly embedded in the sample film. The second step characterizes the steady state in which the release of NPX in the media is governed mainly by a material transfer mechanism.

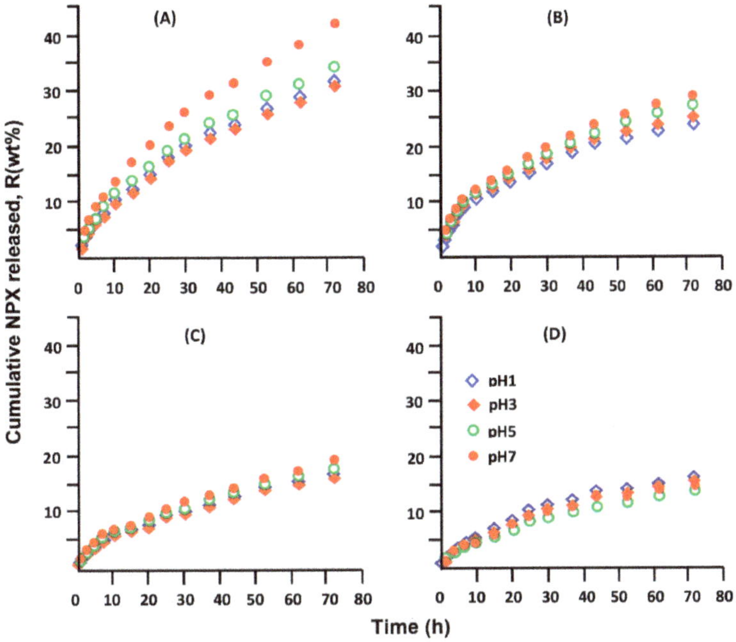

Figure 14. Release kinetics of NPX from (**A**) NPX/pHEMA2, (**B**) NPX/pHEMA5 (**C**) NPX/pHEMA7, (**D**) NPX/pHEMA10 drug carrier system at different pHs.

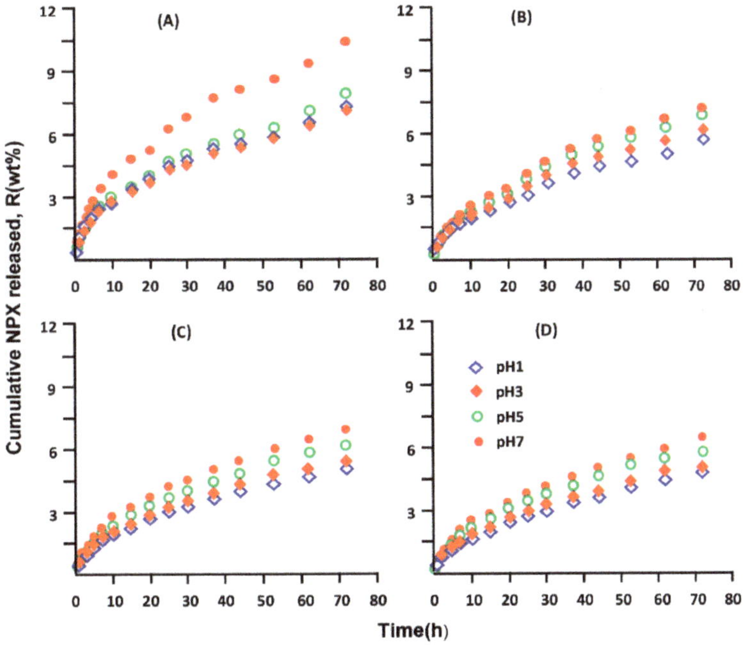

Figure 15. Release kinetics of NPX from (**A**) NPX/pHPMA2, (**B**) NPX/pHPMA5 (**C**) NPX/pHPMA7, (**D**) NPX/pHPMA10 drug carrier system at different pHs.

3.5.2. Enhancement of NPX Solubility

The improvement of the solubility of NPX in water is an integral part among the objectives targeted by this investigation. To reach this goal, a comparison between the solubility of NPX in its powder form and that incorporated in the NPX/pHEMA and NPX/pHPMA drug carrier systems was carried out at 37 °C. An excess amount of NPX powder was dissolved under continuous stirring in a known volume of water maintained at 37 °C until the appearance of a stable precipitate, indicating the supersaturation of the solution. The solution was then filtered through a Whatman filter number 1. The solubility of NPX was determined by means of UV-visible spectroscopic analysis. Two solutions of pH 1 and 7 were prepared, and the results obtained are gathered in Table 6. The maximum NPX amount dissolved in these media was deducted from the maximum release of this medication from the NPX/pHEMA and NPX/pHPMA drug carrier systems, and the results obtained are also grouped for comparison in this table. The comparison of the maximum solubility data obtained reveals that the pHEMA is much more efficient than the pHPMA used as supports. These data also reveal an enhancement of the solubility of NPX in pH media 1 and 7, in which the NPX/pHEMA system was able to dissolve 2.60-fold that of that of because the NPX amount dissolved from the NPX/pHEMA10 system is more than 1.34 times in pH medium 1 and 2.60 times in neutral pH from the NPX/pHPMA system. The comparison of the NPX solubility data obtained by direct dissolution of the powder with that deduced from the release process involving these two polymers reveals a marked improvement when this drug is incorporated in the molecular state in one of these two polymers. These results also show that the pHEMA used as a support is more efficient than the pHPMA in increasing the solubility of NPX in water. For example, in neutral pH medium, pHEMA was able to improve the solubility of this medication by 3.32 times that of its direct dissolution as powder and 2.33 times in pH medium 1, while pHPMA increased this solubility by only 1.28 and 1.74 times in pH media 7 and 1, respectively.

Table 6. Maximum solubility of NPX dissolved in pH media 1 and 7 at 37 °C.

System	pH 1	pH 7
NPX powder	29.51	61.68
NPX/pHEMA10	68.84	205.04
NPX/pHPMA10	51.28	78.77

3.5.3. Surface Morphology

Figure 16 groups the micrographs of NPX powder, virgin pHEMA, NPX/pHEMA2, and NPX/pHEMA10 film samples before and after the release process in media pH 1 and 7 chosen among the most significant images. The NPX image shows crystal particles aggregated into defined geometric shapes resembling piles of rubble from houses destroyed by an earthquake. These aggregates, which are sized between 3 μm × 3 μm × 2 μm and 50 μm × 25 μm × 6 μm, show smooth and homogeneous morphology surfaces, while the micrograph of the virgin pHEMA film presents roughness on the surface, which is probably due to the film preparation. NPX/pHEMA systems containing 2 wt% and 10 wt% NPX contents before the release process show comparable morphology surfaces, in which the observed obliquely aligned parallel grooves mark the surface of the mold where they were prepared. These same samples observed after the NPX release process in pH 1 and 7 media exhibit surface morphologies very marked by the very hollow relief and cavities, thus revealing the large amount of NPX released and also show a significant degree of swelling of the film notably in pH medium 1. Regarding the NPX/pHPMA drug carrier system, as shown in Figure 17, the images are practically comparable to those of the system involving pHEMA as carrier are observed. Indeed, the surfaces of the samples before the drug release process as for the blank carrier show the same type of grooves, except that with 10 wt% NPX, in which they are less marked. This reveals that the surfaces of the two carriers involved in the drug carrier system behave substantially the same during the drug release and show no particular mark distinguishing one or the other polymer.

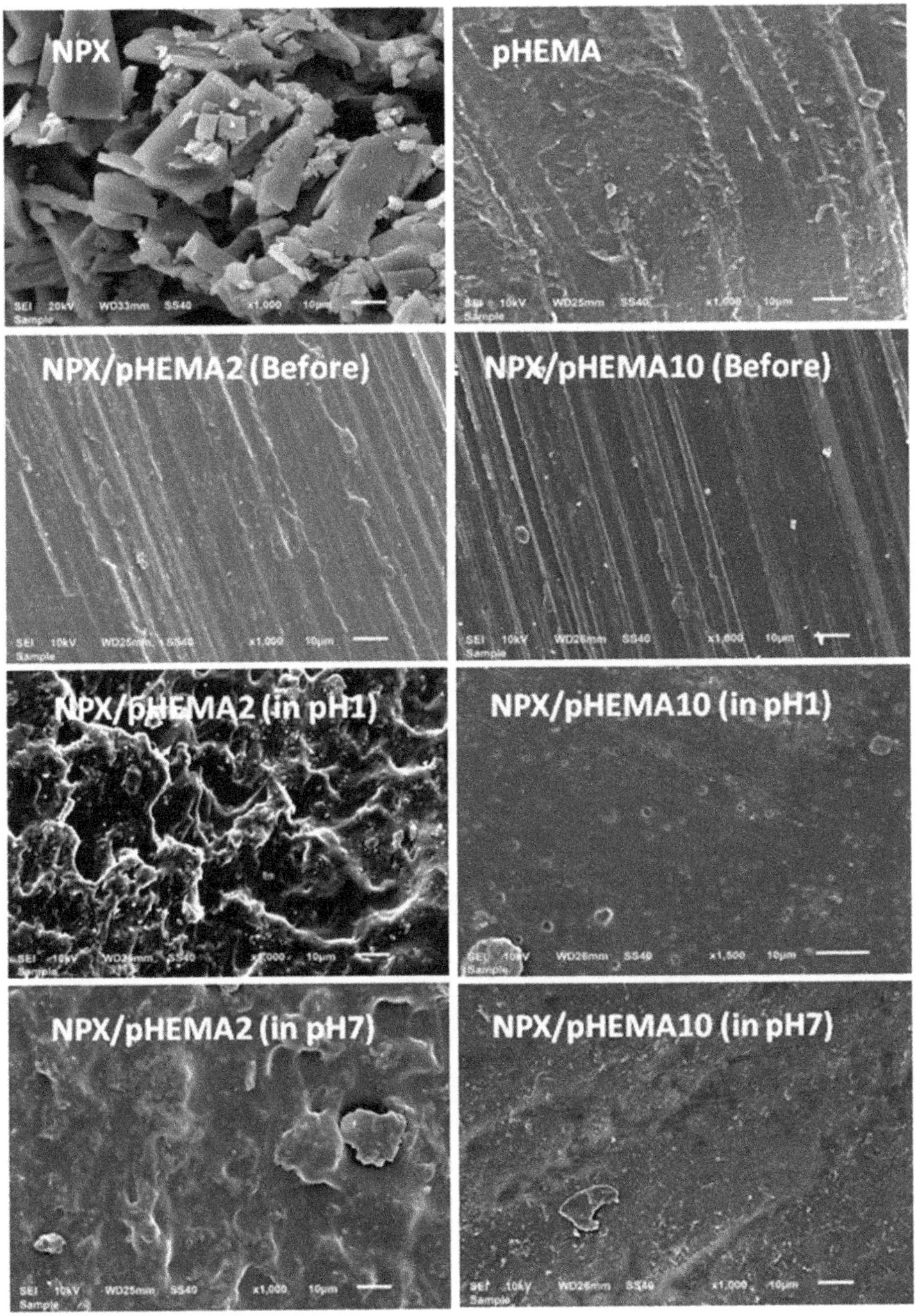

Figure 16. SEM images of NPX powder and surface morphology of virgin pHEMA, NPX/pHEMA2, and NPX/pHEMA10 film samples before and after the NPX release process in media pH 1 and 7.

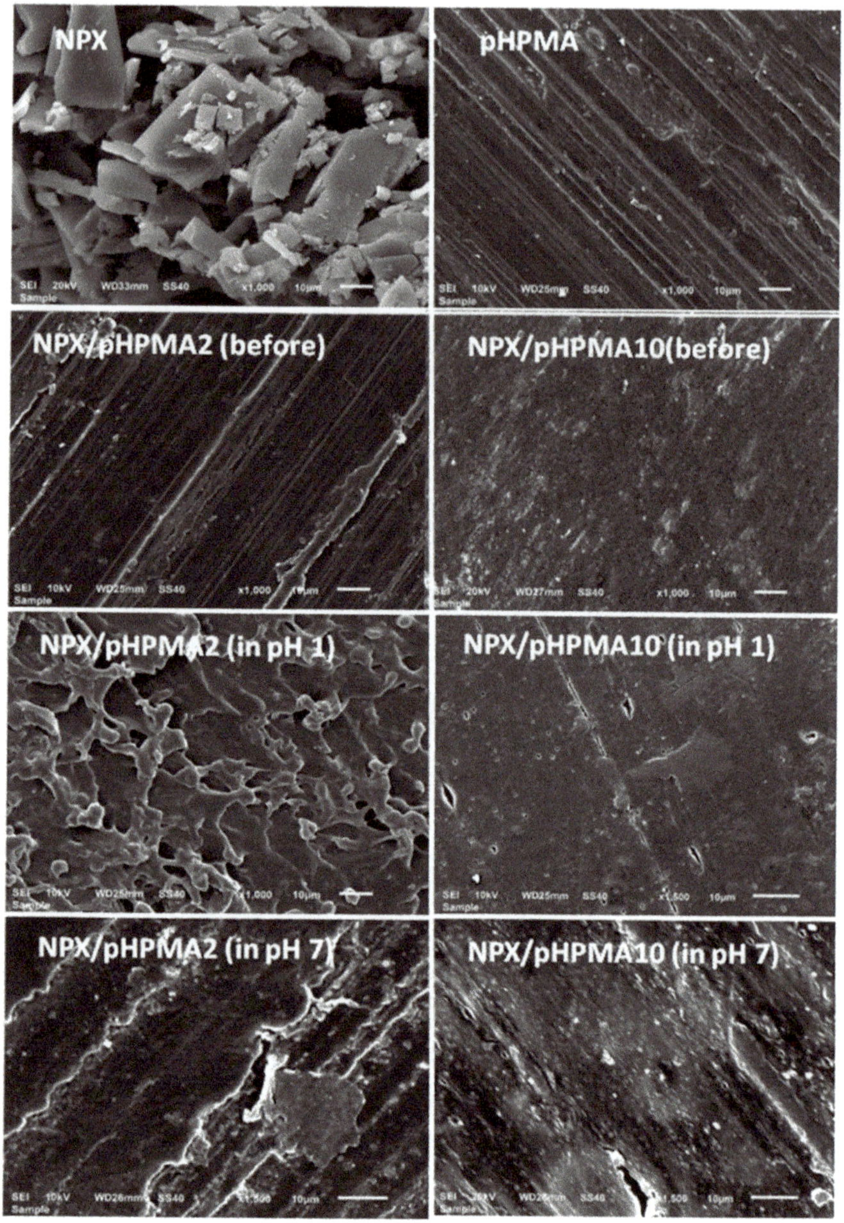

Figure 17. SEM images of NPX powder and surface morphology of virgin pHPMA, NPX/pHPMA2, and NPX/pHPMA10 films samples before and after the NPX release process in media pH 1 and 7.

3.5.4. Diffusion Behavior of NPX

The diffusion behavior of NPX from NPX/pHEMA and NPX/pHPMA drug carrier systems was investigated. According to Lin et al. [52], for a percentage less than 60 wt% of a substance released from the initial amount incorporated into a material, the diffusion of this substance in its liquid state through this material follows a Fickian model, as long as, in this investigation, the limit of the percentage of NPX released is far from being reached whatever the drug carrier system and the composition. Fick model is therefore applicable to describe the diffusion behavior of NPX from the polymer matrix [53]. The equation resulting from the Fickian model is given by Equation (12) [54–56]

$$\frac{m_t}{m_o} = k\sqrt{t}, \tag{12}$$

where m_t/m_o is the fraction of drug released, t is the release time, and k is a constant characteristic of each sample.

If the drug released from the drug carrier system obeyed the Fick diffusion model, the graph showing the change in the fraction of drug released m_t/m_o versus the square root of time would give a straight line with a slope k. Under these conditions, the value of the diffusion coefficient (D) will then be deduced from Equation (13) [57]:

$$k = '\sqrt{\frac{D}{\pi \times l^2}}', \tag{13}$$

where l is thickness of film, from which the drug is released. The k and D values were calculated from the data of Figures 18 and 19 using Equations. (11) and (12) and the results obtained are gathered in Table 7.

Table 7. Summary of diffusion data NPX through NPX/pHEMA and NPX/pHPMA drug carrier systems.

pH	System	$k{\cdot}10^2$ ($h^{-1/2}$)	R^2	$D{\cdot}10^3$ ($mm^2{\cdot}h^{-1}$)	System	k ($h^{-1/2}$)	R^2	$D\ 10^3$ ($mm^2{\cdot}h^{-1}$)
1	NPX/pHEMA2	3.56	0.9988	6.161	NPX/pHPMA2	0.85	0.9991	0.386
3	NPX/pHEMA2	3.54	0.9976	6.642	NPX/pHPMA2	0.82	0.9994	0.359
5	NPX/pHEMA2	3.96	0.9983	8.830	NPX/pHPMA2	0.91	0.9996	0.401
7	NPX/pHEMA2	4.76	0.9990	13.239	NPX/pHPMA2	1.22	0.9995	0.636
1	NPX/pHEMA5	2.71	0.9950	3.628	NPX/pHPMA5	00.66	0.9996	0.270
3	NPX/pHEMA5	2.79	0.9950	3.678	NPX/pHPMA5	0.70	0.9994	0.262
5	NPX/pHEMA5	3.09	0.9977	4.240	NPX/pHPMA5	0.79	0.9997	0.278
7	NPX/pHEMA5	3.25	0.9978	5.343	NPX/pHPMA5	0.81	0.9994	0.396
1	NPX/pHEMA7	1.93	0.9998	1.724	NPX/pHPMA7	0.61	0.9997	0.172
3	NPX/pHEMA7	1.96	0.9994	2.446	NPX/pHPMA7	0.65	0.9994	0.198
5	NPX/pHEMA7	2.26	0.9985	3.017	NPX/pHPMA7	0.73	0.9998	0.262
7	NPX/pHEMA7	2.36	0.9941	3.597	NPX/pHPMA7	0.81	0.9999	0.348
1	NPX/pHEMA10	1.80	0.9980	1.338	NPX/pHPMA10	0.0032	0.9996	0.147
3	NPX/pHEMA10	1.81	0.9987	1.455	NPX/pHPMA10	0.0032	0.9996	0.147
5	NPX/pHEMA10	1.75	0.9976	1.254	NPX/pHPMA10	0.0036	0.9996	0.167
7	NPX/pHEMA10	1.80	0.9977	1.6778	NPX/pHPMA10	0.0046	0.9997	0.205

As it can be seen from these data, all the R^2 values are close to unity. This indicates that the data correspond well to the linear regression of these curve profiles. These results also indicate that the NPX release behavior from both NPX/pHEMA and NPX/pHPMA systems follows a Fickian model with an order of 0.5. The higher the value of k, the higher the diffusion coefficient and, therefore, the faster the rate of the drug diffusion through the carrier. In general, the k and D values increased with the pH medium regardless of the drug carrier system used. This can be explained by the solubility of NPX, which becomes more soluble in media of higher pH. Knowing that the pKa of Naproxen is equal to 4.19 [58], the solubility of this drug increases with increasing pH of the medium due to the passage of the carboxylic acid group towards the carboxylate salt group that is more soluble in water, notably when the pH medium becomes equal to or greater than the pKa. In addition, it can also be seen in Figures 18 and 19, as the amount of NPX increases, the D value decreases. This can be explained by two main factors that can intervene simultaneously in the management of the drug release process: (i) the increase in the viscosity of the medium, which reduces the rate of diffusion and (ii) the presence of an NPX excess not soluble in the polymer matrix, which hinders the passage of soluble molecules during the diffusion process.

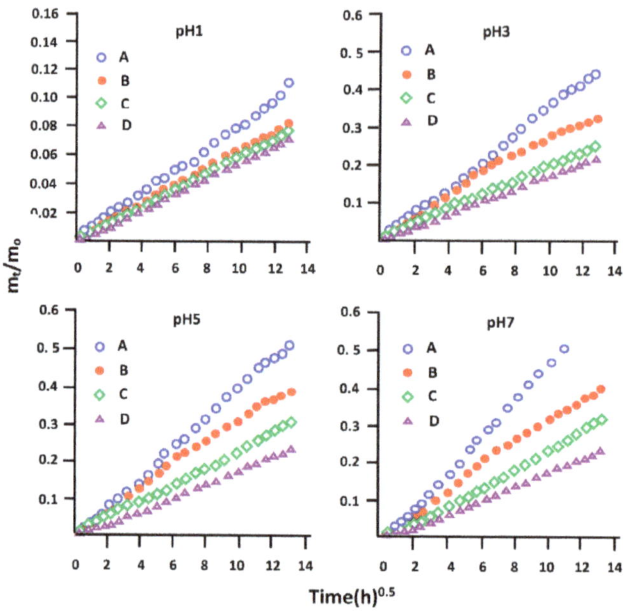

Figure 18. Diffusion behavior of NPX through (A) NPX/pHEMA2, (B) NPX/pHEMA5 (C) NPX/pHEMA7, (D) NPX/pHEMA10drug carrier system with different NPX contents in different pH media.

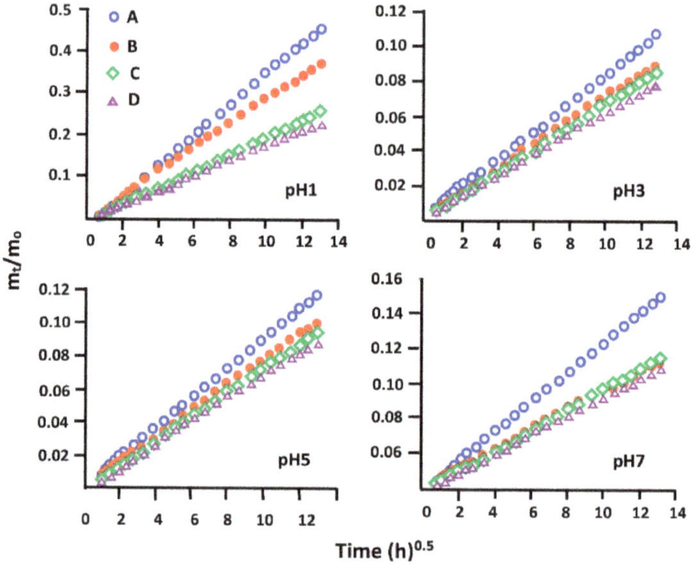

Figure 19. Diffusion behavior of NPX through (A) NPX/pHPMA2, (B) NPX/pHPMA5 (C) NPX/pHPMA7, (D) NPX/pHPMA10drug carrier system with different NPX contents in different pH media.

3.5.5. Effect of the Initial NPX Amount

The influence of the initial NPX amount loaded in pHEMA and pHPMA carriers on the dynamic release of this medication from the corresponding drug-carrier systems was studied at a selected period of 72 h of the release process. The results obtained for NPX/pHEMA and NPX/pHPMA systems are plotted for comparison in Figure 20. As can be observed from these curve profiles, the two drug carrier systems have practically the same trends, in which the release dynamic decreased, passing through a reflection point at 6.0 wt% of NPX common for all samples and then stabilizes or tends to stabilize when the percentage of NPX is greater than 7.0 wt%.

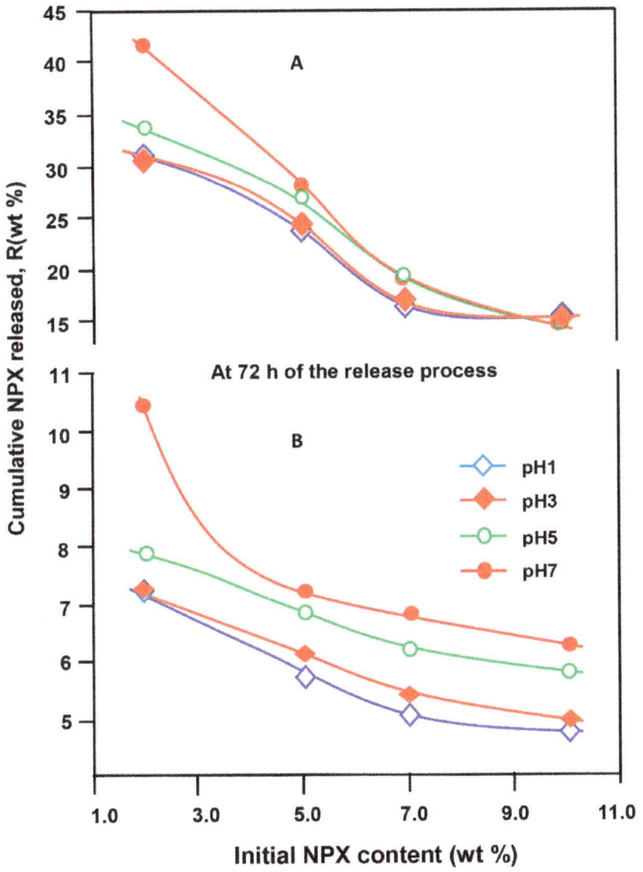

Figure 20. Cumulative NPX released from (A) NPX/pHEMA and (B) NPX/pHPMA drug carri-er systems at 72 h of the release process.

A more rapid decrease in the release dynamic is also observed on these profiles in neutral pH medium when the initial NPX loaded in the drug carrier systems was less than 5 wt%. The decrease in the NPX released observed in all pH media, when the drug content in the polymer matrix increased, is mainly due to the limited solubility of this drug inside the drug carrier system and to the increase of the viscosity of the solution inside the polymer matrix. Indeed, dissolving loads greater than 5 wt% seems to be difficult, especially in acidic pH media.

3.5.6. Effect of pH Medium

The impact of the pH of the medium on the release dynamic of NPX from NPX/pHEMA and NPX/pHPMA systems was carried out at 72 h of the release process, and the results obtained are plotted in Figure 21. These curve profiles reveal comparable dynamics of the NPX released by the two drug-carrier systems regardless of the time period. Pseudo-stability of the release dynamics is observed for all samples in very acidic media (pH 1 and 3), then a slight increase or decrease depending on the initial NPX amount incorporated in the polymer matrix is observed at higher pH (5 and 7), except that containing the lowest NPX load (2 wt%), in which the release dynamic rapidly increased. The pseudo-reproducibility of the behavior of the drug release dynamics at different periods for these two systems shows that the transfer of NPX from the polymer material is mainly handled by a stable, purely mechanical process. The increase in the release dynamic with the pH of the medium is mainly due to the increase of the solubility of Naproxen in neutral pH media inside the polymer matrix. Indeed, as previously revealed from the results of Table 6, the solubility of NPX increased dramatically when the media pH increased. These results were also observed by Kumar et al. [59] and attribute the low solubility of Naproxen in lower pH media to its unionization. These same authors add that the unionization of the drug can facilitate its permeability through the polymer material, but drug solubility is the limiting factor.

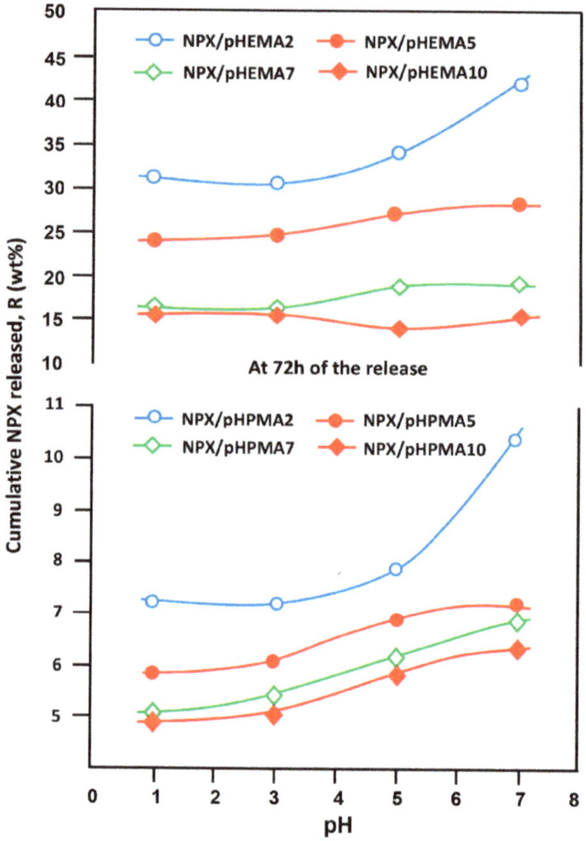

Figure 21. Variation of the NPX released from NPX/pHEMA and NPX/pHPMA drug carrier systems versus the pH of medium taken at 72 h of the release process.

3.5.7. Performance of XPN/pHEMA and NPX/pHPMA Drug Carrier Systems

As it was noted in Section 3.5.1, for both the drug carrier systems, it was revealed that the release behavior of NPX versus time followed two main stages regardless of the composition and the pH of the medium. Each stage is characterized by a zone, in which the release dynamic of NPX passes by pseudo stability. The rate of NPX released during the corresponding period was taken from the slope of the pseudo linear curve, and the data obtained are illustrated for NPX/pHEMA and NPX/pHPMA systems in Tables 8 and 9, respectively, noting that the cumulative percentage of the drug released during each period was calculated by multiplying the rate by time. Knowing that, for a system to be effective in the field of drug delivery, it must be able to uniformly deliver an appropriate amount of this drug in the intestines (neutral pH) and in the stomach (pH = 1–3). On this basis, the performance of these two systems on the NPX release was founded, and the results obtained are summarized for NPX/pHEMA in Table 8 and for NPX/pHPMA in Table 9. These data reveal that the two systems containing 2% by weight NPX appear to be the most effective of all the others because NPX/pHEMA2 drug-carriers were able to release 28.68 wt% of NPX uniformly during 65 h of the release process into a neutral medium with a release rate of 0.441 wt%·h^{-1}. In contrast, only 21.32 wt% was released uniformly (0.328 wt%·h^{-1}) during this same period into the acidic medium (pH = 1). On the other hand, during the same period, the drug carrier system involving the pHPMA2 was able to release uniformly 6.83 wt% of NPX in neutral pH with a rate of 0.102 wt%·h^{-1}; at the same time, only 4.62 wt% was released in medium with pH = 1 with a constant rate of 0.069 wt·h^{-1}. In general, the comparison of the performances of these two systems reveals that involving the pHEMA appears to be the most efficient.

Table 8. Percentage of NPX released and instantaneous release rate of NPX from NPX/pHEMA system with different compositions.

System	pH	SZ (h)	RNR (wt%·h^{-1})	CNR (wt%)	LR (R^2)	System	SZ (h)	RNR (wt%·h^{-1})	CNR (wt%)	LR (R^2)
NPX/pHEMA2	1	0–7	1.001	07.01	0.978	NPX/pHEMA7	0–4	0.776	3.104	0.985
		7–72	0.328	21.32	0.976		4–72	0.171	11.29	0.979
	3	0–7	0.964	0.138	0.972		0–4	0.803	3.21	0.982
		7–72	0.336	21.84	0.975		4–72	0.172	11.35	0.973
	5	0–7	1.124	7.87	0.967		0–4	0.783	3.13	0.985
		7–72	0.366	23.79	0.977		4–72	0.209	13.79	0.981
	7	0–7	1.290	9.03	0.980		0–4	0.636	2.54	0.988
		7–72	0.441	28.67	0.982		4–72	0.223	14.72	0.987
NPX/pHEMA5	1	0–7	0.969	6.783	0.977	NPX/pHEMA10	0–4	0.687	2.75	0.951
		7–72	0.229	14.89	0.938		4–72	0.185	12.58	0.921
	3	0–7	0.993	6.95	0.986		0–4	0.668	2.67	0.973
		7–72	0.233	15.15	0.943		4–72	0.180	6.80	0.933
	5	0–7	1.081	7.57	0.975		0–4	0.718	2.87	0.977
		7–72	0.275	17.88	0.950		4–72	0.161	10.95	0.967
	7	0–7	1.013	7.09	0.971		0–4	0.605	2.42	0.990
		7–72	0.300	19.50	0.939		4–72	0.174	11.83	0.961

SZ: stability zone; RNR: rate of the NPX release; CNR: cumulative NPX released; LR: linear regression.

As it can be seen from the results of Table 9, the NPX/pHPMA2 drug carrier system appeared to be the best performing system in terms of the percentage of NPX released into the medium at neutral pH over the longest period. Indeed, this system was capable to release uniformly the greatest percentage of NPX (28.67% by weight) in the medium at neutral pH (intestines) with a release rate of 0.44 wt%·h^{-1} for 67 h of the release process. During this time, only 21.32 wt% of this drug was released into the medium at pH = 1 (similar to that of in the stomach), with a constant rate of 0.33 wt%·h^{-1}. Concerning the system involving the pHPMA as a carrier, as in the case of that of the NPX/pHEMA system, the most efficient is that initially containing 2 wt% of NPX (NPX/pHPMA2). Indeed,

6.83 wt% NPX was released uniformly (0.102 wt%·h^{-1}) from this system in the neutral pH medium and 4.62 wt% slowly (0.069 wt%·h^{-1}) in acidic medium (pH1) during the same period. Thus, regardless of the polymer used as a carrier in this work, the most efficient system is the one that contains the least NPX load. Finally, the addition of a methylene group on the substituent of the hydroxyl ethyl methacryloyl unit of pHEMA had the effect of reducing by more than four times the percentage of NPX released, as well as its release rate in the various media invested. This can be attributed to the reduction in the hydrophilicity of the polymer upon switching from pHEMA to pHPMA.

Table 9. Percentage of NPX released and instantaneous release rate of NPX from NPX/pHPMA system with different compositions.

System	pH	SZ (h)	RNR (wt%·h^{-1})	CNR (wt%)	LR (R^2)	System	SZ (h)	RDR (wt%·h^{-1})	CNR (wt%)	LR (R^2)
NPX/pHPMA2	1	0–5	0.330	1.65	0.966	NPX/pHPMA7	0–5	0.285	1.43	0.955
		5–72	0.069	4.62	0.977		5–72	0.076	5.09	0.975
	3	0–5	0.311	1.56	0.971		0–5	0.193	0.97	0.953
		5–72	0.069	4.62	0.980		5–72	0.052	3.48	0.973
	5	0–5	0.330	1.65	0.985		0–5	0.209	1.05	0.963
		5–72	0.077	5.162	0.982		5–72	0.057	3.82	0.976
	7	0–5	0.511	2.56	0.990		0–5	0.244	1.22	0.981
		5–72	0.102	6.83	0.972		5–72	0.064	4.29	0.977
NPX/pHPMA5	1	0–5	0.244	1.22	0.968	NPX/pHPMA10	0–5	0.184	0.92	0.929
		5–72	0.061	4.09	0.977		5–72	0.050	3.35	0.974
	3	0–5	0.250	1.25	0.960		0–5	0.193	0.97	0.952
		5–72	0.066	4.42	0.973		5–72	0.052	3.48	0.977
	5	0–5	0.279	1.40	0.963		0–5	0.224	1.12	0.944
		5–72	0.072	4.82	0.971		5–72	0.060	4.02	0.977
	7	0–5	0.285	1.43	0.955		0–5	0.240	1.20	0.954
		5–72	0.076	5.09	0.975		5–72	0.066	4.42	0.978

SZ: stability zone; RNR: rate of the NPX release; CNR: cumulative NPX released; LR: linear regression.

3.5.8. Distribution of NPX Released on Target Organs

According to Belzer et al. [60], the mean total gastrointestinal transit time (GITT) is between 53 and 88 h divided into three main stages: (i) gastric transit (pH 1.5–3, 5), which lasts between one and 4 h; (ii) intestinal transit (pH 7–9), which varies between 4 and 12 h; (iii) transit in the colon (pH 5–7), which lasts between 48 and 72 h. Taking into account the pH of the medium and the GITT, it was possible to estimate approximately from the data in Tables 8 and 9 the distribution of the percentages of cumulative NPX released in different organs and the mean stomach/digestive organ ratio (SDOR) (Equation (14)), independently of the effects of enzymes and microorganisms.

$$SDOR(wt\%) = \frac{r_s}{r_{si} + r_c} \times 100 \quad (14)$$

where r_s, r_{si}, and r_c are the percentages of NPX released in the stomach, small intestine, and colon, respectively, during a certain transit time.

The results obtained are gathered for comparison in Table 10. These data reveal that both the drug carrier systems containing 2 wt% of NPX are the most efficient because the NPX/pHEMA2 drug carrier systems are able to reduce the NPX amount released in the stomach to 3.18 wt% of the total amount released for the fast GITTs and 14.85 wt% for the slow GITTs, and 4.41 wt% and 14.83 wt% for the NPX/pHPMA2 system.

Table 10. Estimated distribution of the cumulative NPX released from NPX/pHEMA and NPX/pHPMA drug carrier systems on the principal digestive organs timed, according to Belzer approach.

Drug Carrier System	Stomach Transit (wt%)		Small Intestine Transit (wt%)		Colon Transit (wt%)		SDOR (wt %)	
Transit Time	Min (1 h)	Max (4 h)	Min (4 h)	Max (12 h)	Min (48 h)	Max (72 h)	Min (48)	Max (72)
NPX/pHEMA2	0.98	3.93	5.16	7.94	24.7	17.2	3.18	14.85
NPX/pHEMA5	0.98	3.92	4.05	5.44	14.4	17.1	5.04	14.46
NPX/pHEMA7	0.79	3.16	2.54	2.68	10.7	12.04	5.63	17.67
NPX/pHEMA10	0.68	5.42	2.42	2.09	8.35	11.83	5.94	28.02
NPX/pHPMA2	0.32	1.28	2.04	1.64	4.9	5.71	4.41	14.83
NPX/pHPMA5	0.25	0.99	1.14	1.13	3.65	5.09	4.96	13.73
NPX/pHPMA7	0.24	0.96	0.96	0.96	3.07	4.29	5.62	15.46
NPX/pHPMA10	0.22	0.87	0.96	0.97	3.17	4.42	5.06	13.9

4. Conclusions

To conclude this work, we can say that the objectives of this work have been achieved. Indeed, the comparison between the physicochemical properties of pHPMA with those of pHEMA revealed properties slightly inferior to those of pHEMA necessary for the admission of pHPMA as a carrier in the drug delivery domain. The miscibility of NPX with pHEMA and pHPMA binary systems, in which the NPX is distributed uniformly in its molecule state, are proven in all compositions by the FTIR method through the presence of hydrogen bonds between their components. This miscibility was also confirmed by the DSC method through the shift toward the low temperatures of the *Tg* of the polymer, the disappearance of the melting temperature of NPX in the mixture, and by XRD through the disappearance of the signals characterizing the crystalline structure of NPX.

The cell adhesion essay and cytotoxicity test of pure polymers and drug carrier systems revealed that the NPX/PHPMA system, as well as the NPX/pHEMA system with compositions, generally exhibit good adhesion compared to the negative and positive controls used in this study. In addition, these two systems, as well as their pure polymers, induce low cytotoxicity compared to the negative and positive controls.

The swelling study of pHEMA and pHPMA carriers revealed that the presence of additional methylene group in the substituent of the HPMA unit of pHPMA caused the swelling capacity to drop to half that of pHEMA. The determination of the Flory–Huggins interaction parameters of the NPX/pHEMA and NPX/pHPMA binary systems reveals greater interactions between the components of NPX/pHEMA system at compositions equal to or less than 5 wt% NPX; on the other hand, they are greater for NPX/pHPMA at compositions greater than 5 wt% NPX.

The "in vitro" study of the release dynamic of Naproxen from NPX/pHEMA and NPX/pHPMA drug carrier systems revealed that the higher percentage of NPX released was obtained from each polymer carrier in neutral pH medium, and the diffusion of water and NPX solution trough these polymer matrices also obeys the Fickian model with a kinetics order close to 0.5, regardless of the pH of the medium. It was also found that the less the mass percent of NPX in the composites, the better its release will be. The comparison between the two drug carrier systems revealed that the pHEMA leads to the best performance in the release dynamic of NPX.

Regarding the Naproxen solubility in water, the results deducted from the "in vitro" study of NPX/pHEMA10 and NPX/pHPMA10 drug carrier systems reveal a very significant improvement in the solubility of NPX in media pH1 (2.33 times, 1.43 times) and 7 (3.32 times, 2.60 times), respectively, compared to those obtained by direct dissolution of Naproxen powder.

According to Belzer, the approximate estimation of the distribution of the percentages of cumulative NPX released in different organs and the mean stomach/digestive organ ratio, independently of the effects of enzymes and microorganisms, revealed that both drug carrier systems containing 2 wt% of NPX are the most efficient because the NPX/pHEMA2 drug carrier systems are able to reduce the NPX amount released in the stomach to 3.18 wt% of the total amount released for the fast GITTs, 14.85 wt% for the slow GITTs, and 4.41 wt% and 14.83 wt% for the NPX/pHPMA2 system. Although pHEMA seems to be the more performing carrier of the two polymers when administered orally (requiring a relatively large amount of drug absorbed at neutral pH), pHPMA combined with a small amount of medication (2 wt%) can also be used if the purpose is the application on the skin surface or as contact lenses to treat certain diseases of the surface of eyes caused by viruses, bacteria, parasites, and fungi because the eyes absorb only a tiny amount of the drug dissolved in a neutral medium. In this case, a regular release of small amounts of drug for as long as possible is desirable in order to limit the frequency of administration of the drug by this route, providing more comfort to the patient.

Author Contributions: Data curation, A.A., S.M.S.A. and A.S.; Formal analysis, A.A., T.S.A.-G., A.S. and T.A.; Funding acquisition, S.M.S.A. and W.S.S.; Investigation, T.A.; Methodology, A.A., S.M.S.A., T.S.A.-G. and T.A.; Project administration, S.M.S.A. and T.A.; Resources, S.M.S.A.; Software, A.A., S.M.S.A., T.S.A.-G. and W.S.S.; Supervision, S.M.S.A.; Visualization, T.S.A.-G., W.S.S. and A.S.; Writing—original draft, A.A., S.M.S.A. and T.A.; Writing—review & editing, T.A. All authors have read and agreed to the published version of the manuscript

Funding: This research received no external funding.

Institutional Review Board Statement: Not applicable.

Informed Consent Statement: Not applicable.

Data Availability Statement: The data presented in this study are available on request from the corresponding author.

Acknowledgments: Authors extend their appreciation to Researchers Supporting Project (RSP2022R475) King Saud University, Riyadh, Saudi Arabia.

Conflicts of Interest: The authors declare no conflict of interest.

References

1. Peterson, B.; Weyers, M.; Steenekamp, J.H.; Steyn, J.D.; Gouws, C.; Hamman, J.H. Drug bioavailability enhancing agents of natural origin (bioenhancers) that modulate drug membrane permeation and pre-systemic metabolism. *Pharmaceutics* **2019**, *11*, 33. [CrossRef] [PubMed]
2. Golan, D.E.; Armstrong, E.J.; Armstrong, A.W. *Principles of Pharmacology: The Pathophysiologic Basis of Drug Therapy*, 4th North American ed.; LWW: Philadelphia, PA, USA, 2016; p. 1024.
3. Krishnaiah, Y. Pharmaceutical Technologies for Enhancing Oral Bioavailability of Poorly Soluble Drugs. *J. Bioequivalence Bioavailab.* **2010**, *2*, 28–36. [CrossRef]
4. Brunton, L.L. *Goodman and Gilman's The Pharmacological Basis of Therapeutics*, 13th ed.; Hilal-Dandan, R., Knollmann, B.C., Eds.; McGraw-Hill Education: New York, NY, USA, 2017; p. 1440.
5. Kim, M.T.; Sedykh, A.; Chakravarti, S.K.; Saiakhov, R.D.; Zhu, H. Critical evaluation of human oral bioavailability for pharmaceutical drugs by using various cheminformatics approaches. *Pharm. Res.* **2014**, *31*, 1002–1014. [CrossRef] [PubMed]
6. Chillistone, S.; Hardman, J.G. Factors affecting drug absorption and distribution. *Anaesth. Intensive Care Med.* **2017**, *18*, 335–339. [CrossRef]
7. Savjani, K.T.; Gajjar, A.K.; Savjani, J.K. Drug solubility: Importance and enhancement techniques. *Int. Sch. Res. Not.* **2012**, *2012*, 195727. [CrossRef]
8. Roderick, P.; Wilkes, H.; Meade, T. The gastrointestinal toxicity of aspirin: An overview of randomised controlled trials. *Br. J. Clin. Pharmacol.* **1993**, *35*, 219–226. [CrossRef]
9. Holliday, W.M.; Berdick, M.; Bell, S.A.; Kiritsis, G.C. Sustained relief analgesic composition. U.S. Patent 3488418A, 6 January 1970.
10. Zheng, J.; Luan, L.; Wang, H.; Xi, L.; Yao, K. Study on ibuprofen/montmorillonite intercalation composites as drug release system. *Appl. Clay Sci.* **2007**, *36*, 297–301. [CrossRef]
11. Carreras, N.; Acuña, V.; Martí, M.; Lis, M. Drug release system of ibuprofen in PCL-microspheres. *Colloid Polym. Sci.* **2013**, *291*, 157–165. [CrossRef]
12. Mangindaan, D.; Chen, C.-T.; Wang, M.-J. Integrating sol–gel with cold plasmas modified porous polycaprolactone membranes for the drug-release of silver-sulfadiazine and ketoprofen. *Appl. Surf. Sci.* **2012**, *262*, 114–119. [CrossRef]

13. Saudi, S.; Bhattarai, S.R.; Adhikari, U.; Khanal, S.; Sankar, J.; Aravamudhan, S.; Bhattarai, N. Nanonet-nano fiber electrospun mesh of PCL–chitosan for controlled and extended release of diclofenac sodium. *Nanoscale* **2020**, *12*, 23556–23569. [CrossRef]
14. Todd, P.A.; Clissold, S.P. Naproxen. *Drugs* **1990**, *40*, 91–137. [CrossRef] [PubMed]
15. Drug Bank, R. DB14761. Available online: http://www.drugbank.ca/drugs. (accessed on 25 May 2020).
16. Lee, B.-J.; Lee, J.-R. Enhancement of solubility and dissolution rate of poorly water-soluble naproxen by complexation with 2-hydroxypropyl-β-cyclodextrin. *Arch. Pharmacal Res.* **1995**, *18*, 22–26. [CrossRef]
17. Branchu, S.; Rogueda, P.G.; Plumb, A.P.; Cook, W.G. A decision-support tool for the formulation of orally active, poorly soluble compounds. *Eur. J. Pharm. Sci.* **2007**, *32*, 128–139. [CrossRef] [PubMed]
18. Netti, P.; Shelton, J.; Revell, P.; Pirie, G.; Smith, S.; Ambrosio, L.; Nicolais, L.; Bonfield, W. Hydrogels as an interface between bone and an implant. *Biomaterials* **1993**, *14*, 1098–1104. [CrossRef]
19. IKADA, Y. Application of biomedical engineering to neurosurgery. *Neurol. Med.-Chir.* **1998**, *38*, 772–779. [CrossRef] [PubMed]
20. Hsiue, G.-H.; Guu, J.-A.; Cheng, C.-C. Poly(2-hydroxyethyl methacrylate) film as a drug delivery system for pilocarpine. *Biomaterials* **2001**, *22*, 1763–1769. [CrossRef]
21. Senol, S.; Akyol, E. Synthesis and characterization of hydrogels based on poly(2-hydroxyethyl methacrylate) for drug delivery under UV irradiation. *J. Mater. Sci.* **2018**, *53*, 14953–14963. [CrossRef]
22. Bettencourt, A.; Almeida, A.J. Poly(methyl methacrylate) particulate carriers in drug delivery. *J. Microencapsul.* **2012**, *29*, 353–367. [CrossRef]
23. Shaked, E.; Shani, Y.; Zilberman, M.; Scheinowitz, M. Poly(methyl methacrylate) particles for local drug delivery using shock wave lithotripsy: I n vitro proof of concept experiment. *J. Biomed. Mater. Res. B Appl. Biomater.* **2015**, *103*, 1228–1237. [CrossRef]
24. Clemons, T.D. Applications of Multifunctional Poly(glycidyl methacrylate)(PGMA) Nanoparticles in Enzyme Stabilization and Drug Delivery. Ph.D. Thesis, University of Western Australia, Crawley, Australia, 2013.
25. Shohraty, F.; Moghadam, P.N.; Fareghi, A.R.; Movagharnezhad, N.; Khalafy, J. Synthesis and Characterization of New pH-Sensitive Hydrogels Based on Poly(glycidyl methacrylate-*co*-maleic anhydride). *Adv. Polym. Technol.* **2018**, *37*, 120–125. [CrossRef]
26. Zare, M.; Bigham, A.; Zare, M.; Luo, H.; Rezvani Ghomi, E.; Ramakrishna, S. pHEMA: An Overview for Biomedical Applications. *Int. J. Mol. Sci.* **2021**, *22*, 6376. [CrossRef] [PubMed]
27. Smetana, K., Jr.; Štol, M.; Korbelář, P.; Novak, M.; Adam, M. Implantation of p(HEMA)-collagen composite into bone. *Biomaterials* **1992**, *13*, 639–642. [CrossRef]
28. Zavřel, V.; Štol, M. p(HEMA) composite as allografting material during therapy of periodontal disease: Three case reports. *Biomaterials* **1993**, *14*, 1109–1112. [CrossRef]
29. Filmon, R.; Basle, M.; Barbier, A.; Chappard, D. Poly(2-hydroxy ethyl methacrylate)-alkaline phosphatase: A composite biomaterial allowing in vitro studies of bisphosphonates on the mineralization process. *J. Biomater. Sci. Polym. Ed.* **2000**, *11*, 849–868. [CrossRef] [PubMed]
30. Filmon, R.; Baslé, M.; Atmani, H.; Chappard, D. Adherence of osteoblast-like cells on calcospherites developed on a biomaterial combining poly(2-hydroxyethyl) methacrylate and alkaline phosphatase. *Bone* **2002**, *30*, 152–158. [CrossRef]
31. Orienti, I.; Bertasi, V.; Zecchi, V. Influence of physico-chemical parameters on the release kinetics of ketoprofen from Poly (HEMA) crosslinked microspheres. *J. Pharm. Belg.* **1992**, *47*, 309–315.
32. Horák, D.; Červinka, M.; Půža, V. Hydrogels in endovascular embolization: VI. Toxicity tests of poly(2-hydroxyethyl methacrylate) particles on cell cultures. *Biomaterials* **1997**, *18*, 1355–1359. [CrossRef]
33. Lesný, P.; De Croos, J.; Přádný, M.; Vacık, J.; Michalek, J.; Woerly, S.; Syková, E. Polymer hydrogels usable for nervous tissue repair. *J. Chem. Neuroanat.* **2002**, *23*, 243–247. [CrossRef]
34. Perlovich, G.L.; Kurkov, S.V.; Kinchin, A.N.; Bauer-Brandl, A. Thermodynamics of solutions III: Comparison of the solvation of (+)-naproxen with other NSAIDs. *Eur. J. Pharm. Biopharm.* **2004**, *57*, 411–420. [CrossRef]
35. Semlali, A.; Beji, S.; Ajala, I.; Rouabhia, M. Effects of tetrahydrocannabinols on human oral cancer cell proliferation, apoptosis, autophagy, oxidative stress, and DNA damage. *Arch. Oral Biol.* **2021**, *129*, 105200. [CrossRef]
36. Semlali, A.; Contant, C.; Al-Otaibi, B.; Al-Jammaz, I.; Chandad, F. The curcumin analog (PAC) suppressed cell survival and induced apoptosis and autophagy in oral cancer cells. *Sci. Rep.* **2021**, *11*, 11701. [CrossRef] [PubMed]
37. Contant, C.; Rouabhia, M.; Loubaki, L.; Chandad, F.; Semlali, A. Anethole induces anti-oral cancer activity by triggering apoptosis, autophagy and oxidative stress and by modulation of multiple signaling pathways. *Sci. Rep.* **2021**, *11*, 13087. [CrossRef] [PubMed]
38. Iza, M.; Stoianovici, G.; Viora, L.; Grossiord, J.; Couarraze, G. Hydrogels of poly(ethylene glycol): Mechanical characterization and release of a model drug. *J. Control. Release* **1998**, *52*, 41–51. [CrossRef]
39. Guan, Y.; Zhang, Y.; Zhou, T.; Zhou, S. Stability of hydrogen-bonded hydroxypropylcellulose/poly (acrylic acid) microcapsules in aqueous solutions. *Soft Matter* **2009**, *5*, 842–849. [CrossRef]
40. Yang, S.; Zhang, Y.; Wang, L.; Hong, S.; Xu, J.; Chen, Y.; Li, C. Composite thin film by hydrogen-bonding assembly of polymer brush and poly(vinylpyrrolidone). *Langmuir* **2006**, *22*, 338–343. [CrossRef] [PubMed]
41. Saritha, D.; Bose, P.S.C.; Reddy, P.S.; Madhuri, G.; Nagaraju, R. Improved dissolution and micromeritic properties of naproxen from spherical agglomerates: Preparation, in vitro and in vivo characterization. *Braz. J. Pharm. Sci.* **2012**, *48*, 683–690. [CrossRef]
42. Akbari, J.; Saeedi, M.; Morteza-Semnani, K.; Rostamkalaei, S.S.; Asadi, M.; Asare-Addo, K.; Nokhodchi, A. The design of naproxen solid lipid nanoparticles to target skin layers. *Colloids Surf. B Biointerfaces* **2016**, *145*, 626–633. [CrossRef]

43. Morita, S. Hydrogen-bonds structure in poly(2-hydroxyethyl methacrylate) studied by temperature-dependent infrared spectroscopy. *Front. Chem.* **2014**, *2*, 10. [CrossRef]
44. Brogden, R.; Finder, R.; Sawyer, P.R.; Speight, T.; Avery, G. Naproxen: A review of its pharmacological properties and therapeutic efficacy and use. *Drugs* **1975**, *9*, 326–363. [CrossRef]
45. Keshavarz, M.H.; Esmaeilpour, K.; Taghizadeh, H. A new approach for assessment of glass transition temperature of acrylic and methacrylic polymers from structure of their monomers without using any computer codes. *J. Therm. Anal. Calorim.* **2016**, *126*, 1787–1796. [CrossRef]
46. Comyn, J. Introduction to Polymer Permeability and the Mathematics of Diffusion. In *Polymer Permeability*; Springer: Berlin/Heidelberg, Germany, 1985; pp. 1–10.
47. Masaro, L.; Zhu, X. Physical models of diffusion for polymer solutions, gels and solids. *Prog. Polym. Sci.* **1999**, *24*, 731–775. [CrossRef]
48. Burchard, W. Solution Thermodynamics of Non-Ionic Water-Soluble Polymers. In *Chemistry and Technology of Water-Soluble Polymers*; Finch, C.A., Ed.; Springer: Berlin/Heidelberg, Germany, 1983; pp. 125–142.
49. Potter, C.B.; Davis, M.T.; Albadarin, A.B.; Walker, G.M. Investigation of the Dependence of the Flory–Huggins Interaction Parameter on Temperature and Composition in a Drug–Polymer System. *Mol. Pharm.* **2018**, *15*, 5327–5335. [CrossRef] [PubMed]
50. Aragon, D.M.; Pacheco, D.P.; Ruidiaz, M.A.; Sosnik, A.D.; Martinez, F. Método extendido de Hildebrand en la predicción de la solubilidad de naproxeno en mezclas cosolventes etanol+ agua. *Vitae* **2008**, *15*, 113–122.
51. Thakral, S.; Thakral, N.K. Prediction of drug–polymer miscibility through the use of solubility parameter based Flory–Huggins interaction parameter and the experimental validation: PEG as model polymer. *J. Pharm. Sci.* **2013**, *102*, 2254–2263. [CrossRef] [PubMed]
52. Lin, M.; Wang, H.; Meng, S.; Zhong, W.; Li, Z.; Cai, R.; Chen, Z.; Zhou, X.; Du, Q. Structure and release behavior of PMMA/silica composite drug delivery system. *J. Pharm. Sci.* **2007**, *96*, 1518–1526. [CrossRef]
53. Reinhard, C.S.; Radomsky, M.L.; Saltzman, W.M.; Hilton, J.; Brem, H. Polymeric controlled release of dexamethasone in normal rat brain. *J. Control. Release* **1991**, *16*, 331–339. [CrossRef]
54. Cypes, S.H.; Saltzman, W.M.; Giannelis, E.P. Organosilicate-polymer drug delivery systems: Controlled release and enhanced mechanical properties. *J. Control. Release* **2003**, *90*, 163–169. [CrossRef]
55. Frank, A.; Rath, S.K.; Venkatraman, S.S. Controlled release from bioerodible polymers: Effect of drug type and polymer composition. *J. Control. Release* **2005**, *102*, 333–344. [CrossRef]
56. Dilmi, A.; Bartil, T.; Yahia, N.; Benneghmouche, Z. Hydrogels based on 2-hydroxyethylmethacrylate and chitosan: Preparation, swelling behavior, and drug delivery. *Int. J. Polym. Mater. Polym. Biomater.* **2014**, *63*, 502–509. [CrossRef]
57. Peppas, N.A.; Narasimhan, B. Mathematical models in drug delivery: How modeling has shaped the way we design new drug delivery systems. *J. Control. Release* **2014**, *190*, 75–81. [CrossRef]
58. Sevelius, H.; Runkel, R.; Segre, E.; Bloomfield, S. Bioavailability of naproxen sodium and its relationship to clinical analgesic effects. *Br. J. Clin. Pharmacol.* **1980**, *10*, 259–263. [CrossRef] [PubMed]
59. Kumar, L.; Suhas, B.; Pai, G.; Verma, R. Determination of saturated solubility of naproxen using UV visible spectrophotometer. *Res. J. Pharm. Technol.* **2015**, *8*, 825–828. [CrossRef]
60. Belzer, C.; De Vos, W.M. Microbes inside—from diversity to function: The case of Akkermansia. *ISME J.* **2012**, *6*, 1449–1458. [CrossRef] [PubMed]

polymers

MDPI

Article

Casein Micelles as Nanocarriers for Benzydamine Delivery

Nikolay Zahariev [1,2], Maria Marudova [3], Sophia Milenkova [3], Yordanka Uzunova [2,4] and Bissera Pilicheva [1,2,*]

1 Department of Pharmaceutical Sciences, Faculty of Pharmacy, Medical University of Plovdiv, 15A Vassil Aprilov Blvd, 4002 Plovdiv, Bulgaria; nikolay.zahariev@mu-plovdiv.bg
2 Research Institute, Medical University of Plovdiv, 15A Vassil Aprilov Blvd, 4002 Plovdiv, Bulgaria; yordanka.uzunova@mu-plovdiv.bg
3 Faculty of Physics and Technology, University of Plovdiv "Paisii Hilendarski", 24 Tsar Asen Str., 4000 Plovdiv, Bulgaria; marudova@uni-plovdiv.net (M.M.); sophiamilenkowa@gmail.com (S.M.)
4 Department of Bioorganic Chemistry, Faculty of Pharmacy, Medical University of Plovdiv, 15A Vassil Aprilov Blvd, 4002 Plovdiv, Bulgaria
* Correspondence: bisera.pilicheva@mu-plovdiv.bg

Abstract: The aim of the present work was to optimize the process parameters of the nano spray drying technique for the formulation of benzydamine-loaded casein nanoparticles and to investigate the effect of some process variables on the structural and morphological characteristics and release behavior. The obtained particles were characterized in terms of particle size and size distribution, surface morphology, production yield and encapsulation efficiency, drug-polymer compatibility, etc., using dynamic light scattering, scanning electron microscopy, differential scanning calorimetry, and Fourier transformed infrared spectroscopy. Production yields of the blank nanoparticles were significantly influenced by the concentration of both casein and the crosslinking agent. The formulated drug-loaded nanoparticles had an average particle size of 135.9 nm to 994.2 nm. Drug loading varied from 16.02% to 57.41% and the encapsulation efficiency was in the range 34.61% to 78.82%. Our study has demonstrated that all the investigated parameters depended greatly on the polymer/drug ratio and the drug release study confirmed the feasibility of the developed nanocarriers for prolonged delivery of benzydamine.

Keywords: benzydamine; casein; biopolymers; nanoparticles; nano spray drying; nano micelles; drug delivery

Citation: Zahariev, N.; Marudova, M.; Milenkova, S.; Uzunova, Y.; Pilicheva, B. Casein Micelles as Nanocarriers for Benzydamine Delivery. *Polymers* **2021**, *13*, 4357. https://doi.org/10.3390/polym13244357

Academic Editors: Ariana Hudita and Bianca Gălățeanu

Received: 24 November 2021
Accepted: 10 December 2021
Published: 13 December 2021

1. Introduction

Over recent decades there has been a growing scientific interest towards the use of naturally occurring materials for drug delivery purposes. This is mainly due to their numerous advantages over synthetic materials, namely biocompatibility, biodegradability, and low immunogenicity [1]. Moreover, natural materials produce non-toxic metabolites, unlike synthetic polymers which can be contaminated with unreacted toxic monomers and crosslinkers [2]. Due to their specific structure and corresponding features, naturally occurring materials such as polysaccharides and peptides are widely used for the formulation of micro- and nanoparticulate drug delivery systems. These carriers can provide controlled and targeted release, thus improving the therapeutic performance of the encapsulated drug and minimizing the risk of side effects [3]. Amongst the potential biopolymers, proteins are preferred as natural drug delivery systems due to the relatively easy preparation processes and production of well-defined structures, which enables surface modification and may provide modified and targeted release [1]. Among proteins, casein (CAS) is considered a suitable biopolymer for the preparation of nanoparticulate drug delivery systems due to its structural and physicochemical characteristics [2].

Casein is a collective term used to define a family of calcium (phosphate)-binding phosphoproteins commonly found in mammalian milk [4]. Casein from bovine milk is composed of four peptides, namely α_{s1}, α_{s2}, β, and k, which differ in the content of amino acids, phosphorus, and carbohydrates, but they are all amphiphilic in nature [5]. Cysteine amino acid residues that allow the formation of disulfide bonds are found only in the polypeptide chains of k-casein. In general, the peptide surface is negatively charged due to phosphorylation [1]. The lack of secondary structures because of the proline-rich amino acid sequence [6] and the tendency for binding amorphous calcium phosphate cause electrostatic, hydrogen, and hydrophobic interactions, leading to self-assembly of the casein peptides into stable agglomerates known as casein micelles [7]. The inner part of the micelle is composed of α_{s1}, α_{s2}, and β caseins, whereas the outer layer that stabilizes the micelle contains glycosylated k-casein [8]. Casein micelles exhibit pH-dependent behavior. Their structure tightens when the negative surface charge of casein molecules decreases, and expands with increasing surface charge, which leads to electrostatic repulsion between the molecules [9–11]. Given the amphiphilic properties and pH-dependent behavior of casein, and its ability to participate in hydrophobic and hydrophilic interactions, it is clear why this biopolymer has found a place in scientific research as a potential nanoparticle drug delivery carrier.

Various methods have been reported for the preparation of casein nanoparticles for drug delivery, including pH-shifting [12], high pressure homogenization [13–18], electrostatic complexation [19], solvent displacement [20], emulsification solvent evaporation [21], and spray drying [22–24]. Nano spray drying, a variation of the established spray drying technology used to convert liquids into solid powders, is a relatively new technique adopted for the preparation of nanosized drug delivery systems. The method is based on the use of a revolutionary sprayer developed by the Swiss Büchi Labortechnik AG, which is equipped with a piezoelectric vibrating spray mesh head, allowing the formation of fine droplets, which are dried and electrostatically collected [25]. As a result, spherical submicron structures of particle size below 1000 nm with improved biopharmaceutical behavior are obtained [26–33]. Although spray drying of proteins has been reported in numerous scientific papers [34–37], no data on nano spray drying of casein have been found in the literature. The technology was therefore a research challenge. For the present study, benzydamine hydrochloride (BZ) was used as a model drug.

Benzydamine hydrochloride is a nonsteroidal anti-inflammatory drug with local anesthetic and analgesic properties for pain relief and treatment of inflammatory conditions of the mouth and throat such as oral mucositis, postoperative sore throat and mucosal ulcers. The mechanism of the anti-inflammatory effect of benzydamine has not yet been fully understood. According to Quane et al. [38], the anti-inflammatory activity of benzydamine may be due to its membrane-stabilizing or inhibitory effect of the synthesis of TNF-α. Unlike NSAIDs, which have acidic properties, benzydamine is a weak base, highly lipid-soluble in its unionized form [39].

According to Beckett et al. [40] and Bickel et al. [41], only a limited amount of weak, basic, lipid-soluble drugs is absorbed into buccal tissue via mouthwash application. The small degree of absorption into buccal tissue is confirmed by the poor systemic availability (5%) [42]. To enhance absorption and thus bioavailability, benzydamine hydrochloride was incorporated into nanoparticles. Due to the specific binding properties and pH-dependent drug release, casein is considered a promising biopolymer for the preparation of benzydamine loaded casein nanoparticles.

The aim of the present work was to optimize the process parameters of the nano spray drying technique for the formulation of BZ-loaded casein nanoparticles. Furthermore, an investigation of the effect of process variables on structural and morphological characteristics and release behavior was conducted.

2. Materials and Methods

Benzydamine hydrochloride (Mw 345.87 g/mol), sodium caseinate (from bovine milk) and $CaCl_2 \cdot 2H_2O$ (Mw 147.01 g/mol) were purchased from Sigma-Aldrich (St. Louis, MO, USA). All other reagents were of analytical grade.

2.1. Preparation of Blank Casein Nanoparticles and BZ-Loaded Casein Nanoparticles

The blank casein nanoparticles where prepared via coacervation followed by spray drying using nano spray dryer Büchi B-90 (Büchi Labortechnik AG, Flawil, Switzerland), as previously reported by Gandhi et al. [2]. A certain amount of sodium caseinate was dissolved in 100 mL deionized water, previously adjusted to pH 2 with 1M hydrochloric acid. Then, the crosslinking agent $CaCl_2{\cdot}2H_2O$ (2 µL/mL) was added dropwise to the casein solution under high-speed homogenization at 25,000 rpm (Miccra MiniBatch D-9, MICCRA GmbH, Heitersheim, Germany) for 15 min and casein micelles were produced. The obtained nanosuspension was then stirred on a magnetic stirrer at 500 rpm for 30 min) to allow effective crosslinking of casein molecules. Finally, the suspension was spray dried using nano spray dryer Büchi B-90 under the following predetermined conditions: mesh size of 4.0 µm, inlet temperature 40 °C, solution feed rate 50%, spray intensity 70%, drying gas speed 120 L/min, pressure 30 nbar. To study the effect of different formulation variables on the produced particles, 3^2 full factorial design was applied. Nine batches of formulations were prepared at varied protein and crosslinker concentrations (Table 1).

Table 1. Composition and characteristics of blank casein nanoparticles (n = 3, PDI (polydispersity index), Dv10, Dv50 and Dv90 (10, 50 and 90% of the total volume of particles, respectively, are with size below the indicated value).

Sample Code	Variables		Dv10 ± SD (nm)	Dv50 ± SD (nm)	Dv90 ± SD (nm)	PDI	ζ ± SD (mV)	Yeild ± SD (%)
	Polymer (%)	Crosslinker (M)						
Cas1-Ca1	0.5	0.5	2885.0 ± 5.26	3470.0 ± 22.3	4020.0 ± 6.6	109.20	−15.4 ± 0.4	37.87 ± 9.26
Cas1-Ca2	0.5	1.0	48.5 ± 2.04	174.5 ± 4.1	256.1 ± 2.7	8.44	−22.5 ± 0.8	43.54 ± 2.04
Cas1-Ca3	0.5	1.5	46.1 ± 7.25	133.2 ± 1.6	2156.0 ± 4.7	34.47	−19.0 ± 0.6	50.30 ± 7.25
Cas2-Ca1	1.0	0.5	89.8 ± 2.83	138.1 ± 3.5	463.2 ± 3.0	22.68	−17.0 ± 0.7	49.57 ± 2.83
Cas2-Ca2	1.0	1.0	51.3 ± 4.55	4190.0 ± 20.2	6050.0 ± 3.2	543.90	−13.3 ± 0.7	40.88 ± 4.55
Cas2-Ca3	1.0	1.5	36.5 ± 3.02	104.1 ± 8.5	191.0 ± 9.0	1.21	−23.6 ± 0.6	64.80 ± 3.02
Cas3-Ca1	1.5	0.5	91.8 ± 2.57	956.2 ± 14.3	5556.0 ± 5.1	6.18	−17.9 ± 0.7	51.04 ± 2.57
Cas3-Ca2	1.5	1.0	115.0 ± 4.50	212.9 ± 2.9	1362.0 ± 7.9	8.06	−15.1 ± 0.5	35.04 ± 4.50
Cas3-Ca3	1.5	1.5	90.1 ± 2.69	149.2 ± 2.6	2944.0 ± 4.0	13.21	−16.3 ± 0.5	57.68 ± 2.69

BZ-loaded casein nanoparticles were prepared following the methodology described in the previous paragraph. Briefly, protein aqueous solution (1.5% *w*/*v*) was prepared by dissolving a certain amount of sodium caseinate in deionized water, previously acidified to pH 2 using 1N hydrochloric acid. Then, BZ was added to the solution, followed by protein crosslinking with $CaCl_2{\cdot}2H_2O$ (2 µL/mL) at a stirring rate of 25,000 rpm. The procedure continued as previously described. Four batches of drug-leaded formulations were developed at varied drug–polymer ratios. The composition of the batches is presented in Table 2.

Table 2. Composition of BZ-loaded casein nanoparticles of various batches.

Sample Code	Sodium Caseinate (%)	Benzydamine HCl (mg)	Variables	
			Polymer: Drug Ratio	
Cas2-Ca3-BZ-1	1.5	1.500	1	1
Cas2-Ca3-BZ-2	1.5	0.750	2	1
Cas2-Ca3-BZ-4	1.5	0.375	4	1
Cas2-Ca3-BZ-6	1.5	0.250	6	1

2.2. Characterization

2.2.1. Production Yield, Drug Loading and Entrapment Efficiency

The production yields of the nanoparticles from different batches were calculated using the weight of the spray dried nanoparticles with respect to the initial quantity of the drug and polymer, according to the following equation:

$$\text{Production yield } (\%) = \frac{\text{Spray dried nanoparticles (mg)}}{\text{Drug (mg)} + \text{Polymer (mg)}} \tag{1}$$

Drug loading and drug entrapment efficiency of BZ-loaded CAS nanoparticles were determined spectrophotometrically. BZ-loaded nanoparticles were dispersed into 10 mL previously acidified at pH 2 deionized water and were stirred for 60 min until complete swelling of casein micelles occurred, allowing BZ extraction in the aqueous medium. Then, the blend was centrifuged at 5000 rpm and filtered (0.22 μm, Chromafil®, Macherey-Nagel, Düren, Germany). Drug concentration was determined after proper dilution using an Evolution 3000 Pro UV/Visible spectrophotometer (Thermo Scientific, Waltham, MA, USA) at a wavelength of 306 nm. The drug loading (DL) was calculated according to Equation (2), and entrapment efficiency (EE) was calculated according to Equation (3):

$$DL\ (\%) = \frac{\text{Amount of drug in the formulation}}{\text{Total amount of nanoparticles}} \times 100 \quad (2)$$

$$EE\ (\%) = \frac{\text{Actual drug content}}{\text{Theoretical drug content}} \times 100 \quad (3)$$

2.2.2. Particle Size Analysis, Size Distribution and Zeta Potential

Particle size of the obtained CAS nanoparticles was determined by dynamic light scattering method using Nanotrac particle size analyzer (Microtrac, York, PA, USA). The system is equipped with 3 mW helium/neon laser at 780 nm wavelength and measures the particle size with noninvasive backscattering technology, performing particle size analysis in the range of 0.8 nm to 6.5 μm. The equipment allows determination of ζ-potential in the range from −200 mv to +200 mV. The samples were prepared by dispersing a small amount of dry nanoparticles in purified water, and the dispersions (refractive index 1.33, average viscosity 0.87 ± 0.05 cP) were stirred on a magnetic stirrer and then analysed for particle size and zeta potential. All the measurements were performed at 25.0 °C at 20-s intervals and were repeated three times.

2.2.3. SEM Analysis

Imaging of the obtained CAS nanoparticles was performed using scanning electron microscopy (Prisma E SEM, Thermo Scientific, Waltham, MA, USA). The samples were loaded on a copper sample holder and sputter coated with carbon followed by gold using vacuum evaporator (BH30). The images were recorded at 15 kV acceleration voltage at various magnifications using DBS (back-scattered electrons) detector.

2.2.4. FTIR Spectroscopy

The samples were evaluated for drug/polymer interactions by Fourier transformed infrared spectroscopy (FTIR). The spectra were collected using a Nicolet iS 10 FTIR spectrometer (Thermo Fisher Scientific, Pittsburgh, PA, USA), equipped with a diamond attenuated total reflection (ATR) accessory, operating in the range from 600 cm^{-1} to 4000 cm^{-1} with a resolution 4 nm and 16 scans. The obtained spectra were analysed with OMNIC® software package (Version 7.3, Thermo Electron Corporation, Madison, WI, USA).

2.2.5. Differential Scanning Calorimetry (DSC)

Thermal analysis of the CAS nanoparticles was performed using DSC 204F1 Phoenix (Netzsch Gerätebau GmbH, Selb, Germany) based on the heat flux principle and cooled with a an intracooler. An indium standard (T_m = 156.6 °C, ΔH_m = 28.5 J/g) was used for the temperature and heat flow calibration. The samples were hermetically sealed in aluminum sample pans. An empty pan, identical to the sample pan, was used as reference. The measurements were performed under argon atmosphere at a heating rate of 10 °C/min.

2.2.6. In Vitro Drug Release

In vitro release study was carried out by diffusion using dialysis bag. The dialysis membrane (Sigma, MWCO 12,000 Da) was cut into equal pieces (6 × 2.5 cm^2) and soaked in distilled water for 24 h before use. An accurately weighed amount of nanoparticles (equivalent to 10 mg BZ) was dispersed in 2 mL of PBS buffer (pH = 7.4) and transferred into the dialysis bag. Each bag was placed into a beaker containing 20 mL dissolution media (PBS buffer, pH 7.4) and kept on an electromagnetic stirrer at 50 rpm and 37 ± 0.5 °C. Samples of 2 mL were taken at predetermined time intervals and replaced with equivalent volume of fresh media. The samples were then filtered (0.45 μm Chromafil® syringe filter, Macherey-Nagel, Düren, Germany) and analyzed for drug content as mentioned above. Mean results of triplicate measurements and standard deviation were reported.

3. Results and Discussion

3.1. Synthesis and Characterization of Blank Casein Nanoparticles

Casein concentration and the crosslinking agent ($CaCl_2{\cdot}2H_2O$, Mw = 147.01 g/mol) concentration were varied at three different levels according to the applied 3^2 full factorial design. Three different concentrations of casein solution were used: low concentration 0.05% (1), medium concentration 0.1% (2) and high concentration 0.15% (3). The concentration of the crosslinking agent was also set at three levels: low concentration 0.5 M (1), medium concentration 1.0 M (2) and high concentration 1.5 M (3). The other process parameters were kept constant as described in the Materials and Methods section. The dependent variables were particle size and production yields. The composition of the obtained casein nanoparticles with different formulation variables is shown in Table 1.

Nine batches of blank casein nanoparticles were obtained by spray drying technique. Mean particle sizes (Dv50, Table 1) varied in a wide range between 74.5 nm and 4.19 μm. The smaller particles showed extremely high degree of aggregation, leading to the formation of larger clusters, as evidenced from the scanning electron micrographs. Analysis of Dv10 was found to be far more representative for the particle size range of the formulated structures. According to the results, a tendency for reduction of the particle size was observed when the crosslinker concentration increased. According to the data published in the literature [42], the larger the amount of the crosslinker, the stronger the packing of the structure and the denser the micelle, resulting in particles of a smaller size range. Our results confirmed this relation, but the impact was not significant. As for the aggregation, clusters of nanoparticles occurred within each of the three groups of batches and no dependences could be outlined. For that reason, the combined effect of the two factors—the concentration of casein and the crosslinker—on the particle size was investigated, and the plot is shown in Figure 1. Since our goal was to produce particles of the smallest possible size, the batches revealing practically no or little degree of aggregation were considered for further investigation (samples prepared at casein concentration 1.5% and crosslinker concentration of 0.5 M proved to be unsatisfying).

Production yields, on the other hand, gradually increased when higher casein concentrations were used. The yields obtained varied in the range from 35.04% to 64.80%. Production yields were determinative for the selection of optimum models for drug loading and further investigation, therefore the combined effect of the two variables on the yields was studied. The plot is presented in Figure 2. The highest values were obtained when 1.5 M calcium chloride solution was used. Among all the formulated batches, Cas2-Ca3 (casein 1%, calcium chloride 1.5 M) was determined to be optimal in terms of production yield and desired particle size range.

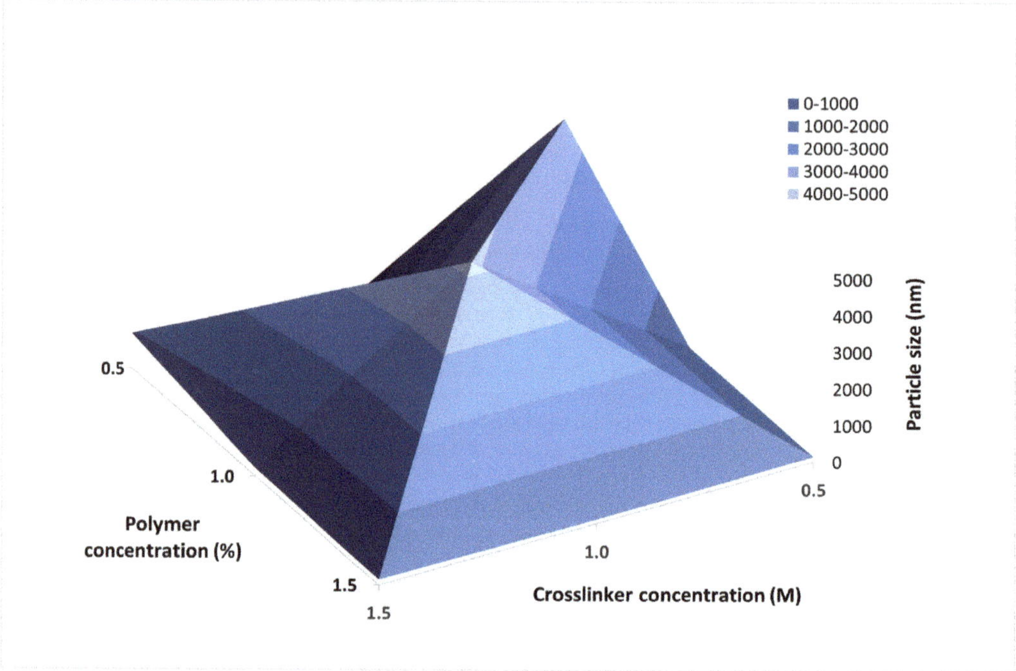

Figure 1. 3D plot representing the impact of the concentration of the polymer and the crosslinking agent on the mean particle size of blank casein spray dried nanoparticles.

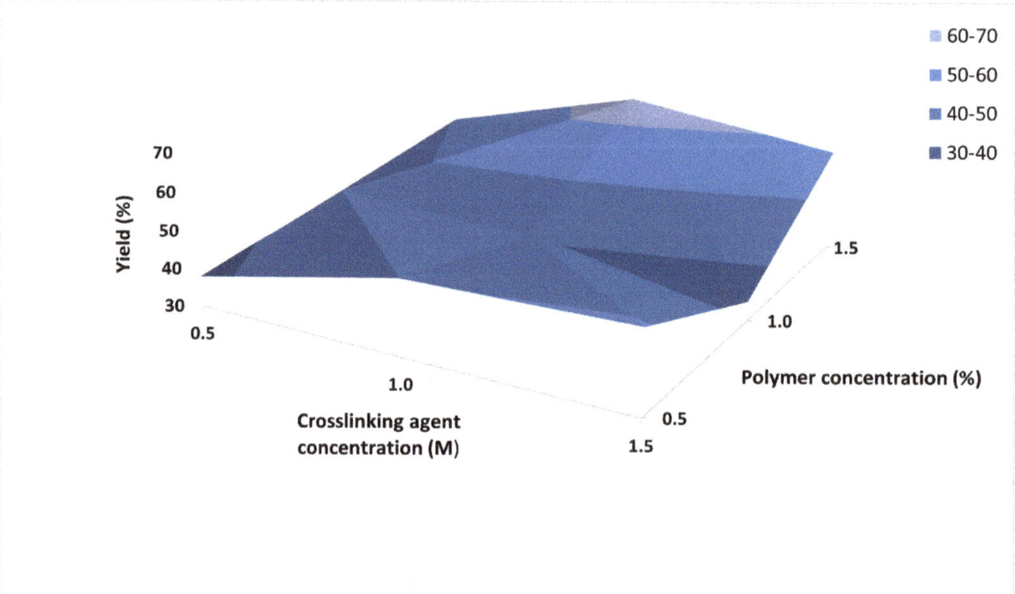

Figure 2. 3D plot representing the impact of the concentration of the polymer (%) and the crosslinking agent (M) on the particles production yields.

3.2. Synthesis and Characterization of BZ Loaded Casein Nanoparticles

BZ-loaded CAS nanoparticles were prepared via coacervation method, followed by spray drying. In order to investigate the effect of polymer and drug concentration over the production yield, particle size, surface morphology, drug entrapment efficiency and release behavior, four batches of drug-loaded nanoparticles were prepared based on the optimized formulation of blank nanoparticles (sample Cas2-Ca3, prepared at 1.0% casein concentration, 1.5 M $CaCl_2 \cdot 2H_2O$) and varying the polymer/drug ratio (1:1, 2:1, 4:1, 6:1) (Table 2). The results of the study are summarized in Table 3.

3.2.1. Drug Loading and Entrapment Efficiency

Drug loading of the developed BZ-loaded casein nanoparticles varied in a wide range from 16.02% to 57.41%. A tendency for decrease in drug loading was observed with increase of polymer/drug ratio, which was not surprising regarding the amount of polymer used for the formulation of the model particles. Entrapment efficiency was substantial, varying from 76.23% to 78.82% for the samples prepared at 2:1, 4:1 and 6:1 ratio. Significantly lower entrapment efficiency was determined for the sample Cas2-Ca3-BZ-1, prepared at 1:1 polymer/drug ratio. It could be suggested that the polymer had a limited capacity to incorporate drug molecules during nanoparticles formulation. For the above sample, the amount of the polymer was probably not sufficient to entrap and retain the drug and form a stable structure.

Our hypothesis was confirmed by morphological analysis of the samples using scanning electron microscopy (Figure 4). The lack of clearly defined nanostructures in model Cas2-Ca3-BZ-1 was evidenced by the obtained scanning electron micrographs in contrast to the other samples. In addition, the larger amount of BZ in this sample probably led to displacement of calcium phosphate and disruption of micellar integrity. The phenomenon has been observed in other studies and has been thoroughly described in the literature [43]. With an increase of the polymer/drug ratio from 1:1 to 2:1, a double increase of the EE was observed (Table 3). Higher amounts of casein led to more efficient incorporation of benzydamine in the nanoparticles, which is probably due to the enhanced hydrophobic effect favoring micellar solubilization of the drug [19]. A further increase in the polymer/drug ratio (4:1 and 6:1) did not lead to a significant change in the drug entrapment efficiency.

3.2.2. Production Yield

Production yields were high, ranging from 58.23% to 74.71% except for the batch produced at 1:1 polymer/drug ratio (34.61%). The increase in the amount of casein in the formulations, relative to BZ, led to a slight reduction of production yields, which was probably due to the enhanced viscosity of the feeding suspension, which made it difficult to pass through the spray mesh. On the other hand, batch Cas2-Ca3-BZ-1, although expected to provide the highest yield, refuted our suggestions. A possible explanation for this could be the disruption of micellar integrity due to displacement of calcium phosphate and the formation of precipitate prior to spray drying [44].

Table 3. Characteristics of the spray dried BZ-loaded casein nanoparticles (n = 3). DL = drug loading, EE = entrapment efficiency.

Sample Code	Particle Size ± SD (nm)	ζ ± SD (mV)	DL ± SD (%)	EE ± SD (%)	Yield ± SD (%)
Cas2-Ca3-BZ-1	994.2 ± 2.21	18.11 ± 0.86	57.41 ± 0.27	34.61 ± 0.23	30.42 ± 4.28
Cas2-Ca3-BZ-2	243.6 ± 2.47	16.33 ± 0.55	35.04 ± 034	78.82 ± 0.39	74.71 ± 5.41
Cas2-Ca3-BZ-4	159.8 ± 2.43	15.24 ± 0.58	26.21 ± 0.22	76.23 ± 0.28	68.76 ± 5.01
Cas2-Ca3-BZ-6	1359 ± 1.73	14.23 ± 0.66	16.02 ± 0.31	77.44 ± 0.57	58.23 ± 5.08

3.2.3. Particle Size and Size Distribution

Particle size and size distribution were analyzed by dynamic light scattering and the results are presented in Table 3 and Figure 3. The median particle size ranged from 135.9 nm to 994.2 nm with a clear tendency for size reduction with increase in casein concentration. Bimodal particle size distribution was observed in batch Cas2-Ca3-BZ-1, suggesting a high aggregation tendency. However, no clearly distinguished structures were observed under scanning electron microscope, corresponding to the results obtained for production yield, drug loading and entrapment efficiency. Probably, nanoparticle formation could not be accomplished at 1:1 polymer/drug ratio, whereas the samples prepared at 2:1, 4:1 and 6:1 polymer/drug ratio were clearly distinguished and less cohesive, with minimal degree of aggregation.

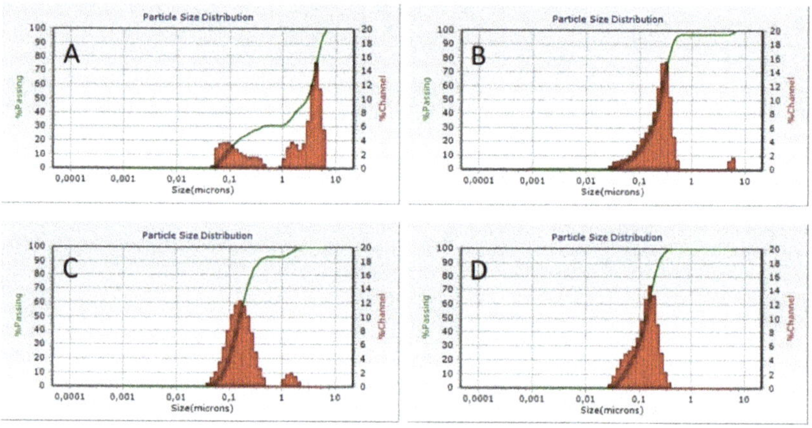

Figure 3. Dynamic light scattering histograms of BZ loaded casein nanoparticles of batches Cas2-Ca3-BZ-1 (**A**), Cas2-Ca3-BZ-2 (**B**), Cas2-Ca3-BZ-4 (**C**) and Cas2-Ca3-BZ-6 (**D**).

3.2.4. Surface Morphology

Surface morphology evaluation of the four batches of nanoparticles was performed using scanning electron microscopy. The micrographs are presented in Figure 4. Three different patterns of surface morphology were observed: rough spherical particles, wrinkled spherical particles and wrinkled irregularly shaped particles. A tendency towards increased surface roughness was observed with raising casein concentrations. Irregular, wrinkled, fragmented, and highly aggregated structures with an average particle size of about 994 nm were observed at drug/polymer ratio 1:1 (Figure 4A, batch Cas2-Ca3-BZ-1). It is well known that inlet temperature plays a key role in the spray drying process, significantly affecting the surface morphology of the dry particles. According to Both et al. [45], spray drying at high inlet temperatures generally results in the formation of less wrinkled particles with a large, hollow core. Therefore, it could be assumed that the higher viscosity of the feeding suspension together with the low inlet temperature (40 °C) might be associated with increased stickiness and subsequent agglomeration of these particles. As for the other three batches of nanoparticles, they all had a rounded shape and a wrinkled surface. In addition, as the concentration of the polymer raised and the percentage of drug diminished relative to the casein content, the rugosity degree of particles increased. It could be assumed that the lower drug content per unit mass led to the formation of loose matrix structures. Upon drying, these structures shrink, leading to the formation of smaller particles with multiple surface invaginations. Our hypothesis was confirmed by particle size analysis.

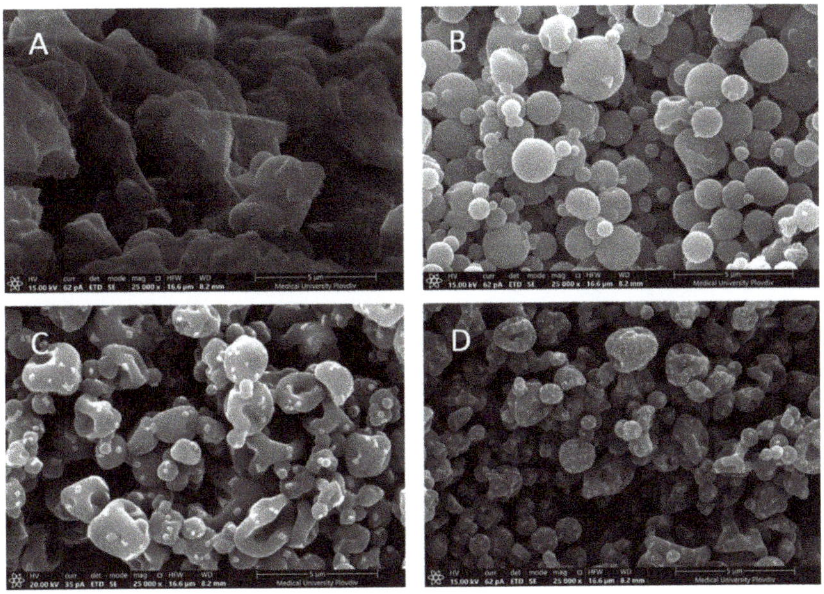

Figure 4. SEM micrographs of BZ-loaded casein nanoparticles of batches Cas2-Ca3-BZ-1 (**A**), Cas2-Ca3-BZ-2 (**B**), Cas2-Ca3-BZ-4 (**C**) and Cas2-Ca3-BZ-6 (**D**) at 25,000× magnification.

3.2.5. Differential Scanning Calorimetry (DSC)

The phase state of BZ incorporated into the spray dried casein nanoparticles was analyzed using differential scanning calorimetry (DSC). The obtained thermograms are presented in Figure 5. The thermogram of casein revealed a broad endothermic peak at 84.8 °C, which could be attributed to the evaporation of water present in casein micelles. Benzydamine hydrochloride, on the other hand, being solid crystalline, showed a characteristic peak at 166.5 °C, which corresponds to its melting point. In drug-loaded samples, a gradual decrease in peak intensity was observed, with an increase in the polymer/drug ratio from 1:1 to 6:1, as shown in Figure 5C–F. It can be assumed that changes occurred in the degree of crystallinity of BZ during spray drying and the drug was partially transformed into its amorphous phase depending on the drug content in the formulated nanoparticles.

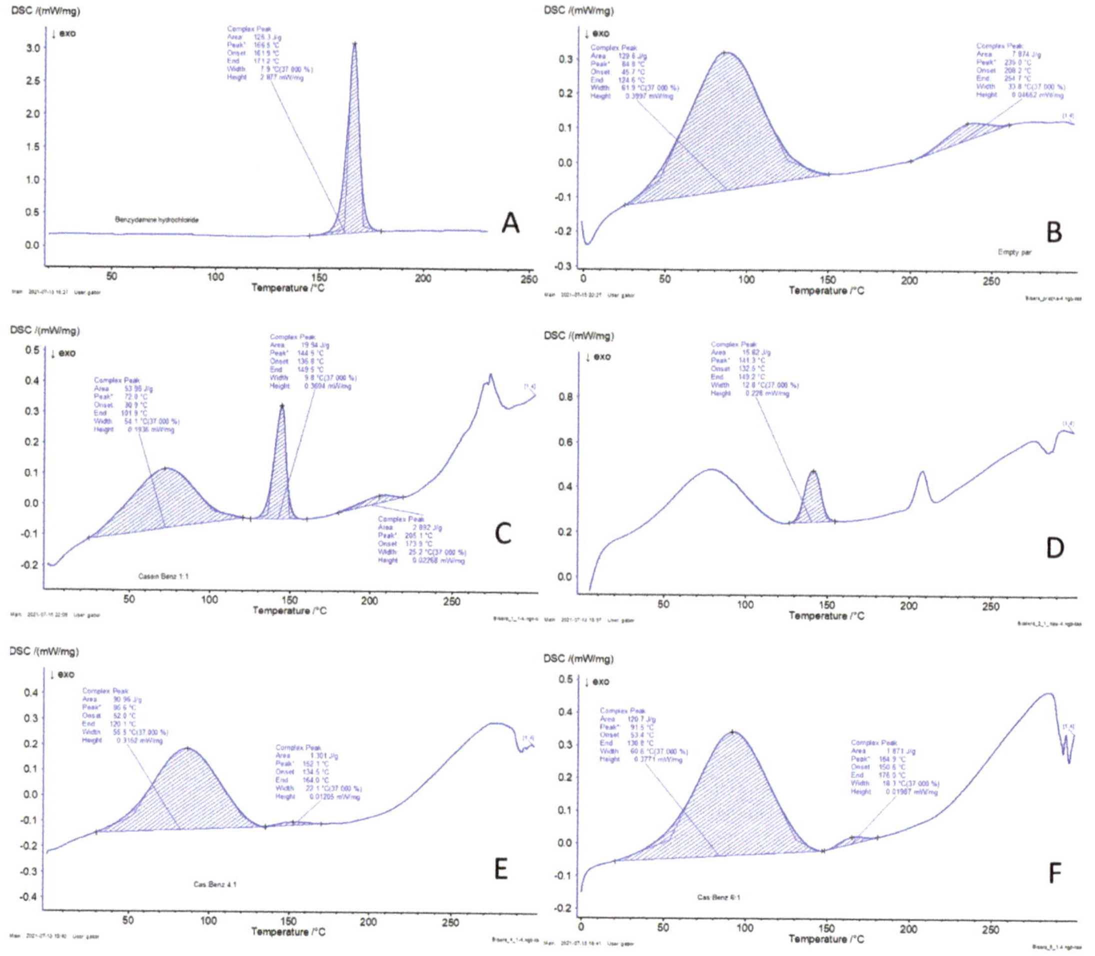

Figure 5. DSC thermograms of Benzydamine hydrochloride (**A**), blank casein nanoparticles (**B**), BZ-loaded casein nanoparticles of batches Cas2-Ca3-BZ-1 (**C**), Cas2-Ca3-BZ-2 (**D**), Cas2-Ca3-BZ-4 (**E**) and Cas2-Ca3-BZ-6 (**F**).

3.2.6. Fourier-Transform Infrared Spectroscopy (FTIR)

The spectra for casein (Figure 6) show peaks at 1646 cm^{-1} in the amide I region and 1530 cm^{-1} in the amide II region, which could be assigned to the stretching of the carbonyl group (C=O) and to the symmetric stretching of N-C=O bonds, respectively. Casein shows a band at 1077 cm^{-1}, suggesting interactions of monocationic phosphates with Na^+. The peaks in the amide I and amide II regions also appear in the crosslinked nanoparticles and in the drug loaded nanoparticles. The band at 977 cm^{-1}, attributed to bionic phosphate, has very low intensity on the casein spectrum and increased intensity in the crosslinked systems, suggesting interaction with Ca^{2+}. Characteristic bands for stretching of aromatic C=C at 1497 cm^{-1} of benzydamine hydrochloride can be seen in the spectra of the drug-loaded nanoparticles. The band at 1357 cm^{-1} is attributed to C-N vibrations of the heterocyclic ring.

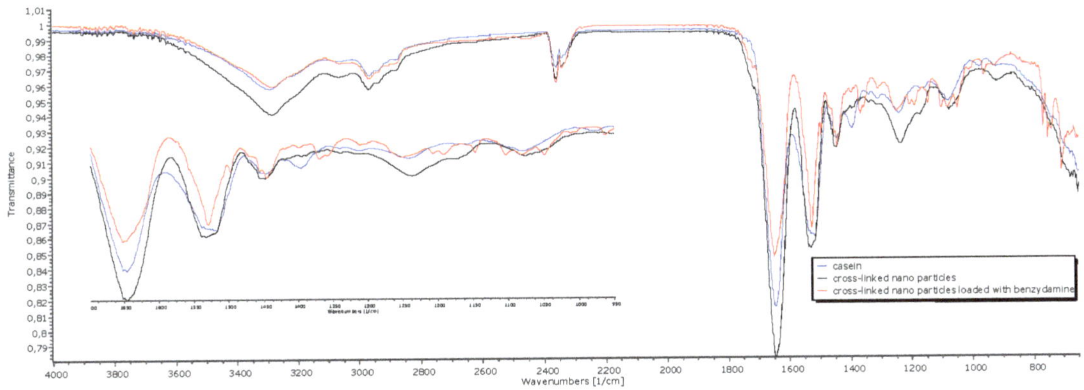

Figure 6. ATR-FTIR spectra of casein, crosslinked placebo nanoparticles and crosslinked BZ-loaded nanoparticles.

3.2.7. In Vitro Drug Release

The dissolution profiles of BZ from casein nanoparticles are presented in Figure 7. The percentage of released drug during 5-h study was incomplete, varying from 73.30% (Cas2-Ca3-BZ-4) to 91.81% (Cas2-Ca3-BZ-1). Initial burst effect was observed in models Cas2-Ca3-BZ-1 and Cas2-Ca3-BZ-2 in the first 60 min, releasing more than 50% of the encapsulated drug. This was probably due to the higher drug loading and the accumulation of BZ in the periphery of the nanoparticles during the process of spray drying. The batches Cas2-Ca3-BZ-4 and Cas2-Ca3-BZ-6 demonstrated prolonged drug release over time, releasing almost 75% of the incorporated benzydamine. It was probably the greater amount of casein per unit mass in these two batches that refrained the drug from free diffusion from the particle core to the periphery, despite the larger surface area available for dissolution.

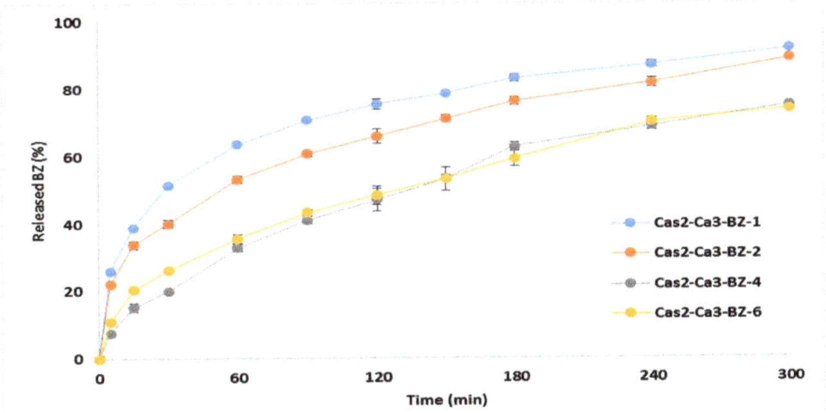

Figure 7. In vitro BZ release from spray dried casein nanoparticles prepared at different polymer/drug ratios (n = 3).

4. Conclusions

In this study, self-assembled casein nanocarriers were produced by nano spray drying. The process parameters were investigated, and an optimized model of blank casein nanostructures was outlined. Furthermore, four batches of BZ-loaded nanoparticles with a particle size from 135.9 nm to 994.2 nm were developed. BZ loading in the nanoparticles depended on the polymer/drug ratio. BZ was transformed from crystalline into amorphous during spray drying, which implies an increased dissolution rate. The drug release study confirmed the feasibility of the developed nanocarriers for prolonged delivery of benzydamine.

Author Contributions: Conceptualization, B.P. and M.M.; methodology, N.Z., S.M. and Y.U.; investigation, N.Z., B.P. and M.M.; writing—original draft preparation, N.Z.; writing—review and editing, B.P.; visualization, N.Z., Y.U. and M.M.; supervision, B.P.; project administration, M.M.; funding acquisition, M.M. and B.P. All authors have read and agreed to the published version of the manuscript.

Funding: The authors appreciate the financial support of the Bulgarian National Science Fund (BNSF) via Project KP-06-N38/3 for ensuring the chemicals and consumables.

Institutional Review Board Statement: Not applicable.

Informed Consent Statement: Not applicable.

Data Availability Statement: The data presented in this study are available on request from the corresponding author.

Acknowledgments: The authors acknowledge Project BG05M2OP001-1.002-0005-C01 Centre of competence for Personalised Innovative Medicine PERIMED for providing access to the scientific infrastructure and sophisticated equipment for this work.

Conflicts of Interest: The authors declare no conflict of interest.

References

1. Sundar, S.; Kundu, J.; Kundu, S.C. Biopolymeric nanoparticles. *Sci. Technol. Adv. Mater.* **2010**, *11*, 014104. [CrossRef]
2. Gandhi, S.; Roy, I. Doxorubicin-loaded casein nanoparticles for drug delivery: Preparation, characterization and In Vitro evaluation. *Int. J. Biol. Macromol.* **2019**, *121*, 6–12. [CrossRef]
3. Couvreur, P.; Gref, R.; Andrieux, K.; Malvy, C. Nanotechnology for drug delivery: Aplication to cancer and autoimmune diseases. *Prog. Solid State Chem.* **2016**, *2–4*, 231–235. [CrossRef]
4. Głąb, T.K.; Boratyński, J. Potential of casein as a carrier for biologically active agents. *Top. Curr. Chem.* **2017**, *375*, 71. [CrossRef]
5. Elzoghby, A.O.; El-Fotoh, W.S.; Elgindy, N.A. Casein-based formulations as promising controlled release drug delivery systems. *J. Control. Release* **2011**, *153*, 206–216. [CrossRef]
6. Bhat, M.Y.; Dar, T.A.; Singh, L.R. Casein Proteins: Structural and Functional Aspects. In *Milk Proteins—From Structure to Biological Properties and Health Aspects*; Gigly, I., Ed.; InTech: London, UK, 2016. [CrossRef]
7. Fox, P.; Brodkorb, A. The casein micelle: Historical aspects, current concepts and significance. *Int. Dairy J.* **2008**, *18*, 677–684. [CrossRef]
8. Dalgleish, D.G. Casein micelles as colloids: Surface structures and stabilities. *J. Dairy Sci.* **1998**, *81*, 3013–3018. [CrossRef]
9. Liu, Y.; Guo, R. pH-dependent structures and properties of casein micelles. *Biophys. Chem.* **2008**, *136*, 67–73. [CrossRef]
10. Liu, Z.; Juliano, P.; Williams, R.P.; Niere, J.; Augustin, M.A. Ultrasound effects on the assembly of casein micelles in reconstituted skim milk. *J. Dairy Res.* **2014**, *81*, 146–155. [CrossRef]
11. Liu, C.; Yao, W.; Zhang, L.; Qian, H.; Wu, W.; Jiang, X. Cell-penetrating hollow spheres based on milk protein. *Chem. Commun.* **2010**, *46*, 7566–7568. [CrossRef] [PubMed]
12. Pan, K.; Luo, Y.; Gan, Y.; Baek, S.J.; Zhong, Q. pH-driven encapsulation of curcumin in self-assembled casein nanoparticles for enhanced dispersibility and bioactivity. *Soft Matter* **2014**, *10*, 6820–6830. [CrossRef]
13. Rahimi Yazdi, S.; Bonomi, F.; Iametti, S.; Miriani, M.; Brutti, A.; Corredig, M. Binding of curcumin to milk proteins increases after static high pressure treatment of skim milk. *J. Dairy Res.* **2013**, *80*, 152–158. [CrossRef]
14. Roach, A.; Dunlap, J.; Harte, F. Association of triclosan to casein proteins through solvent-mediated high-pressure homogenization. *J. Food Sci.* **2009**, *74*, N23–N29. [CrossRef]
15. Elzoghby, A.O.; Helmy, M.W.; Samy, W.M.; Elgindy, N.A. Micellar delivery of flutamide via milk protein nanovehicles enhances its anti-tumor efficacy in androgen-dependent prostate cancer rat model. *Pharm. Res.* **2013**, *30*, 2654–2663. [CrossRef]
16. Menéndez-Aguirre, O.; Stuetz, W.; Grune, T.; Kessler, A.; Weiss, J.; Hinrichs, J. High pressure-assisted encapsulation of vitamin D_2 in reassembled casein micelles. *High Press. Res.* **2011**, *31*, 265–274. [CrossRef]
17. Menéndez-Aguirre, O.; Kessler, A.; Stuetz, W.; Grune, T.; Weiss, J.; Hinrichs, J. Increased loading of vitamin D_2 in reassembled casein micelles with temperature-modulated high pressure treatment. *Food Res. Int.* **2014**, *64*, 74–80. [CrossRef] [PubMed]
18. Cohen, Y.; Ish-Shalom, S.; Segal, E.; Nudelman, O.; Shpigelman, A.; Livney, Y.D. The bioavailability of vitamin D3, a model hydrophobic nutraceutical, in casein micelles, as model protein nanoparticles: Human clinical trial results. *J. Funct. Foods* **2017**, *30*, 321–325. [CrossRef]
19. Chen, L.; Wei, J.; An, M.; Zhang, L.; Lin, S.; Shu, G.; Yuan, Z.; Lin, J.; Peng, G.; Liang, X.; et al. Casein nanoparticles as oral delivery carriers of mequindox for the improved bioavailability. *Colloids Surf. B Biointerfaces* **2020**, *195*, 111221. [CrossRef] [PubMed]

20. Chu, B.S.; Ichikawa, S.; Kanafusa, S.; Nakajima, M. Preparation and characterization of beta-carotene nanodispersions prepared by solvent displacement technique. *J. Agric. Food Chem.* **2007**, *55*, 6754–6760. [CrossRef] [PubMed]
21. Chu, B.S.; Ichikawa, S.; Kanafusa, S.; Nakajima, M. Preparation of protein-stabilized β-carotene nanodispersions by emulsification-evaporation method. *J. Am. Oil Chem. Soc.* **2007**, *84*, 1053–1062. [CrossRef]
22. Elzoghby, A.O.; Helmy, M.W.; Samy, W.M.; Elgindy, N.A. Spray-dried casein-based micelles as a vehicle for solubilization and controlled delivery of flutamide: Formulation, characterization, and In Vivo pharmacokinetics. *Eur. J. Pharm. Biopharm.* **2013**, *84*, 487–496. [CrossRef]
23. Hartini, N.; Ponrasu, T.; Wu, J.-J.; Sriariyanun, M.; Cheng, Y.-S. Microencapsulation of curcumin in crosslinked jelly fig pectin using vacuum spray drying technique for effective drug delivery. *Polymers* **2021**, *13*, 2583. [CrossRef]
24. Rodrigues, S.; da Costa, A.M.R.; Flórez-Fernández, N.; Torres, M.D.; Faleiro, M.L.; Buttini, F.; Grenha, A. Inhalable spray-dried chondroitin sulphate microparticles: Effect of different solvents on particle properties and drug activity. *Polymers* **2020**, *12*, 425. [CrossRef]
25. Pan, K.; Zhong, Q.; Baek, S.J. Enhanced dispersibility and bioactivity of curcumin by encapsulation in casein nanocapsules. *J. Agric. Food Chem.* **2013**, *61*, 6036–6043. [CrossRef]
26. Penalva, R.; Esparza, I.; Agüeros, M.; Gonzalez-Navarro, C.J.; Gonzalez-Ferrero, C.; Irache, J.M. Casein nanoparticles as carriers for the oral delivery of folic acid. *Food Hydrocoll.* **2015**, *44*, 399–406. [CrossRef]
27. Heng, D.; Lee, S.H.; Ng, W.K.; Tan, R.B. The nano spray dryer B-90. *Expert Opin. Drug Deliv.* **2011**, *8*, 965–972. [CrossRef]
28. Li, X.; Anton, N.; Arpagaus, C.; Belleteix, F.; Vandamme, T.F. Nanoparticles by spray drying using innovative new technology: The Büchi nano spray dryer B-90. *J. Control. Release* **2010**, *147*, 304–310. [CrossRef] [PubMed]
29. Schmid, K.; Arpagaus, C.; Friess, W. Evaluation of the nano spray dryer B-90 for pharmaceutical applications. *Pharm. Dev. Technol.* **2011**, *16*, 287–294. [CrossRef] [PubMed]
30. Lee, S.H.; Heng, D.; Ng, W.K.; Chan, H.-K.; Tan, R.B. Nano spray drying: A novel method for preparing protein nanoparticles for protein therapy. *Int. J. Pharm.* **2011**, *403*, 192–200. [CrossRef]
31. Durli, T.L.; Dimer, F.A.; Fontana, M.C.; Pohlmann, A.R.; Beck, R.C.; Guterres, S.S. Innovative approach to produce submicron drug particles by vibrational atomization spray drying: Influence of the type of solvent and surfactant. *Drug Dev. Ind. Pharm.* **2014**, *40*, 1011–1020. [CrossRef]
32. Anton, N.; Benoit, J.P.; Saulnier, P. Design and production of nanoparticles formulated from nano-emulsion templates—A review. *J. Control. Release* **2008**, *128*, 185–199. [CrossRef]
33. Marante, T.; Viegas, C.; Duarte, I.; Macedo, A.S.; Fonte, P. An overview on spray-drying of protein-loaded polymeric na-noparticles for dry powder inhalation. *Pharmaceutics* **2020**, *12*, 1032. [CrossRef]
34. Moslehi, M.; Mortazavi, S.A.R.; Azadi, A.; Fateh, S.; Hamidi, M.; Foroutan, S.M. Preparation, optimization and characteri-zation of chitosan-coated liposomes for solubility enhancement of furosemide: A model BCS IV drug. *Iran. J. Pharm. Res.* **2020**, *19*, 366–382. [PubMed]
35. Gu, B.; Linehan, B.; Tseng, Y.C. Optimization of the Büchi B-90 spray drying process using central composite design for preparation of solid dispersions. *Int. J. Pharm.* **2015**, *491*, 208–217. [CrossRef]
36. Bürki, K.; Jeon, I.; Arpagaus, C.; Betz, G. New insights into respirable protein powder preparation using a nano spray dryer. *Int. J. Pharm.* **2011**, *408*, 248–256. [CrossRef] [PubMed]
37. Harsha, S.; Al-Dhubiab, B.E.; Nair, A.B.; Attimarad, M.; Venugopala, K.N.; Sa, K. Pharmacokinetics and tissue distribution of microspheres prepared by spray drying technique: Targeted drug delivery. *Biomed. Res.* **2017**, *28*, 3387–3396.
38. Quane, P.A.; Graham, G.G.; Ziegler, J.B. Pharmacology of benzydamine. *Inflammopharmacology* **1998**, *6*, 95–107. [CrossRef]
39. Hansch, C.; Sammes, P.G.; Taylor, J.B. *Comprehensive Medicinal Chemistry. The Rational Design, Mechanistic Study and Therapeutic Application of Chemical Compounds*; Six Volumes; Pergamon: Oxford, UK, 1990; ISBN 0-08-032530-0.
40. Beckett, A.H.; Triggs, E.J. Buccal absorption of basic drugs and its application as an In Vivo model of passive drug transfer through lipid membranes. *J. Pharm. Pharmacol.* **1967**, *19*, 31S–41S.
41. Bickel, M.H.; Weder, H.J. Buccal absorption and other properties of pharmacokinetic importance of imipramine and its metabolites. *J. Pharm. Pharmacol.* **1969**, *3*, 160–168. [CrossRef]
42. Baldock, G.A.; Brodie, R.R.; Chasseaud, L.F.; Taylor, T.; Walmsley, L.M.; Catanese, B. Pharmacokinetics of benzydamine after intravenous, oral, and topical doses to human subjects. *Biopharm. Drug Dispos.* **1991**, *12*, 481–492. [CrossRef]
43. Li, M.; Wang, K.; Wang, Y.; Han, Q.; Ni, Y.; Wen, X. Effects of genipin concentration on cross-linked β-casein micelles as nanocarrier of naringenin: Colloidal properties, structural characterization and controlled release. *Food Hydrocoll.* **2020**, *108*, 105989. [CrossRef]
44. Rose, D.; Tessier, H. Effect of various salts on the coagulation of casein. *J. Dairy Sci.* **1959**, *42*, 989–997. [CrossRef]
45. Both, E.M.; Boom, R.M.; Schutyser, M.A.I. Particle morphology and powder properties during spray drying of maltodextrin and whey protein mixtures. *Powder Technol.* **2020**, *363*, 519–524. [CrossRef]

polymers

MDPI

Article

Impacts of Blended *Bombyx mori* Silk Fibroin and Recombinant Spider Silk Fibroin Hydrogels on Cell Growth

Chavee Laomeephol [1], Apichai Vasuratna [2], Juthamas Ratanavaraporn [1,3,4], Sorada Kanokpanont [1,3,5], Jittima Amie Luckanagul [1,6], Martin Humenik [7], Thomas Scheibel [7,*] and Siriporn Damrongsakkul [1,3,5,*]

1 Biomaterial Engineering for Medical and Health Research Unit, Faculty of Engineering, Chulalongkorn University, Bangkok 10330, Thailand; papomchavee@gmail.com (C.L.); Juthamas.R@chula.ac.th (J.R.); sorada.k@chula.ac.th (S.K.); jittima.l@pharm.chula.ac.th (J.A.L.)
2 Department of Obstetrics and Gynecology, Faculty of Medicine, Chulalongkorn University, Bangkok 10330, Thailand; apichai.v@chula.ac.th
3 Biomedical Engineering Research Center, Faculty of Engineering, Chulalongkorn University, Bangkok 10330, Thailand
4 Biomedical Engineering Program, Faculty of Engineering, Chulalongkorn University, Bangkok 10330, Thailand
5 Department of Chemical Engineering, Faculty of Engineering, Chulalongkorn University, Bangkok 10330, Thailand
6 Department of Pharmaceutics and Industrial Pharmacy, Faculty of Pharmaceutical Sciences, Chulalongkorn University, Bangkok 10330, Thailand
7 Department of Biomaterials, Faculty of Engineering Science, University of Bayreuth, Prof.-Rüdiger-Bormann Str. 1, 95447 Bayreuth, Germany; martin.humenik@bm.uni-bayreuth.de
* Correspondence: thomas.scheibel@bm.uni-bayreuth.de (T.S.); siriporn.d@chula.ac.th (S.D.); Tel.: +49-921-55-6700 (T.S.); +662-218-6862 (S.D.); Fax: +662-218-6877 (S.D.)

Citation: Laomeephol, C.; Vasuratna, A.; Ratanavaraporn, J.; Kanokpanont, S.; Luckanagul, J.A.; Humenik, M.; Scheibel, T.; Damrongsakkul, S. Impacts of Blended *Bombyx mori* Silk Fibroin and Recombinant Spider Silk Fibroin Hydrogels on Cell Growth. *Polymers* **2021**, *13*, 4182. https://doi.org/10.3390/polym13234182

Academic Editors: Ariana Hudita and Bianca Gălăţeanu

Received: 27 October 2021
Accepted: 25 November 2021
Published: 29 November 2021

Abstract: Binary-blended hydrogels fabricated from *Bombyx mori* silk fibroin (SF) and recombinant spider silk protein eADF4(C16) were developed and investigated concerning gelation and cellular interactions in vitro. With an increasing concentration of eADF4(C16), the gelation time of SF was shortened from typically one week to less than 48 h depending on the blending ratio. The biological tests with primary cells and two cell lines revealed that the cells cannot adhere and preferably formed cell aggregates on eADF4(C16) hydrogels, due to the polyanionic properties of eADF4(C16). Mixing SF in the blends ameliorated the cellular activities, as the proliferation of L929 fibroblasts and SaOS-2 osteoblast-like cells increased with an increase of SF content. The blended SF:eADF4(C16) hydrogels attained the advantages as well as overcame the limitations of each individual material, underlining the utilization of the hydrogels in several biomedical applications.

Keywords: silk fibroin; spider silk; hydrogel; self-assembly; cell culture

1. Introduction

Hydrogels are well suited as scaffolds for tissue engineering due to their characteristics resembling natural extracellular matrices. Hydrogels can be applied in various biomedical fields, such as injectable hydrogels or printable bioinks for space-filling or cell/biological factor delivery [1]. Silk fibroin is a naturally derived fibrous protein which is widely used as a base material in hydrogel fabrication, due to its self-assembly, mechanical stability of the gels, and biocompatibility [2]. In this work, silk fibroin derived from two different sources, *Bombyx mori* silk cocoons and recombinant spider dragline silk proteins, were chosen to form blended hydrogels, and their cytocompatibility was tested in vitro.

Combining two materials is an approach to gain the advantage from both materials as well as to overcome some limitations to achieve products with desired features [3]. Silk fibroin (SF) can be derived at high amounts from silkworms by isolating the SF solution from silk glands or dissolving silk cocoons with the drawback of some inhomogeneities common to all nature-derived materials. However, SF solution can be produced under certified conditions, and SF is already available from several companies, such as Fibrothelium GmbH, Aachen, Germany, Sigma-Aldrich, MO, USA, and Advanced Biomatrix, CA, USA. Glycine-alanine repeats of SF can form beta sheet structures, which is relevant for self-assembly as well as the physical strength of the obtained materials [4]. However, self-gelation of SF is extremely slow (ca. 7–12 days depending on the SF concentration) [5], which is impractical in various applications, especially for cell encapsulation. Several strategies have been applied to accelerate the self-assembly process of SF, including an application of physical or mechanical forces, an addition of chemicals, as well as a simple blending with other polymers. Regarding the works from Mandal BB's group, non-mulberry SF, with its primary structure containing a high ratio of alanine-glycine and poly-alanine sequences, was simply blended with *B. mori* SF, and rapid gelation can be achieved. The biological properties were drastically improved due to the presence of arginine-glycine-aspartic acid (RGD) motifs in non-mulberry SF [6–9].

The recombinant spider silk protein (eADF4(C16)), derived from ADF4 of the dragline silk protein of *Araneus diadematus* containing poly-alanine sequence, can be spontaneously assembled into hydrogels within hours depending on the protein concentration or the ionic strength of the solution [10,11]. The advantage of this recombinant protein is its large availability with continuous properties. eADF4(C16) is a commercially available spider silk protein from AMSilk GmbH, Martinsried, Germany. Major limitation of eADF4(C16) is cell adhesion and proliferation unless the protein is genetically modified, e.g., with a tag comprising the cell adhesion motif RGD [12]. Hence, binary blending of SF and eADF4(C16) could expectedly be beneficial in enhancing the interaction with cells, as well as accelerating the gelation process within a range suitable for practical uses.

Herein, the gelation of SF and eADF4(C16) blends was evaluated. Since the hydrogels were proposed to serve as cell-loaded substrates, the interaction with primary cells and cell lines was tested. Physico-chemical properties of the hydrogels, which could affect the cellular behavior, namely micromorphology, hydrophobicity, and protein diffusivity, were also identified. This work as a proof-of-concept study provides information of blended SF:eADF4(C16) hydrogels for further applications e.g., cell-encapsulation for cell delivery or injectable or printable materials.

2. Materials and Methods

2.1. Material

Thai *Bombyx mori* silk cocoons were received from Queen Sirikit Sericulture Center, Nakhon Ratchasima province, Thailand. Recombinant protein eADF4(C16) based on dragline silk protein ADF-4 of *Araneus diadematus* was produced according to the published protocol [13]. Briefly, a bacterial expression plasmid containing gene corresponding to 16 repeats of module C of ADF4 protein (sequence: GSSAAAAAAAASGPGGYG PENQGPSGPGGYGPGGP) was induced in *E. coli* strain BLR(DE3) using isopropyl β-D-1-thiogalactopyranoside (IPTG). Approximately 3–4 h after the induction, cells were harvested, lysed and eADF4(C16) was isolated by precipitation using 30% ammonium sulfate, before redissolution in 6 M guanidine thiocyanate (GdnSCN) and lyophilization. All reagents used in this study were of analytical grade and supplied from Sigma-Aldrich, MO, USA, unless otherwise stated.

To extract SF from silk cocoons, 40 g of the cocoons were boiled in 1 L of 0.02 M Na_2CO_3 for 20 min twice to remove sericin, before leaving to dry. Four gram of dried silk fiber was dissolved in 16 mL of 9.3 M LiBr (1:4 weight-to-volume ratio), and incubated at 60 °C for 4 h. After that, LiBr was eliminated by dialyzing the SF solution against deionized water for 48 h using a dialysis tube with molecular weight cut-off (MWCO) of 12–16 kDa [14]. The concentration of the protein solution was determined from dry solid weight. To prepare the sterile SF solution, the solution was autoclaved at 121 °C for 20 min and stored in a refrigerator until usage.

eADF4(C16) powder was dissolved in 6 M GdnSCN at a concentration of 4 mg/mL and incubated at 37 °C for 1 h. The solution was then dialyzed against 10 mM Tris-HCl buffer (pH 7.4) using a dialysis tube with MWCO of 6–9 kDa. Subsequently, the protein solution was concentrated using dialysis against 20% *w*/*v* polyethylene glycol (PEG; M_n = 35 kDa), and the concentration was determined from the absorbance at 280 nm [10]. The eADF4(C16) was UV-irradiated for 20 min for sterilization.

2.2. Gelation Kinetics

Gelation of the protein solutions and blends was investigated using the change of turbidity, as the gelation is associated with fibril assembly causing light scattering [5,10]. 2% and 3% *w*/*v* SF and eADF4(C16) solutions were prepared and mixed at the volume ratio of SF:eADF4(C16) of 10:0, 7:3, 5:5, 3:7 and 0:10. The effects of different buffers, including Dulbecco's Modified Eagle's Medium (DMEM), phosphate buffer saline (PBS), and normal saline solution (NSS), on the gelation kinetics were investigated by supplementing the protein mixtures with 10X concentrated solutions. The samples were prepared from the sterile stock solutions using aseptic techniques to avoid microbial contamination. 100 µL of the mixtures were transferred to a 96-well plate, and the change of visible light absorption at 550 nm was measured using a microplate reader (FLUOstar Omega, BMG Labtech, Ortenberg, Germany). The temperature was controlled at 37 °C, and the microplate was sealed to prevent water evaporation. The measurement was conducted every 15 min for 40 h.

2.3. Visualization of Microstructures

The morphologies of hydrogels were observed upon freeze-drying using a field emission scanning electron microscope (FESEM; JSM-7610F, Jeol, Tokyo, Japan). The samples were prepared as above described by incubating the mixtures in tight-sealed vials at 37 °C until achieving the complete gelation. The samples were immediately frozen using liquid nitrogen for 30 min and at −80 °C overnight before lyophilization. Flash freezing the samples in liquid nitrogen was performed to preserve the microstructure of the hydrogels.

To visualize the micromorphological structure using FESEM, the freeze-dried samples were cut and coated with platinum. The FESEM was operated with an acceleration voltage of 5 kV.

2.4. Quantitative Determination of Secondary Structures

The structures of freeze-dried hydrogels were determined using Fourier-transform infrared (FTIR) spectroscopy in an attenuated total reflection (ATR) mode (Nicolet™ iS™ 5, Thermo Fisher Scientific, Waltham, MA, USA). The absorbance spectra within the range of 4000 to 800 cm^{-1} were collected with 1.0 cm^{-1} resolution. The secondary structures were quantified using Fourier self-deconvolution (FSD) and curve-fitting techniques according to the established protocol [15]. In brief, the deconvolution of amide I spectrum (1725–1575 cm^{-1}) was obtained using Omnic 8.0 software (Thermo Fisher Scientific, Waltham, MA, USA) by fitting the Voigt line shape with a half-bandwidth of 25 cm^{-1} and an enhancement factor of 2.5. Subsequently, the deconvoluted spectrum was fitted with the Gaussian function using Origin Pro 9.0 software (OriginLab, MA, USA). The content of beta sheet structure was obtained from the peak area between 1616–1637 cm^{-1}. Other structures, namely random coil, alpha-helix and beta turn, were determined from the peaks at 1638–1655, 1656–1662 and 1663–1696 cm^{-1}, respectively [16].

2.5. Protein Adsorptivity of Blended Hydrogels

Fetal bovine serum (FBS), a mixture of soluble proteins which is widely used in in vitro biological experiments, was selected to study the protein adsorptivity of the SF:eADF4(C16) hydrogels. 2% Hydrogels were prepared in a silicone mold, cut into a disc shape with a diameter of 5 mm and a thickness of 1 mm, and immersed in 10% FBS in PBS buffer (pH 7.4) at 37 °C for a particular period. After that, the hydrogels were washed with PBS to remove redundant proteins and incubated in 1 mL of PBS at 4 °C overnight with gentle shaking to extract the protein. Supernatants obtained from the samples immersed in PBS overnight without FBS were used as blanks. Protein content in the extracted solution was determined using Bio-Rad protein assay (Bio-Rad Laboratories, Hercules, CA, USA) according to the manufacturer's instruction. 80 µL of the supernatant was mixed with 20 µL of the dye, incubated at room temperature for 15 min, and the absorbance was measured at 595 nm.

2.6. Cell Preparation

Three different cell types, including human adipose-derived stromal cells (hASCs), L929 mouse fibroblasts, and SaOS-2 human osteoblast-like cells, were chosen to investigate their activities while culturing on the developed SF:eADF4(C16) hydrogels. Primary hASCs were used as a model for tissue engineering applications. L929 and SaOS-2 were selected to represent normal and tumor cells, respectively.

Human ASCs were isolated from subcutaneous fat tissues collected from female participants enrolled to Chulalongkorn Memorial Hospital, Thailand for laparotomy with an approval from institutional ethic committee on human research, Faculty of Medicine, Chulalongkorn University (project no. 416/61). Isolation and culture procedures were conducted following established protocols [17,18]. Briefly, 10–15 g fat tissue was washed with PBS before enzymatically digested using 0.1% collagenase type II (Gibco, New York, NY, USA) supplemented with 1% bovine serum albumin at 37 °C with continuous shaking for 1 h. The digested specimen was then centrifuged, and the upper oil layer was removed. The bottom dark brown layer, known as stromal vascular fraction (SVF), was collected, resuspended with PBS and centrifuged. After that, SVF was resuspended with culture medium (DMEM/F12 + 10% FBS + 1% antibiotics) and transferred to a T-75 tissue culture flask. The culture was maintained at 37 °C with fed air supplemented with 5% CO_2.

After initial plating for 48 h, cells were washed with PBS to remove unattached cells and refed with the new medium. Typically, hASCs reached 80–90% confluency within 2 weeks. Subculture was performed using TrypLE Express enzyme (Thermo Fisher Scientific, Waltham, MA, USA) according to the manufacturer's advice with a subculture ratio of 1:2.

L929 and SaOS-2 were cultured in DMEM supplemented with 10% FBS and 1% antibiotics. The cells were maintained in a CO_2-incubator, and the subculture was treated using trypsin (Thermo Fisher Scientific, Waltham, MA, USA) according to the manufacturer's protocol.

2.7. Evaluation of Cell Attachment and Proliferation on the Silk Hydrogels

Hydrogels were prepared as mentioned above using autoclaved SF and UV-irradiated eADF4(C16) solutions, and the overall concentration of protein mixtures was fixed at 2% *w*/*v*. 100 µL of the protein mixtures were transferred to a tissue culture-treated 48-well pate and incubated at 37 °C under humidified atmosphere for at least 48 h to allow complete gelation. For pure SF solutions, gelation was accelerated using sonication at 40% amplitude for 30 s [19].

All hydrogels were hydrated with the complete media for 24 h prior to cell seeding. hASCs at a density of 5000 cell/cm^2, L929 and SaOS-2 at a density of 10,000 cell/cm^2 were seeded on the hydrogels. Seeding density was based on the proliferation profile of each cell, of which its logarithm growth phase should be achieved within 1 to 5 days. Cells cultured on the tissue culture-treated plate (TCP) were used as controls. The cell culture media, DMEM/F-12 + 10%FBS + 1% antibiotics and DMEM + 10%FBS + 1% antibiotics were used for culturing hASCs and L929 or SaOS-2, respectively, and the samples were stored in a CO_2 incubator at 37 °C and 5% CO_2 supplementation.

On day 1, 3, 5 and 7 after cell seeding, the cell proliferation was determined using the 3-(4,5-dimethylthiazol-2-yl)-2,5-diphenyltetrazolium bromide (MTT) assay (Thermo Fisher Scientific, Waltham, MA, USA) according to the manufacturer's protocol. Briefly, cells were washed with PBS, treated with 0.5 mg/mL MTT solution, and incubated at

37 °C in the dark for 30 min. After that, the MTT dye was removed and replaced with dimethyl sulfoxide (DMSO) to extract the precipitated formazan. The blue solution was then retrieved, and its absorbance at 570 nm was measured with a visible-light background correction at 650 nm.

Cell morphology was observed using a phase contrast imaging as well as fluorescent live-dead staining with calcein AM and propidium iodide (PI) dyes (Thermo Fisher Scientific, Waltham, MA, USA). At the designated time-points, cells were washed and stained with the fluorescent dyes. Bright-field and fluorescent images were obtained using a fluorescence microscope (Nikon Eclipse 80i, Nikon, Tokyo, Japan) with green and red filters to visualize calcein AM stained (lived) and PI stained (dead) cells, respectively.

2.8. Statistical Analysis

Statistical analysis was performed using IBM® SPSS® Statistics software version 22 under the license of Chulalongkorn University. Data was analyzed using one-way analysis of variance (ANOVA) with Bonferroni post-hoc tests at the significant level of 0.05.

3. Results

3.1. Gelation Time of SF:eADF14(C16) Blends

The gelation of fibroin solutions was associated with the formation of heterogenous microstructures, which affected the light scattering degree in the visible range [5]. Therefore, the gelation can be noticed from the point at which an abrupt change of the visible-light absorbance value occurs. Figure 1 demonstrates the gelation time of 2% and 3% SF:eADF14(C16) blends. Only eADF4(C16) showed gel formation within less than a day, while a blending with SF significantly prolonged the gelation time. Increasing protein concentrations resulted in faster gelation. The SF:eADF4(C16) samples at a ratio 3:7 underwent the gelation in less than 40 h. Furthermore, a supplementation with DMEM, PBS, and NSS significantly reduced the gelation time, especially for blended samples (5:5 and 3:7 ratio). DMEM addition yielded slower gelation kinetics in case of blended 5:5 and 3:7 SF:eADF4(C16) solutions, when compared to those in presence of PBS and NSS.

3.2. Micromorphology of the Freeze-Dried Hydrogels

The microstructures of freeze-dried hydrogels were visualized using FESEM (Figure 2). The microstructures of all samples presented a high porosity with a ridge- or wall-like structure, and the fracture surfaces displayed an accumulation of nanofibers.

3.3. Secondary Structures of the Hydrogels

Figure 3A,B display the FTIR spectra and the secondary structure content of the freeze-dried SF:eADF4(C16) hydrogels, respectively. Comparing in sol and gel state, the peak shift from approximately 1650 cm^{-1} of the sol groups toward a lower wavenumber (1625 cm^{-1}) of the gel group can be noticed, indicating a transition of the predominated random coil structure in the sol state to a beta sheet structure after gelation. The results were in accordance with the amount of secondary structures quantified using FSD and curve-fitting. A reduction in random coil and an increase of beta sheet structure could be clearly observed, especially in samples containing SF. The samples with higher eADF4(C16) content possessed a higher beta sheet and lower random coil content in the sol state, which slightly changed after gelation.

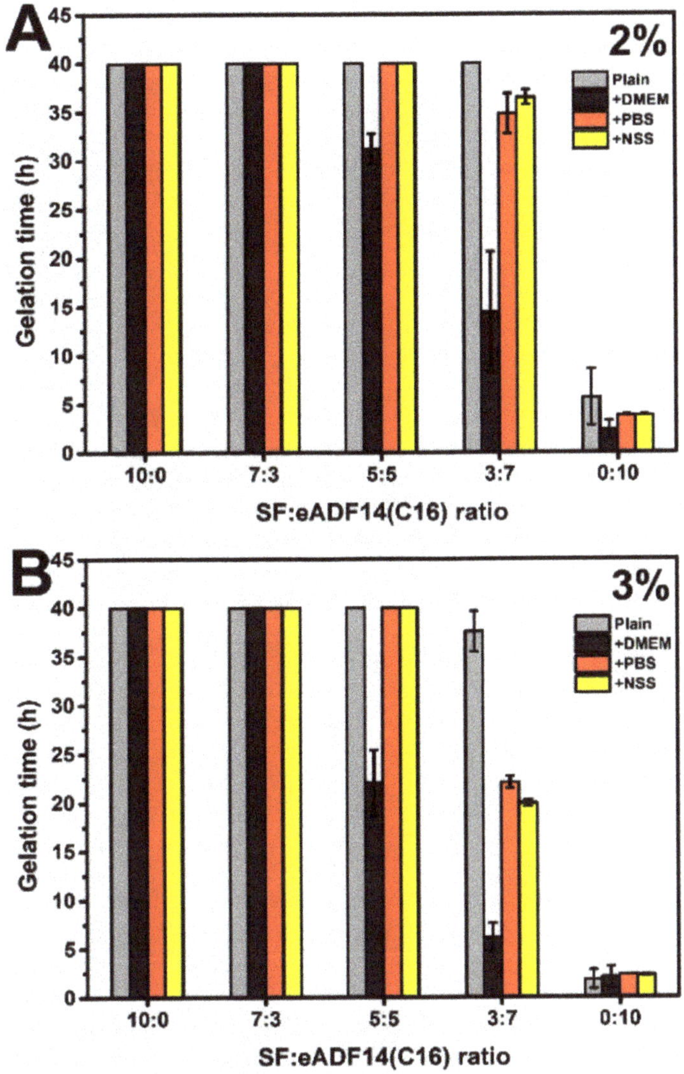

Figure 1. Gelation time of (**A**) 2% and (**B**) 3% w/v of SF:eADF4(C16) solutions with different ratios as indicated and upon addition of Dulbecco's Modified Eagle's Medium (DMEM), phosphate buffer saline (PBS) and normal saline solution (NSS) at 37 °C. "Plain" refers to SF:eADF4(C16) blends in the absence of salts. The gelation time was interpreted from the time-point at which the absorbance values reached the half-maxima. The experimental time was limited to 40 h. The ratio indicates the volume ratio of the respective 2% and 3% protein solutions.

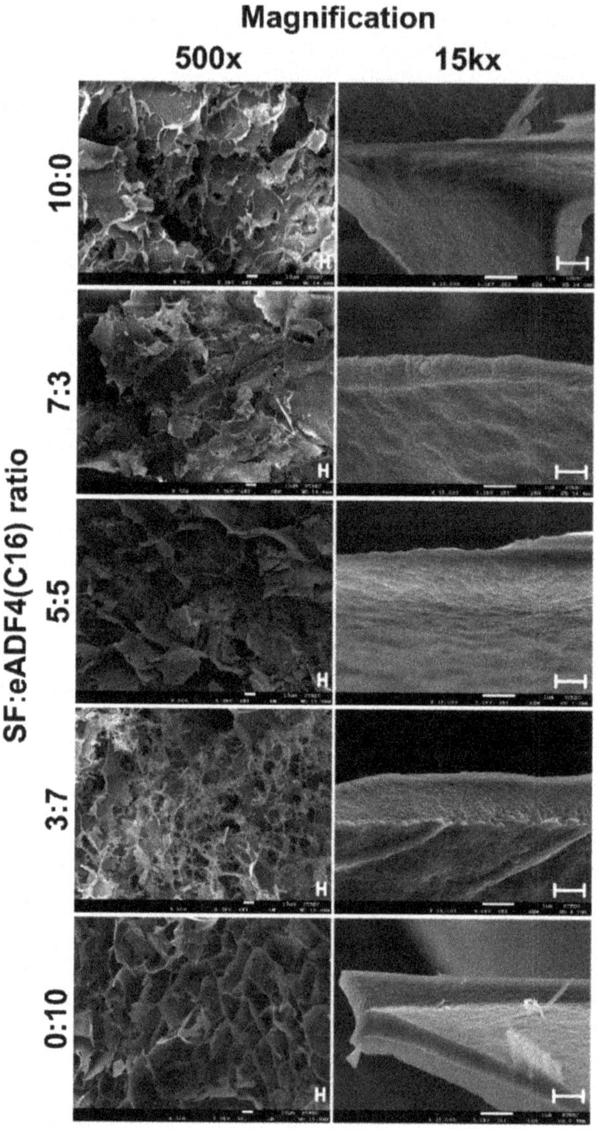

Figure 2. FESEM images of freeze-dried hydrogels. The number on the left indicates the SF:eADF4(C16) volume ratio. The scale bars of 500× and 15 kx magnification are 10 and 1 μm, respectively.

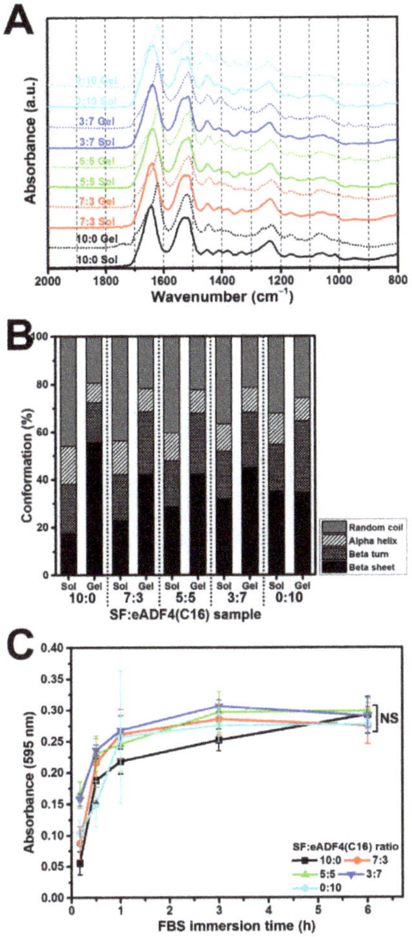

Figure 3. Physico-chemical properties of the SF:eADF4(C16). (**A**) FTIR spectra of freeze-dried hydrogels, (**B**) The content of protein conformation, quantified from the amide I region of FTIR spectra, and (**C**) Adsorption of proteins in the hydrogels after immersion in 10% FBS. NS: non-significant difference.

3.4. Protein Adsorptivity of the Hydrogels

Adsorption of soluble proteins in SF:eADF4(C16) hydrogels was determined using FBS as a model (Figure 3C). For all samples, FBS can be rapidly adsorbed onto the hydrogels in the first hour of immersion before maintaining a plateau. The adsorption of the proteins in the SF hydrogel (10:0) was slower than in the others, but the identical protein level could be achieved within 6 h.

3.5. Proliferation of Cells Cultured on the Silk Hydrogels

Due to the very long gelation time, gelation of pure SF was accelerated using ultrasonication, and the 7:3 SF:eADF4(C16) hydrogel was omitted from the cell culture experiment. Cell proliferation was determined from MTT assay (Figure 4). For primary cells, such as hASC, the results showed no significant difference among all samples. Cells on TCP presented the growth phase in the first five days before plateauing on day 7 (population doubling time (PDT) = 61.3 h), while those on the silk hydrogels showed low proliferation rates after 3 days (Figure 4A). The proliferation of two cell lines, L929 and SaOS-2 was depended on the sample composition, as samples with higher SF content showed enhanced cellular activities. L929 cells (PDT on TCP = 21.0 h) cultured on hydrogels containing SF (10:0, 5:5, 3:7 SF:eADF4(C16)) showed a similar behavior, while the growth of cells on the eADF4(C16) (0:10) hydrogel was significantly lower (Figure 4B). Similar results were noticed for SaOS-2 (PDT on TCP = 25.9 h) (Figure 4C). The cells on 2% SF hydrogels presented the highest growth rate, which was not significantly different from those on TCP for all time-points. The absorbance values were lower depending on an increasing eADF4(C16) content, with the lowest cell activity for 2% eADF4(C16).

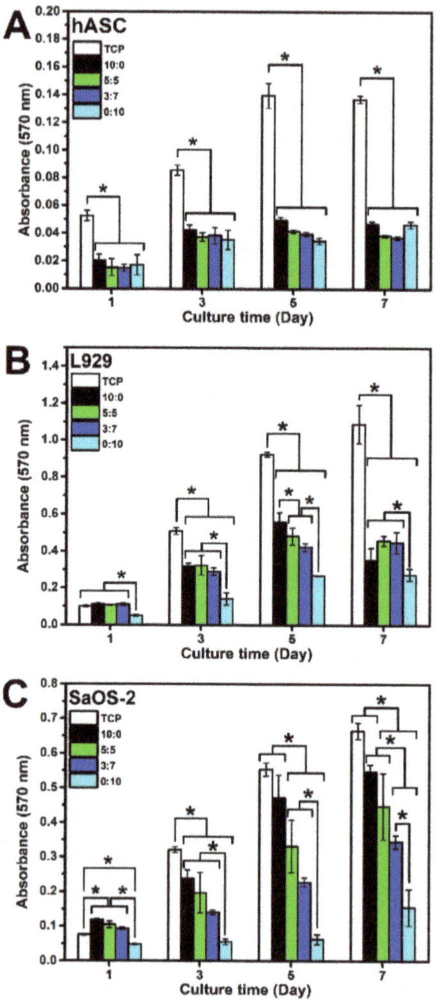

Figure 4. Cell proliferation determined using the MTT assay of (**A**) human adipose-derived stromal cells (hASC), (**B**) mouse L929 fibroblasts, and (**C**) human SaOS-2 osteosarcoma cells cultured on 2% SF:eADF4(C16) hydrogels for 7 days. Cells cultured on tissue culture. The asterisk (*) indicates the statistical difference at p-value ≤ 0.05.

3.6. *Morphology of Cells on the Hydrogels*

Bright-field images of hASC, L929, and SaOS-2 (Figure 5) cultured on silk hydrogels for 1 and 5 days showed the cell morphology after the initial attachment and in the exponential growth phase, respectively. Fluorescent images of hASCs (Figure 6) demonstrated live and dead cells, stained by calcein AM and PI dyes, respectively. The number of cells visualized from the images were in accordance with the cell proliferation results. It can be recognized that cells attached to hydrogels containing SF within 1 day after seeding. In contrast, cells on hydrogels with an increasing content of eADF4(C16) were less stretched and presented a lower attachment. On pure 2% eADF4(C16), cells could not attach well and preferably formed cell aggregates.

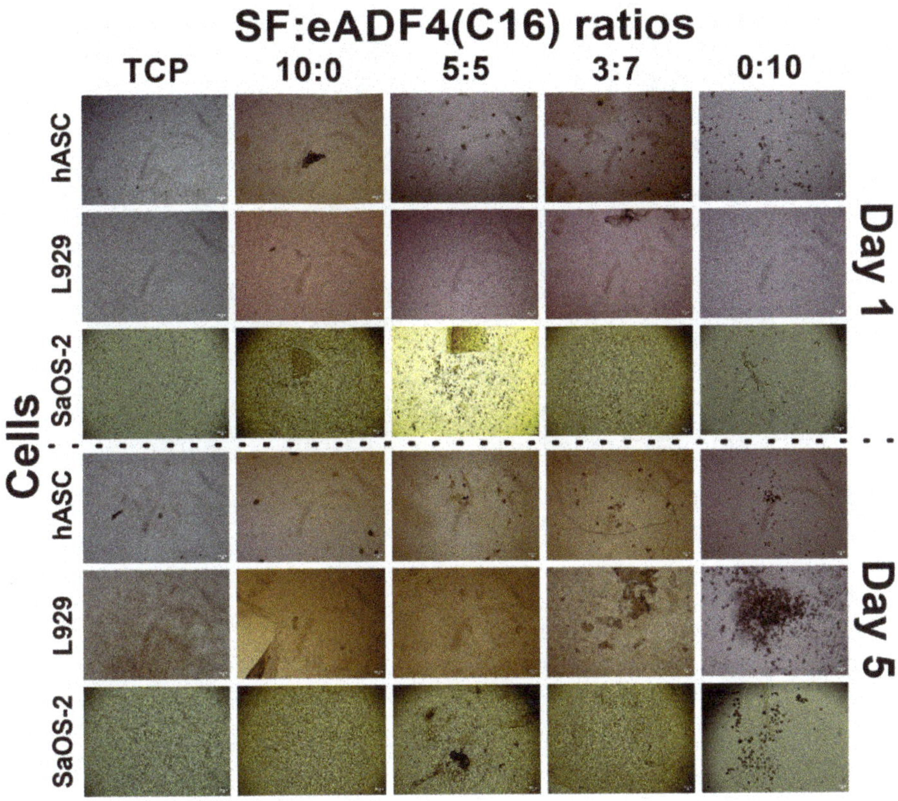

Figure 5. Bright-field images of hASC, L929 and SaOS-2 cells cultured on 2% SF:eADF4(C16) hydrogels on day 1 (**top** panel) and day 5 (**bottom** panel) (scale bar = 50 μm).

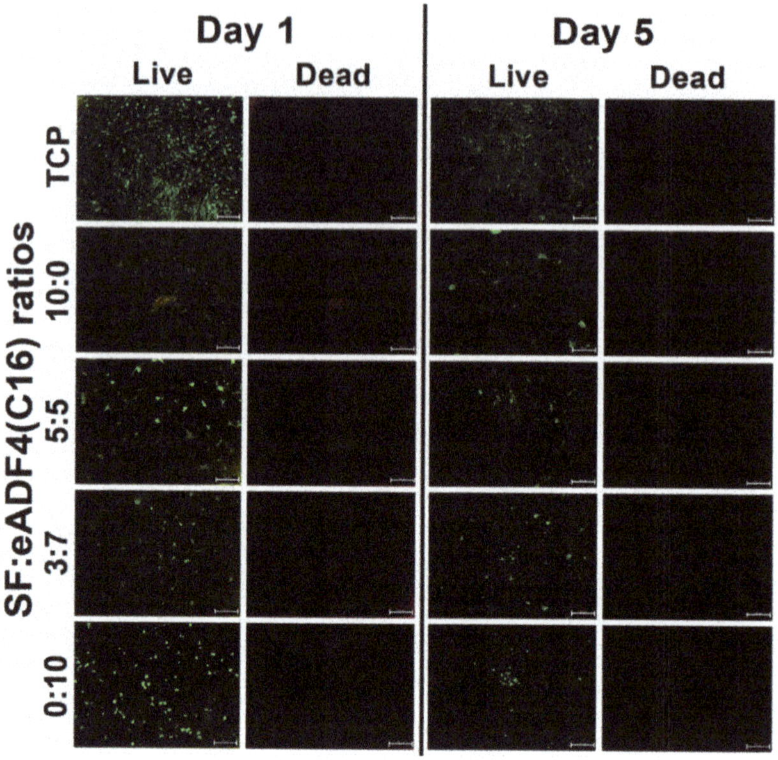

Figure 6. Fluorescence images of calcein AM (Live) and PI (Dead) stained hASC cells cultured on 2% SF:eADF4(C16) hydrogels on day 1 and day 5 (scale bar = 500 μm).

4. Discussion

Blended SF:eADF4(C16) hydrogels were fabricated in order to combine the advantages of eADF4(C16) in facilitating rapid gelation and enhanced interactions with cells provided by SF. As shown in Figure 1, it can be clearly seen that the addition of eADF4(C16) can induce faster gelation of SF, i.e., the gelation of 5:5 and 3:7 hydrogels occurred within 48 h. The gelation of pure 2% and 3% SF solutions were not observed after one week, which was in agreement with a previous study [20]. As expected, an increase of protein concentration from 2% to 3% led to a shorter gelation time, due to a higher opportunity in chain-chain interactions [5,10]. Additionally, the presence of cations, especially the mixtures of monovalent and divalent cations as in DMEM, significantly shortened the gelation time of SF:eADF4(C16) blends. This finding confirmed the previous study in which the addition of DMEM and divalent cations, such as Ca^{2+}, triggered a faster eADF4(C16) hydrogel formation [11]. Also, accelerated gelation of SF was achieved in presence of the divalent cation Ca^{2+}, but not the monovalent K^{+} [21]. However, our results showed that mixtures of monovalent cations, such as in PBS, and NSS, were able to affect the gelation of SF:eADF4(C16). We propose that the presence of cations mostly influences assembly of eADF4(C16) rather than that of SF, due to the polyanionic characteristics of eADF4(C16). Interactions with counterions result in a decrease of chain repulsion and facilitates molecular interactions.

Typically, the sol-gel transition of silk proteins, either by spontaneous gelling or physical intervention, relates to the self-assembly process [22]. Conformational changes from random coil in sol state to the highly ordered beta sheet structure in gel state were in accordance with the gelation of SF and eADF4(C16). The hydrogen bonding and hydrophobic interaction between hydrophobic glycine-alanine repeats or poly-alanine sequences result in a formation of beta sheet stacks and a subsequent gel formation [5,10]. Our findings showed a reduction of random coil structure as well as an increased beta sheet content after the gelation of all samples (Figure 3B). Notably, the eADF4(C16) showed a lower degree of beta sheet structures than that of SF. As the heavy chain of SF, which mainly directs the formation of beta sheets, possesses a molecular weight of 391 kDa [4], the presence of smaller eADF4(C16) chains with a molecular weight of 47.7 kDa [23] could disturb the molecular organization, resulting in the detected reduction of beta sheet formation upon an increase of eADF4(C16) content.

Cell interactions of SF:eADF4(C16) hydrogels were evaluated using the primary cell, hASC, and two cell lines, L929 mouse fibroblasts and SaOS-2 human osteoblast-like cells. For all cells, those cultured on pure eADF4(C16) hydrogels were rounded, loosely attached to the hydrogel surface, and preferably formed cell aggregates (Figure 5). The surface hydrophilicity, determined from water contact angle measurement of blended films (Figure S1), and the protein adsorptivity of the hydrogels (Figure 3C) could not strongly influence the cellular adhesion. In addition, as the isoelectric point of SF and eADF4(C16) was approximately 3.9 and 3.5, respectively [24,25], the negatively charged materials at the physiological pH (pH 7.4) could not well support cellular attachment through electrostatic interaction with negative-charged cell surface.

The proliferation of hASC primary cells on all hydrogels was slower, and a lower cell number was obtained than that on the TCP control (Figure 4A). SF and eADF4(C16) hydrogels were biocompatible, as seen from no dead cells in all samples (Figure 6). However, the lack of cell adhesion motifs in eADF4(C16), such as RGD, resulted in a low cellular adhesion to the material [26,27]. Furthermore, cells maintained their spherical forms and preferably formed micro-aggregates due to the loose attachment on the hydrogel surface [28]. Therefore, an extended initial lag phase of primary cells with low proliferation potential could be observed, and any growth phase could not be observed within the experimental period.

For L929 fibroblastic cells, proliferation rate depended on the type of samples, as hydrogels with high SF content supported cellular activities better (Figure 4B). Apart from the reason that eADF4(C16) cannot support proper cellular adhesion [12,29], Yamada et al. reported the presence of fibroblast growth-promoting peptides, VITTDSDGNE and NINDFDED, at the N-terminus of SF heavy chain [30]. Hence, the higher content of SF in the blended materials could be beneficial in promoting the proliferation of L929 fibroblasts.

As shown in Figure 4C, the growth of SaOS-2 on 10:0 hydrogels were statistically similar to the control (TCP) on day 5 and 7, and the values were proportionally reduced with increasing eADF4(C16) content. Furthermore, as noticed from Figure 6, the hASC could spread more on hydrogels with higher SF content. It can be presumed that the material acts as a physical cue directing the cellular activities including cytoskeleton organization and cell morphology [31]. Cells cultured on the stiffer materials exhibit a stretch morphology, generating a greater force on actin cytoskeleton, which favor an osteogenic expression [32,33].

The blended hydrogels, especially the 7:3 and 5:5 SF:eADF4(C16) formulations, exhibited accelerated gelation kinetics together with an enhanced cellular activity. Our results showed that the physical and biological properties of the hydrogels are tunable depending on the blending ratio. The simple blending of two different silk proteins reflects a simple preparation route to obtain the hydrogels with required properties such as biological activity of mechanical features for an intended cell biological application.

5. Conclusions

Blended SF:eADF4(C16) solutions show accelerated sol-gel transition in combination with the enhanced cell binding of the resulting hydrogels. Faster gelation was noticed with an increment of eADF4(C16) content to at least 50%. At the ratio of 7:3 and 5:5 SF:eADF4(C16), an enhanced cellular adhesion as well as cell proliferation have been noticed. The developed blended hydrogels supported viability of hASC primary cells. Proliferation of cell lines depended on SF content, as an increasing SF content enhanced cell proliferation. The different behavior of each cell type on the blended SF:eADF4(C16) hydrogels could serve as the fundamental data to design the applications of such hydrogels in the future. In addition, for further works, the mechanical properties of the SF:eADF4(C16) should be evaluated, both the modulus and the thixotropic properties of the hydrogels, to investigate their utilization as a printable bioink. Since the high crystallization of SF could limit the shear thinning and self-recovery of the hydrogels, the addition of eADF4(C16) which is known for its thixotropy [11] could ameliorate the printability of the developed hydrogels.

Supplementary Materials: The following are available online at https://www.mdpi.com/article/10.3390/polym13234182/s1, Figure S1: Water contact angle of blend SF:eADF4(C16) films.

Author Contributions: Conceptualization, C.L., J.R., M.H., T.S. and S.D.; Data curation, C.L.; Formal analysis, C.L.; Funding acquisition, J.R., S.K. and S.D.; Investigation, C.L.; Methodology, C.L., M.H., T.S. and S.D.; Project administration, T.S. and S.D.; Resources, A.V., J.R., Jittima Luckanagul, M.H., T.S. and S.D.; Supervision, A.V., J.R., S.K., J.A.L., M.H., T.S. and S.D.; Validation, J.R., S.K., J.A.L., M.H., T.S. and S.D.; Visualization, C.L.; Writing—original draft, C.L.; Writing—review & editing, A.V., J.R., S.K., J.A.L., M.H., T.S. and S.D. All authors have read and agreed to the published version of the manuscript.

Funding: This study was supported by the project "The enhancement of engineering academic collaboration with the Federal Republic of Germany", Ratchadapisek Somphot Fund, Chulalongkorn University. Authors also acknowledged the financial supports from the National Research Council of Thailand (NRCT5-RSA63001-20) and Chulalongkorn University (RES_64_117_21_014).

Institutional Review Board Statement: The study was conducted according to the guidelines of the Declaration of Helsinki and approved by the Institutional Review Board of Faculty of Medicine, Chulalongkorn University (IRB No.416/61, Date of Approval: September 10, 2019).

Informed Consent Statement: Informed consent was obtained from all subjects involved in the study.

Data Availability Statement: Data is contained within the article and Supplementary Materials.

Conflicts of Interest: T.S. is co-founder, shareholder and consultant of AMSilk GmbH, Martinsried, Germany.

References

1. Hoffman, A.S. Hydrogels for biomedical applications. *Adv. Drug Deliv. Rev.* **2012**, *64*, 18–23. [CrossRef]
2. Kapoor, S.; Kundu, S.C. Silk protein-based hydrogels: Promising advanced materials for biomedical applications. *Acta Biomater.* **2016**, *31*, 17–32. [CrossRef]
3. Abbasian, M.; Massoumi, B.; Mohammad-Rezaei, R.; Samadian, H.; Jaymand, M. Scaffolding polymeric biomaterials: Are naturally occurring biological macromolecules more appropriate for tissue engineering? *Int. J. Biol. Macromol.* **2019**, *134*, 673–694. [CrossRef] [PubMed]
4. Murphy, A.R.; Romero, I.S. 8—iochemical and biophysical properties of native Bombyx mori silk for tissue engineering applications. In *Silk Biomaterials for Tissue Engineering and Regenerative Medicine*; Kundu, S.C., Ed.; Woodhead Publishing: Sawston, UK, 2014; pp. 219–238.
5. Matsumoto, A.; Chen, J.; Collette, A.L.; Kim, U.-J.; Altman, G.H.; Cebe, P.; Kaplan, D.L. Mechanisms of Silk Fibroin Sol−Gel Transitions. *J. Phys. Chem. B* **2006**, *110*, 21630–21638. [CrossRef] [PubMed]
6. Bhunia, B.K.; Mandal, B.B. Exploring Gelation and Physicochemical Behavior of in Situ Bioresponsive Silk Hydrogels for Disc Degeneration Therapy. *ACS Biomater. Sci. Eng.* **2019**, *5*, 870–886. [CrossRef]
7. Kumar, M.; Gupta, P.; Bhattacharjee, S.; Nandi, S.K.; Mandal, B.B. Immunomodulatory injectable silk hydrogels maintaining functional islets and promoting anti-inflammatory M2 macrophage polarization. *Biomaterials* **2018**, *187*, 1–17. [CrossRef]
8. Chouhan, D.; Lohe, T.-u.; Samudrala, P.K.; Mandal, B.B. In Situ Forming Injectable Silk Fibroin Hydrogel Promotes Skin Regeneration in Full Thickness Burn Wounds. *Adv. Healthc. Mater.* **2018**, *7*, 1801092. [CrossRef]
9. Singh, Y.P.; Bandyopadhyay, A.; Mandal, B.B. 3D Bioprinting Using Cross-Linker-Free Silk–Gelatin Bioink for Cartilage Tissue Engineering. *ACS Appl. Mater. Interfaces* **2019**, *11*, 33684–33696. [CrossRef]
10. Schacht, K.; Scheibel, T. Controlled Hydrogel Formation of a Recombinant Spider Silk Protein. *Biomacromolecules* **2011**, *12*, 2488–2495. [CrossRef]

11. DeSimone, E.; Schacht, K.; Scheibel, T. Cations influence the cross-linking of hydrogels made of recombinant, polyanionic spider silk proteins. *Mater. Lett.* **2016**, *183*, 101–104. [CrossRef]
12. Wohlrab, S.; Müller, S.; Schmidt, A.; Neubauer, S.; Kessler, H.; Leal-Egaña, A.; Scheibel, T. Cell adhesion and proliferation on RGD-modified recombinant spider silk proteins. *Biomaterials* **2012**, *33*, 6650–6659. [CrossRef]
13. Huemmerich, D.; Helsen, C.W.; Quedzuweit, S.; Oschmann, J.; Rudolph, R.; Scheibel, T. Primary Structure Elements of Spider Dragline Silks and Their Contribution to Protein Solubility. *Biochemistry* **2004**, *43*, 13604–13612. [CrossRef]
14. Vachiraroj, N.; Ratanavaraporn, J.; Damrongsakkul, S.; Pichyangkura, R.; Banaprasert, T.; Kanokpanont, S. A comparison of Thai silk fibroin-based and chitosan-based materials on in vitro biocompatibility for bone substitutes. *Int. J. Biol. Macromol.* **2009**, *45*, 470–477. [CrossRef]
15. Hu, X.; Kaplan, D.; Cebe, P. Determining Beta-Sheet Crystallinity in Fibrous Proteins by Thermal Analysis and Infrared Spectroscopy. *Macromolecules* **2006**, *39*, 6161–6170. [CrossRef]
16. Laomeephol, C.; Guedes, M.; Ferreira, H.; Reis, R.L.; Kanokpanont, S.; Damrongsakkul, S.; Neves, N.M. Phospholipid-induced silk fibroin hydrogels and their potential as cell carriers for tissue regeneration. *J. Tissue Eng. Regen. Med.* **2020**, *14*, 160–172. [CrossRef]
17. Bunnell, B.A.; Flaat, M.; Gagliardi, C.; Patel, B.; Ripoll, C. Adipose-derived stem cells: Isolation, expansion and differentiation. *Methods* **2008**, *45*, 115–120. [CrossRef]
18. Katz, A.J.; Tholpady, A.; Tholpady, S.S.; Shang, H.; Ogle, R.C. Cell Surface and Transcriptional Characterization of Human Adipose-Derived Adherent Stromal (hADAS) Cells. *Stem Cells* **2005**, *23*, 412–423. [CrossRef]
19. Wang, X.; Kluge, J.A.; Leisk, G.G.; Kaplan, D.L. Sonication-induced gelation of silk fibroin for cell encapsulation. *Biomaterials* **2008**, *29*, 1054–1064. [CrossRef]
20. Laomeephol, C.; Ferreira, H.; Yodmuang, S.; Reis, L.R.; Damrongsakkul, S.; Neves, M.N. Exploring the Gelation Mechanisms and Cytocompatibility of Gold (III)-Mediated Regenerated and Thiolated Silk Fibroin Hydrogels. *Biomolecules* **2020**, *10*, 466. [CrossRef]
21. Kim, U.-J.; Park, J.; Li, C.; Jin, H.-J.; Valluzzi, R.; Kaplan, D.L. Structure and Properties of Silk Hydrogels. *Biomacromolecules* **2004**, *5*, 786–792. [CrossRef]
22. Jin, H.-J.; Kaplan, D.L. Mechanism of silk processing in insects and spiders. *Nature* **2003**, *424*, 1057–1061. [CrossRef] [PubMed]
23. Elsner, M.B.; Herold, H.M.; Müller-Herrmann, S.; Bargel, H.; Scheibel, T. Enhanced cellular uptake of engineered spider silk particles. *Biomater. Sci.* **2015**, *3*, 543–551. [CrossRef] [PubMed]
24. Ayub, Z.H.; Arai, M.; Hirabayashi, K. Mechanism of the Gelation of Fibroin Solution. *Biosci. Biotechnol. Biochem.* **1993**, *57*, 1910–1912. [CrossRef]
25. Hofer, M.; Winter, G.; Myschik, J. Recombinant spider silk particles for controlled delivery of protein drugs. *Biomaterials* **2012**, *33*, 1554–1562. [CrossRef]
26. Acharya, C.; Ghosh, S.K.; Kundu, S.C. Silk fibroin protein from mulberry and non-mulberry silkworms: Cytotoxicity, biocompatibility and kinetics of L929 murine fibroblast adhesion. *J. Mater. Sci. Mater. Med.* **2008**, *19*, 2827–2836. [CrossRef]
27. Schacht, K.; Jüngst, T.; Schweinlin, M.; Ewald, A.; Groll, J.; Scheibel, T. Biofabrication of Cell-Loaded 3D Spider Silk Constructs. *Angew. Chem. Int. Ed.* **2015**, *54*, 2816–2820. [CrossRef]
28. Leal-Egaña, A.; Scheibel, T. Interactions of cells with silk surfaces. *J. Mater. Chem.* **2012**, *22*, 14330–14336. [CrossRef]
29. Borkner, C.B.; Wohlrab, S.; Möller, E.; Lang, G.; Scheibel, T. Surface Modification of Polymeric Biomaterials Using Recombinant Spider Silk Proteins. *ACS Biomater. Sci. Eng.* **2017**, *3*, 767–775. [CrossRef]
30. Yamada, H.; Igarashi, Y.; Takasu, Y.; Saito, H.; Tsubouchi, K. Identification of fibroin-derived peptides enhancing the proliferation of cultured human skin fibroblasts. *Biomaterials* **2004**, *25*, 467–472. [CrossRef]
31. Han, F.; Zhu, C.; Guo, Q.; Yang, H.; Li, B. Cellular modulation by the elasticity of biomaterials. *J. Mater. Chem. B* **2016**, *4*, 9–26. [CrossRef]
32. Lee, J.; Abdeen, A.A.; Huang, T.H.; Kilian, K.A. Controlling cell geometry on substrates of variable stiffness can tune the degree of osteogenesis in human mesenchymal stem cells. *J. Mech. Behav. Biomed. Mater.* **2014**, *38*, 209–218. [CrossRef]
33. Engler, A.J.; Sen, S.; Sweeney, H.L.; Discher, D.E. Matrix Elasticity Directs Stem Cell Lineage Specification. *Cell* **2006**, *126*, 677–689. [CrossRef]

polymers

MDPI

Article

Ethyl Cellulose and Hydroxypropyl Methyl Cellulose Blended Methotrexate-Loaded Transdermal Patches: In Vitro and Ex Vivo

Muhammad Shahid Latif [1,†], Abul Kalam Azad [2,*,†], Asif Nawaz [1], Sheikh Abdur Rashid [1], Md. Habibur Rahman [3], Suliman Y. Al Omar [4], Simona G. Bungau [5], Lotfi Aleya [6] and Mohamed M. Abdel-Daim [7,*]

1 Department of Pharmaceutics, Faculty of Pharmacy, Gomal University, Dera Ismail Khan 29050, Pakistan; shahidlatif1710@gmail.com (M.S.L.); asifnawaz676@gmail.com (A.N.); sheikhabdurra-shid11@gmail.com (S.A.R.)
2 Advanced Drug Delivery Laboratory, Pharmaceutical Technology Department, Faculty of Pharmacy, International Islamic University Malaysia, Kuantan 25200, Pahang, Malaysia
3 Department of Global Medical Science, Wonju College of Medicine, Yonsei University, Seoul 26426, Gangwon-do, Korea; pharmacisthabib@gmail.com
4 Department of Zoology, College of Science, King Saud University, P.O. Box 2455, Riyadh 11451, Saudi Arabia; syalomar@ksu.edu.sa
5 Department of Pharmacy, Faculty of Medicine and Pharmacy, University of Oradea, 410087 Oradea, Romania; simonabungau@gmail.com
6 Chrono-Environnement Laboratory, UMR CNRS 6249, Bourgogne, Franche-Comté University, CEDEX, F-25030 Besançon, France; lotfi.aleya@univ-fcomte.fr
7 Pharmacology Department, Faculty of Veterinary Medicine, Suez Canal University, Ismailia 41522, Egypt
* Correspondence: aphdukm@gmail.com (A.K.A.); abdeldaim.m@vet.suez.edu.eg (M.M.A.-D.)
† These authors contributed equally to this work.

Citation: Latif, M.S.; Azad, A.K.; Nawaz, A.; Rashid, S.A.; Rahman, M.H.; Al Omar, S.Y.; Bungau, S.G.; Aleya, L.; Abdel-Daim, M.M. Ethyl Cellulose and Hydroxypropyl Methyl Cellulose Blended Methotrexate-Loaded Transdermal Patches: In Vitro and Ex Vivo. *Polymers* **2021**, *13*, 3455. https://doi.org/10.3390/polym13203455

Academic Editors: Ariana Hudita and Bianca Gălățeanu

Received: 15 September 2021
Accepted: 4 October 2021
Published: 9 October 2021

Abstract: Transdermal drug delivery systems (TDDSs) have become innovative, fascinating drug delivery methods intended for skin application to achieve systemic effects. TDDSs overcome the drawbacks associated with oral and parenteral routes of drug administration. The current investigation aimed to design, evaluate and optimize methotrexate (MTX)-loaded transdermal-type patches having ethyl cellulose (EC) and hydroxypropyl methyl cellulose (HPMC) at different concentrations for the local management of psoriasis. In vitro release and ex vivo permeation studies were carried out for the formulated patches. Various formulations (F1–F9) were developed using different concentrations of HPMC and EC. The F1 formulation having a 1:1 polymer concentration ratio served as the control formulation. ATR–FTIR analysis was performed to study drug–polymer interactions, and it was found that the drug and polymers were compatible with each other. The formulated patches were further investigated for their physicochemical parameters, in vitro release and ex vivo diffusion characteristics. Different parameters, such as surface pH, physical appearance, thickness, weight uniformity, percent moisture absorption, percent moisture loss, folding endurance, skin irritation, stability and drug content uniformity, were studied. From the hydrophilic mixture, it was observed that viscosity has a direct influence on drug release. Among all formulated patches, the F5 formulation exhibited 82.71% drug release in a sustained-release fashion and followed an anomalous non-Fickian diffusion. The permeation data of the F5 formulation exhibited about a 36.55% cumulative amount of percent drug permeated. The skin showed high retention for the F5 formulation (15.1%). The stability study indicated that all prepared formulations had very good stability for a period of 180 days. Therefore, it was concluded from the present study that methotrexate-loaded transdermal patches with EC and HPMC as polymers at different concentrations suit TDDSs ideally and improve patient compliance for the local management of psoriasis.

Keywords: transdermal drug delivery system (TDDS); hydroxypropyl methyl cellulose (HPMC); ethyl cellulose (EC); methotrexate; patches

1. Introduction

Psoriasis is a chronic inflammatory disease affecting about 1–3% of the world's population [1]. This lifelong disease has an equal gender distribution, and its incidence rate may vary from 50 to 140 new cases per 100,000 cases per year [2]. Its mortality risk is increased exponentially in terms of severe psoriasis when compared to the general population, even though it is usually not life threatening [3]. Psoriasis leads to decreased patient quality of life due to its link with high levels of morbidity and ailment. The management protocols of psoriasis vary depending upon the severity index of the disease [4].

Topical agents constitute first-line therapy, typically sufficient for active management of the disease to combat mild to moderate types of psoriasis [5]. Phototherapy and systemic management are crucial to consider when either suboptimal effects arise from topical therapies or when the intensity of the psoriasis limits the use of topical agents [6]. Currently available systemic tools comprise biological and nonbiological therapies, which are utilized as monotherapy or in combination with other modalities to manage moderate to severe psoriasis [7].

Methotrexate, orally administered retinoid, and cyclosporine represent prominent nonbiological systemic agents. Methotrexate administered via oral and parenteral routes presents an excellent therapeutic strategy to treat psoriasis owing to its epidermal cell proliferation inhibition, as well as anti-inflammatory actions at low doses [8]. However, a large number of reported toxicities due to methotrexate systemic administration, such as liver impairment, and gastric side effects, including diarrhea vomiting and stomatitis, appear [9]. Methotrexate is a folate antagonist, and it displays prominent antineoplastic activity, as well as having a use for psoriasis management [10]. It competitively inhibits the enzyme dihydrofolate reductase, leading to DNA inhibition synthesis. Methotrexate, when delivered to the psoriatic site by means of transdermal drug delivery, has the potential to reduce side effects associated with this drug and avoid first-pass metabolism [11]. A major problem with methotrexate is that the drug is hydrosoluble and available in ionized form at physiological pH (7.4), leading to limited capacity for passive diffusion [12].

To minimize the likelihood of side effects, as well as skin permeation, and to maintain a therapeutic concentration in the target tissues, numerous approaches have been proposed, such as liposomes, polymeric nanoparticles, microspheres, solid lipid nanoparticles, nanoemulsions and nanoemulsion gel and patch formulations [13]. However, transdermal patches offer numerous advantages in terms of ease of preparation, high loading capacity for hydrophilic and lipophilic drugs and long-term stability with improved dermal delivery [14]. A transdermal patch is used to deliver a specific dose of medication through the skin and into the bloodstream. Transdermal delivery provides controlled, consistent drug administration and produces continuous drug input. It has a short biological half-life and eliminates pulsed entry into the systemic circulation [15]. It is convenient and especially evident in patches that require application only once a week. Such a simple dosing regimen could enhance patient compliance with drug therapy [16].

Polymers are widely used in modern pharmaceutical technologies, and they play a vital role in drug delivery advancements. Polymers act as carriers in targeted therapies and offer controlled drug delivery while reducing the bitter taste of drugs [17]. Hydroxypropyl methylcellulose (HPMC) is a derivative of cellulose of hydrophilic nature [18]. It is widely used in controlled-release formulations due to its swelling, gelling and thickening properties. Furthermore, HPMC is nontoxic in nature, and its swelling and easy compression properties make it convenient for use in the preparation of controlled drug delivery systems [19].

Ethylcellulose (EC) is a derivative of cellulose of hydrophobic nature. It is a white to light free-flowing powder used widely in the manufacturing of controlled drug delivery systems. EC has very limited side effects; hence, it is considered safe to employ in tablets, oral capsules, ocular or vaginal preparations and topical preparations [20]. EC is an inert, hydrophobic polymer that exhibits certain properties such as good stability during storage, lack of toxicities and good compressibility, which are suitable for designing controlled drug delivery systems [21].

This study was undertaken to develop methotrexate-loaded matrix-type patches by employing a combination of hydrophilic and hydrophobic polymers, hydroxypropyl methylcellulose (HPMC) and ethyl cellulose (EC), for the pertinent and effective dermal treatment of psoriasis, with improved cutaneous deposition of methotrexate to enhance its local effect. The success associated with methotrexate through dermal application via patch formulations could also be represented by increased patient compliance due to the topical administration of therapeutic substances, representing a less invasive and more comfortable and convenient route of administration.

2. Materials and Methods

2.1. Materials

Methotrexate was kindly gifted by Wilsons Pharmaceutical (Pvt.) Ltd., Pakistan. Ethyl cellulose (EC) and hydroxypropyl methyl cellulose (HPMC) (Dow Chemical Company, 693 Washington St, #627, Midland, MI, 48640, USA, +1-989-636-1000) were used as rate-controlling polymers. Ethanol (Sigma-Aldrich Inc., P.O. Box 14508, St. Louis, MO, 63178, USA, +1-314-771-5750), PEG-400 (Sigma-Aldrich Co., 3050 Spruce Street, St. Louis, MO, 63103, USA, +1-314-771-5765), sodium hydroxide (NaOH) (Sigma-Aldrich Chemie GmbH, Riedstrasse 2, 89555 Steinheim, Germany, +49-7329-970), dichloromethane and calcium chloride (Dow Chemical Co., 2030 Dow Ctr, Midland, MI, 48674, USA, +1-989-638-8173) were used in the preparation of patches and buffers. Analytical-grade chemicals were used in this study.

2.2. Preparation of Transdermal Patch

The solvent evaporation technique was used for the formulation of methotrexate-loaded patches, with EC and HPMC as rate-controlling polymers at different concentrations (Table 1). The polymers were weighted accurately using an analytical weighing balance (Shimadzu AX 200, Kyoto, Japan), placed in a solvent system (15 mL) comprising ethanol and dichloromethane (1:1) and allowed to swell for 6 h. The plasticizer used was PEG-400. A 100 μL volume of ethanolic hydrochloric acid was taken in a beaker, and a proper amount of methotrexate was added. Dichloromethane and ethanol (1:1) were taken in a separate beaker and placed in a sonicator (Elma D-78224, Germany) for 2 min. The drug and polymers were mixed homogeneously by slow stirring. A uniform dispersion was poured in Petri dishes with an area of 19.5 cm^2. The Petri dishes were placed in an oven (Memmert, Germany) at 40 °C for 12 h.

Table 1. Composition of methotrexate transdermal patches.

Batch	Amount of MTX (mg)	Total Amount of Polymers			Plasticizer PEG-400 (%)	Amount of Solvents (*v/v*) mL	
		EC (mg)	HPMC (mg)	Combination EC/HPMC		Dichloromethane	Ethanol
F1 (Control)	5	100	100	1:1	25	20	20
F2	5	100	200	1:2	25	20	20
F3	5	100	300	1:3	25	20	20
F4	5	100	400	1:4	25	20	20
F5	5	100	500	1:5	25	20	20
F6	5	200	100	2:1	25	20	20
F7	5	300	100	3:1	25	20	20
F8	5	400	100	4:1	25	20	20
F9	5	500	100	5:1	25	20	20

2.3. ATR–FTIR Analysis (Preformulation Study)

ATR–FTIR analysis was carried out on the pure drug (methotrexate) and various physical mixtures of patch formulations (F1 to F9) to investigate possible interactions. A total of 32 scans were observed for each spectrum at a resolution of 4 cm^{-1} from 4000 to 600 cm^{-1}.

2.4. Physicochemical Evaluation of Patches

The physicochemical properties of the formulated patches were evaluated using the following parameters.

2.4.1. Surface pH

The surface pH of the formulated patches was evaluated by placing a 1 cm^2 portion of a patch in 1 mL of distilled water for 2 h at room temperature (25 ± 2 °C) in a test tube. Excess water from the test tube was removed by the filtration process. A pH meter (InoLab®, Xylem Analytics, Dr. Karl Slevogt Street 1. Weilheim 82362, Germany) was used for the identification of the surface pH of the formulated patches. The pH meter was placed at the swollen part of the patch, and three readings were recorded for the average (mean ± SD) result [22].

2.4.2. Physical Appearance

All formulated patches were physically inspected for smoothness, color, clarity, transparency, and homogeneity.

2.4.3. Thickness

The uniformity of thickness was evaluated for all formulated patches. A vernier caliper (Germany) was used for the evaluation of the thickness of the formulated patches. The thickness of patches was evaluated at 6 different places, and then the average was calculated [23].

2.4.4. Weight Uniformity

All formulated patches were weighed individually for weight uniformity. An analytical weighing balance (Shimadzu AX 200, Kyoto, Japan) was used for the determination of weight. Individual weight was compared with average weight [24].

2.4.5. Folding Endurance

The efficacy of the plasticizer was investigated with the folding endurance test. The folding of a patch at the same point until a break or crack appears shows the folding endurance capacity of a patch. At the same point, a patch was folded several times without cracking or breaking defines the value of folding endurance. The folding endurance test was conducted for all formulated patches [24].

2.4.6. Moisture Uptake

The formulated patches were weighed accurately for the determination of percent moisture uptake. Aluminum chloride and the patches were placed in a desiccator to maintain humid conditions. After 3 days, the patches were taken out of the desiccator. The patches were weighed again. The difference between the initial and final weights of the patches gave the value of percent moisture uptake. Finally, average percent moisture uptake was calculated [25].

$$\%\ \text{Moisture Uptake} = (\text{wf} - \text{wi})/\text{wi} \times 100 \quad (1)$$

where wf is the final patch weight, and wi is the initial patch weight.

2.4.7. Moisture Loss

All formulated patches were weighed individually for the determination of moisture loss. The patches, along with anhydrous calcium chloride, were placed in the desiccator at 37 °C in order to maintain dry conditions. After 3 days, the patches were taken out of the desiccator. The patches were weighed again. The difference between the initial and final weights of the patches gave the value of percent moisture loss. Finally, average percent moisture loss was calculated [14].

$$\% \text{ Moisture Loss} = (\text{wi} - \text{wf})/\text{wi} \times 100 \quad (2)$$

where wi is the initial weight, and wf is the final weight.

2.4.8. Moisture Content

The formulated patches were weighed accurately for the determination of moisture content. The patches, along with silica, were placed in the desiccator at room temperature for 24 h. The patches were taken out of the desiccator and weighed again until a constant weight was calculated. Percent moisture content was calculated using the following equation [26].

$$\% \text{ Moisture Loss} = (\text{wi} - \text{wf})/\text{wi} \times 100 \quad (3)$$

whereas wi is the initial patch weight, and wf is the final patch weight

2.4.9. Tensile Strength and Percent Elongation at Break

The mechanical properties of the formulated patches were determined using a pulley system. A scale was used for the identification of the initial patch length. One end of the patch was tied with a thread, while the second end was tied with a rope crossing over the pulley. A weighing pan was attached to the hanging side of the thread. Gradually, weight was added until a crack or break appeared in the patch. The total weight in the pan was calculated for the tensile strength. The thread pointer indicated percent elongation of work. From the following equation, the total amount of force (tensile strength, kg/cm^2) required to break a patch was calculated.

$$\text{Tensile Strength} = \text{F}/(\text{a.b}(1+\text{L}/\text{I})) \quad (4)$$

Where,

F is the force needed to break a patch, a is the patch width (cm) and b is patch thickness (cm).

L is patch length (cm), and I is patch elongation before patch breakage (cm). The percent elongation of the patches was determined from the following equation [27].

$$\% \text{ Elongation} = (\text{Lf} - \text{Li})/\text{Li} \times 100 \quad (5)$$

where Lf is the final patch length before breaking, and Li is the initial patch length.

2.4.10. Drug Content Uniformity

Drug content uniformity was evaluated for the formulated patches. A patch was placed in a volumetric flask filled with phosphate buffer (pH = 7.4) and then placed in a sonicator for 8 h. After sonication, the solutions were filtered. A double-beam UV–visible spectrophotometer (Shimadzu 1601, Kyoto, Japan) was used for the identification of drug content using a 303 nm wavelength [11].

2.4.11. Water Vapor Transmission Rate

Oven-dried and properly washed equal-diameter glass vials were used as transmission cells. In the transmission cells, 1 g of anhydrous calcium chloride was kept. At brim, the formulated patches were fixed. Transmission cells were weighed and then placed in the desiccator. Potassium chloride solution was kept closed in the desiccator to maintain 84% humidity. After predetermined time intervals, i.e., 6, 12, 24, 36, 48 and 72 h, cells were removed from the desiccator. Then, the cells were weighed again for the identification of the water vapor transmission rate [28].

2.5. Stability Studies

Stability studies were carried out for a period of 180 days for all formulated patches. The incubator was maintained at 37 ± 0.05 °C and 75 ± 5 RH, and the patches were kept for stability studies. After a regular interval of 30 days, drug content and physical appearance were evaluated for the formulated patches. Drug content determination was carried out according to the procedure described in Section 2.4.1.

2.6. Skin Irritation Studies

For skin irritation studies, healthy male albino rabbits (2–2.5 kg) were used, and proper NOC was taken from the Research Ethical Review Board of Gomal Center of Pharmaceutical Sciences, Faculty of Pharmacy, Gomal University, Dera Ismail Khan, KPK, Pakistan. The rabbits used in this study were given standard food at least three days before administration of the formulations. The standard food was prepared according to a published recipe composed of 10% white fish meat, 18% middlings, 20% grass meal and 40% bran [29]. Water was also allowed ad libitum. All rabbits were housed in a temperature (25 ± 2 °C)- and relative-humidity-controlled (50 ± 10%) room. The sparse hairs on the abdomen of each rabbit were carefully shaved a day before the scheduled experiment with an electrical clipper without damaging the stratum corneum. The application area was swept with dry cotton. The Draize patch test was used for skin irritation studies. Five groups, each containing 3 rabbits, were selected for skin irritation (hypersensitivity) reactions. Group 1 served as the nontreated group, and Group II served as the control group with USP adhesive tape. Group III was served with methotrexate-loaded patches, Group IV was served with 0.8% v/v aqueous solution of formalin (which is a standard irritant) and Group V was served with blank patches. The skin irritation study was carried out for a period of 1 week. A visual scoring scale was used for the identification of skin irritation grades. Skin irritation was graded as follows: "0" indicated no skin irritation, "1" indicated slight skin irritation, "2" indicated well-defined skin irritation, "3" indicated moderate-type skin irritation and "4" indicated scar formation on the skin [29].

2.7. In Vitro Drug Release Studies

In Vitro drug release studies were carried out using a Franz diffusion cell apparatus (model $_{\gamma}$9-CB (71026), PermeGear, Hellertown, PA, USA). A Tuffryn membrane was used as a synthetic membrane for drug release from the formulated patch of methotrexate with EC and HPMC at different concentrations. Between the donor and receptors compartments, the Tuffryn membrane was placed. A 1 cm^2 area of the formulated patch was placed over the Tuffryn membrane. The formulated patch piece was placed in such a manner that the drug-releasing surface faced the Tuffryn membrane. Phosphate buffer (pH = 5.5) was used in the receptor compartment. The receptor compartment is surrounded by water jackets in which water circulates. The receptor fluid temperature was maintained at 32 ± 0.5 °C. Magnetic beads were used for the stirring of receptor fluids. At predetermined time intervals of 0.5, 1, 1.5, 2, 4, 8, 12, 16, 20 and 24 h, a 2 mL sample was taken from the receptor fluid. In order to maintain sink conditions, fresh receptor fluid of an equal volume was added to the receptor compartment. For the determination of drug content, these samples were analyzed spectrophotometrically at a 303 nm wavelength [30].

Kinetic Model Profiling

The drug release data were fitted into a Korsmeyer–Peppas kinetic model to investigate the mechanism of drug release [31,32]. The power-law equation is shown below

$$Mt/M\infty = Kt\hat{}n \tag{6}$$

where Mt and M∞ are the fractions of drug released after time t.

K represents the constant rate.

n represents the exponential release value.

When n = 0.5, it is a quasi-Fickian diffusion mechanism.

When n > 0.5, drug release occurred by an anomalous non-Fickian, Case II or zero-order release mechanism

When n = 0, it indicates a zero-order release mechanism.

2.8. Ex Vivo Permeation Studies

Preparation of Rabbit Skin

For skin irritation studies, healthy male albino rabbits (2–2.5 kg) were used. For ex vivo permeation studies, proper NOC was taken from the Research Ethical Review Board of Gomal Center of Pharmaceutical Sciences, Faculty of Pharmacy, Gomal University, Dera Ismail Khan. The rabbits were allowed to roam freely and take food (standard) and water (ad libitum) at their own will. After administering an overdose of ketamine and xylazine injections, the rabbits were sacrificed. Hairs from the abdominal region of each rabbit were surgically removed. The skin was then dipped in warm water at 60 °C for 45 s for adhering fats. Excised skin was placed in distilled water, washed at –20 °C and stored until further use [30].

For the determination of drug (methotrexate) permeation from the formulated patches across rabbit skin, a Franz diffusion apparatus was used for this study. Before starting the experiment, excised skin that was kept at −20 °C was hydrated for at least 1 h. Then, between the donor and receptor compartments, the skin was placed. The stratum corneum (SC) side of the skin was placed facing the donor compartment of the Franz cell apparatus. Then, a 1 cm^2 piece of the formulated patch was placed over the rabbit skin. The drug-releasing surface of the formulated patch faced the SC of the rabbit skin. In the receptor compartment, phosphate buffer (pH 7.4) was used. The temperature of the receptor fluid was maintained at 37 ± 0.5 °C by means of water circulating in the water jackets around the receptor compartment. The receptor fluids were stirred by means of magnetic beads. At predetermined time intervals of 0.5, 1, 1.5, 2, 4, 8, 12, 16, 20 and 24 h, a 2 mL sample was taken from the receptor fluid. In order to maintain sink conditions, fresh receptor fluid of an equal volume was added to the receptor compartment. For the determination of drug content, these samples were analyzed spectrophotometrically at a 303 nm wavelength [30].

2.9. Drug Retention Study

The drug retention study was carried out after the permeation experiment. Skin from the Franz cell apparatus was removed carefully and cleaned with phosphate buffer solution, then dried and cut into small pieces. The skin pieces were then stirred in phosphate buffer (pH = 7.4) overnight. Retained drug from the skin was extracted using methanol, and samples were centrifuged. The supernatant was filtered through a 0.45 μm cellulose acetate filter. The filtrate was analyzed with a UV–visible spectrophotometer at a 303 nm wavelength.

2.10. Statistical Analysis

One-way ANOVA was used as the statistical tool. A value of $p < 0.05$ was considered significant. All experiments were performed in triplicate, and the result was expressed as the mean value ± standard deviation.

3. Results

TDDS is user friendly, painless, and convenient, and it usually leads to enhanced patient compliance. Transdermal patches control drug delivery by employing different combinations of polymers. In the current study, various polymers were used to prepare methotrexate-loaded transdermal patches. The polymers used were ethyl cellulose (EC) and hydroxypropyl methyl cellulose (HPMC) at different ratios.

3.1. Drug Excipient Compatibility Studies (ATR–FTIR Analysis)

The methotrexate spectrum was compared with the spectrum of EC and HPMC polymers formulations at different concentrations.

The methotrexate ATR–FTIR spectrum shows its characteristic absorption band as a broad signal at 3450 cm^{-1} (O–H being stretched from the carboxyl group overlaying with O–H being stretched from crystallized water). At 3080 cm^{-1} (primary amine, N–H stretched), 1670–1600 cm^{-1} is allocated to C=O stretching (–C=O stretched from the carboxylic group and C=O stretched from the amidic group (Figure 1). Hence, the C=O band is split into a double in the methotrexate sample). The corresponding N-H band from the amidic group appears in the spectral range of 1550–1500 cm^{-1}. It is partly overlapped with the aromatic C=C stretching. The carboxylic group band appears in the range of 1400–1200 cm^{-1}, corresponding to –C–O stretching. The molecular structure of the entire formulated patch of methotrexate indicates that it is in good agreement, which confirms its purity. The ATR–FTIR spectra of the physical mixture of methotrexate and polymers show major peaks. The FTIR spectrum of methotrexate with EC and HPMC polymers at different concentrations was compared with the methotrexate spectra. The major peaks of methotrexate and polymers were found preserved.

3.2. Physicochemical Assessment of Methotrexate-Loaded Transdermal Patches

All fabricated patch formulations of methotrexate were tested for physicochemical characterization. The results of various physicochemical tests revealed that all formulated patches were clear, smooth, transparent, flexible and nonsticky in appearance. The surface pH of all formulated patches (F1–F9) was found to be within the acceptable range, i.e., 5–5.9. Hence, skin irritation did not occur. Patch (F1–F9) thickness ranged between 0.50 and 0.60 nm, which showed uniform thickness. The weight ranged between 72.25 $\pm$ 0.08 and 78.67 $\pm$ 0.004 mg, which showed that the formulated patch weights were almost similar (Table 2). All formulated patches (F1–F9) passed the folding endurance test. The folding endurance of all formulated patches was more than the predefined range of folding endurance, i.e., $\geq$30. Hence, all formulated patches proved to be the best and acceptable dosage forms used transdermally. However, there was an increase in moisture content with an increase in hydrophilic polymers. This may be due to the higher affinity of water for hydrophilic polymers than hydrophobic polymers. The percent moisture uptake was found to be higher in the patch containing EC because it absorbs moisture. The low moisture content in the prepared formulations helped them to remain stable and free from being completely dry and brittle patches. The moisture loss ranged from 6.8 $\pm$ 0.38 to 8.28 $\pm$ 0.85. Similarly, the moisture absorption observed was found to be satisfactory and ranged from 8.45 $\pm$ 1.22 to 12.79 $\pm$ 1.46. The tensile strength and elongation values of the formulated patches ranged between 9.36 kg/cm^2 and 12.75 kg/cm^2, and proper flexibility was observed, as indicated by the high values. All formulated patches showed uniformity in drug content that was quite good and ranged between 97.42% and 99.13%. The results of this study show that the formulated patches could produce transdermal matrix-type patches with uniform drug contents. The plasticizer used was PEG-400 to reduce the brittleness of the patches. The current study indicates that the addition of PEG-400 at 25% *w/w* of polymers produces uniform, flexible and smooth patches. Patches formulated with the addition of PEG-400 as plasticizer were found to be best for tensile strength and folding endurance properties.

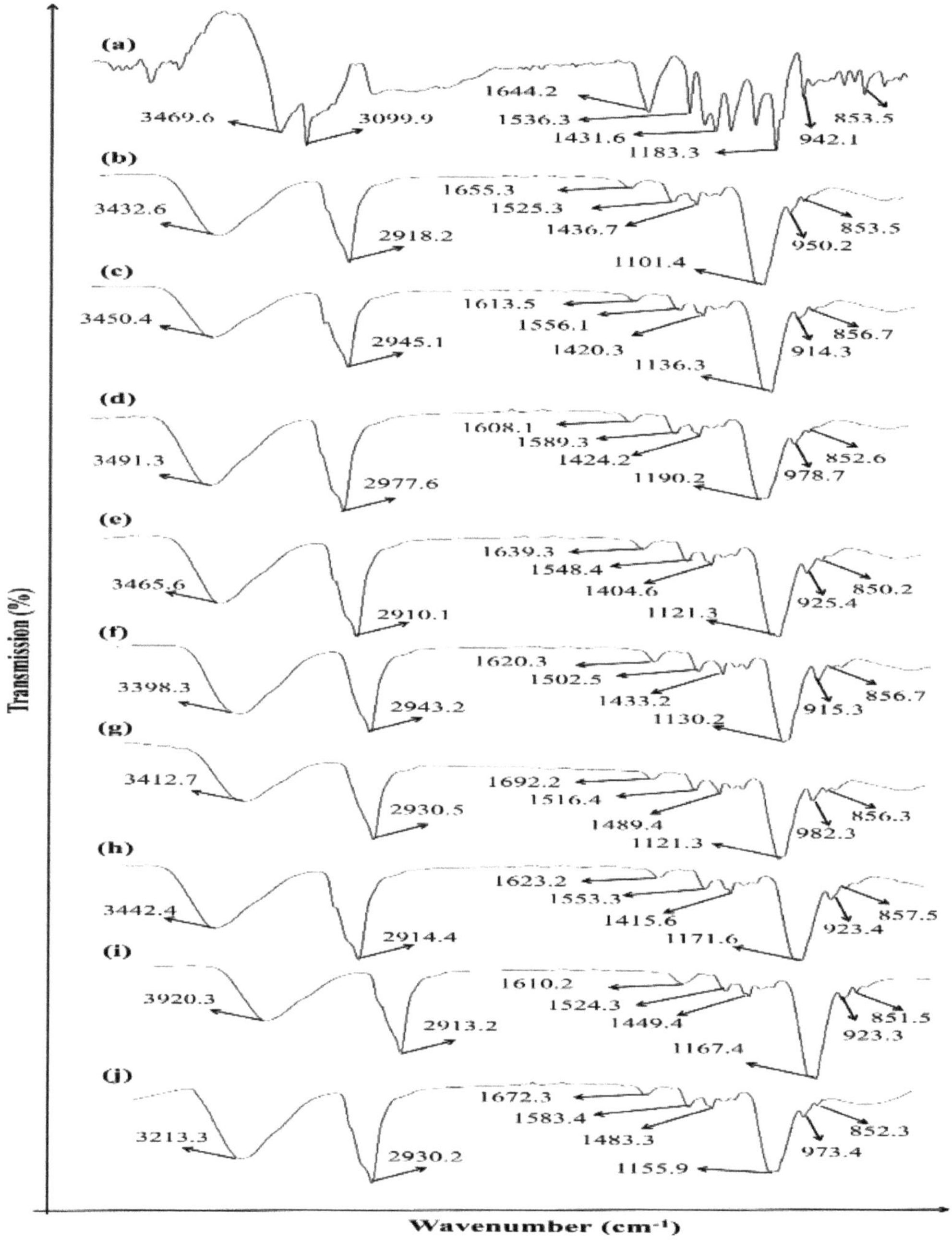

Figure 1. ATR–FTIR spectra: (**a**) MTX; (**b**) F1; (**c**) F2; (**d**) F3; (**e**) F4; (**f**) F5; (**g**) F6; (**h**) F7; (**i**) F8; (**j**) F9.

Table 2. Characterization of MTX-loaded transdermal patches (F1–F9). Data were expressed as mean ± SD, *n* = 3.

Formulation Code	Characteristics							
	Thickness (mm)	Weight Uniformity (mg)	% Moisture Absorbance	% Moisture Loss	% Drug Content	Water Transmission Rate	Folding Endurance	Tensile Strength, Kg/cm^2
F1	0.51 ± 0.03	73.86 ± 0.05	9.25 ± 1.62	6.28 ± 0.85	97.17 ± 3.21	3.58 ± 0.23	83 ± 2.03	10.43 ± 0.71
F2	0.52 ± 0.02	74.37 ± 0.03	10.81 ± 1.12	7.31 ± 0.21	99.13 ± 2.34	3.77 ± 0.65	92 ± 1.21	9.36 ± 0.83
F3	0.54 ± 0.07	76.55 ± 0.08	11.36 ± 1.32	7.82 ± 0.38	96.97 ± 2.24	3.89 ± 0.34	78 ± 3.24	11.35 ± 0.85
F4	0.55 ± 0.04	77.15 ± 0.05	11.56 ± 0.73	8.25 ± 0.22	97.11 ± 3.23	3.97 ± 0.54	86 ± 2.32	12.75 ± 0.72
F5	0.56 ± 0.04	78.67 ± 0.04	12.79 ± 1.46	8.94 ± 0.62	95.42 ± 2.23	4.23 ± 0.37	89 ± 2.54	10.93 ± 0.76
F6	0.50 ± 0.05	72.75 ± 0.08	8.45 ± 1.22	6.22 ± 0.15	99.54 ± 2.56	4.13 ± 0.41	64 ± 4.62	9.45 ± 0.81
F7	0.51 ± 0.02	73.57 ± 0.04	8.99 ± 1.44	6.58 ± 0.45	97.25 ± 3.43	3.93 ± 0.18	58 ± 5.12	12.34 ± 0.77
F8	0.52 ± 0.04	73.48 ± 0.08	9.67 ± 0.52	7.10 ± 0.23	98.64 ± 1.65	3.69 ± 0.32	52 ± 5.32	10.87 ± 0.83
F9	0.53 ± 0.09	74.86 ± 0.04	10.24 ± 0.97	7.31 ± 0.57	97.67 ± 4.36	3.84 ± 0.69	56 ± 5.35	11.66 ± 0.74

3.3. Stability Studies

Stability studies of all formulated patches are shown in Table 3. All formulated patches showed almost similar drug content data, as observed at the beginning of the study. All formulated patches showed acceptable, flexible and elasticity properties at the beginning Figure 2a, and end of this study in Figure 2b, thus ensuring the stability of the formulated patches.

Table 3. Stability studies of MTX-loaded transdermal patches (F1–F9), Data were expressed as mean ± SD, *n* = 3. Significant compared to formalin ($p < 0.05$).

Evaluation Parameters	F. Code	30 Days	60 Days	90 Days	120 Days	150 Days	180 Days
Drug content (%)	F1	97.17 ± 3.21	97.02 ± 1.99	96.89 ± 1.11	96.72 ± 3.22	96.61 ± 1.21	96.55 ± 2.11
	F2	99.13 ± 2.01	99.04 ± 1.65	98.85 ± 1.28	98.77 ± 2.14	98.58 ± 3.11	98.49 ± 2.01
	F3	96.97 ± 2.11	96.86 ± 2.56	96.71 ± 1.32	96.59 ± 2.14	96.42 ± 3.01	96.32 ± 2.46
	F4	97.11 ± 3.10	96.89 ± 2.35	96.73 ± 2.12	96.64 ± 2.98	96.59 ± 2.76	96.47 ± 3.23
	F5	95.42 ± 1.23	95.33 ± 2.27	95.09 ± 2.32	94.89 ± 2.65	94.76 ± 2.35	94.66 ± 2.87
	F6	98.92 ± 2.11	98.14 ± 3.11	97.76 ± 2.46	96.56 ± 2.45	95.78 ± 2.26	95.23 ± 3.43
	F7	99.19 ± 3.01	98.89 ± 1.02	97.93 ± 2.75	97.45 ± 3.46	96.69 ± 2.33	96.21 ± 3.10
	F8	97.25 ± 2.76	96.87 ± 3.15	95.87 ± 2.46	95.22 ± 3.25	94.76 ± 3.11	94.32 ± 2.33
	F9	96.63 ± 1.90	96.34 ± 2.32	95.82 ± 3.56	95.25 ± 2.31	94.78 ± 2.10	94.21 ± 2.19
Appearance	F1 F2 F3 F4 F5 F6 F7 F8 F9	No change	No change	No change	No change	No change	No change

3.4. Skin Irritation Study

The skin irritation study revealed that no irritation erythema or edema occurred. During the study period and after the removal of transdermal patches, no edema or erythema was found, which indicates that the formulations were nonirritant, while formalin (standard irritant) produced severe erythema and edema. The Draize test was negative, indicating that no skin irritation occurred if the score of the tests was less than 2 showed in Table 4.

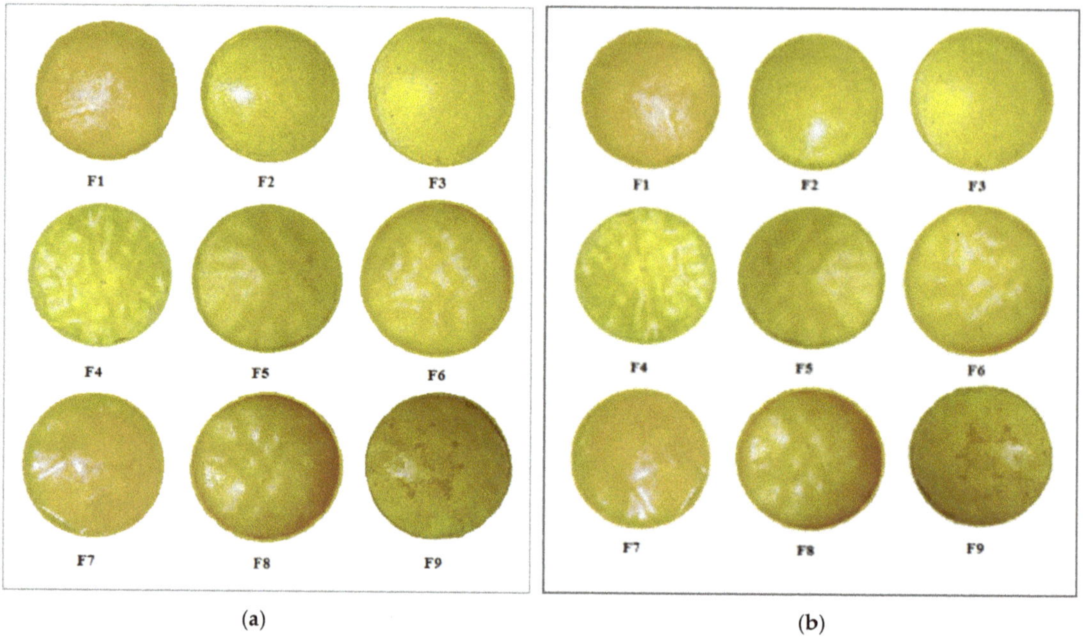

Figure 2. Physical appearance of formulated (F1–F9) patches shown at Day 1 (**a**) and after 180 days (**b**).

Table 4. Results of skin irritation on rabbits. Data were expressed as mean ± SD, $n = 3$. Significant compared to formalin ($p < 0.05$).

Rabbit Groups	Visual Observation	
	Erythema	Edema
	Result of Skin Irritation Studies	
Control	0.00 ± 0.00	0.00 ± 0.00
Adhesive tape	0.46 ± 0.62	1.02 ± 0.19
Blank patch	1.12 ± 0.46	1.06 ± 0.31
F1 Patch	0.83 ± 0.28	1.25 ± 0.16
F2 Patch	0.93 ± 0.67	1.08 ± 0.24
F3 Patch	1.03 ± 0.82	0.83 ± 0.69
F4 Patch	1.25 ± 0.71	1.12 ± 0.33
F5 Patch	0.37 ± 0.54	0.72 ± 0.26
F6 Patch	0.85 ± 0.43	1.05 ± 0.16
F7 Patch	0.94 ± 0.83	1.22 ± 0.35
F8 Patch	1.04 ± 0.74	1.06 ± 0.66
F9 Patch	0.98 ± 0.64	0.86 ± 0.27
Formalin	3.05 ± 0.23	3.21 ± 0.51

3.5. In Vitro Drug Release Study

In vitro drug release studies are needed for predicting the reproducibility of the rate and duration of drug release. The results from the in vitro drug release studies show that the release from the formulated patches increases with an increase in the concentration of the hydrophilic polymer (HPMC). The formulations from F1 to F5 showed higher release, while formulations from F6 to F9 showed lower release over a time period of 24 h in Figure 3. Hence, from the drug release profile of all formulations, F5 showed the best controlled-release profile of 82.71%. This may be attributed to the presence of the hydrophobic polymer (EC) and hydrophilic polymer (HPMC) in the ratio of 1:5. It was noted that the hydrophilic polymer released the drug at a faster rate than the hydrophobic polymer. The F5 formulation is an optimized formulation. The diffusion mechanism is responsible for drug release for transdermal drug delivery systems. This mechanism involves drug transport from the polymer matrix into the respective medium based on the concentration gradient. The variation in the concentration gradient leads to drug release, as well as a greater distance for diffusion. This could be the most probable reason for the comparatively slower rate of drug diffusion when the distance for diffusion increases.

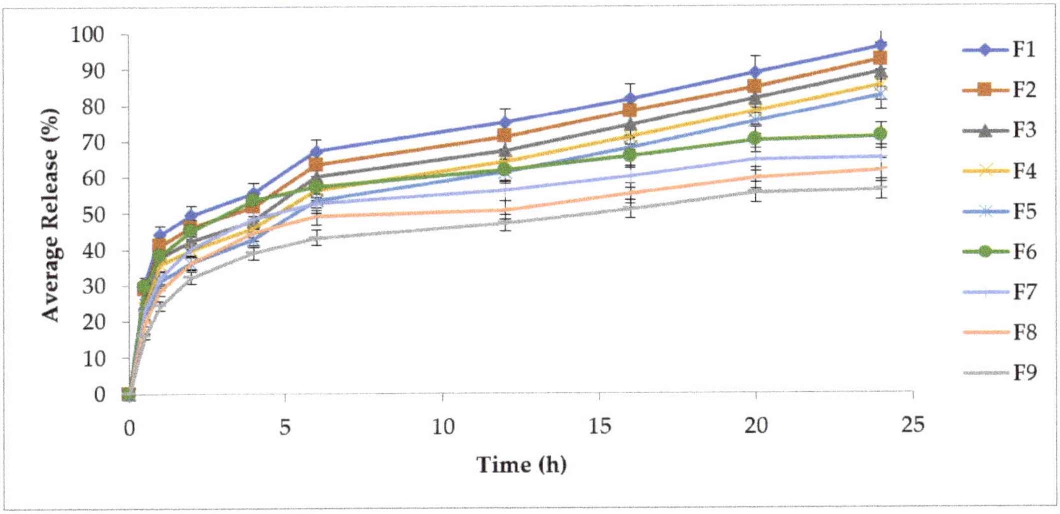

Figure 3. Release profile of MTX from F1 to F9, data were expressed as mean ± SD, *n* = 3.

3.6. Drug Release Kinetics

All the data obtained from the formulated patches were fitted to the Korsmeyer–Peppas model for the confirmation of the exact drug release behavior showed in Table 5. In our present study, the F5 formulation showed the best fit with the Korsmeyer–Peppas equation (R^2 = 0.974), showing an anomalous or non-Fickian diffusion mechanism of drug release (n = 0.50131). There was complete and controlled drug release of methotrexate found over a period of 24 h. The optimized formulation for the current study was F5. Thus, F5 releases the drug at the predefined rate for a prolonged period of time into the systemic circulation, leading to minimal dose frequency and adverse effects.

Table 5. Drug release kinetics (F1–F9), Data were expressed as mean ± SD, *n* = 3.

Formulations	K ± SO	R^2		N
F1	0.001 ± 2.8337	0.943	0.307	Fickian diffusion
F2	0.001 ± 5.0820	0.949	0.317	Fickian diffusion
F3	0.002 ± 0.0001	0.940	0.342	Fickian diffusion
F4	0.001 ± 0.0002	0.941	0.345	Fickian diffusion
F5	0.001 ± 0.0005	0.974	0.501	Anomalous non-Fickian diffusion
F6	0.001 ± 0.0001	0.929	0.301	Fickian diffusion
F7	0.001 ± 0.0001	0.924	0.301	Fickian diffusion
F8	0.002 ± 0.0003	0.933	0.303	Fickian diffusion
F9	0.001 ± 0.0006	0.967	0.303	Fickian diffusion

3.7. *Ex Vivo Drug Permeation Study*

The ex vivo permeation result of methotrexate-loaded patches having EC and HPMC at different concentrations showed in Figure 4. The F5 formulation exhibited a maximum percent cumulative amount of drug permeation (34.68%), producing a significant difference ($p < 0.05$) compared to the F1 formulation (25.11%) in 24 h. The F1 formulation produced 9.14 μg/h/cm^2, which is less than the required flux of 20.11 μg/h/cm^2. An increase in HPMC concentration increases flux values; thus, the F5 formulation, containing a greater amount of HPMC, produced about 1.5-fold greater flux compared to the target flux. Similarly, the F9 formulation exhibited a cumulative drug permeation of 21.68% compared to the F1 formulation, which exhibited a cumulative drug permeation of 25.11%. The flux value of F9 was found to be 9.23 μg/h/cm^2.

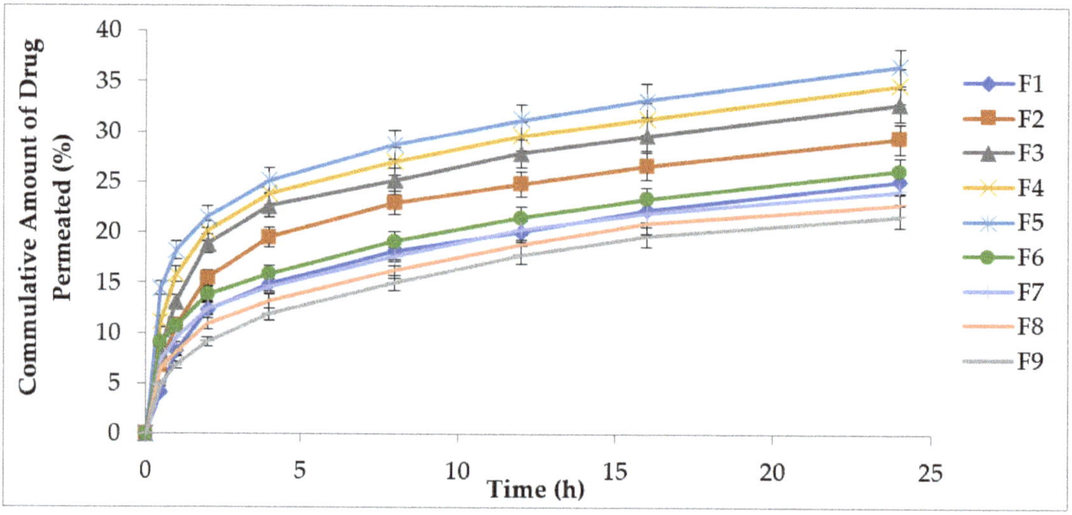

Figure 4. Percent cumulative amount of MTX permeated (F1–F9), Data were expressed as mean ± SD, $n = 3$. Significant compared to formalin ($p < 0.05$).

3.8. *Drug Retention Analysis*

The drug retention study (see Figure 5), revealed that the retention of methotrexate is more in the deep layers of the skin. In the case of F5, when compared to other formulations, there is a statistically significant difference in drug retention primarily in the epidermis and dermis (ANOVA, $p < 0.05$).

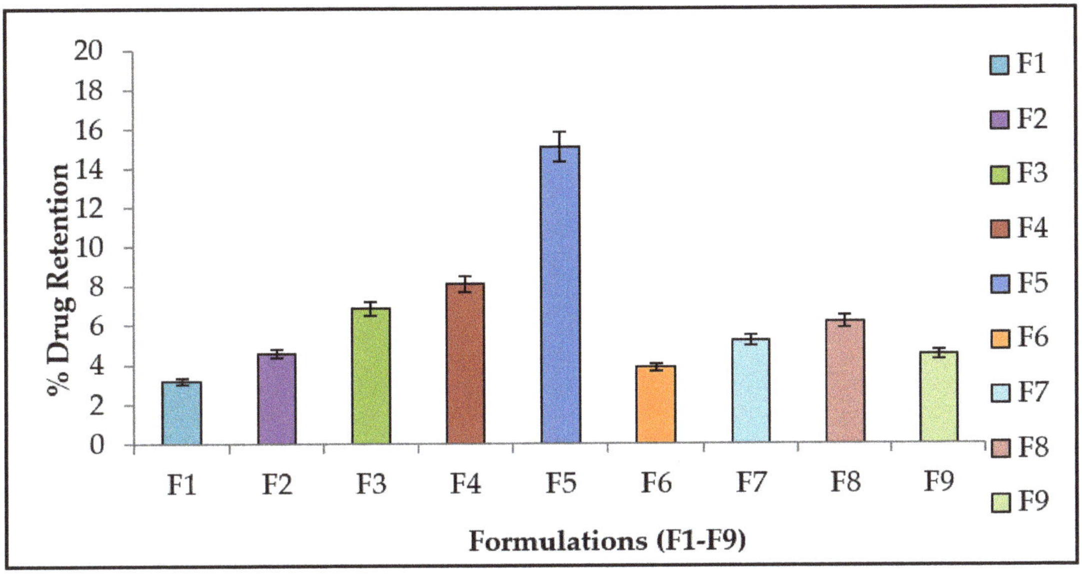

Figure 5. Skin drug retention analysis of methotrexate patches (F1–F9), Data were expressed as mean ± SD, $n = 3$. Significant compared to formalin ($p < 0.05$).

4. Discussion

Preformulation studies play a very crucial role in successfully creating a formulation. In order to determine drug excipient compatibility, ATR–FTIR analysis was carried out. ATR–FTIR is a nondestructive and quick technique used for obtaining the IR spectrum of a pure drug (methotrexate), as well as various patch formulations (F1 to F9). It is used for the identification and characterization of interactions between drugs and synthetic, semisynthetic, and native macromolecules. The ATR–FTIR spectra revealed several peaks in the final formulation, which confirmed the chemical structure retained within the drug with efficient loading into the formulation. The current study showed no chemical interaction between methotrexate and the physical mixtures with polymers (EC, HPMC) used [33]. The solvent evaporation technique was employed for the formulation of methotrexate-loaded transdermal-type patches having HPMC and EC at different concentrations [34]. The surface pH of the formulated patches was determined to investigate the possibilities of irritation during in vivo studies. This test is of utmost importance during transdermal drug delivery because alkaline or acidic pH causes irritation to the skin [35]. Uniform weight measurements and thicknesses were observed, which were evident due to low standard deviation values [26]. The folding endurance value is important during the formulation of transdermal patches. This test indicates that formulated patches can integrate with skin folding and do not break during use [24]. The drug release pattern of transdermal matrix-type patches can be affected by moisture uptake and moisture loss. In this study, the values of moisture content and moisture uptake were low. This indicates that the patches remain stable during long-term storage, and brittleness is reduced. Formulated patches are protected from microbial contaminations, having low moisture uptake and reduced bulkiness [8]. Tensile strength and elongation are related to the effectiveness of the plasticizer used in formulations [36]. It is necessary for the drug distribution to be homogenous and uniform because it helps in the evaluation of sustained drug deliveries from formulated patches. Uniform drug content data were observed from the formulated patches, which were revealed by low standard deviation values. Plasticizer use in the formulation of transdermal patches is necessary to improve the patch-forming properties and physical appearance of the patches. This prevents the patches from cracking and breaking and increases patch flexibility for obtaining the desired mechanical properties [37]. The successfully formulated patch formulations were screened for their irritation potential, as withdrawal of the patch from the site of application resulted in irritation in the form of erythema and edema, so it is necessary for patch formulations to completely lack irritation potential in order to obtain patient acceptability and a therapeutic outcome [3]. The in vitro drug release profile of the methotrexate-loaded patches showed an initial burst release, followed by a gradually approaching plateau, giving an indication of the controlled-release behavior of the matrix formulations. This burst release might be due to the release of superficially adhered drug contents. Drug release could be prolonged by adjusting adequate ratios of EC and HPMC [38]. There was a decrease in the release rate with an increasing concentration of HPMC in the formulations. This is because increased proportions of HPMC in the matrix resulted in a proportionate decrease in the amount of water uptake, leading to lesser drug release. EC, owing to its hydrophobic nature, produced a retarded drug release from the matrix [16]. The optimized F5 formulation showed a controlled-release pattern. This is advantageous in the case of chronic conditions such as psoriasis because rapid or burst release is not useful owing to its toxicity potential due to greater drug release and faster absorption of the drug in inflamed psoriatic lesions. The Korsmeyer–Peppas model was used for kinetic drug profiling. Data were fitted into this model in order to investigate the Fickian or non-Fickian diffusion pattern, followed by formulated patches. The value of "n" determines whether the drug release mechanism that weathers the release pattern is Fickian or non-Fickian diffusion. If the value of "n" is equal to 0.5, the diffusion is said to be Fickian. If the value of "n" ranges between 0.5 and 1, the diffusion is said to be anomalous diffusion. When the value of "n" is equal to 1, the diffusion is said to be Case II transport behavior [39]. The in vitro skin permeation profile is considered a significant tool to exclude the risks of unfortunate drug effects. In vitro skin permeation experiments describe the rate and mechanism of the percutaneous absorption of drugs. Studies have shown that diffusion rate is affected due to the physicochemical characteristics of the formulations. These include hydrogen bonding, drug loading capacity, surface charge and mode of application [16]. The skin permeation experiment concluded that patch formulations play a significant role in controlling methotrexate release, as well as drug targeting to the skin. Studies have shown that diffusion rate is affected due to the physicochemical characteristics of the formulation. During skin retention studies, greater retention was observed. This might be due to the strong interaction of the drug with keratinocytes. It is also expected that more drug is retained in the dermis compared to the epidermis, as revealed by skin anatomy studies, which show the dermis layer is thicker than that of the epidermis. Regardless, the accumulation of more amounts of drug in the deeper layer of skin is advantageous, as these layers are mainly affected by psoriasis [40]. Stability studies performed for the formulated patches were carried out for 180 days. After specific intervals of 30 days, the formulated patches showed optimum stability with no obvious physicochemical changes [41].

5. Conclusions

The results of the current study indicate that methotrexate, having EC/HPMC polymers at different concentrations, has the best and excellent patch-forming abilities. All formulated patches (F1–F9) were evaluated, and the F5 formulation exhibited the best in vitro drug release pattern and ex vivo drug permeation ability, having the highest deposition of methotrexate compared to other formulated patches. This greater retention on the deeper layer of the skin is significant for targeting psoriasis because this chronic ailment prevails in the epidermis and dermis layers. Thus, F5 showed the best potential for transdermal drug delivery. The controlled and slow release of the drug showed that the F5 formulation is suitable for transdermal patches. FTIR studies showed no interaction between the drug (methotrexate) and polymers (EC/HPMC) used. Methotrexate was distributed uniformly in the formulated patches and was of an amorphous nature.

Author Contributions: Conceptualization, A.K.A.; Data curation, A.K.A. and A.N.; Formal analysis, A.K.A.; Funding acquisition, M.M.A.-D.; Investigation, M.S.L. and A.K.A.; Methodology, A.K.A. and A.N.; Resources, M.M.A.-D.; Supervision, A.N.; Validation, A.K.A.; Visualization, A.K.A.; Writing—original draft, M.S.L.; Writing—review & editing, A.K.A., A.N., S.A.R., M.H.R., S.Y.A.O., S.G.B., L.A. and M.M.A.-D. All authors surely contributed to this work. All authors have read and agreed to the published version of the manuscript.

Funding: This project was supported by the Researchers Supporting Project (RSP-2021/35), King Saud University, Riyadh, Saudi Arabia.

Institutional Review Board Statement: The study was conducted according to the guidelines of the Declaration of Helsinki and approved by the Institutional Review Board/Ethics Committee of GOMAL UNIVERSITY (protocol code No: 116/ERB/GU and 26/02/2021).

Informed Consent Statement: Not applicable.

Data Availability Statement: Not applicable.

Acknowledgments: This project was supported by the Researchers Supporting Project (RSP-2021/35), King Saud University, Riyadh, Saudi Arabia.

Conflicts of Interest: The authors declare no conflict of interest.

References

1. Munguía-Calzada, P.; Drake-Monfort, M.; Armesto, S.; Reguero-del Cura, L.; López-Sundh, A.E.; González-López, M.A. Psoriasis flare after influenza vaccination in COVID-19 era: A report of four cases from a single center. *Dermatol. Ther.* **2021**, *34*, e14684. [CrossRef]
2. Mehrmal, S.; Uppal, P.; Nedley, N.; Giesey, R.L.; Delost, G.R. The global, regional, and national burden of psoriasis in 195 countries and territories, 1990 to 2017: A systematic analysis from the Global Burden of Disease Study 2017. *J. Am. Acad. Dermatol.* **2021**, *84*, 46–52. [CrossRef]
3. Gottlieb, A.B.; Merola, J.F. Axial psoriatic arthritis: An update for dermatologists. *J. Am. Acad. Dermatol.* **2021**, *84*, 92–101. [CrossRef] [PubMed]
4. Dabholkar, N.; Rapalli, V.K.; Singhvi, G. Potential herbal constituents for psoriasis treatment as protective and effective therapy. *Phytother. Res.* **2021**, *35*, 2429–2444. [CrossRef] [PubMed]
5. Zhang, B.; Lai, R.C.; Sim, W.K.; Choo, A.B.H.; Lane, E.B.; Lim, S.K. Topical application of mesenchymal stem cell exosomes alleviates the imiquimod induced psoriasis-like inflammation. *Int. J. Mol. Sci.* **2021**, *22*, 720. [CrossRef] [PubMed]
6. Elmets, C.A.; Korman, N.J.; Prater, E.F.; Wong, E.B.; Rupani, R.N.; Kivelevitch, D.; Armstrong, A.W.; Connor, C.; Cordoro, K.M.; Davis, D.M.; et al. Joint AAD–NPF Guidelines of care for the management and treatment of psoriasis with topical therapy and alternative medicine modalities for psoriasis severity measures. *J. Am. Acad. Dermatol.* **2021**, *84*, 432–470. [CrossRef] [PubMed]
7. Sudhakar, K.; Fuloria, S.; Subramaniyan, V.; Sathasivam, K.V.; Azad, A.K.; Swain, S.S.; Sekar, M.; Karupiah, S.; Porwal, O.; Sahoo, A.; et al. Ultraflexible Liposome Nanocargo as a Dermal and Transdermal Drug Delivery System. *Nanomaterials* **2021**, *11*, 2557. [CrossRef]
8. Dehshahri, A.; Kumar, A.; Madamsetty, V.S.; Uzieliene, I.; Tavakol, S.; Azedi, F.; Fekri, H.S.; Zarrabi, A.; Mohammadinejad, R.; Thakur, V.K. New horizons in hydrogels for methotrexate delivery. *Gels* **2021**, *7*, 2. [CrossRef] [PubMed]
9. Nam, S.; Mooney, D. Polymeric tissue adhesives. Chemical Reviews. *Chem. Rev.* **2021**, *121*, 11336–11384. [CrossRef] [PubMed]
10. Ezhilarasan, D. Hepatotoxic potentials of methotrexate: Understanding the possible toxicological molecular mechanisms. *Toxicology* **2021**, *458*, 152840. [CrossRef]
11. Biswasroy, P.; Pradhan, D.; Kar, B.; Ghosh, G.; Rath, G. Recent Advancement in Topical Nanocarriers for the Treatment of Psoriasis. *AAPS PharmSciTech* **2021**, *22*, 1–27. [CrossRef] [PubMed]
12. Giri, B.R.; Kim, J.S.; Park, J.H.; Jin, S.G.; Kim, K.S.; Choi, H.G.; Kim, D.W. Improved Bioavailability and High Photostability of Methotrexate by Spray-Dried Surface-Attached Solid Dispersion with an Aqueous Medium. *Pharmaceutics* **2021**, *13*, 111. [CrossRef] [PubMed]
13. Khan, A.; Qadir, A.; Ali, F.; Aqil, M. Phytoconstituents loaded based nanomedicines for the management of psoriasis. *J. Drug Deliv. Sci. Technol.* **2021**, *64*, 102663. [CrossRef]
14. Imtiaz, M.S.; Shoaib, M.H.; Yousuf, R.I.; Ali, F.R.; Saleem, M.T.; Khan, M.Z.; Sikandar, M. Formulation development and evaluation of drug-in-adhesive-type transdermal patch of metoclopramide HCl. *Polym. Bull.* **2021**, *2*, 1–14.
15. Patel, D.; Chaudhary, S.A.; Parmar, B.; Bhura, N. Transdermal drug delivery system: A review. *Pharma Innov.* **2012**, *1*, 66. [CrossRef]
16. Bernardes, M.T.C.P.; Agostini, S.B.N.; Pereira, G.R.; da Silva, L.P.; da Silva, J.B.; Bruschi, M.L.; Novaes, R.D.; Carvalho, F.C. Preclinical study of methotrexate-based hydrogels versus surfactant based liquid crystal systems on psoriasis treatment. *Eur. J. Pharm. Sci.* **2021**, *165*, 105956. [CrossRef]

17. Malviya, R.; Sundram, S.; Fuloria, S.; Subramaniyan, V.; Sathasivam, K.V.; Azad, A.K.; Sekar, M.; Kumar, D.H.; Chakravarthi, S.; Porwal, O.; et al. Evaluation and Characterization of Tamarind Gum Polysaccharide: The Biopolymer. *Polymers* **2021**, *13*, 3023. [CrossRef]
18. Zaidul, I.S.; Fahim, T.K.; Sahena, F.; Azad, A.K.; Rashid, M.A.; Hossain, M.S. Dataset on applying HPMC polymer to improve encapsulation efficiency and stability of the fish oil: In vitro evaluation. *Data Brief* **2020**, *32*, 106111. [CrossRef]
19. Hu, M.; Yang, J.; Xu, J. Structural and biological investigation of chitosan/hyaluronic acid with silanized-hydroxypropyl methylcellulose as an injectable reinforced interpenetrating network hydrogel for cartilage tissue engineering. *Drug Deliv.* **2021**, *28*, 607–619. [CrossRef]
20. Rekhi, G.S.; Jambhekar, S.S. Ethylcellulose-a polymer review. *Drug Dev. Ind. Pharm.* **1995**, *21*, 61–77. [CrossRef]
21. Wasilewska, K.; Winnicka, K. Ethylcellulose–a pharmaceutical excipient with multidirectional application in drug dosage forms development. *Materials* **2019**, *12*, 3386. [CrossRef] [PubMed]
22. Kharia, A.; Singhai, A.K.; Gilhotra, R. Formualtion and evalaution of transdermal patch for the treatment of inflammation. *J. Pharm. Sci. Res.* **2020**, *12*, 780–788.
23. Yang, X.; Tang, Y.; Wang, M.; Wang, Y.; Wang, W.; Pang, M.; Xu, Y. Co-delivery of methotrexate and nicotinamide by cerosomes for topical psoriasis treatment with enhanced efficacy. *Int. J. Pharm.* **2021**, *605*, 120826. [CrossRef] [PubMed]
24. Ullah, W.; Nawaz, A.; Akhlaq, M.; Shah, K.U.; Latif, M.S.; Alfatama, M. Transdermal delivery of gatifloxacin carboxymethyl cellulose-based patches: Preparation and characterization. *J. Drug Deliv. Sci. Technol.* **2021**, *66*, 102783. [CrossRef]
25. Sahu, K.; Pathan, S.; Khatri, K.; Upmanyu, N.; Shilpi, S. Development, characterization, in vitro and ex vivo evaluation of antiemetic transdermal patches of ondansetron hydrochloride and dexamethasone. *GSC Biol. Pharm. Sci.* **2021**, *14*, 067–078. [CrossRef]
26. Jan, S.U.; Gul, R.; Jalaludin, S. Formulation and evaluation of transdermal patches of pseudoephedrine HCL. *Int. J. Pharm.* **2020**, *5*, 121–127.
27. Kulkarni, S. Formulation and evaluation of transdermal patch for atomoxetine hydrochloride. *J. Drug Deliv. Ther.* **2019**, *9*, 32–35.
28. Sakhare, A.D.; Biyani, K.R.; Sudke, S.G. Design and Evaluation of Transdermal Patches of Carvedilol. *J. Curr. Pharm. Res.* **2019**, *9*, 3124–3137.
29. Oxley, J.A.; Ellis, C.F.; McBride, E.A.; McCormick, W.D. A survey of rabbit handling methods within the United Kingdom and the Republic of Ireland. *J. Appl. Anim. Welf. Sci.* **2019**, *22*, 207–218. [CrossRef] [PubMed]
30. Ramadon, D.; McCrudden, M.T.; Courtenay, A.J.; Donnelly, R.F. Enhancement strategies for transdermal drug delivery systems: Current trends and applications. *Drug Deliv.* **2021**, 1–34.
31. Azad, A.K.; Al-Mahmood, S.M.; Kennedy, J.F.; Chatterjee, B.; Bera, H. Electro-hydrodynamic assisted synthesis of lecithin-stabilized peppermint oil-loaded alginate microbeads for intestinal drug delivery. *Int. J. Biol. Macromol.* **2021**, *185*, 861–875. [CrossRef] [PubMed]
32. Bera, H.; Abbasi, Y.F.; Gajbhiye, V.; Liew, K.F.; Kumar, P.; Tambe, P.; Azad, A.K.; Cun, D.; Yang, M. Carboxymethyl fenugreek galactomannan-g-poly (N-isopropylacrylamide-co-N, N′-methylene-bis-acrylamide)-clay based pH/temperature-responsive nanocomposites as drug-carriers. *Mater. Sci. Eng. C* **2020**, *110*, 110628. [CrossRef] [PubMed]
33. Rashid, S.A.; Bashir, S.; Ullah, H.; Ullah Shah, K.; Khan, D.H.; Shah, P.A.; Danish, M.Z.; Khan, M.H.; Mahmood, S.; Sohaib, M.; et al. Development, characterization and optimization of methotrexate-olive oil nano-emulsion for topical application. *Pak. J. Pharm. Sci.* **2021**, *34*, 205–215. [PubMed]
34. Sivasankarapillai, V.S.; Das, S.S.; Sabir, F.; Sundaramahalingam, M.A.; Colmenares, J.C.; Prasannakumar, S.; Rajan, M.; Rahdar, A.; Kyzas, G.Z. Progress in natural polymer engineered biomaterials for transdermal drug delivery systems. *Mater. Today Chem.* **2021**, *19*, 100382. [CrossRef]
35. Yadav, K.; Soni, A.; Singh, D.; Singh, M.R. Polymers in topical delivery of anti-psoriatic medications and other topical agents in overcoming the barriers of conventional treatment strategies. *Prog. Biomater.* **2021**, *10*, 1–17. [CrossRef]
36. Latha, A.V.S.; Ravikiran, T.N.; Kumar, J.N. Formulation, Optimization and Evaluation of Glibenclamide Transdermal Patches by using chitosan Polymer. *Asian J. Pharm. Technol.* **2019**, *9*, 1–7. [CrossRef]
37. Pünnel, L.C.; Lunter, D.J. Film-forming systems for dermal drug delivery. *Pharmaceutics* **2021**, *13*, 932. [CrossRef]
38. Azad, A.K.; Al-Mahmood, S.M.; Chatterjee, B.; Wan Sulaiman, W.M.; Elsayed, T.M.; Doolaanea, A.A. Encapsulation of black seed oil in alginate beads as a ph-sensitive carrier for intestine-targeted drug delivery: In vitro, in vivo and ex vivo study. *Pharmaceutics* **2020**, *12*, 219. [CrossRef]
39. Altun, E.; Yuca, E.; Ekren, N.; Kalaskar, D.M.; Ficai, D.; Dolete, G.; Ficai, A.; Gunduz, O. Kinetic Release Studies of Antibiotic Patches for Local Transdermal Delivery. *Pharmaceutics* **2021**, *13*, 613. [CrossRef]
40. Haroon, M.; Batool, S.; Asif, S.; Hashmi, F.; Ullah, S. Combination of Methotrexate and Leflunomide Is Safe and Has Good Drug Retention Among Patients with Psoriatic Arthritis. *J. Rheumatol.* **2021**, *48*, 1–2. [CrossRef]
41. Sabir, F.; Qindeel, M.; Rehman, A.U.; Ahmad, N.M.; Khan, G.M.; Csoka, I.; Ahmed, N. An efficient approach for development and optimisation of curcumin-loaded solid lipid nanoparticles' patch for transdermal delivery. *J. Microencapsul.* **2021**, *38*, 233–248. [CrossRef] [PubMed]

polymers

MDPI

Article

Surface Characterization and Physiochemical Evaluation of P(3HB-*co*-4HB)-Collagen Peptide Scaffolds with Silver Sulfadiazine as Antimicrobial Agent for Potential Infection-Resistance Biomaterial

Sevakumaran Vigneswari [1], Tana Poorani Gurusamy [2], Wan M. Khairul [1], Abdul Khalil H.P.S. [3], Seeram Ramakrishna [4] and Al-Ashraf Abdullah Amirul [2,5,6,*]

1 Faculty of Science and Marine Environment, Universiti Malaysia Terengganu, Kuala Terengganu 21030, Terengganu, Malaysia; vicky@umt.edu.my (S.V.); wmkhairul@umt.edu.my (W.M.K.)
2 School of Biological Sciences, Universiti Sains Malaysia, Gelugor 11800, Penang, Malaysia; purani_guru@yahoo.com
3 School of Industrial Technology, Universiti Sains Malaysia, Gelugor 11800, Penang, Malaysia; akhalilhps@gmail.com
4 Center for Nanofibers and Nanotechnology, Department of Mechanical Engineering, National University of Singapore, Singapore 117581, Singapore; seeram@nus.edu.sg
5 Centre for Chemical Biology, Universiti Sains Malaysia, Bayan Lepas 11900, Penang, Malaysia
6 Malaysian Institute of Pharmaceuticals and Nutraceuticals, NIBM, Gelugor 11700, Penang, Malaysia
* Correspondence: amirul@usm.my

Citation: Vigneswari, S.; Gurusamy, T.P.; Khairul, W.M.; H.P.S., A.K.; Ramakrishna, S.; Amirul, A.-A.A. Surface Characterization and Physiochemical Evaluation of P(3HB-*co*-4HB)-Collagen Peptide Scaffolds with Silver Sulfadiazine as Antimicrobial Agent for Potential Infection-Resistance Biomaterial. *Polymers* **2021**, *13*, 2454. https://doi.org/10.3390/polym13152454

Academic Editors: Ariana Hudita and Bianca Gălăţeanu

Received: 4 July 2021
Accepted: 20 July 2021
Published: 26 July 2021

Abstract: Poly(3-hydroxybutyrate-*co*-4-hydroxybutyrate) [P(3HB-*co*-4HB)] is a bacterial derived biopolymer widely known for its unique physical and mechanical properties to be used in biomedical application. In this study, antimicrobial agent silver sulfadiazine (SSD) coat/collagen peptide coat-P(3HB-*co*-4HB) (SCCC) and SSD blend/collagen peptide coat-P(3HB-*co*-4HB) scaffolds (SBCC) were fabricated using a green salt leaching technique combined with freeze-drying. This was then followed by the incorporation of collagen peptides at various concentrations (2.5–12.5 wt.%) to P(3HB-*co*-4HB) using collagen-coating. As a result, two types of P(3HB-*co*-4HB) scaffolds were fabricated, including SCCC and SBCC scaffolds. The increasing concentrations of collagen peptides from 2.5 wt.% to 12.5 wt.% exhibited a decline in their porosity. The wettability and hydrophilicity increased as the concentration of collagen peptides in the scaffolds increased. In terms of the cytotoxic results, MTS assay demonstrated the L929 fibroblast scaffolds adhered well to the fabricated scaffolds. The 10 wt.% collagen peptides coated SCCC and SBCC scaffolds displayed highest cell proliferation rate. The antimicrobial analysis of the fabricated scaffolds exhibited 100% inhibition towards various pathogenic microorganisms. However, the SCCC scaffold exhibited 100% inhibition between 12 and 24 h, but the SBCC scaffolds with SSD impregnated in the scaffold had controlled release of the antimicrobial agent. Thus, this study will elucidate the surface interface-cell interactions of the SSD-P(3HB-*co*-4HB)-collagen peptide scaffolds and controlled release of SSD, antimicrobial agent.

Keywords: P(3HB-*co*-4HB); silver sulfadiazine; collagen peptide; infection-resistance scaffolds

1. Introduction

Biomaterial scaffolds are materials which have been engineered to interact with our biological system in providing three-dimensional structure and mimicking an extracellular matrix (ECM). Therefore, it is crucial to design biologically active scaffolds with well interconnected configuration and surface chemistry to enhance the cellular interactions on the scaffold interface [1,2]. The scaffold interface would enhance and facilitate the cell infiltration, proliferation and differentiation of cell lines, and eventually contribute to the tissue regeneration.

Polyhydroxyalkonates (PHAs) are insoluble granules accumulated in cell cytoplasm as carbon and energy storage compounds under stress conditions [3–5]. PHAs are a biodegradable thermoplastic which exhibit similar thermo-mechanical properties to synthetic polymers [6]. Among the variety of PHAs, copolymer P(3HB-*co*-4HB) is widely used in biomedical applications due to the non-toxic biodegradation products, wide range of physical and mechanical properties, non-carcinogenic effects and biocompatibility [7]. It possesses exceptional properties for medical and pharmaceutical fields [8,9]. Moreover, P(3HB-*co*-4HB) has Food and Drug Administration (FDA) clearance for clinical usages among all the other PHAs available [3]. The P(3HB-*co*-4HB) was biosynthesized by bacterium *Cupriavidus necator* (formally *Ralstonia eutropha*) from structurally related sources such as 4-hydroxybutyric acid (4HBA), 4-chlorobutyric and γ-butyrolactone [7].

However, P(3HB-*co*-4HB) lacks active functional sites for cell attachment which limits the applications for regenerative medicine. Many studies have been carried out in this direction to overcome this limitation. Therefore, surface modification is carried out by incorporating natural polymers, such as collagen, gelatin, pullulan and chitosan, in enhancing the hydrophilicity of the scaffolds [10]. Nevertheless, the desirability and wide applicability of collagen is often attributed to its abundance in the human body as the key structural fibrous protein of the ECM [11]. Hence, collagen peptide was used as the biomolecules to enhance the hydrophilicity of the scaffolds fabricated in our study. Collagen peptide is a biomolecule which not only has the ability to improve the hydrophilicity of the scaffold but has the natural ability to interact with host cells [12,13].

Biomaterial scaffold-affiliated microbial infections are an emerging threat in clinical practices, which cause serious infection and impact healing. Therefore, designing scaffolds with antimicrobial efficacy have extensively gained priority in resolving biomaterial-associated infections [14]. Silver sulfadiazine (SSD) is an antibacterial agent that exhibits broad-spectrum antibacterial activity against Gram-positive and Gram-negative bacteria, as well as fungi, even at very low concentrations [15–17]. SSD is a much preferred antibacterial agent of choice due to the ability of SSD to reduce early infections at low concentration. However, currently available formulations of antimicrobial agents lack the ability to control the release of antimicrobial properties [18,19]. There are many scaffolds developed with antimicrobial properties and Table 1 lists common examples of antimicrobial biopolymer incorporated with SSD.

Table 1. List of common examples of various antimicrobial scaffolds incorporated with SSD.

Biopolymer/Materials	Fabrication of Scaffolds	Applications	References
Collagen/SSD	Facile blending	Wound dressings	[19]
Collagen/SSD	Electrospinning	Wound healing applications	[15]
Collagen/SSD	Blending with SSD-loaded alginate microspheres	Conventional burn dressings in second-degree burns	[16]
Polycaprolactone (PCL)/SSD	Electrospinning	Antibacterial scaffold	[20]
P(3HB-*co*-4HB)/collagen peptide/SSD	Aminolysis	Potential wound healing	[9]
Polycaprolactone (PCL) and Polyvinyl alcohol (PVA)/SSD	Electrospinning	Antimicrobial wound dressing	[21]
Poly(lactic acid) (PLA)/SSD	Electrospinning, structural reconstruction	Antimicrobial wound dressing	[22]

Following the aforementioned background, in the present work, the surface architecture of P(3HB-*co*-4HB) was enhanced by incorporating collagen peptides and silver sulfadiazine (SSD) as the antimicrobial mechanism agent. Two different scaffolds, namely SSD coat/collagen peptide coat-P(3HB-*co*-4HB) [SSCC] and SSD blend/collagen peptides coat-P(3HB-*co*-4HB) [SBCC] scaffold, were fabricated by the combination of salt leaching and freeze-drying techniques which are low cost and apply green technology to fabricate the scaffolds. The study provides evidence for increased hydrophilicity due to the incorporation of collagen peptide. This elucidates surface interface-cell interactions of the modified P(3HB-*co*-4HB) scaffolds and release mechanism of the antimicrobial agent from the scaffolds, thus driving the research effort forward for emerging infection-resisting biomaterials in tissue engineering and regenerative medicine in the future.

2. Materials and Methods

2.1. Biosynthesis of P(3HB-co-95 mol% 4HB) Copolymer

The bacteria strains used in this study were *Cupriavidus malaysiensis* USMAA1020 transformant harbouring additional PHA synthase gene from *Cupriavidus malaysiensis* USMAA2–4 to produce P(3HB-*co*-95 mol% 4HB) copolymer. The biosynthesis was carried out as previously described [23]. A preculture of 5% (*v/v*) of the working volume was transferred into 20 L fermenter (Biostat® C plus, Sartorius Stedim, German) containing mineral salts medium (MSM) with carbon precursors (1,4-butanediol and 1,6-hexanediol in the 1:5 ratio). The fermentation was carried out at 30 °C with an agitation speed of 200 rpm, the aeration rate of 1 vvm and controlled pH of 7 for 108 h. Sampling was done at intervals of every 12 h. The composition of PHA produced was determined by gas chromatography (GC) using Shimadzu Gas Chromatography GC-2014 according to methods previously described [24]. Endotoxin removal was carried out on extracted P(3HB-*co*-95 mol% 4HB) copolymer as previously described. The extracted polymer was characterized based on the molecular weight using Shimadzu GPC-2014 and tensile test using tensile testing machine (GoTech Al-3000, Shimadzu, Japan) [24].

2.2. Surface Functionalization of SSD/Collagen Peptide-P(3HB-co-4HB) Scaffolds

Surface functionalization of P(3HB-*co*-4HB) was carried out by salt leaching and solvent casting technique followed by freeze-drying method. Briefly, P(3HB-*co*-4HB) copolymer was dissolved in chloroform (5.5% w/v) and sodium bicarbonate ($NaHCO_3$) particles sieved with known mesh sizes (200 μm) were added as porogen with mass ratio of salt:polymer at 6:1. The resulting polymer matrix was washed with deionized water to leach out the porogens. The scaffolds were freeze-dried for 24 h and later vacuum-dried for 48 h (BINDER GmbH, Tuttlingen, Germany) to remove any remaining solvent.

There were two types of scaffolds prepared using the various functionalization combination methods by incorporating different concentration of collagen peptide (2.5 wt.%, 5 wt.%, 7.5 wt.%, 10 wt.%, 12.5 wt.%) and 0.04% (w/v) of SSD. Collagen peptide powder from Tilapia fish skin with high purity (95%) and molecular weight of less than 3000 Da was used (Hainan Zhongxin Chemical Co. Ltd., Haikou, China).

The SSD coat/collagen peptide coat-P(3HB-*co*-4HB) scaffold (SCCC) was prepared by coating different concentration of collagen peptide in the silver (I) sulfadiazine (Sigma Aldrich) dispersed in hydrochloric acid solution (1.0 mM, pH 3.0).

The preparation of SSD blend/collagen peptide coat-P(3HB-*co*-4HB) scaffold (SBCC) was prepared with SSD added into the dissolved P(3HB-*co*-4HB) with $NaHCO_3$ porogen and then solvent cast, as mentioned above.

Cross-linking was carried out using GA vapor-phase technique where the scaffolds were placed in an airtight desiccator containing 25% aqueous GA solution heated to 100 °C. Subsequently, the samples were washed for 24 h to remove GA, and then dried in vacuum for 24 h [8,25]. The scaffolds will be known as SCCC and SBCC from here on.

2.3. Characterization of Scaffolds

The functional group present in the scaffolds fabricated were determined and analyzed using FTIR-ATR spectrophotometer (Model RX1, PerkinElmer, Buckinghamshire, UK). The spectra of the samples were obtained in the range of wave number between 650 cm^{-1} and 4000 cm^{-1}. The spectrum of the FTIR was recorded in transmittance mode as function of wave number and the results were computed after 4 automated scans [24]. The surface morphology of the scaffolds coated with gold were mounted on aluminium stump and was observed using scanning electron microscopy (SEM) (Leo Supra 50 VP Field Mission SEM, Carl-Ziess SMT, Oberkochen, Germany). The scaffolds were cut into 1 cm × 1 cm. The dry weight before immersion (m_o) was used as the initial weight of the scaffolds. The scaffolds were immersed in distilled water for 24 h. In order to obtain the wet weights (m_f), the immersed scaffolds were removed from the solution, gently wiped with absorbent paper and air-died for 15 s before weighing. Water uptake was calculated using the formula below:

$$Water\ uptake = \left(m_o - m_f\right)\ /\ m_o \times 100\% \quad (1)$$

The contact angle of the fabricated scaffolds was conducted by using sessile drop method (KSV CM200 Contact Angle) to determine their wettability properties. The scaffolds were cut into 1 cm × 1 cm pieces. The scaffolds were placed on the instrument and the droplet of water was then deposited on the polymer surface by a specialized microsyringe. The water droplet was observed from the computer screen and the contact angle was calculated. The porosity of the scaffolds was calculated using Image Analyser Software (Olympus Co. Ltd., Tokyo, Japan). The values of 100 different spots were analyzed and averaged [8].

2.4. Antimicrobial Activity

Four bacterial strains, which include *Bacillus licheniformis*, *Staphylococcus aureus* ATCC 12600, *Escherichia coli* ATCC 11303 and *Pseudomonas aeruginosa* ATCC 17588, were used. Briefly, the tested bacterial suspensions (1.5×10^6 CFU/mL) were transferred in sterilized nutrient broth. Then, 20 µL of the bacteria suspension (7.5×10^5 CFU/mL) was added to each antimicrobial coated porous scaffold. The incubation is done under suitable conditions for varied time intervals (0, 6, 12 and 24 h). In every 6 h interval, the scaffold with bacteria adhesion was dissolved in 10 mL of distilled water and vortexed. After that, 100 µL of the bacterial suspension was spread on nutrient agar to observe the colonization of bacteria. The percentage of dead cells is calculated relatively to the growth control by determining the number of living cells (CFU/mL) of each scaffold using the agar plate count method. The percentages of inhibition were calculated using following Equation:

$$C\% = (Co - Ce)\ /\ Co \times 100\% \quad (2)$$

where $C\%$ is percentage of inhibition, Ce is CFU after incubation period and Co is initial CFU before incubation period.

2.5. Biocompatibility and Cell Proliferation Evaluation

Mouse fibroblast cell culture (L929, ATCC) was cultured in cell culture flasks containing Modified Eagle Medium (MEM) supplemented with 2 mM L-glutamine, 1.5 g/L sodium bicarbonate, 1 mM of sodium pyruvate, 1000 U/mL penicillin-streptomycin and 10% (*v/v*) of bovine calf serum, which were incubated at 37 °C in 5% (*v/v*) CO_2 for 2–3 days. The various scaffolds fabricated and its positive control (P(3HB-*co*-4HB) without collagen were cut in size (6 mm in diameter) fitting the 96-well flat bottom culture plate and sterilized under UV cross-linker (Spectrolinker™, XL-1000 UV Cross-linker, Westbury, NY, USA) at 1200 µJ/cm^2 for 30 min [8,9]. The scaffolds were then placed in the 96-well flat bottom culture plate. Suspension of the mouse fibroblast cell lines (L929) [2.5×10^4 cells/mL] were directly cultivated onto the scaffolds and film. The seeded scaffolds and film were incubated in a 5% (*v/v*) CO_2 incubator at 37 °C for 96 h. The cells viability and proliferation were assayed with MTS[3-(4,5-dimethylthiazol-2-yl)-5-(3-carboxymethoxyphenyl)-2-(4-sulfophenyl)-2H-tetrazolium/PMS (phenazinmethosulfate). MTS and PMS solution were used to evaluate the biocompatibility of the fabricated of scaffolds. Standard curve was plot based on the cell density from the range of 1×10^3 to 5×10^5 cells/mL. The media was used as the positive control and scaffolds without any incorporation of collagen peptide were used as negative control. The absorbance values were plotted against the counted cell numbers, and thus a standard curve was established [9].

2.6. Statistical Analysis

The qualitative results were presented as means and standard deviation (s.d). The qualitative data were analyzed using ANOVA and Tukey's HSD test with SPSS 20.0 software. All p values < 0.05 were considered significant.

3. Results and Discussion

3.1. Biosynthesis of P(3HB-co-4HB) via Batch Fermentation

The biosynthesis of P(3HB-*co*-4HB) copolymer was carried out using *Cupriavidus malaysianesis* USMAA1020 transformant, which possessed an excess copy of the phaC gene. This cultivation regulated 4HB molar fraction to achieve 95 mol% of P(3HB-*co*-4HB) with PHA content of 78 wt.% and its concentration at 17.3 g/L in 20 L bioreactor, the mixed substrates of 1,6-hexanediol and 1,4-butanediol at 1:5 ratio. The high 4HB monomers are favored for implantable medical products. This was in agreement with the previous study [23], the 1,6-hexanediol and 1,4-butanediol were utilized as carbon sources as 4-hydroxybutyryl-CoA was initially formed and converted to 4-hydroxybutyrate. The copolymer was recovered by the chloroform extraction method and subjected to physical properties. Based on the results obtained, as summarized Table 2, the average molecular weight (MW) of the polymer was 585 kDa while the polydispersity index was in the range 3.2. Besides, the tensile strength of the polymer was recorded at about 23 MPa with the elongation at break around 611%.

Table 2. Physical and mechanical properties of P(3HB-*co*-4HB).

Copolymer	Tensile Strength (MPa) [a]	Elongation at Break (%) [a]	Young Modulus (Mpa) [a]	M_w (kDa) [b]	M_n (kDa) [b]	PDI [b]
P(3HB-*co*-95 mol% 4HB)	23.2 ± 4	611.8 ± 1	226.6 ± 20	585 ± 8	132 ± 11	3.2 ± 0.5

Values are mean ± SD of three replicates; [a] Determined using Gotech Al-3000 tensile Machine; [b] Calculated from GPC analysis, M_n: number-average molecular weight; M_w: weight average molecular weight; M_w/M_n: polydispersity index.

3.2. Fabrication of SBCC and SCCC Scaffolds

In this study a three-dimensional, porous scaffold was successfully engineered with the use of a combination of techniques, namely particle leaching and freeze-drying. Figure 1 shows a schematic of the fabrication of porous antimicrobial SSD-P(3HB-*co*-4HB)-collagen peptide scaffolds termed as SBCC and SCCC scaffolds. The system contained two phases in developing a highly porous, well interconnected pore structure of the scaffold. The first phase involved the particle leaching using $NaHCO_3$ (200 µm), followed by the freeze-drying technique. The combination of methods has shown many advantages over other methods as it is easier to control pore structures. This will produce porous scaffolds with open surface pores and interconnected bulk pores which will facilitate cell seeding and homogeneous cell distribution and promote tissue regeneration [26,27]. Despite the homogenous pores' structures, the surface properties of these polymers are hydrophobic which will possibly inhibit the infiltration of cell suspension into the scaffolds preventing smooth cell seeding.

Therefore, it is crucial to modify the surface characteristic from hydrophobic to hydrophilic to facilitate cell seeding. In this case, the surface of the porous P(3HB-*co*-4HB) scaffolds was coated with hydrophilic collagen peptide to increase the hydrophilicity of the surface, thus improving cell interaction [28–30]. In this study, apart from the surface modification of P(3HB-*co*-4HB) scaffold with collagen layer, incorporation of antimicrobial agent, SSD was carried out. This was executed by introducing the SSD either through blending (SBCC) or coating of the scaffolds (SCCC). Fabrication of scaffolds that release the antimicrobial agents or respond to infections is crucial in developing biomaterials in tissue engineering [31–33].

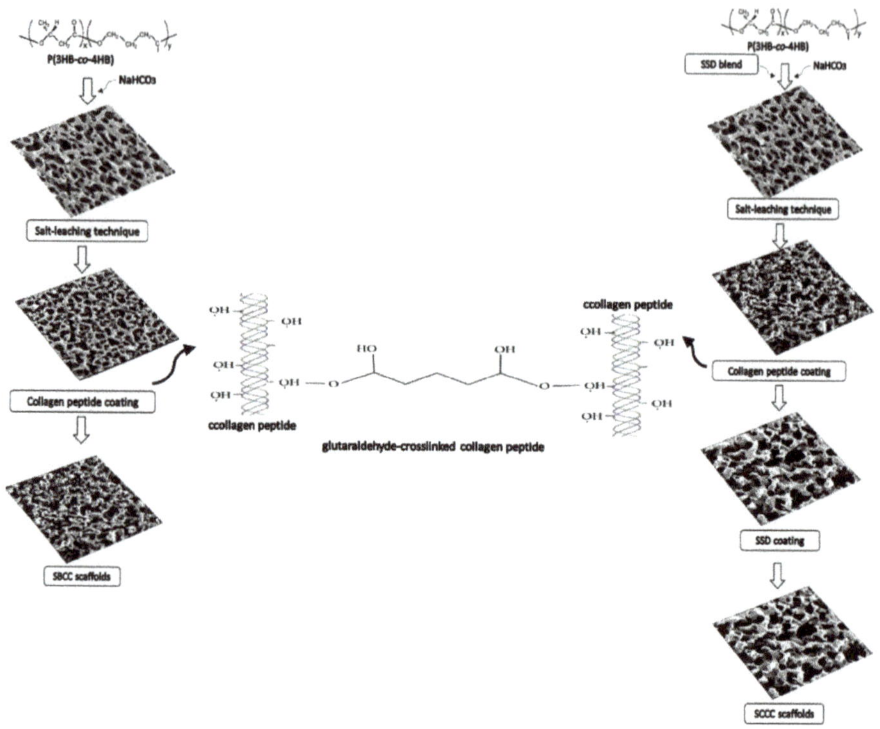

Figure 1. Schematic of fabrication of SBCC and SCCC scaffolds using a combination of salt leaching modification and freeze-drying technique with collagen peptide coating and cross-linking with glutaraldehyde.

Besides, surface morphology is crucial in developing biomaterials as this determines the cell-matrix interface interactions. As seen in Figure 2, the SEM micrographs reveal the formation of the three-dimensional interconnected porous structure of SBCC and SCCC scaffolds. Interestingly, the pore sizes observed using SEM were much smaller than the range of porogen sizes ($NaHCO_3$) used to create them. This could be attributed to the combination of techniques used, mainly freeze-drying. Hence, combining salt leaching with freeze-drying may enhance pore interconnectivity and assist the formation of homogenous pores ranging from 100 to 200 μm [34]. However, SCCC scaffolds exhibited rougher appearance with less interconnection and possessed numerous macropores as compared to SBCC. Basically, the porous-based connectivity surface is favored to enhance the ECM architecture and provide a larger space to induce cell-material interactions [35–39]. Additionally, both the scaffolds created similar morphology with generally amorphous pores with smooth edge.

The fabricated scaffolds differed in terms of their construction. SSD coated onto the porous SCCC scaffolds showed solubility in aqueous medium in contrast to the impregnated collagen in SBCC. Hence, the collagen peptide coated P(3HB-*co*-4HB) porous scaffold were then cross-linked via GA vapor phase. The dissolution analysis on scaffolds of cross-linking and uncross-linking scaffolds were shown in Figure 3. The percentage of dissolutions was significantly higher for uncross-linked SCCC scaffolds with the amount retained only between 15% and 45%. On the contrary, the crosslinked SSD/collagen peptide-coated P(3HB-*co*-4HB) scaffold exhibited collagen retain percentage from up to 80%

to 90%. This demonstrates that cross-linking with GA enhanced the scaffolds resistance to dissolution. After GA vapor cross-linking, the membranes became visibly yellowish and shrunk dimensionally. The aldimine linkages (CH=N) between the free amine groups of protein and GA attributes to the color change, whereas the covalent bond formed between the aldehyde groups of GA caused shrinkage [40,41]. The aldimine linkage was a reflection of Schiff base reaction, whereby the carbon in the aldehyde group of GA was attacked by nucleophilic nitrogen in the amino group of collagen peptides, and hence replaced the oxygen in the aldehyde group and eliminated water molecule [30]. Nonetheless, GA cross-linking with vapor phase methodology showed low or no detectable cytotoxic effects [25,42]. As described by Teixera et al. (2021), this will enable a sustainable approach in achieving green methodology and the lowest environment impact possible at all stages of fabrication for biomedical application [43].

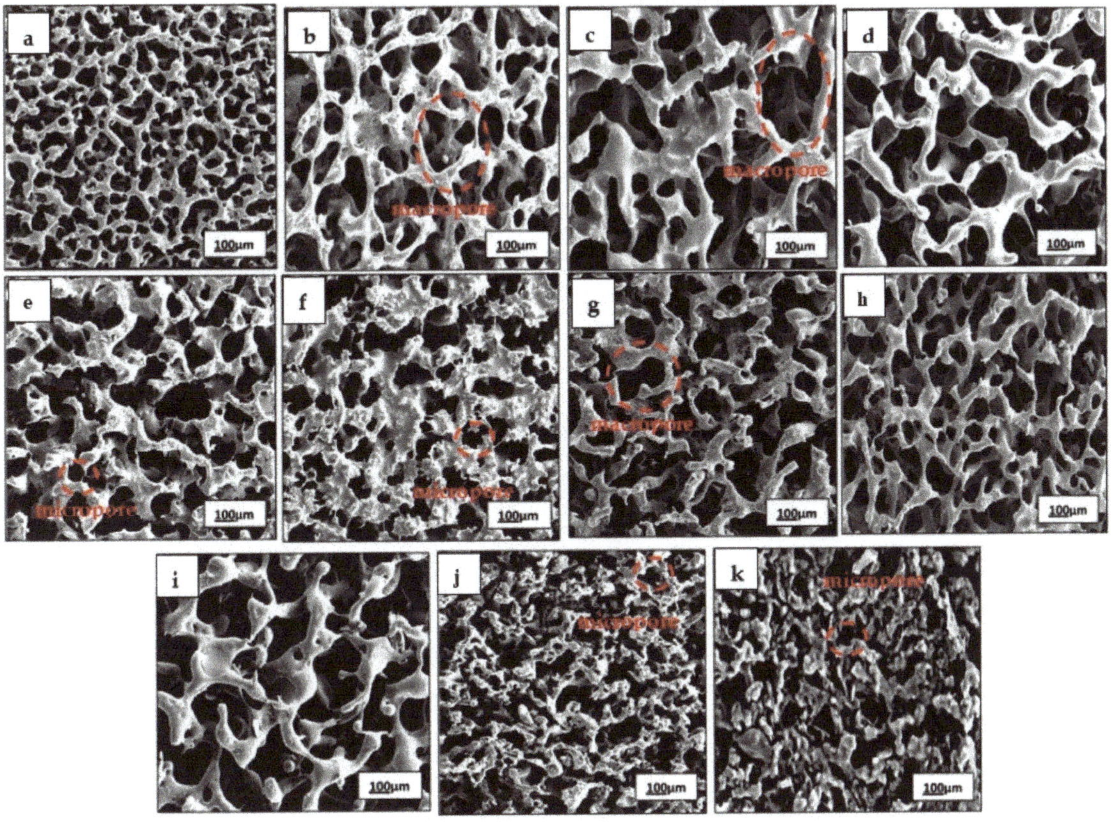

Figure 2. Micrograph of porous structure of (**a**) control-P(3HB-*co*-4HB), (**b**) SCCC 2.5 wt.%, (**c**) SCCC 5 wt.%, (**d**) SCCC 7.5 wt.%, (**e**) SCCC 10 wt.%, (**f**) SCCC 12.5 wt.%, (**g**) SBCC 2.5 wt.%, (**h**) SBCC 5 wt.%, (**i**) SBCC 7.5 wt.%, (**j**) SBCC 10 wt.% and (**k**) SBCC 12 wt.%.

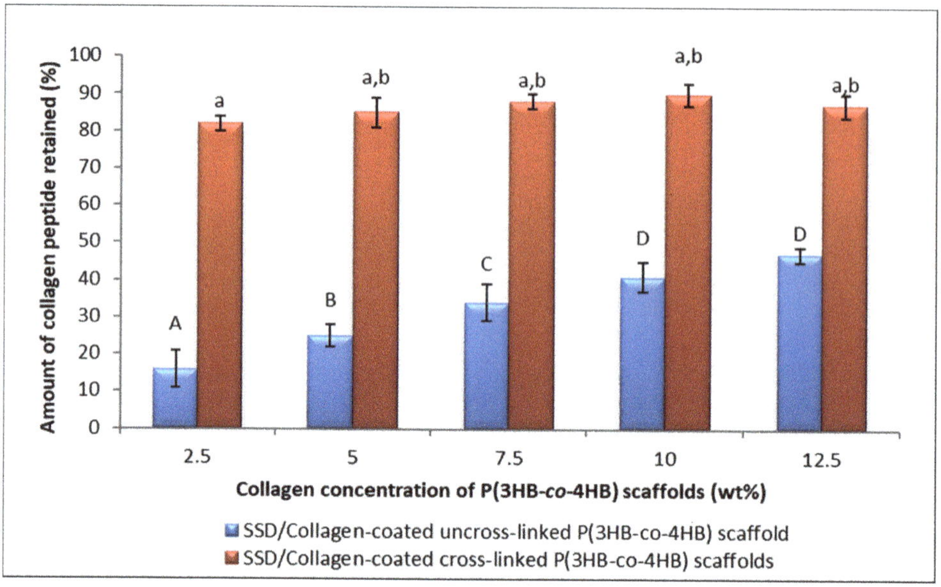

Figure 3. The percentage of collagen retained on cross-linked SCCC and uncross-linked SCCC scaffold. Data represent means ± SD (n = 3). Mean data accompanied by different alphabets as of cross-linked SCCC scaffolds (a–b); and uncross-linked SCCC (A–D) indicates significant difference within each respective group (Tukey's HSD test, $p < 0.05$).

3.3. Functional Group Identification Using FTIR Analysis

FTIR analysis shown in Figure 4 was carried as an evidential analysis to determine and analyze the characteristic bands that correlate to functional groups of the fabricated scaffolds. The FTIR spectrum for collagen (a) showed symmetric and unsymmetric stretching of the primary amine (NH_2) bands at 3275 cm^{-1} and 3150 cm^{-1}, respectively. The hydroxyl (OH) from carboxylic acid portion also is expected to be overlapped with the symmetric amine at 3275 cm^{-1}. The moderate peaks at 2937 cm^{-1} represent CH_3 (bend) and CH_2 (stretch) of the alkanes' substructure. A strong band at 1633 cm^{-1} represents (C=O) from the amide moiety. Another strong peak can be seen at 1531 cm^{-1} and represents NH_2 bending [44–48]. The peaks of the (C=C) bands of the aromatic portion also can be clearly observed between the peaks of 1531 cm^{-1} to 1449 cm^{-1}. In addition, a moderate peak at 920.89 cm^{-1} would represent a C-H (out-of-plane) band from the aromatics.

In the case of P(3HB-*co*-4HB) polymer (b), moderate peaks at 2963 cm^{-1} and 2899 cm^{1} represent CH_3 (bend), CH_2 (stretch) and CH of the alkanes' substructure. A strong band at 1633 cm^{-1} represents (C=O) and another strong band at 1161 cm^{-1} exhibits the (C-O) band [8,49].

Comparatively, the FTIR spectra of SBCC (c) and SCCC (d) are rather comparable to each other as they exhibit all the expected bands and peaks of the designated collagen, P(3HB-*co*-4HB) polymer and pure SSD. The major characteristic absorption peaks in both FTIR spectra of SBCC (c) and SCCC (d) ca. 3283, 3150, 2900, 1719, 1630, 1540, 1450 and 1164 cm^{-1}. The absorption peak at 3283 and 3150 cm^{-1} are assigned to NH_2 symmetric and asymmetric stretching, respectively. A distinctive peak at 2900 cm^{-1} represents CH_3 (bend) and CH_2 (stretch) of the alkanes' substructure. Whilst the strong peak at 1719 cm^{-1} represents (C=O) peak. The absorption peak at 1630 cm^{-1} corresponds to NH_2 bending. The peaks at 1540 cm^{-1}, 1450 cm^{-1} belong to the peaks of the (C=C) bands of the aromatic portion. The peaks of asymmetric stretching vibration of (SO_2) group cannot be resolved in these spectra as the band of (C-O) can be dominantly seen in this fingerprint region at 1164 cm^{-1} [50].

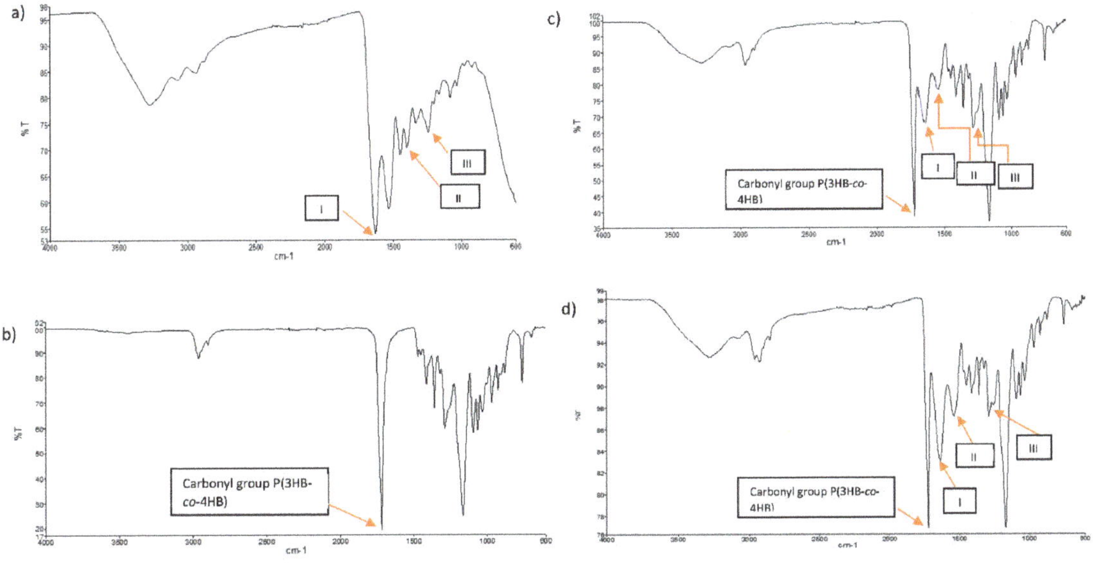

Figure 4. FTIR spectra of scaffolds (**a**) collagen, (**b**) P(3HB-*co*-4HB), (**c**) SBCC, and (**d**) SCCC. Arrows I, II, III indicate amide I, amide II and amide III, respectively.

It was observed that the prominent characteristic peaks of SBCC (c) and SCCC (d) with a few bands shift in comparison to each other with the dominant characteristics are from the P(3HB-*co*-4HB) polymer (b) which are indicative of the reservation of the chemical aspect of these blends production. It can be concluded that the SSD did not engage with its active groups in any chemical interaction with any of the components of SBCC (c) and SCCC (d) built up. From the FTIR spectra of the two, there is also no evidence of electrostatic interaction nor chemical reaction have taken place between all the materials that made up the blend due to very little shift of all the vibrational wavenumbers (i.e., less than 5 cm^{-1}) throughout the major bands of interest.

3.4. Porosity Analysis

The pores in scaffolds are imperative as they provide an ideal framework for cells to bind, proliferate and form extracellular matrix. As such, here the porosity was determined with six different collagen concentrations of the scaffolds. The fabricated scaffolds exhibited a gradual drop of the pore size from 145 to 53 µm with increasing collagen concentrations (Figure 5). Similarly, the porosity of the SBCC declined by 50% from the control scaffold. This decrease in porosity could have been due to the larger collagen layer deposits on the surface of scaffolds [51–55]. Based on various studies, pore sizes above 100 µm are ideal for cell infiltration and migration. Interestingly, 10 wt.% scaffolds resulted in a desirable pore size despite the higher concentration of collagen peptide. In developing biomaterial, pore structures of scaffolds play a crucial role in facilitating cell seeding, cell penetration and distribution in the scaffolds. Thus, the adhesion of cells and formation of new tissues and organs occurs [56–58]. It is emphasized that an ideal scaffold depends on biomaterial source, fabrication technique and the pore geometry. As such, it is vital to develop a scaffold with specific porosity properties for potential application in tissue engineering and regenerative medicine [59,60].

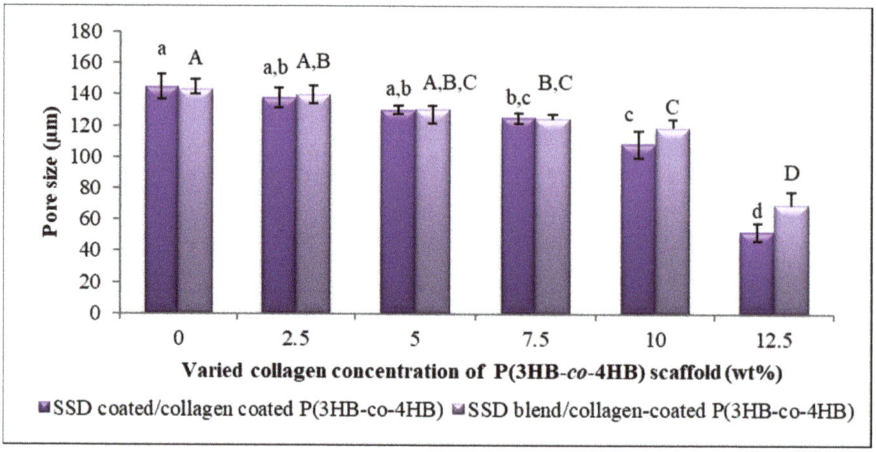

Figure 5. Porosity analysis of SSD coated/collagen coated P(3HB-*co*-4HB); SCCC and SSD blend/collagen coated P(3HB-*co*-4HB); SBCC scaffolds. Data represent means ± SD ($n = 3$). Mean data accompanied by different alphabets as of SCCC scaffolds (a–d) and SBCC scaffolds (A–D) indicates significant difference within each respective group (Tukey's HSD test, $p < 0.05$).

3.5. Hydrophilicity of Fabricated Scaffolds

The hydrophilicity of the SCCC and SBCC scaffolds was determined using water contact angle analysis (Table 3). The graph clearly showed a decline in the contact angle as the concentration of collagen peptide increases, thus indicating the increase of hydrophilicity. Ideally, a contact angle of less than 90° indicates that the surface is wet-prone, hence being categorized as a hydrophilic surface [61–63]. Whole wetting was observed with the water droplet becoming a flat puddle with 0° contact angle on SBCC and SCCC with 10 wt.% and 12.5 wt.% collagen peptides. Additionally, the collagen peptide coating enhanced the surface wettability of sample scaffolds. The significant hydrophilicity enhancing effect of collagen peptide could be associated with the amino groups in collagen [64,65].

The wettability analysis of different sample collagen concentrations is demonstrated in Figure 6. A steady rise of water uptake percentage with the increment of collagen peptide concentrations can be observed. Water uptake ability elucidates the hydrophilicity of fabricated scaffolds which will increase the efficiency of absorption of essential supplements required for cell attachment. Overall, the collagen peptide coated P(3HB-*co*-4HB) scaffold absorbed a larger amount of water, exceeding 100% (*v*/*v*) of the total volume of the scaffold even at the low concentrations of collagen peptide (2.5 wt.%). As anticipated, the results pointed out that the hydrophilicity of both SCCC and SBCC scaffolds have similar water uptake ability. The water uptake ability properties of scaffolds are crucial in order to enhance the proliferation of a cell. The optimal design of a scaffold strongly depends on both materials and the surface treatment in modulating cell seeding and proliferation [60,66].

3.6. Evaluation of Cell Proliferation of Fibroblast Cells on Scaffolds

In general, a functional scaffold requires the ability to support attachment and promote proliferation of cultured cells [67]. In line with it, the L929 fibroblasts cells behavior towards SCCC and SBCC scaffolds with different collagen concentrations was investigated as shown in Figure 7. Cells adhered well with progressive growth and by day three, the scaffold surfaces supported high cell density. The cell proliferation was spotted to increase significantly on scaffold coated with 2.5 wt.% until it reaches 10 wt.% as compared to the collagen free scaffold. However, the number of fibroblast cells decreased (10.6×10^5 cells/mL) at the highest collagen peptide concentration (12.5 wt.%). This may be attributed to the reduction of pore size, which caused less pore accessibility and proliferation [8,30,68,69].

In the current study, 10 wt.% collagen coated scaffold with pore size around 108.6 ± 8.7 µm demonstrated highest proliferation rate (12.4 × 10^5 cells/ mL), as shown in Figure 8, in comparison to control, as well as 2.5 wt.%, 5 wt.% and 7.5 wt.% collagen coated scaffolds. In short, scaffolds fabricated using combined techniques displayed the highest cell proliferation. These findings clearly implied the enhancement of cell proliferation attributes to the effects of collagen on cell viability. In short, these findings clearly demonstrated the process of incorporating collagen layer on the scaffold is an efficient way to initiate cell attachment and supports cell growth [63,64].

Table 3. Water contact angle of scaffolds with various collagen peptide concentration.

Collagen Peptide (wt.%)	Types of Scaffolds	
	Coat/Coat	Blend/Coat
0	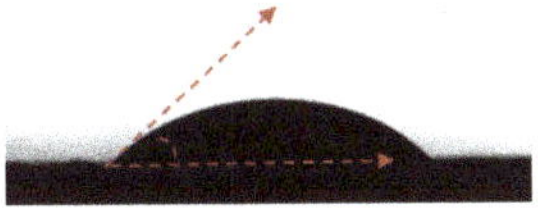 49.9 ± 2.7	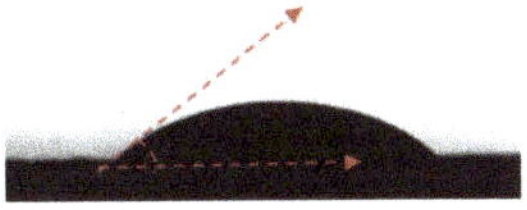 73.3 ± 1.4
2.5	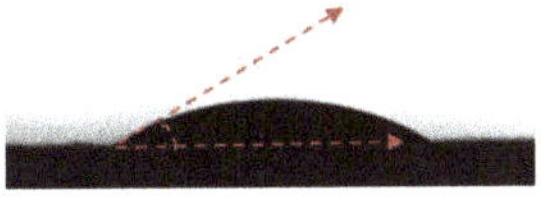 32.5 ± 2.8	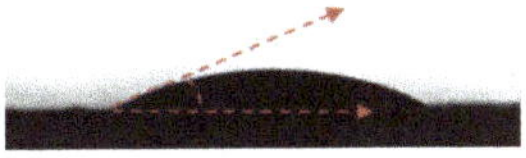 45.2 ± 2.4
5.0	 15.3 ± 3.3	 25 ± 5.4
7.5	 8.58 ± 0.8	 14.81 ± 1.2
10.0	 0	 0
12.5	 0	 0

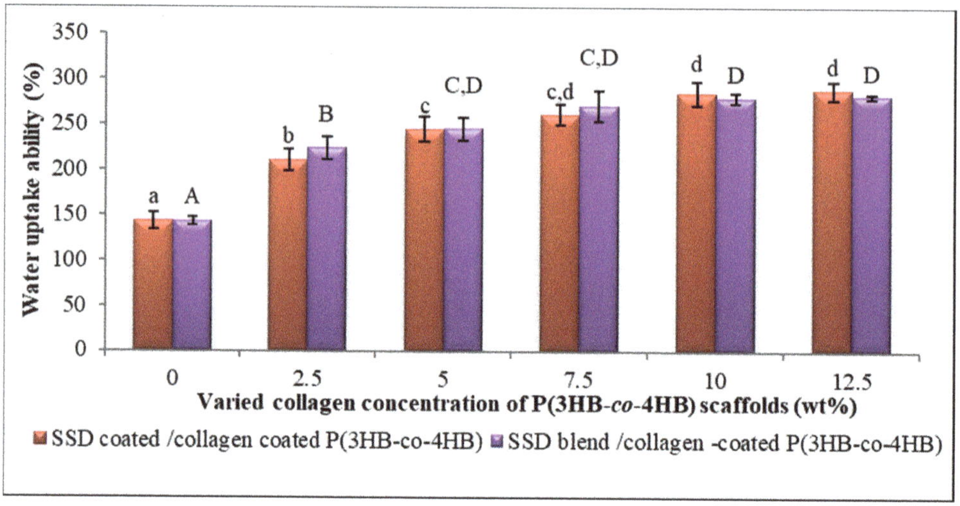

Figure 6. Water uptake analysis of SSD coated/collagen coated P(3HB-*co*-4HB); SCCC and SSD blend/collagen coated P(3HB-*co*-4HB); SBCC scaffolds. Data represent means ± SD (n = 3). Mean data accompanied by different alphabets as of SCCC scaffolds (a–d) and SBCC scaffolds (A–D) indicates significant difference within each respective group (Tukey's HSD test, $p < 0.05$).

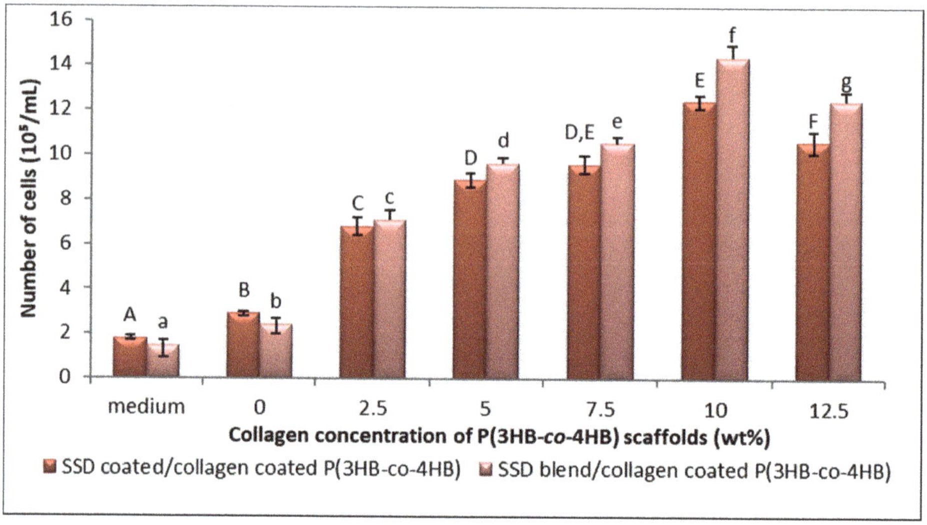

Figure 7. Proliferation of L929 cells on the SSD coated/collagen coated P(3HB-*co*-4HB); SCCC and SSD blend/collagen coated P(3HB-*co*-4HB); SBCC scaffolds. Data represent means ± SD (n = 3). Mean data accompanied by different alphabets as of SCCC scaffolds (A–F) and SBCC scaffolds (a–g) indicates significant difference within each respective group (Tukey's HSD test, $p < 0.05$).

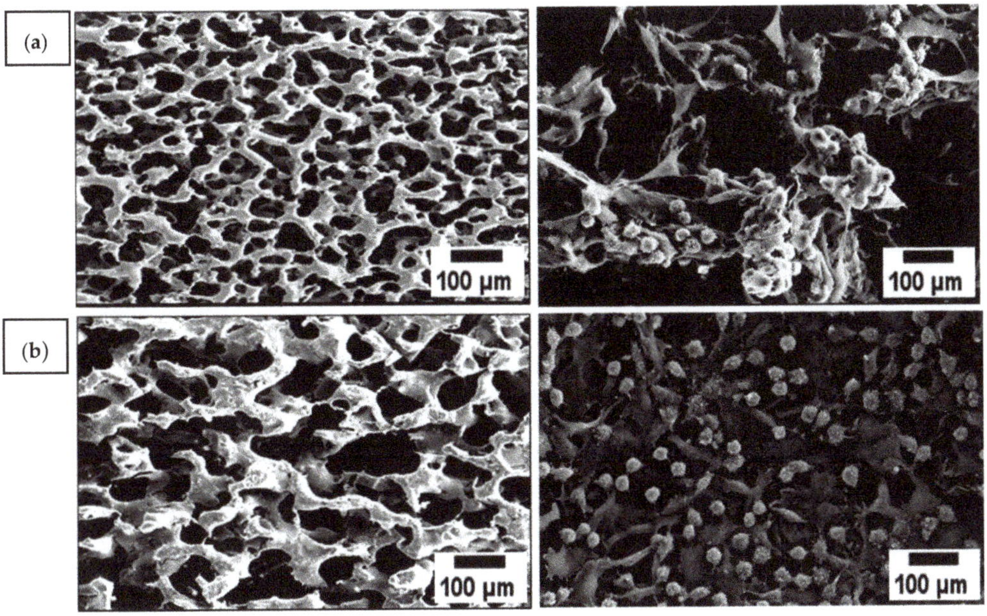

Figure 8. Micrograph of proliferation of L929 cells on (**a**) control-P(3HB-*co*-4HB), SCCC scaffolds (**b**) SCCC 10 wt.%. Data represent means ± SD ($n = 5$).

3.7. Antimicrobial Analysis of SCCC and SBCC Scaffolds

Antimicrobial analysis was carried out using the colonization test as summarised in Table 4. Antimicrobial substance, silver sulfadiazine (SSD), was incorporated in the scaffolds. Silver compounds, especially (SSD), has been widely used as an antibacterial agent in various biomedical applications [69,70]. Based on the results obtained, both SCCC and SBCC scaffolds revealed desirable antimicrobial effects. However, SBCC scaffolds required 48 h to inhibit certain pathogenic microorganisms which was due to the elution of silver sulfurdiazine impregnated with SSD possessed, whereby Ag ions were physically entrapped in the scaffolds where controlled release of antimicrobial agent occurred [70]. Meanwhile, the results revealed that in SCCC with scaffolds, the silver ion was continuously released directly leading to almost 100% inhibition for most of the microorganism within 12 h. Both scaffolds showed different functionality according to the releasing rate of silver ion. The schematic of the antimicrobial release of both the scaffolds is illustrated in Figure 9. The SCCC scaffolds, which rapidly release SSD, are thus appropriate for further work towards dermal application, especially skin damage to the epidermis and the upper dermis that can be regenerated spontaneously and healed in relatively shorter periods [71–74]. On the condition of chronic wounds, such as diabetic ulcers, long-term release of antimicrobials is highly suggested since regeneration occurs at the edges of injuries [75]. Therefore, the SBCC scaffold can be beneficial for such cases. The antimicrobial effect of SBCC scaffold is effective by the significantly prolonged release of silver ion, which continues to kill microbes after the release system is exhausted. The release of silver ions is accompanied by the contact killing of the layer that contains silver ion gradually released by diffusion and scaffold degradation [69]. Furthermore, according to Heo and coworkers [73], silver sulfadiazine binds with microbial DNA and releases the sulfonamide, interfering with the intermediary metabolic pathway [76].

Table 4. Antimicrobial test of SCCC and SBCC scaffolds against various microorganisms.

Time (h)	Inhibition of Microorganisms (%)							
	6		12		24		48	
	SCCC	**SBCC**	**SCCC**	**SBCC**	**SCCC**	**SBCC**	**SCCC**	**SBCC**
Staphylococus aerus ATCC 12600	65 ± 5	13 ± 1	85 ± 3	36 ± 5	100 ± 0	83 ± 8	NA	100 ± 0
Escherichia coli ATCC 11303	79 ± 8	34 ± 5	100 ± 0	51 ± 9	100 ± 0	92 ± 6	NA	100 ± 0
Pseudomonas aeruginosa ATCC 17588	85 ± 7	43 ± 9	100 ± 0	45 ± 5	100 ± 0	87 ± 12	NA	100 ± 0
Bacillus licheniformis	98 ± 2	65 ± 10	100 ± 0	95 ± 5	100 ± 0	100 ± 0	NA	100 ± 0
Candida albicans	93 ± 7	33 ± 6	100 ± 0	71 ± 10	100 ± 0	94 ± 6	NA	100 ± 0

Values are mean ± SD of three replicates; NA denotes not applicable.

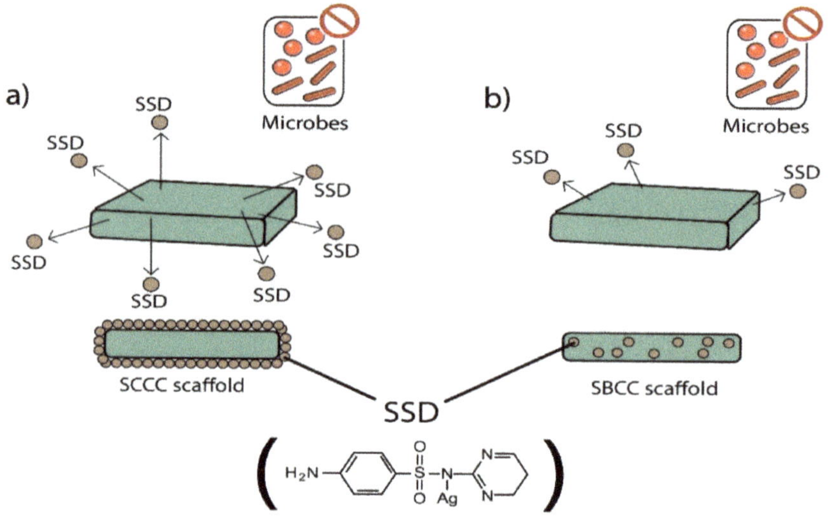

Figure 9. Schematic represents the releasing rate of silver ion from (**a**) SCCC scaffolds which rapidly release SSD and (**b**) the slow release of silver ion impregnated in the SBCC scaffolds.

4. Conclusions

In this study, we demonstrated that a combination of a simple and green approach to fabricate collagen and SSD incorporated P(3HB-*co*-4HB) scaffolds using porogen leaching and freeze-drying techniques. In comparing the SCCC and SBCC scaffolds, both the scaffolds differed in the incorporation of antimicrobial agent. Biomaterial based microbial infections pose serious concerns in the biomedical field. This study focuses on the development of highly efficient potential biomaterials that release the antimicrobial agents. This is in response to the limitations caused by some biomaterials with antimicrobial properties that inhibit microbial infections but slow down the cell seeding and tissue integration. Here, both the SCCC and SBCC scaffolds enhanced cell seeding and proliferation of L929 cells. Nonetheless, SCCC has higher antibacterial efficiency within the first 24 h, whereby the antibiotic is rapidly released as compared to the controlled release of the antimicrobial properties in SBCC scaffolds. Entrapment of SSD in P(3HB-*co*-4HB), as in SBCC, resulted in a reduced burst release of SSD as compared to SCCC. Nonetheless, both the SCCC and SBCC scaffolds could be an excellent candidate to inhibit microbial colonization based on the biomaterial application without causing antibiotic resistance. The study provides evidence and elucidates the surface interface-cell interactions of the modified P(3HB-*co*-4HB) scaffolds and release of the antimicrobial agent from the scaffolds, thus paving the way in developing infection-resistance biomaterials in the biomedical field in the future.

Author Contributions: Conceptualization, A.-A.A.A. and S.V.; methodology, T.P.G.; validation, T.P.G., A.-A.A.A., S.V. and W.M.K.; formal analysis, T.P.G. and S.V.; investigation, T.P.G.; resources, A.-A.A.A.; data curation, T.P.G., W.M.K., S.V.; writing—original draft preparation, S.V.; writing—review and editing, S.V. and S.R.; visualization, A.-A.A.A., A.K.H.P.S. and S.R.; supervision, A.-A.A.A.; project administration, T.P.G.; funding acquisition, A.-A.A.A. All authors have read and agreed to the published version of the manuscript.

Funding: We would like to acknowledge Universiti Sains Malaysia (USM) for the research fund provided (311.PCCB.411954, 'USM-Strategic Initiative-Ten Q1-Q2').

Institutional Review Board Statement: Not applicable.

Informed Consent Statement: Not applicable.

Data Availability Statement: The data presented in this study is openly available.

Conflicts of Interest: The authors declare no conflict of interest.

References

1. Pina, S.; Ribeiro, V.P.; Marques, C.F.; Maia, F.R.; Silva, T.H.; Reis, R.L.; Oliveira, J.M. Scaffolding Strategies for Tissue Engineering and Regenerative Medicine Applications. *Materials* **2019**, *12*, 1824. [CrossRef] [PubMed]
2. Abdulghani, S.; Mitchell, G.R. Three-dimensional Scaffolds for Tissue Engineering Applications: Role of Porosity and Pore Size. *Biomolecules* **2019**, *9*, 750. [CrossRef] [PubMed]
3. Salim, Y.S.; Sharon, A.; Vigneswari, S.; Mohd Ibrahim, M.N.; Amirul, A.A. Environmental degradation of microbial polyhydroxyalkanoates and oil palm-based composites. *Appl. Bichem. Biotechnol.* **2012**, *167*, 314. [CrossRef] [PubMed]
4. Winnacker, M. Polyhydroxyalkanoates: Recent Advances in Their Synthesis and Applications. *Eur. J. Lipid Sci. Technol.* **2019**, *121*, 1900101. [CrossRef]
5. Trakunjae, C.; Boondaeng, A.; Apiwatanapiwat, W.; Kosugi, A.; Arai, T.; Sudesh, K.; Vaithanomsat, P. Enhanced Polyhydroxybutyrate (PHB) Production by Newly Isolated Rare Actinomycetes *Rhodococcus* sp. strain BSRT1-1 using Response Surface Methodology. *Sci. Rep.* **2021**, *11*, 1896. [CrossRef] [PubMed]
6. Akinmulewo, A.B.; Nwinyi, O.C. Polyhydroxyalkanoate: A Biodegradable Polymer (a mini review). *J. Phys. Conf. Ser.* **2019**, *1378*, 042007. [CrossRef]
7. Faezah, A.N.; Rahayu, A.; Vigneswari, S.; Majid, M.I.A.; Amirul, A.A. Regulating the molar fraction of 4-hydroxybutyrate in Poly(3-hydroxybutyrate-*co*-4-hydroxybutyrate) by biological fermentation and enzymatic degradation. *World J. Microbiol. Biotechnol.* **2011**, *27*, 2455. [CrossRef]
8. Vigneswari, S.; Chai, J.M.; Kamarudin, K.H.; Amirul, A.A.; Focarate, M.L.; Ramakrishna, S. Elucidating the Surface Functionality of Biomimetic RGD Peptides Immobilized on Nano-P(3HB-*co*-4HB) for H9c2 Myoblast Cell Proliferation. *Front. Bioeng. Biotechnol.* **2020**, *8*, 567693. [CrossRef]
9. Vigneswari, S.; Gurusamy, T.P.; Abdul Khalil, H.P.S.; Ramakrishna, S.; Amirul, A.A. Elucidation of Antimicrobial SSD blend/poly (3-hydroxybutyrate-*co*-4-hydroxybutyrate) Immobilised with Collagen Peptide As Potential Biomaterial. *Polymer* **2020**, *12*, 2979. [CrossRef]
10. Sun, F.; Guo, J.; Liu, Y.; Yu, Y. Preparation and Characterization of Poly (3-hydroxybutyrate-*co*-4-hydroxybutyrate)/Pullulan-Gelatin Electrospun Nanofibers with Shell-Core Structure. *Biomed. Mater.* **2020**, *15*, 045023. [CrossRef]
11. Lim, Y.S.; Ok, Y.J.; Hwang, S.Y.; Kwak, J.Y.; Yoon, S. Marine Collagen as A Promising Biomaterial for Biomedical Applications. *Mar. Drugs* **2019**, *17*, 467. [CrossRef]
12. Hernandez-Gordillo, V.; Chmielewski, J. Mimicking the Extracellular Matrix with Functionalized, Metal-Assembled Collagen Peptide Scaffolds. *Biomaterials* **2014**, *35*, 7363–7373. [CrossRef]
13. Yamada, S.; Yamamoto, K.; Ikeda, T.; Yanagiguchi, K.; Hayashi, Y. Potency of Fish Collagen as a Scaffold for Regenerative Medicine. *BioMed Res. Int.* **2014**, *2014*, 302932. [CrossRef]
14. Xiao, X.; Zhao, W.; Liang, J.; Sauer, K.; Libera, M. Self-Defensive Antimicrobial Biomaterial Surfaces. *Colloids Surf. B Biointerfaces* **2020**, *192*, 110989. [CrossRef]
15. Ilomuanya, M.O.; Adebona, A.C.; Wang, W.; Sowemimo, A.; Eziegbo, C.L.; Silva, B.O.; Adeosun, S.O.; Joubert, E.; De Beer, D. Development and Characterization of Collagen-Based Electrospun Scaffolds Containing Silver Sulphadiazine and *Aspalathus linearis* Extract for Potential Wound Healing Applications. *SN Appl. Sci.* **2020**, *2*, 881. [CrossRef]
16. Mehta, M.A.; Shah, S.; Ranjan, V.; Sarwade, P.; Philipose, A. Comparative Study of Silver-Sulfadiazine-Impregnated Collagen Dressing Versus Conventional Burn Dressings in Second-Degree Burns. *J. Fam. Med. Prim. Care* **2019**, *8*, 215–219.

17. Banerjee, J.; Seetharaman, S.; Wrice, N.L.; Christy, R.J.; Natesan, S. Delivery of Silver Sulfadiazine and Adipose Derived Stem Cells Using Fibrin Hydrogel Improves Infected Burn Wound Regeneration. *PLoS ONE* **2019**, *14*, e0217965. [CrossRef]
18. Ceresa, C.; Fracchia, L.; Marchetti, A.; Rinaldi, M.; Bosetti, M. Injectable Scaffolds Enriched with Silver to Inhibit Bacterial Invasion in Tissue Regeneration. *Materials* **2019**, *12*, 1931. [CrossRef]
19. Shanmugasundaram, N.; Sundaraseelan, J.; Uma, S.; Selvaraj, D.; Babu, M. Design and Delivery of Silver Sulfadiazine from Alginate Microspheres-Impregnated Collagen Scaffold. *J. Biomed. Mater. Res. B Appl. Biomater.* **2006**, *77*, 378–388. [CrossRef]
20. Nejaddehbashi, F.; Hashemitabar, M.; Bayati, V.; Moghimipour, E.; Movaffagh, J.; Orazizadeh, M.; Abbaspour, M.R. Incorporation of Silver Sulfadiazine into an Electrospun Composite of Polycaprolactone as An Antibacterial Scaffold for Wound Healing in Rats. *Cell J.* **2020**, *21*, 379–390.
21. Mohseni, M.; Shamloo, A.; Aghababaei, Z.; Vossoughi, M.; Moravvej, H. Antimicrobial Wound Dressing Containing Silver Sulfadiazine with High Biocompatibility: In Vitro Study. *Artif. Organs* **2016**, *40*, 765–773. [CrossRef]
22. Malafatti, J.O.D.; Bernardo, M.P.; Moreira, F.K.V.; Ciol, H.; Inada, N.M.; Mattoso, L.H.C.; Paris, E.C. Electrospun Poly (lactic acid) Nanofibers Loaded with Silver Sulfadiazine/[Mg–Al]-Layered Double Hydroxide As An Antimicrobial Wound Dressing. *Polym. Adv. Technol.* **2020**, *31*, 1377–1387. [CrossRef]
23. Huong, K.H.; Azuraini, M.J.; Aziz, N.A.; Amirul, A.A. Pilot scale production of poly (3-hydroxybutyrate-*co*-4-hydroxybutyrate) biopolymers with high molecular weight and elastomeric properties. *J. Biosci. Bioeng.* **2017**, *124*, 76–83. [CrossRef]
24. Norhafini, H.; Thinagaran, L.; Shantini, K.; Huong, K.H.; Syafiq, I.M.; Bhubalan, K.; Amirul, A.A. Synthesis of Poly (3-hydroxybutyrate-*co*-4-hydroxybutyrate) with High 4HB Composition and PHA Content Using 1,4-butanediol and 1,6-hexanediol for Medical Application. *J. Polym. Res.* **2017**, *24*, 24–189. [CrossRef]
25. Destaye, A.G.; Lin, C.K.; Lee, C.K. Glutaraldehyde Vapor Cross-Linked Nanofibrous PVA Mat with In Situ Formed Silver Nanoparticles. *ACS Appl. Mater. Interfaces* **2013**, *5*, 4745–4752. [CrossRef]
26. Aramwit, P.; Ratanavaraporn, J.; Ekgasit, S.; Tongsakul, D.; Bang, N.A. Green Salt-Leaching Technique to Produce Sericin/ PVA/Glycerin Scaffolds with Distinguished Characteristics for Wound-Dressing Applications. *J. Biomed. Mater. Res. B* **2015**, *103*, 915–924. [CrossRef]
27. Wang, Y.; Ke, Y.; Ren, L.; Wu, G.; Chen, X.; Zhao, Q. Surface engineering of PHBV by covalent collagen immobilization to improve cell compatibility. *J. Biomed. Mater. Res. A* **2009**, *88*, 616–627. [CrossRef]
28. Köse, G.T.; Kenar, H.; Hasirci, N.; Hasirci, V. Macroporous poly (3-hydroxybutyrate-*co*-3-hydroxyvalerate) matrices for bone tissue engineering. *Biomaterials* **2003**, *24*, 1949–1958. [CrossRef]
29. Ismail, I.; Gurusamy, T.P.; Ramachandran, H.; Amirul, A.A. Enhanced production of poly (3-hydroxybutyrate-co-4-hydroxybutyrate) copolymer and antimicrobial yellow pigmentation from *Cupriavidus* sp. USMAHM13 with antibiofilm capability. *Prep. Biochem. Biotechnol.* **2017**, *47*, 388–396. [CrossRef]
30. Vigneswari, S.; Murugaiyah, V.; Kaur, G.; Abdul Khalil, H.P.S.; Amirul, A.A. Biomacromolecule Immobilization: Grafting of Fish-Scale Collagen Peptides onto Aminolyzed P(3HB-*co*-4HB) Scaffolds as A Potential Wound Dressing. *Biomed. Mater.* **2016**, *68*, 1927–1934. [CrossRef]
31. Ahmed, W.; Zhai, Z.; Gao, C. Adaptive Antibacterial Biomaterial Surfaces and Their Applications. *Mater. Today Bio* **2019**, *2*, 100017. [CrossRef] [PubMed]
32. Wieszczycka, K.; Staszak, K.; Woźniak-Budych, M.J.; Litowczenko, J.; Maciejewska, B.M.; Jurga, S. Surface Functionalization–The Way for Advanced Applications of Smart Materials. *Coord Chem. Rev.* **2021**, *436*, 213846. [CrossRef]
33. Zhu, Y.; Ke, J.; Zhang, L. Anti-biofouling and antimicrobial biomaterials for tissue engineering. In *Racing for the Surface*; Li, B., Moriarty, T., Webster, T., Xing, M., Eds.; Springer: New York, NY, USA, 2020.
34. Goswami, M.; Rekhi, P.; Debnath, M.; Ramakrishna, S. Microbial Polyhydroxyalkanoates Granules: An Approach Targeting Biopolymer for Medical Applications and Developing Bone Scaffolds. *Molecules* **2021**, *26*, 860. [CrossRef] [PubMed]
35. Bružauskaitė, I.; Bironaitė, D.; Bagdonas, E.; Bernotienė, E. Scaffolds and cells for tissue regeneration: Different scaffold pore sizes—Different cell effects. *Cytotechnology* **2016**, *68*, 355–369. [CrossRef] [PubMed]
36. Liao, C.J.; Chen, C.F.; Chen, J.H.; Chiang, S.F.; Lin, Y.J.; Chang, K.Y. Fabrication of Porous Biodegradable Polymer Scaffolds Using a Solvent Merging/Particulate Leaching Method. *J. Biomed. Mater. Res.* **2002**, *59*, 676–681. [CrossRef] [PubMed]
37. Subia, B.; Kundu, J.C.S. Biomaterial Scaffold Fabrication Techniques for Potential Tissue Engineering Applications. *Tissue Eng.* **2010**, *141*, 20–31.
38. Ho, M.H.; Kuo, P.Y.; Hsieh, H.J.; Hsien, T.Y.; Hou, L.T.; Lai, J.Y.; Wang, D.M. Preparation of Porous Scaffolds by Using Freeze-Extraction and Freeze-Gelation Methods. *Biomaterials* **2004**, *25*, 129–138. [CrossRef]
39. Wang, Z.; Qing, Q.; Chen, X.; Liu, C.; Luo, J.; Hu, J.; Qin, T. Effects of Scaffold Surface Morphology on Cell Adhesion and Survival Rate in Vitreous Cryopreservation of Tenocyte-Scaffold Constructs. *Appl. Surf. Sci.* **2016**, *388*, 223–227. [CrossRef]
40. Zhu, B.; Li, W.; Chi, N.; Lewis, R.V.; Osamor, J.; Wang, R. Optimization of Glutaraldehyde Vapor Treatment for Electrospun Collagen/Silk Tissue Engineering Scaffolds. *ACS Omega* **2017**, *2*, 2439–2450. [CrossRef]
41. Campiglio, C.E.; Negrini, N.C.; Farè, S.; Draghi, L. Cross-linking Strategies for Electrospun Gelatin Scaffolds. *Materials* **2019**, *12*, 2476. [CrossRef]
42. Vashisth, P.; Pruthi, V. Synthesis and Characterization of Crosslinked Gellan/PVA Nanofibers for Tissue Engineering Application. *Mater. Sci. Eng. C* **2016**, *67*, 304–312. [CrossRef]

43. Teixeira, M.A.; Antunes, J.C.; Amorimm, M.T.P.; Felgueiras, H.P. Green Optimization of Glutaraldehyde Vapor-Based Crosslinking on Poly (Vinyl Alcohol)/Cellulose Acetate Electrospun Mats for Applications as Chronic Wound Dressings. *Proceedings* **2021**, *69*, 30. [CrossRef]
44. Bettini, S.; Bonfrate, V.; Syrgiannis, Z.; Sannino, A.; Salvatore, L.; Madaghiele, M.; Valli, L.; Giancane, G. Biocompatible Collagen Paramagnetic Scafold For Controlled Drug Release. *Biomacromolecules* **2015**, *16*, 2599–2608. [CrossRef]
45. Bonfrate, V.; Manno, D.; Serra, A.; Salvatore, L.; Sannino, A.; Buccolieri, A.; Serra, T.; Giancane, G. Enhanced Electrical Conductivity of Collagen Films Through Long-Range Aligned Iron Oxide Nanoparticles. *J. Colloid Interface Sci.* **2017**, *501*, 185–191. [CrossRef]
46. Terzi, A.; Storelli, E.; Bettini, S.; Sibillano, T.; Altamura, D.; Salvatore, L.; Madaghiele, M.; Romano, A.; Siliqi, D.; Ladisa, M.; et al. Effects of Processing on Structural, Mechanical and Biological Properties of Collagen-Based Substrates for Regenerative Medicine. *Sci. Rep.* **2018**, *8*, 1429. [CrossRef]
47. Riaz, T.; Zeeshan, R.; Zarif, F.; Ilyas, K.; Muhammad, N.; Safi, S.Z.; Rahim, A.; Rizvi, S.A.A.; Rehman, I.U. FTIR Analysis of Natural and Synthetic Collagen. *Appl. Spectrosc. Rev.* **2018**, *53*, 703–746. [CrossRef]
48. Sotelo, C.G.; Comesaña, M.B.; Ariza, P.R.; Pérez-Martín, R.I. Characterization of Collagen from Different Discarded Fish Species Of The West Coast of the Iberian Peninsula. *J. Aquat. Food Prod. Technol.* **2016**, *25*, 388–399. [CrossRef]
49. Zhijiang, C.; Qin, Z.; Xianyou, S.; Yuanpei, L. Zein/poly (3-hydroxybutyrate-*co*-4-hydroxybutyrate) Electrospun Blend Fiber Scaffolds: Preparation, Characterization and Cytocompatibility. *Mater. Sci. Eng. C Mater. Biol. Appl.* **2017**, *1*, 797–806. [CrossRef]
50. Bayari, S.; Severcan, F. FTIR Study of Biodegradable Biopolymers: P(3HB), P(3HB-*co*-4HB) and P(3HB-*co*-3HV). *J. Mol. Strut* **2005**, *744*, 529–534. [CrossRef]
51. Mighri, N.; Mao, J.; Mighri, F.; Ajji, A.; Rouabhia, M. Chitosan-Coated Collagen Membranes Promote Chondrocyte Adhesion, Growth, and Interleukin-6 Secretion. *Materials* **2015**, *8*, 7673–7689. [CrossRef]
52. Chang, M.C.; Ikoma, T.; Kikuchi, M.; Tanaka, J. Preparation of a porous hydroxyapatite/collagen nanocomposite using glutaraldehyde as a crosslinkage agent. *J. Mater. Sci. Lett.* **2001**, *20*, 1199–1201. [CrossRef]
53. Chen, G.; Ushida, T.; Tateishi, T. Scaffold Design for Tissue Engineering. *Macromol Biosci* **2002**, *2*, 67–77. [CrossRef]
54. Wang, H.M.; Chou, Y.T.; Wen, Z.H.; Wang, Z.R.; Chen, C.H.; Ho, M.L. Novel biodegradable porous scaffold applied to skin regeneration. *PLoS ONE* **2013**, *8*, 118–126. [CrossRef]
55. Karageorgiou, V.; Kaplan, D. Porosity of 3D Biomaterial Scaffolds and Osteogenesis. *Biomaterials* **2005**, *26*, 5474–5491. [CrossRef]
56. Chen, G.; Kawazoe, N. 3.1-Preparation of Polymer Scaffolds by Ice Particulate Method for Tissue Engineering. In *Biomaterials Nanoarchitectonics*; Ebara, M., Ed.; William Andrew Publishing: New York, NY, USA, 2016; pp. 77–95.
57. Venkatesan, J.; Kim, S.; Wong, T.W. Chapter 9-Chitosan and Its Application as Tissue Engineering Scaffolds. In *Nanotechnology Applications for Tissue Engineering*; Thomas, S., Grohens, Y., Ninan, N., Eds.; William Andrew Publishing: New York, NY, USA, 2015; pp. 133–147.
58. Bartoš, M.; Suchý, T.; Foltán, R. Note on the Use of Different Approaches to Determine the Pore Sizes Of Tissue Engineering Scaffolds: What Do We Measure? *BioMed Eng. OnLine* **2018**, *17*, 110. [CrossRef]
59. Han, F.; Wang, J.; Ding, L.; Hu, Y.; Li, W.; Yuan, Z.; Guo, Q.; Zhu, C.; Yu, L.; Wang, H.; et al. Tissue Engineering and Regenerative Medicine: Achievements, Future, And Sustainability in Asia. *Front. Bioeng. Biotechnol.* **2020**, *8*, 83. [CrossRef]
60. Dzobo, K.; Thomford, N.E.; Senthebane, D.A.; Shipanga, H.; Rowe, A.; Dandara, C.; Pillay, M.; Motaung, K.S.C.M. Advances in Regenerative Medicine and Tissue Engineering: Innovation and Transformation of Medicine. *Stem Cells Int.* **2018**, 2495848. [CrossRef]
61. Hebbar, R.S.; Isloor, A.M.; Ismail, A.F. Contact angle measurements in membrane characterization. *Biomaterials* **2017**, *131*, 167–178.
62. Wang, W.; Caetano, G.; Ambler, W.S.; Blaker, J.J.; Frade, M.A.; Mandal, P.; Diver, C.; Bártolo, P. Enhancing the Hydrophilicity and Cell Attachment of 3D Printed PCL/Graphene Scaffolds for Bone Tissue Engineering. *Materials* **2016**, *9*, 992. [CrossRef]
63. Remya, K.R.; Chandran, S.; Mani, S.; John, A.; Ramesh, P. Hybrid polycaprolactone/polyethylene oxide scaffolds with tunable fiber surface morphology, improved hydrophilicity and biodegradability for bone tissue engineering applications. *J. Biomater. Sci. Polym. Ed.* **2018**, *29*, 12. [CrossRef]
64. Rouabhia, M.; Mighri, N.; Mao, J.; Park, H.J.; Mighri, F.; Ajji, A.; Zhang, Z. Surface Treatment with Amino Acids of Porous Collagen-Based Scaffolds to Improve Cell Adhesion and Proliferation. *Can. J. Chem. Eng.* **2018**, *39*, 875–886. [CrossRef]
65. Zhang, D.; Wu, X.; Chen, J.; Lin, K. The development of collagen based composite scaffolds for bone regeneration. *Bioact. Mater.* **2018**, *31*, 129–138. [CrossRef] [PubMed]
66. Hu, Y.; Dan, W.; Xiong, S.; Kang, Y.; Dhinakar, A.; Wu, J.; Gu, Z. Development of Collagen/Polydopamine Complexed Matrix as Mechanically Enhanced and Highly Biocompatible Semi-Natural Tissue Engineering Scaffold. *Acta Biomater.* **2017**, *47*, 135–148. [CrossRef] [PubMed]
67. Liu, Z.; Tamaddon, M.; Gu, Y.; Yu, J.; Xu, N.; Gang, F.; Sun, X.; Liu, C. Cell Seeding Process Experiment and Simulation on Three-Dimensional Polyhedron and Cross-Link Design Scaffolds. *Front. Bioeng. Biotechnol.* **2020**, *8*, 104. [CrossRef] [PubMed]
68. Vigneswari, S.; Majid, M.I.A.; Amirul, A.A. Tailoring the Surface Architecture of Poly (3-hydroxybutyrate-*co*-4-hydroxybutyrate) Scaffolds. *J. Appl. Polym. Sci.* **2011**, *124*, 2777–2788. [CrossRef]
69. Lim, M.M.; Sultana, N. In Vitro Cytotoxicity and Antibacterial Activity of Silver-Coated Electrospun Polycaprolactone/Gelatine Nanofibrous Scaffolds. *Biotech* **2016**, *6*, 211. [CrossRef]

70. Sandri, G.; Bonferoni, M.C.; D'Autilia, F.; Rossi, S.; Ferrari, F.; Grisoli, P.; Caramella, C. Wound Dressings Based on Silver Sulfadiazine Solid Lipid Nanoparticles for Tissue Repairing. *Eur. J. Pharm. Biopharm.* **2013**, *84*, 84–90. [CrossRef]
71. Siedenbiedel, F.; Tiller, J.C. Antimicrobial Polymers in Solution and On Surfaces: Overview and Functional Principles. *Polymers* **2012**, *4*, 46–71. [CrossRef]
72. Heo, D.N.; Yang, D.H.; Lee, J.B.; Bae, M.S.; Kim, J.H.; Moon, S.H.; Chun, H.J.; Kim, C.H.; Lim, H.N.; Kwon, I.K. Burn-Wound Healing Effect of Gelatin/Polyurethane Nanofiber Scaffold Containing Silver-Sulfadiazine. *J. Biomed. Nanotechnol.* **2013**, *9*, 511–515. [CrossRef]
73. Zhong, S.P.; Zhang, Y.Z.; Lim, C.T. Tissue Scaffolds for Skin Wound Healing and Dermal Reconstruction. *Wiley Interdiscip. Rev. Nanomed. Nanobiotechnol.* **2010**, *2*, 510–525. [CrossRef]
74. Kalantari, K.; Mostafavi, E.; Afifi, A.M.; Izadiyan, Z.; Jahangirian, H.; Rafiee-Moghaddam, R.; Webster, T.J. Wound Dressings Functionalized with Silver Nanoparticles: Promises and Pitfalls. *Nanoscale* **2020**, *12*, 2268–2291. [CrossRef]
75. Sripriya, R.; Kumar, M.S.; Ahmed, M.R.; Sehgal, P.K. Collagen Bilayer Dressing with Ciprofloxacin, an Effective System for Infected Wound Healing. *J. Biomater. Sci. Polym. Ed.* **2007**, *18*, 335–351. [CrossRef]
76. Timofeeva, L.; Kleshcheva, N. Antimicrobial polymers: Mechanism of action, factors of activity, and applications. *Appl. Microbiol.* **2011**, *89*, 475–492. [CrossRef]

polymers

MDPI

Article

In Vitro Interaction of Doxorubicin-Loaded Silk Sericin Nanocarriers with MCF-7 Breast Cancer Cells Leads to DNA Damage

Ionuț-Cristian Radu [1], Cătălin Zaharia [1], Ariana Hudiță [2], Eugenia Tanasă [3], Octav Ginghină [4,5], Minodora Marin [1], Bianca Gălățeanu [2,*] and Marieta Costache [2]

1 Advanced Polymer Materials Group, Faculty of Applied Chemistry and Materials Science, University Politehnicaof Bucharest, 1-7 Gh. Polizu Street, Sector 1, 011061 Bucharest, Romania; radu.ionut57@yahoo.com (I.-C.R.); zaharia.catalin@gmail.com (C.Z.); minodora.marin@ymail.com (M.M.)
2 Department of Biochemistry and Molecular Biology, University of Bucharest, 050095 Bucharest, Romania; arianahudita@yahoo.com (A.H.); marietacostache@yahoo.com (M.C.)
3 University Politehnica of Bucharest, 313 Splaiul Independentei, Sector 6, 060042 Bucharest, Romania; eugenia.vasile27@gmail.com
4 Department of Surgery, "Sf. Ioan" Clinical Emergency Hospital, 042122 Bucharest, Romania; octav.ginghina@umfcd.ro
5 Faculty of Dental Medicine, Department II, "Carol Davila" University of Medicine and Pharmacy, 020021 Bucharest, Romania
* Correspondence: bianca.galateanu@bio.unibuc.ro

Abstract: In this paper, *Bombyx mori* silk sericin nanocarriers with a very low size range were obtained by nanoprecipitation. Sericin nanoparticles were loaded with doxorubicin, and they were considered a promising tool for breast cancer therapy. The chemistry, structure, morphology, and size distribution of nanocarriers were investigated by Fourier transformed infrared spectroscopy (FTIR–ATR), scanning electron microscopy (SEM) and transmission electron microscopy (TEM), and dynamic light scattering (DLS). Morphological investigation and DLS showed the formation of sericin nanoparticles in the 25–40 nm range. FTIR chemical characterization showed specific interactions of protein–doxorubicin–enzymes with a high influence on the drug delivery process and release behavior. The biological investigation via breast cancer cell line revealed a high activity of nanocarriers in cancer cells by inducing significant DNA damage.

Keywords: silk sericin; nanoprecipitation; breast cancer; doxorubicin

Citation: Radu, I.-C.; Zaharia, C.; Hudiță, A.; Tanasă, E.; Ginghină, O.; Marin, M.; Gălățeanu, B.; Costache, M. In Vitro Interaction of Doxorubicin-Loaded Silk Sericin Nanocarriers with MCF-7 Breast Cancer Cells Leads to DNA Damage. *Polymers* **2021**, *13*, 2047. https://doi.org/10.3390/polym13132047

Academic Editor: Ki Hyun Bae

Received: 19 May 2021
Accepted: 19 June 2021
Published: 22 June 2021

1. Introduction

Silks are natural protein-like fibers produced by arthropods, such as spiders or silkworms. Domestic-species-producing silks have been used since antiquity. Certain species, such as domesticated silkworm *Bombyx mori*, have a central role within textile industry applications and more recently in biomedical applications [1–3]. *Bombyx mori* proteins have been intensively studied for their biocompatibility, great mechanical properties, tunable biodegradation process, easy processing, and favorable source supply. Silk is composed of two major proteins: silk fibroin (fibrous protein) and silk sericin (globular protein) [4–8]. Silk fibroin is the main protein with large usage in the biomedical field. Silk sericin was originally removed as it was associated with the general immune body response to silks [9–11]. Therefore, silk sericin was largely neglected as a biomaterial for medical applications. However, silk sericin has attracted the particular interest of researchers due to special properties such as antioxidant effect, UV protection, moisture adsorption, or antibacterial protection. Recent studies in the literature revealed that sericin is currently used in cosmetics, pharmaceuticals, wound dressing, drug delivery, or cell culture. The antioxidant properties allow sericin to stand against lipid peroxidation by scavenging reactive species or to suppress tumor genesis by reducing oxidative stress or inflammatory responses [12–14]. Furthermore, the antioxidant properties of silk sericin contributed to cancer applications due to the capability to reduce oxidative stress or suppressing cancer cytokines for skin and colon cancer [15,16]. In recent years, silk sericin was used in the development of scaffolds for regenerative medicine in wound healing or tissue engineering [6,17–19].

In this regard, silk sericin may favor and sustain migration, proliferation, or collagen type I production due to methionine amino acid [20–22]. Silk sericin promoted open wound healing and added to silver-zinc sulfadiazine cream prevented burn wound infection. It was also effective in healing second-degree burn wounds without serious inflammatory reactions [22]. Besides particular properties, which recommended silk sericin for various biomedical applications, the drug-delivery field represents the specific area in which different nanoparticle systems have been developed [23–30]. However, the future concept pathway is used to overcome the current limitations so that it sustains the continuous development of nanoparticle systems. Silk sericin proved the ability for self-assembling capacity by loading various active principles. The unique chemistry favored surface charge modification for DNA binding and active targeting by poly(ethylene glycol) (PEG) and folate for cancer management [24,26,31].

This unique ability offers the possibility to easily prepare nanoparticles for drug/biomolecule delivery. These characteristics are tightly related to sericin chemistry and, in the last decades, studies showed new interesting insights on silk sericin structure. Therefore, silk sericin showed less amphiphilic character but higher hydrophilic character. This aspect was considered an impediment for self-assembling into nanoparticles, as compared to silk fibroin [32]. However, the synergistic effect of a proper preparation method, such as nanoprecipitation, and a specific precipitation agent may favor the preparation of silk sericin nanoparticles. Nanoparticles should display the desired characteristics such as size, morphology, and size distribution. Nanoprecipitation is a simple and fast method to produce nanoparticles from various types of polymers [33–38].

Doxorubicin is an anthracycline with therapeutic effects in a wide range of solid tumors, which still plays a major role in chemotherapy that induces apoptosis by causing DNA damage [39]. A great effort has been made to develop targeted nanodrug delivery systems due to their high therapeutic efficacy in cancer management. Polymeric nanoparticles are promising systems for drug delivery based on their nanometric size, high surface-area-to-volume ratio, favorable drug release profiles, and targeting features that can promote their preferential accumulation in tumor tissue. Various preparation routes have been addressed in the literature in an attempt to show the interaction of doxorubicin with various polymeric systems [40,41]. In this regard, doxorubicin-loaded RGD-conjugated polypeptide nanoformulation was developed by an emulsion solvent evaporation method [42]. These zwitterionic biodegradable drug-loaded polypeptide vesicles showed great potential for cancer treatment having high drug loading content (45%) and loading efficiency (95%) [42]. An interesting approach was related to the encapsulation of doxorubicin by polymerization-induced self-assembly methods [41]. Photopolymerization of various monomers in the presence of photocatalytic doxorubicin hydrochloride proved an interesting method to prepare drug-loaded polymeric nanoparticles with higher polymerization rates and good doxorubicin encapsulation efficiency [41]. Protein nanoparticles were also prepared by various routes to easily entrap and release various anticancer drugs [43]. They are prepared within biological systems, require fewer production steps, and show high biocompatibility and biodegradability, as compared to synthetic polymers. Protein nanoparticles have been prepared from various proteins including water-soluble proteins (bovine and human serum albumin, silk sericin) and partially soluble or insoluble proteins (silk fibroin, zein, and gliadin) [44]. TRAIL and Dox-loaded albumin nanoparticles showed great potential for synergistic apoptosis-based anticancer therapy [45].

This research study emphasizes the possibility to easily develop silk sericin nanoparticles by an optimized nanoprecipitation method. The novelty of the study arises from the development of sericin nanoparticles with a lower size distribution with respect to the literature data. Furthermore, the nanoparticles' preparation is based on an optimized procedure that can be easily transferred being useful for other similar systems.

The present research study reports on the preparation and complex characterization of self-assembled silk sericin nanoparticles in the presence of acetone with secondary conformational changes. Doxorubicin loading in sericin nanoparticles and drug release behaviors were studied in neutral, acidic, and enzymatic media. DLS, SEM, and TEM were employed for morphological and structural characterization of the nanoparticles. The structural changes showed similar behavior to silk fibroin by revealing a special and stable physical crosslinking structure of nanoparticles. Moreover, we investigated the sericin particles' potential to reduce MCF-7 cells' viability and to induce DNA fragmentation when loaded with doxorubicin.

2. Materials and Methods

2.1. Materials

Bombyx mori silk sericin powder (quality level 200) and all other reagents, including sodium hydroxide (reagent grade, ≥98%, pellets, anhydrous), potassium phosphate (ACS reagent, ≥98%), hydrochloric acid (ACS reagent, 37%), doxorubicin (doxorubicin hydrochloride 98.0–102.0%, HPLC), protease XIV from *Streptomyces griseus* (≥3.5 units/mg solid), *α-Chymotrypsin* from bovine pancreas (≥40 units/mg protein), and nonsolvent acetone (ACS reagent, ≥99.5%) were provided by Sigma Aldrich (3050 SPRUCE Street, St. Louis, MO 63103, USA).

2.2. Preparation of Silk Sericin Nanoparticles

Firstly, silk sericin solutions were obtained by direct dissolution of protein in distilled water considering the hydrophilic character and good water solubility. Briefly, solutions with 0.1, 0.25, 0.5, and 1% (*w*/*v*) were prepared under moderate stirring at room temperature.

Silk sericin nanoparticles were prepared via nanoprecipitation in which silk sericin solutions (0.1, 0.25, 0.5, and 1% *w*/*v*) were added into an organic phase of a nonsolvent, acetone. The nanoprecipitation technique involved the drop-wise addition of protein aqueous solution into acetone under vigorous stirring. The resulted nanoparticles were recovered by water and acetone vaporization. The workflow of nanoparticle preparation is shown in Scheme 1. The nanoparticle formulations were called SER 0.1%–SER 1% following the sericin concentration.

2.3. Drug Loading in Sericin Nanoparticles

Doxorubicin was loaded within sericin nanoparticles via the direct method by dissolution into the aqueous phase. The next step involved the addition of the dissolved drug aqueous phase into the acetone organic phase under vigorous stirring at neutral pH of 7.45 (1.25–5 wt.% doxorubicin in NPs). The drug-loaded nanoparticles were also recuperated by water and acetone vaporization. The drug was loaded in all prepared formulations 0.1, 0.25, 0.5, and 1% (*w*/*v*).

2.4. Drug Release Behavior

In vitro drug release behaviors of doxorubicin-loaded sericin nanoparticles were investigated in different pH conditions and enzymatic medium at 37 °C. Briefly, 5 mL of drug-loaded nanoparticles (4 wt.%) in phosphate buffer saline solution (PBS, pH 7.45) was placed in a tubular cellulose membrane, followed by immersion in flasks with a fixed volume (40 mL). The flasks were further incubated in an orbital mixer (Benchmark Scientific) at 300 rpm, and 37.0 ± 0.5 °C. 5 mL of PBS dialysate was collected at predetermined time intervals and then investigated by UV–VIS spectroscopy (SHIMADZU UV-3600 instrument). To maintain a constant volume, after each collection, 5 mL of fresh PBS were added to every flask. A similar procedure was performed for an acidic medium (pH = 3). Protease type XIV from *Streptomyces griseus* and protease α-chymotrypsin from bovine pancreas were used for enzymatic drug release test. Enzymes were added in a suspension of drug-loaded NPs for 1 h prior to release test, and the drug release behavior followed the above-mentioned protocol. The enzyme activity was 8 U/mL. The release efficiency was calculated as follows:

$$RE\ (\%) = \frac{amount\ of\ released\ DOX}{amount\ of\ the\ loaded\ DOX} \times 100 \quad (1)$$

The encapsulation efficiency was calculated with the following equation:

$$EE\ (\%) = \frac{amount\ of\ the\ loaded\ DOX - amount\ of\ unloaded\ DOX}{amount\ of\ the\ loaded\ DOX} \times 100 \quad (2)$$

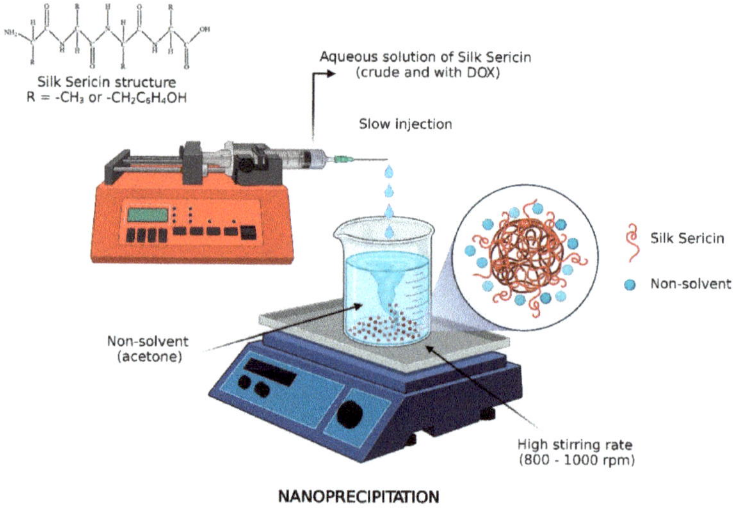

Scheme 1. Workflow of sericin nanoparticle preparation.

2.5. Characterization Methods

2.5.1. FTIR–ATR Analysis

FTIR–ATR investigation was performed using a Bruker Vertex 70 FTIR spectrophotometer with an attenuated total reflectance (ATR) accessory. FTIR spectrophotometer used 32 scans and a resolution of 4 cm^{-1} in mid-IR region 4000–600 cm^{-1}. Sericin nanoparticles SER 1% (w/v), loaded with doxorubicin and aqueous mixtures of enzyme–doxorubicin, were analyzed.

2.5.2. Morphological Characterization

Scanning electron microscopy (SEM) analysis was performed to reveal the main features of sericin nanoparticles, including aggregates' size, shape, or morphology. Silk sericin nanoparticles were investigated by a Quanta Inspect F scanning electron microscopy device equipped with a field emission gun (FEG) with 1.2 nm resolution and with an X-ray energy-dispersive spectrometer (EDS).

Transmission electron microscopy (TEM) analysis was performed using a TECNAI F30 G2 S-TWIN microscope operated at 300 kV in an energy-dispersive X-ray analysis (EDAX) facility. Formulations of sericin nanoparticles with 1% (w/v) and 0.1% (w/v) were evaluated by TEM.

2.5.3. Dynamic Light Scattering (DLS)

The size distribution of the nanoparticles was evaluated by dynamic light scattering in a static domain using a Malvern Zetasizer Nano instrument. Nanoparticles prepared from all sericin concentrations were subjected to DLS investigation (0.1, 0.25, 0.5, and 1% (*w*/*v*)).

The average molecular weight of silk sericin was determined by DLS in a static domain using a Malvern Zetasizer Nano instrument in the molecular weight module. The analysis was performed using glass cuvettes with square aperture. Toluene was used as a reference standard solvent and water as a common solvent for silk sericin. Zeta potential and isoelectric point were also determined by DLS using several protein solutions with pH ranging between acid and neutral (pH 1–7.45). The zeta potential value was considered at a neutral point. The isoelectric point value was considered for zeta potential.

2.5.4. Conformational Analysis by Circular Dichroism (CD)

The secondary structure of silk sericin solution and sericin nanoparticles dispersion was evaluated by a Jasco J-1500 spectrophotometer, Japan (J-1500 Circular Dichroism Spectrophotometer) using a quartz cell of 1 mm path length. During this analysis, the samples were scanned three times at low concentrations in the range of 180–250 nm with a scan rate of 100 nm/min.

2.6. In Vitro Biological Evaluation of Free and Doxorubicin-Loaded Sericin Nanocarriers

2.6.1. Cell Culture Model

MCF-7 human epithelial mammary gland cell line (ATCC® HTB-22™) was employed in this study as in vitro model since it retains several characteristics of differentiated mammary epithelium including the ability to process estradiol via cytoplasmic estrogen receptors and the capability of forming domes. Moreover, the MCF-7 cell line is positive for the estrogen and the progesterone receptor and negative for the HER2 marker. The MCF-7 cells were cultured in Dulbecco's modified Eagle medium (DMEM) supplemented with 10% fetal bovine serum (FBS) and 1% antibiotic–antimycotic solution (ABAM, containing 100 U/mL penicillin, 100 µg/mL streptomycin, and 0.25 µg amphotericin B) and maintained at 37 °C in a humidified air atmosphere of 5% CO_2 throughout the experiment. The medium renewal was carried out every other day.

2.6.2. Cell Viability Assay

The MTT assay was used to measure MCF-7 breast cancer cells' viability after incubation with Ser NPs and Ser NPs + DOX at a final concentration of 20 mg/mL Ser NPs ± Dox. Briefly, MCF-7 cells were seeded at an initial density of 2×10^5 cells/cm^2 in 96–well culture plates and treated with 20 mg/mL Ser NPs and Ser NPs + DOX for 6 and 24 h. At each time point, the culture media was discarded and replaced with a freshly prepared solution of MTT (1 mg/mL). The samples were further incubated for 4 h in standard cell culture conditions to allow the metabolically active cells to form formazan crystals, which were dissolved in DMSO. The absorbance of the resulting solutions was measured at 550 nm using a Flex Station III multimodal reader (Molecular Devices).

2.6.3. Cytoskeleton Investigation and DAPI staining

To evaluate the potential morphological modifications induced in the MCF-7 breast cancer cells by the treatment with Ser NPs + DOX, the cytoskeleton's actin filaments were stained with FITC–phalloidin. In this view, MCF-7 cells were seeded at an initial cell density of 2×10^5 cells/cm^2 in 96–well culture plates and treated with 20 mg/mL Ser NPs and Ser NPs + DOX for 6 and 24 h. At each time point, the MCF-7 monolayers were fixed with a 4% paraformaldehyde solution for 20 min, permeabilized with a 2% BSA/0.1% Triton X100 solution for 1 h and consequently stained with FITC–conjugated phalloidin for 1 h at 37 °C in a humidified environment. In the end, the MCF-7 monolayers were stained with DAPI to highlight cell nuclei and reveal chromatin fragmentation. The samples were imaged using the Olympus IX73 inverted fluorescence microscope (Olympus) and images were captured using CellSense Imaging Software.

2.6.4. Measurement of DNA Damage by Comet Assay

The DNA damage induced by the Ser NPs + DOX in MCF-7 breast cancer cells was quantified at a single cell level using OxiSelect Comet Assay Kit three-well slides (Cell Biolabs) assay. The MCF-7 cells were seeded in six-well plates at an initial density of 0.3×10^6 cells/cm^2 and treated with simple Ser NPs and DOX-loaded Ser NPs. After 6 h and 24 h, the cells were detached from the culture vessels and processed as recommended by the manufacturers' instructions. Briefly, cells were resuspended in agarose in a 1:10 ratio and transferred on the agarose precoated comet assay slides. After gelling, the comet assay slides were immersed in the lysis buffer for 60 min at 40 °C in the dark, followed by alkaline solution treatment for 30 min in the same conditions. Then, the slides were placed in the electrophoresis solution and then transferred to the electrophoresis tank. The electrophoresis was carried out at 1 V/cm for 15 min. Finally, samples were stained with 100 µL Vista Green DNA Dye solution for 15 min at room temperature in the dark. In total, 150 randomly selected cells from each slide were analyzed using a fluorescence microscope (Olympus IX73) and CellSense software. The length of the comet tail was chosen as an indicator of DNA damage.

2.6.5. Statistical Analysis

The data obtained from the MTT assay were statistically analyzed using GraphPad Prism 6 software (San Diego, CA, USA), one-way ANOVA, and the Bonferroni test. Control samples were considered as 100% viability for each time point. All the experiments were performed with three biological replicates and each data set is presented as the average of three replicates (mean ± standard deviation). A value of $p \leq 0.05$ was considered to indicate a statistically significant difference. All experimental controls were represented by MCF-7 cell cultures, where fresh culture media was added instead of nanoparticles treatment and were identically processed as described for each assay separately.

3. Results and Discussions

3.1. FTIR–ATR Analysis

FTIR investigation showed the main characteristic peaks for doxorubicin, enzymes, sericin, and their interactions. The spectrum of doxorubicin had the following peaks: peak at 3517 cm^{-1} was assigned to water molecules bonded within the drug structure; peak at 3323 cm^{-1} was attributed to hydroxyl stretching vibration; peak at 3128 cm^{-1} was attributed to N-H stretching vibration; peaks at 2977 cm^{-1} and 2931 cm^{-1} were assigned to C-H stretching vibration within the ring; peak at 1728 cm^{-1} was assigned to C=O stretching in carbonyl group within vibrating in quinone and ketone; peak at 1584 cm^{-1} was attributed to N-H bending vibration and C-N stretching vibration; peak at 1525 cm^{-1} was assigned to hydroxyl group bending vibration; peak at 1410 cm^{-1} was attributed to methyl bending vibration; peaks at 1285 cm^{-1}, 1207 cm^{-1}, 1115 cm^{-1}, and 1075 cm^{-1} were attributed to C-N and C-O-C stretching vibrations, C-O stretching vibration within the tertiary, and secondary and primary alcohols; peaks at 999 cm^{-1} and 865 cm^{-1} were assigned to the skeletal ring (Figure 1). The sericin spectrum revealed a specific protein spectrum with the following characteristic peaks: peak at 3291 cm^{-1} was attributed to hydroxyl stretching vibration; peak at 3070 cm^{-1} was attributed to N-H stretching vibration; peak at 2972 cm^{-1} was assigned to C-H stretching vibration; peak at 1649 cm^{-1} was assigned to C=O stretching in carbonyl group within amidic backbone (amide I); peak at 1529 cm^{-1} was attributed to N-H bending vibration and C-N stretching vibration (amide II); peak at 1395 cm^{-1} was attributed to methyl, methylene, and methine groups bending vibration; peak at 1242 cm^{-1} was attributed to C-N (amide III); peak at 1068 cm^{-1} was attributed to C-O-C and C-O stretching vibrations (Figure 1).

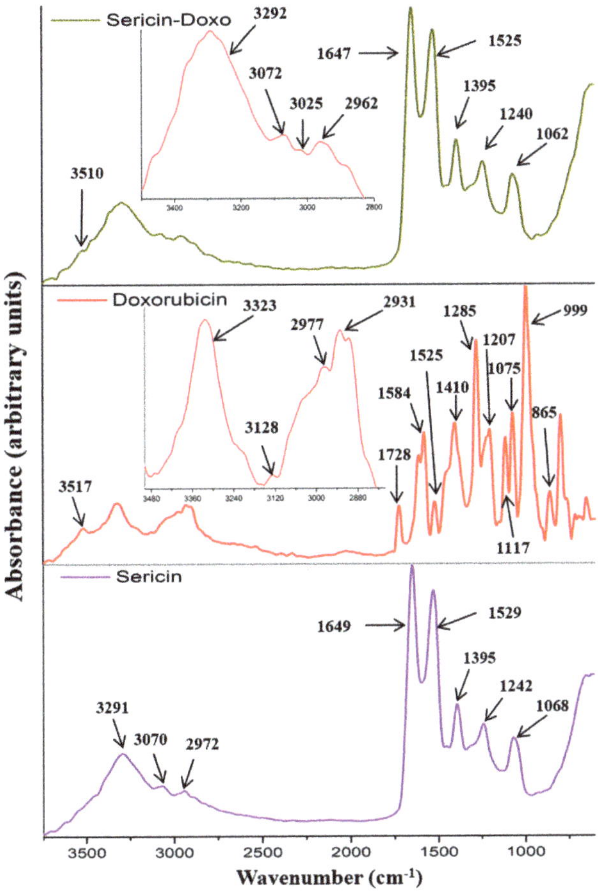

Figure 1. FTIR–ATR spectra of sericin, doxorubicin, and doxorubicin-loaded sericin nanoparticles.

Both enzymes (protease IV and chymotrypsin) showed a typical protein-specific spectrum with amide I, amide II, and amide III. The main peaks are shown in Figure 2a,b. Sericin–doxorubicin interaction revealed a drug-specific peak at 3510 cm^{-1}, assigned to the water molecules bonded within the drug molecules. Another important peak appeared at 3070 cm^{-1}, attributed to unsaturated =C-H stretching vibration characteristic for the aromatic structures similar to doxorubicin structure. Therefore, FTIR analysis showed the presence of doxorubicin within the sericin structure. Furthermore, the analysis revealed no chemical bonds between the sericin and drug, suggesting that the association was induced only by physical interactions (Figure 1).

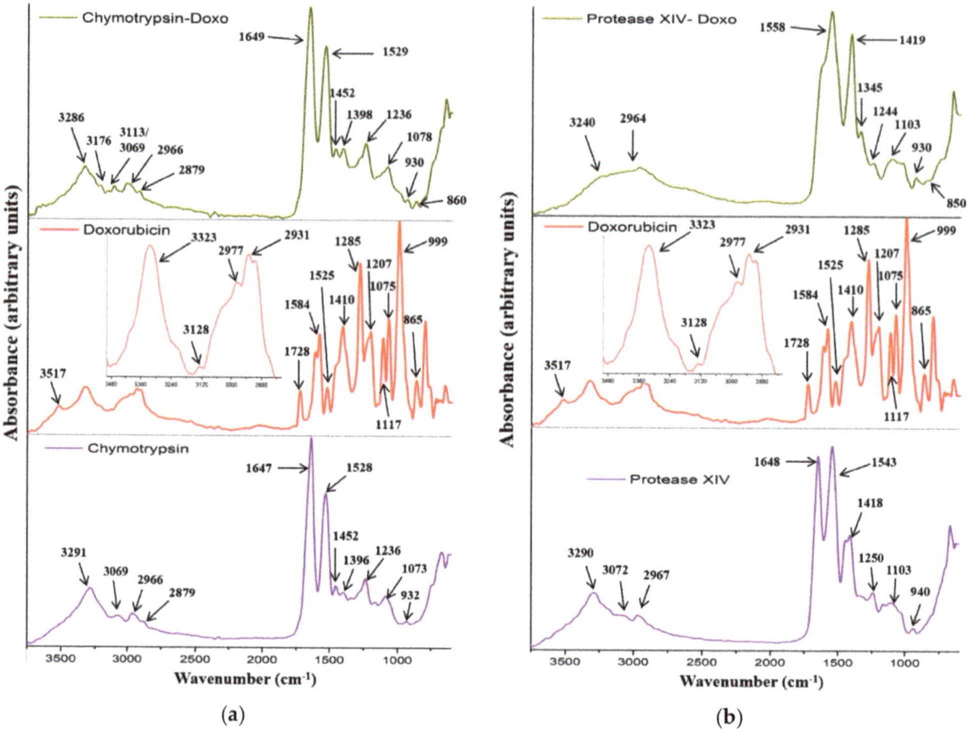

Figure 2. FTIR–ATR spectra of enzymes (protease IV and chymotrypsin), doxorubicin, and association of chymotrypsin/doxorubicin (**a**) and protease IV/doxorubicin (**b**).

Protease IV and chymotrypsin revealed an association with doxorubicin even to a greater extent than sericin. Thus, chymotrypsin–drug association showed two new peaks at 3176 cm^{-1} and 3113 cm^{-1}, attributed to N-H stretching vibration. This fact may suggest that the amidic groups are more visible, as compared to the singular drug or enzyme. A higher contribution to the spectrum can be explained by the fact that these groups manifested supplementary physical interactions. Another important peak at 860 cm^{-1}, attributed to the drug skeletal ring, appeared into the association spectrum (Figure 2a). This approach revealed that the presence of the drug molecules within the chymotrypsin structure was also governed by physical interactions. Protease IV had also some association with doxorubicin. The spectrum (Figure 2b) showed a different overview regarding drug association, as compared to the chymotrypsin approach. Therefore, the spectrum showed the absence of amide I and the presence of a broad peak centered at 1558 cm^{-1} with a significant visible shoulder around 1600 cm^{-1}. This peak probably represents the contributions of amide I and amide II from the protease and peak contribution assigned to hydroxyl group bending vibration (1584 cm^{-1}). The peak at 1419 cm^{-1} increased its intensity, and a new peak appeared at 1345 cm^{-1}. This fact can be attributed to the contribution of the bending vibration for methyl, methylene, and methine groups. Furthermore, the presence of the peak at 850 cm^{-1}, attributed to the skeletal drug structure, confirmed the presence of doxorubicin within the protease XIV structure (physical interactions). One last aspect that can be concluded from protease investigation with respect to sericin highlights the apparent stronger association interactions for the enzymes. This fact can be explained in the case of enzymes by the amide, hydroxyl, and skeletal contributions.

3.2. Drug Release Behavior

3.2.1. Neutral Medium

Doxorubicin release behaviors from sericin nanoparticle formulations were investigated by in vitro tests (PBS: pH 7.45, 37 °C). All nanoparticle formulations had relatively similar behavior with significant differences in release efficiency and time. The results are shown in Figure 3. Both the efficiency and the release time increased for sericin nanoparticles prepared from lower concentrations. The release profile showed a specific behavior with a faster release for the first 100 min with an efficiency of 35%. This behavior cannot be associated with a specific burst release. The release behavior followed a slower release for the next 180 min for SER 1% and for the next 360 min for SER 0.1%. Formulations with SER 0.25% and SER 0.5% followed a slower release time placed within this interval. All nanoparticles SER 0.1%–SER 1% revealed a high entrapment efficiency of 90–95%. Morphological and dimensional analyses showed that the size of the nanoparticle aggregates decreased for those obtained from lower sericin concentrations. These results may suggest that small-size aggregates reached higher efficiency but for a longer release time. The maximum cumulative release efficiency was 74%. This means there was still some entrapped doxorubicin within the mass of sericin nanoparticles. The obtained results in terms of release and encapsulation efficiency can be correlated with other sericin formulations containing low soluble agents [46,47]. Most probably, this behavior was influenced by the fact that the nanoparticles were self-assembled in the presence of doxorubicin. This approach facilitated the formation of sericin–doxorubicin conjugates which are driven by multiple physical bonding [47]. In this situation, the doxorubicin remained entrapped, and only specific environments could act as release stimulus. Therefore, specific media such as pH decrease or enzymatic activity may facilitate further release.

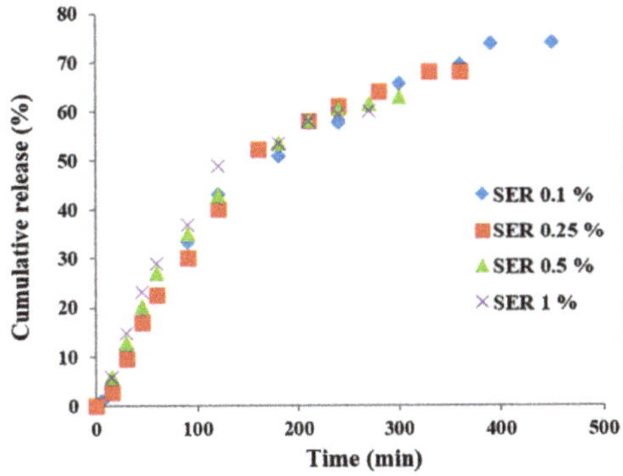

Sample	R.E. Release efficiency, %	E.E. Encapsulation efficiency, %	D.C. Drug content, %
Sericin 0.1 %	74 % ±3	92 %±2	5 % ± 0.092
Sericin 0.25 %	68 % ±2	92 %±2	3 % ±0.062
Sericin 0.5 %	62 % ±2	94 %±3	2 % ±0.055
Sericin 1 %	60 % ±2	93 %±2	1.25 % ±0.025

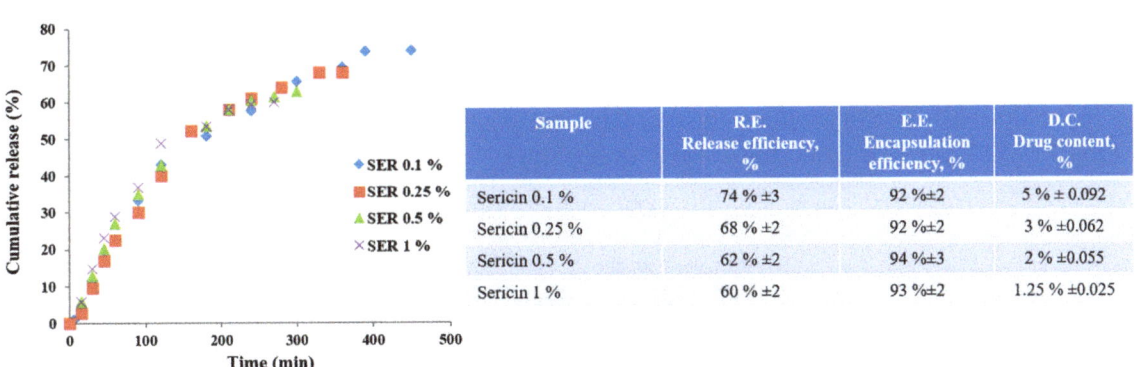

Figure 3. Doxorubicin release profile from sericin nanoparticles in neutral medium.

3.2.2. Acidic Medium

FTIR was performed to investigate the drug–protein interaction. As previously mentioned, there are significant physical interactions that keep the drug inside the nanoparticles. Considering that the main interactions appeared between carboxyl and amino groups, a new in vitro test was performed in an acidic medium. The idea was to protonate the carboxyl negative conjugate form of COO^- within sericin structure to disrupt the interaction with amino positive conjugate form NH^{3+} from doxorubicin. Thus, the entrapped drug could be released. This approach was based on the sericin zeta potential results, which had a considerable negative charge into a neutral medium (pH 7.45). The analysis showed a higher release efficiency within a shorter time period, as compared to the neutral medium. Figure 4 revealed a release efficiency of over 90% for all formulations SER 0.1%–SER 1% in a release time of 30–50 min. The release profile showed different behavior with a faster release for SER 0.1% and SER 0.25% for the first 15 min, followed by a slower release for the next 15 min. SER 0.5% and SER 1% revealed a relatively constant release behavior for longer release time, as compared to SER 0.1% and SER 0.25%. This fact allowed them to reach a higher release.

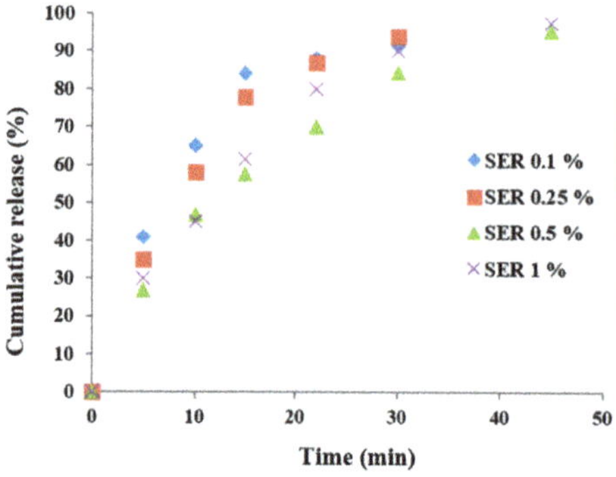

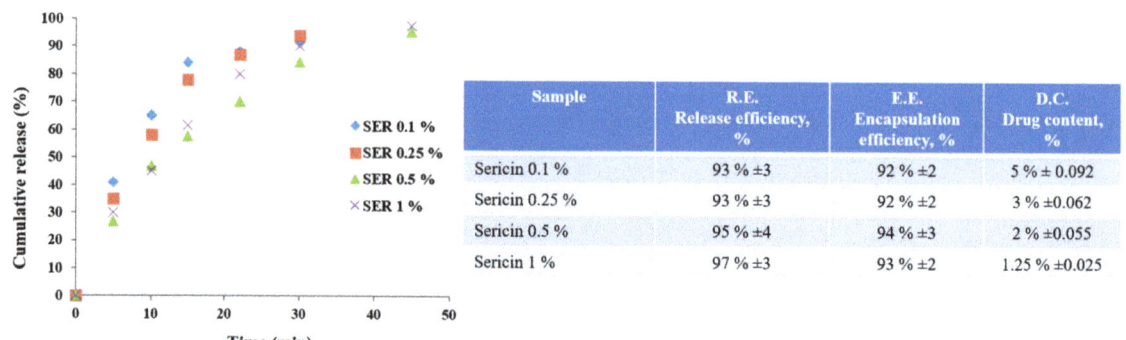

Sample	R.E. Release efficiency, %	E.E. Encapsulation efficiency, %	D.C. Drug content, %
Sericin 0.1 %	93 % ±3	92 % ±2	5 % ± 0.092
Sericin 0.25 %	93 % ±3	92 % ±2	3 % ±0.062
Sericin 0.5 %	95 % ±4	94 % ±3	2 % ±0.055
Sericin 1 %	97 % ±3	93 % ±2	1.25 % ±0.025

Figure 4. Doxorubicin release profile from sericin nanoparticles in acidic medium (pH 3).

3.2.3. Enzymatic Media

The drug–protein interaction was also investigated by enzymatic activity. Two specific enzymes, namely, protease type XIV from *Streptomyces griseus* and protease α-chymotrypsin from bovine pancreas were used. The enzymatic activity involved an enzymatic degradation of sericin protein with the easier release of entrapped drug molecules due to the disruption of physical interactions in drug–protein structure. The results showed an unexpected behavior considering the lower release with respect to the neutral medium. In the case of protease type XIV from *Streptomyces griseus*, the release efficiency was slightly over 20% (Figure 5a), while in the case of protease α-chymotrypsin from bovine pancreas, the release efficiency was slightly under 20% (Figure 5b). Both situations followed a similar profile release and a release time of 180 min. The profile showed a faster release within the first 50 min, followed by a slower release. The poor release efficiency can be explained by the strong physical interactions between the doxorubicin molecule and enzymes' chemical structure. In this case, the drug molecules released from the nanoparticles' mass were further embedded in the structure of the enzymes. These results are in good agreement with the FTIR analysis that showed stronger interactions within drug–enzyme association as compared to sericin nanoparticles.

3.3. Morphological Characterization

3.3.1. SEM Analysis

SER 1% nanoparticles were obtained with a size distribution of 200–300 nanometers with round and specific fusiform shapes (Figure 6a). Higher magnification images revealed closer insights into nanoparticles' morphology with a bunch nanostructure and nanowaved surface (Figure 6b). The nanowaved morphology showed lower sizes of 15–20 nm winding the surface. Individual nanoparticles of 23 nm could be detected, suggesting that the bunch nanostructuring is formed of aggregates of smaller nanoparticles. Morphological characterization continued with formulations SER 0.5% and SER 0.25%. The results for SER 0.5% and SER 0.25% formulations revealed nanoparticle aggregates with a size range size of 100–200 nm (SER 0.5%) and 100–150 nm (SER 0.25%), as shown in Figure 6c,d. Aggregates for SER 0.5% had round and fusiform shapes (Figure 6c), while those for SER 0.25% exhibited only round shapes (Figure 6d). This could be explained by the size range differences since the size increase led to the deviation from the usual round shape. Formulation SER 0.1% was investigated in order to reveal a specific trend of sericin formulations. The lower the sericin concentration was, the smaller the nanoparticle aggregates were obtained. Figure 6e showed aggregates with a size range of 80–130 nm. The aggregates had a round shape like SER 0.25%. The higher magnification image revealed the same bunch nanostructuring with a nanowaved surface (Figure 6f).

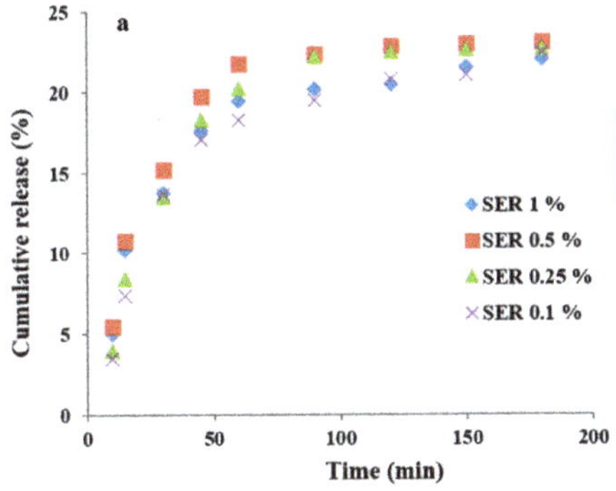

Sample	R.E. Release efficiency, %	E.E. Encapsulation efficiency, %	D.C. Drug content, %
Sericin 0.1 %	23 % ±2	92 % ±2	5 % ± 0.092
Sericin 0.25 %	22 % ±1	92 % ±2	3 % ±0.062
Sericin 0.5 %	22 % ±1	94 % ±3	2 % ±0.055
Sericin 1 %	22 % ±2	93 % ±2	1.25 % ±0.025

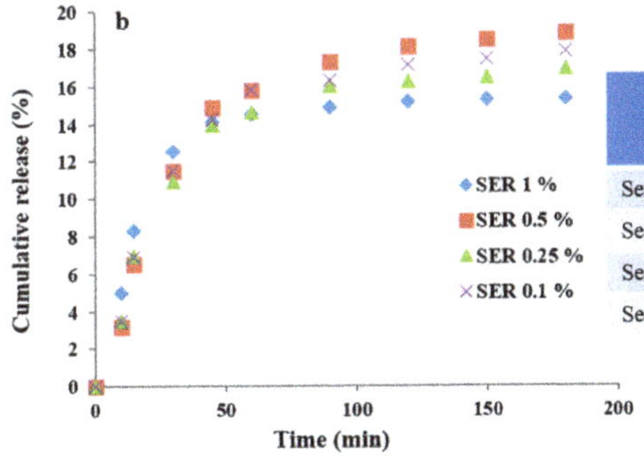

Sample	R.E. Release efficiency, %	E.E. Encapsulation efficiency, %	D.C. Drug content, %
Sericin 0.1 %	18 % ±2	92 % ±2	5 % ± 0.092
Sericin 0.25 %	17 % ±2	92 % ±2	3 % ± 0.062
Sericin 0.5 %	17 % ±1	94 % ±3	2 % ± 0.055
Sericin 1 %	15 % ±2	93 % ±2	1.25 % ± 0.025

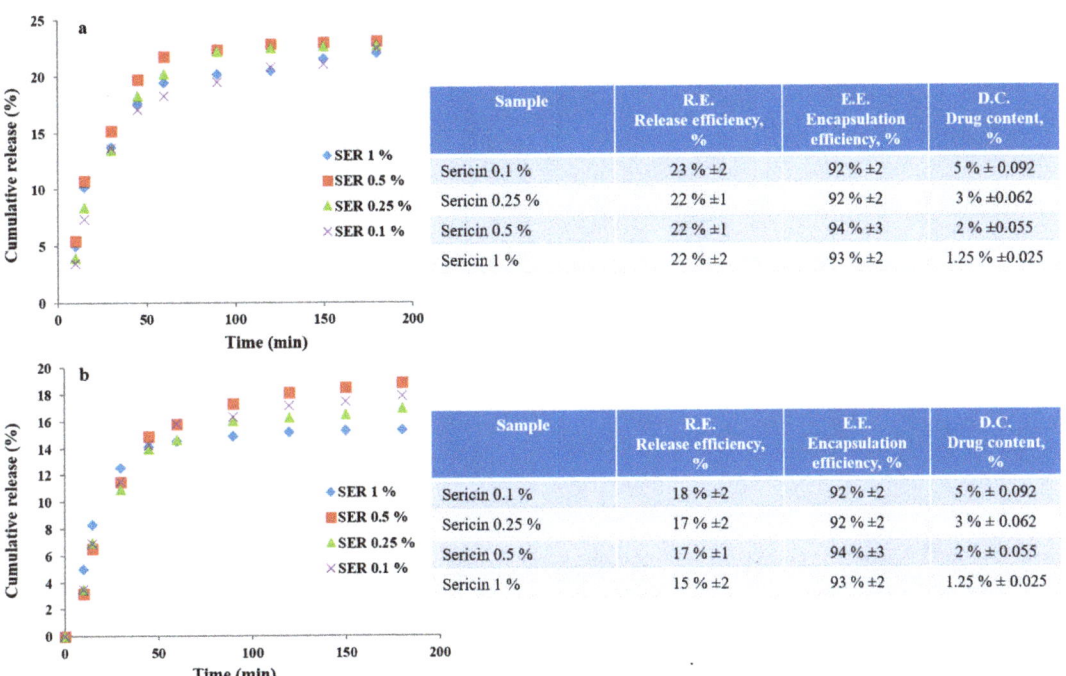

Figure 5. Doxorubicin release profiles from sericin nanoparticles in enzymatic media: protease XIV (**a**); chymotrypsin (**b**).

3.3.2. TEM Analysis

TEM analysis was performed to confirm the SEM results regarding aggregates size, morphology, and nanostructuring. Formulation SER 1% exhibited individualized nanoparticles with a range size between 20 and 35 nm (Figure 7a, higher magnification). The overview image showed also individualized nanoparticles in a significant number (Figure 7b). The nanoparticle aggregates revealed the internal nanostructuring. The aggregates were composed of smaller nanoparticles of 20–35 nm (Figure 7c,d). Besides aggregates, one may notice individualized nanoparticles, suggesting that only a part of them associate with such structures. Formulation SER 0.1% had even smaller individualized nanoparticles with respect to other formulations (15–25 nm, Figure 7e,f). This result confirmed the formulation trend with smaller nanoparticles for lower sericin concentration. This fact can explain the size differences between the aggregates or their shape. The nanoprecipitation appeared as a suitable self-assembling method able to optimize the nanoparticles' features among other nanoformulation methods [30,48,49]. The nanoparticle low average dimension was directly related to some important sericin properties. The mechanism followed a nucleation step, which is typical for the nanoprecipitation method [37,50–52]. This mechanism assured the preparation of a high number of small nanoparticles.

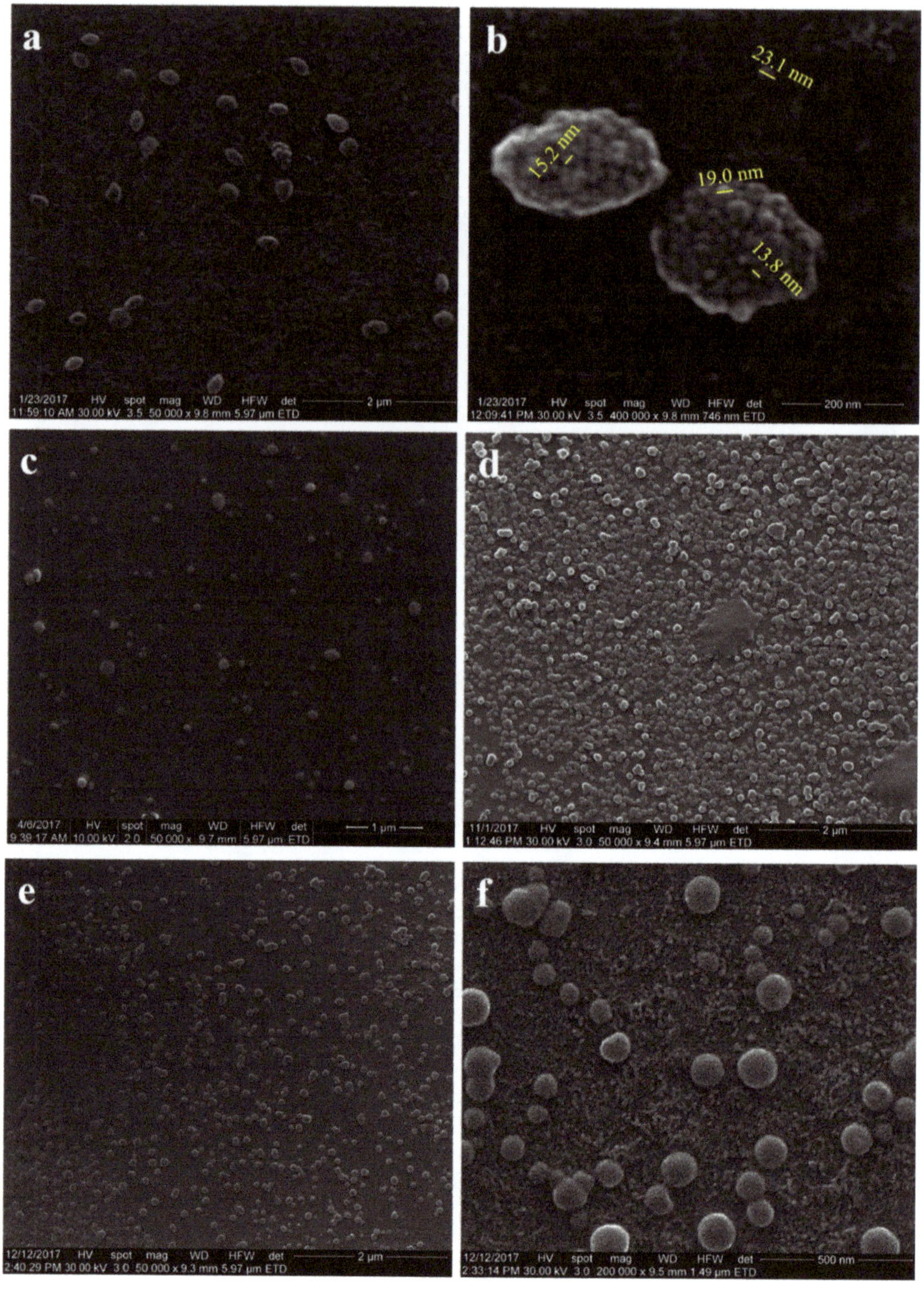

Figure 6. SEM microphotographs for sericin nanoparticles from various sericin concentrations: (**a**,**b**) SER 1%; (**c**,**d**) SER 0.5% and SER 0.25%; (**e**,**f**) SER 0.1%.

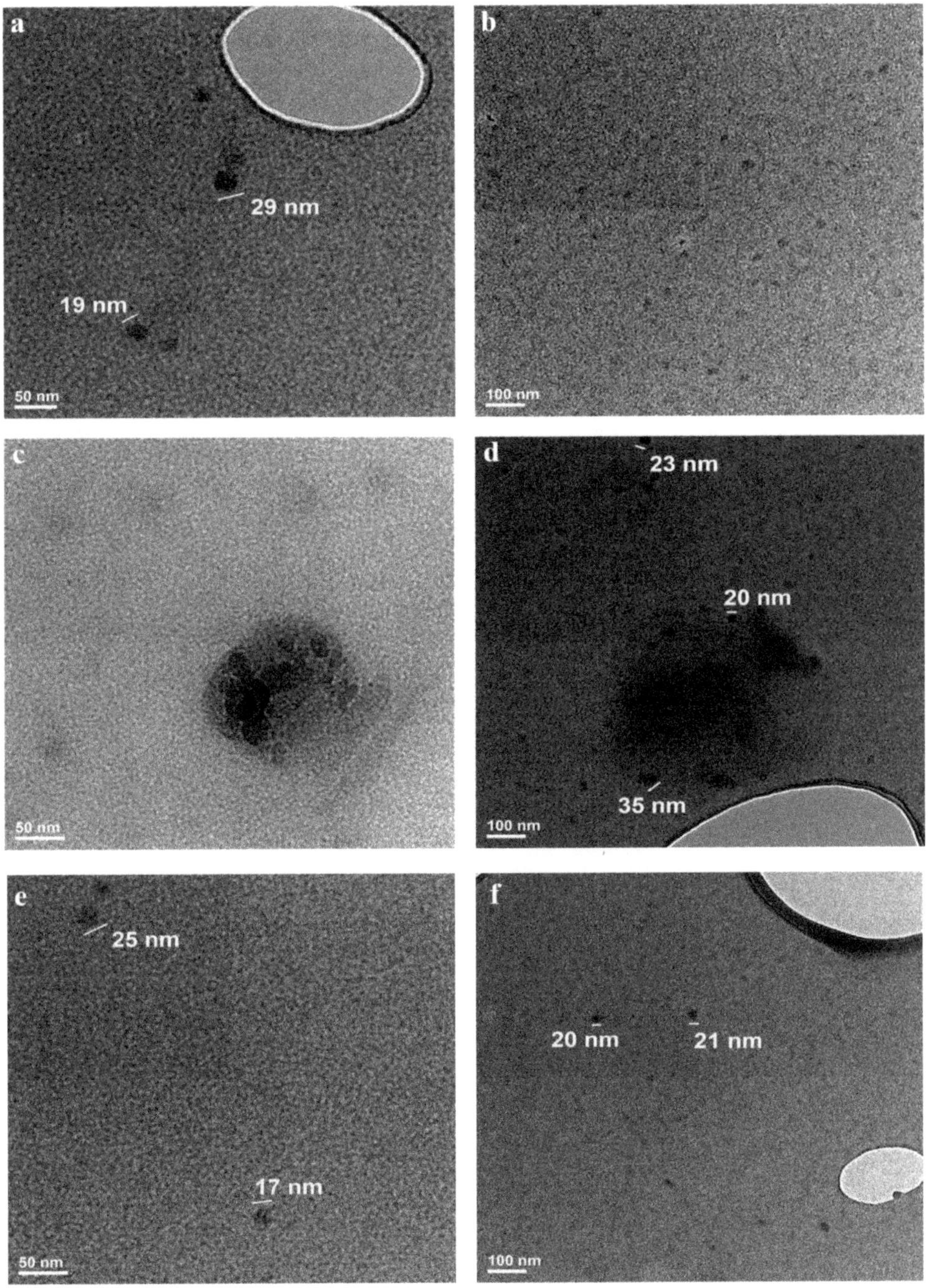

Figure 7. TEM images for individualized nanoparticles obtained from 1% sericin concentration (**a**,**b**); TEM images for nanoparticles aggregates obtained from 1% sericin concentration (SER 1%) (**c**,**d**); TEM images for individualized nanoparticles obtained from 0.1% sericin concentration (SER 0.1%) (**e**,**f**).

3.4. Dynamic Light Scattering, Zeta Potential, and Isoelectric Point

DLS analysis showed more specifically nanoparticles aggregates size. The sericin nanoparticle formulations SER 0.1%–SER 1% revealed an increase of the mean diameter with the increase of the concentration of sericin solutions. All formulations showed a close size distribution profile and close mean size diameter. The size distribution of sericin nanoparticle aggregates and mean diameter are shown in Figure 8a. The results confirmed the morphological investigation by SEM and TEM on aggregates size increasing profile but with a higher size distribution due to the swelling effect.

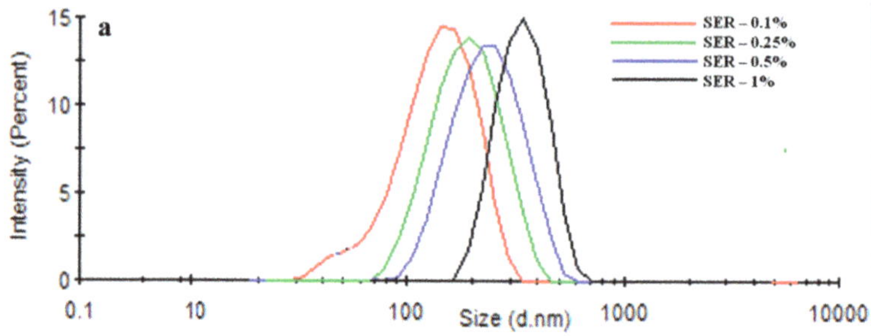

Sample	Mean diameter
SER 0.1%	130±10 nm
SER 0.25%	176±10 nm
SER 0.5%	222±10 nm
SER 1%	342±14 nm

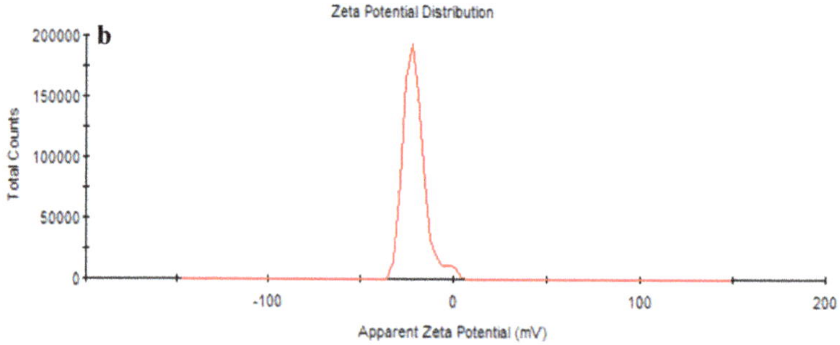

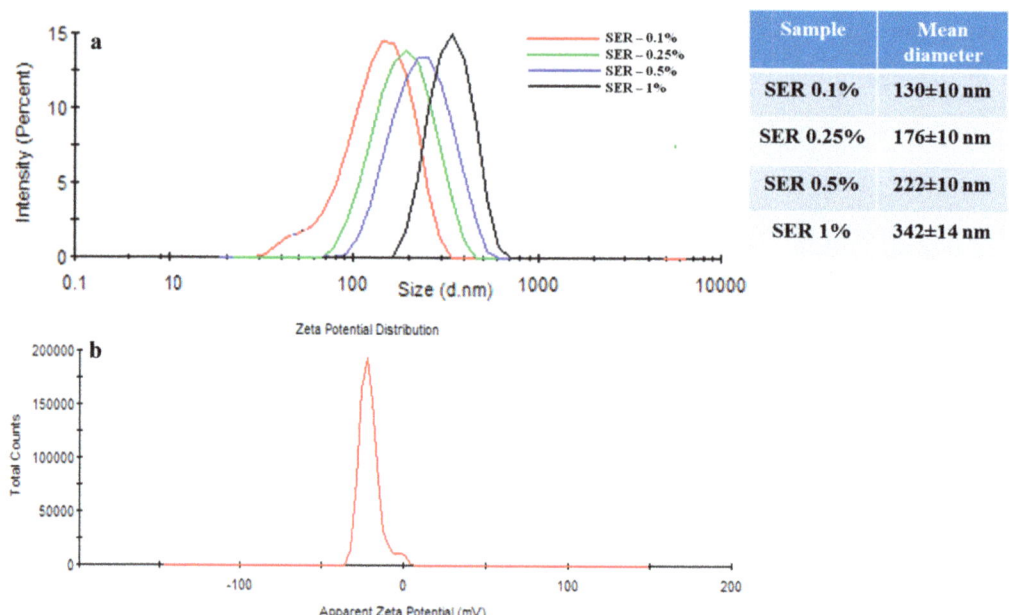

Figure 8. DLS dimensional distribution of sericin nanoparticles with various concentrations (**a**); DLS zeta potential of sericin nanoparticles—SER 0.1% formulation (**b**).

The nanoparticles' surface zeta potential and isoelectric point were used to evaluate the protein surface charging at various pH values. The results revealed a negative surface charging with a zeta potential of −20.2 mV with a standard deviation ±0.9 mV (Figure 8b.) This value represents the zeta potential for neutral pH (7.45). The isoelectric point was established in the pH range 2–2.5. This value of zeta potential can influence the tumor cell line interaction or the mechanism of nanoparticle synthesis.

Molecular Weight Evaluation by DLS

The average molecular weight of silk sericin was evaluated by the detection of light scattering based on the interactions of protein molecule–light. Therefore, the sericin solutions were exposed to a monochromatic wave of light and using multiple detectors. The analysis supposed the investigation of four diluted solutions with various concentrations between 0.2 and 2% (w/v) of silk sericin. An average molecular weight of about 11,700 ± 100 g/mole was determined.

3.5. Conformational Analysis by Circular Dichroism (CD)

CD analysis for sericin showed two positive peaks and one sharp negative peak. The positive peaks at 180 and 186 nm, together with the negative peak at 205 nm, suggest a secondary conformational arrangement dominated by β-sheet and random coil. The sericin nanoparticles had a shifting of the negative peak to lower values and shifting of the positive peaks to higher values. The positive peak was also split into two peaks (Figure 9). This shows some conformational changes of random coil toward the β-sheet structure.

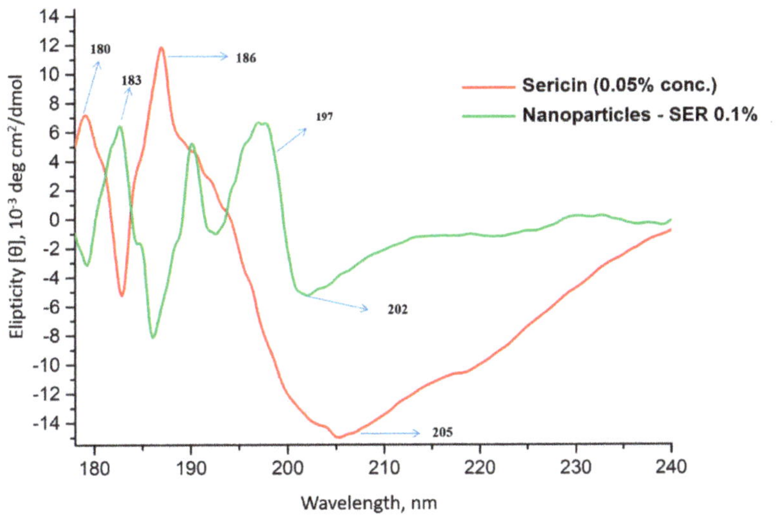

Figure 9. CD curves of native sericin (0.05 wt.% concentration) and sericin nanoparticles (SER-0.1%).

The sericin formulations led to the obtaining of sericin nanoparticles ranging between 15 and 40 nm depending on the solution concentration. These size values are below the usual sizes of polymeric nanoparticles. This fact can be attributed to several important factors including sericin chemistry, molecular weight, concentration, or preparation method. In the case of molecular weight (MW), there are various studies in the literature showing the influence or not of the MW on the size of the nanoparticles for different polymeric systems. Most probably, the molecular weight's influence on nanoparticles size is directly correlated to every studied system. In our study, the low molecular weight fitted the sericin in the oligomeric range (polypeptides), and it could induce such small nanoparticles. The sericin chemistry clearly influenced the hydrophilicity, water solubility, or behavior within the organic phase (acetone). The high water solubility could be also influenced by the low molecular weight. The high water solubility, together with the low molecular weight and chemistry, positively influenced the sericin behavior in acetone dispersion. In contact with acetone, the sericin molecules gather and induce a nucleation process. This is a more controlled process mechanism than instant precipitation. This approach is directly correlated to the preparation method. Nanoprecipitation follows a three-stage process: nucleation, growth, and aggregation [52–54]. Therefore, this approach allowed a supersaturation of sericin molecules per volume with a nucleating process, followed by a growth step. A high sericin concentration led to the generation of nanoparticles with a larger size distribution reaching 35–40 nm, while a low concentration led to the generation of nanoparticles with a narrow distribution. Both high (1% *w*/*v*) and low (0.1 *w*/*v*) concentrations led to the generation of a relatively high number of nanoparticles, suggesting no influence on the number of nanoparticles. Thus, in the initial stage, a high number of nanoparticles were formed, while the size differences appeared in the growth step due to the addition of new molecules on the nuclei surface. Another important issue to be addressed is the aggregation process. The dimensional and morphological investigation showed larger nanoparticle aggregates with a range size between 100 and 300 nm. These nanoparticle aggregates were formed along with individual nanoparticles. Therefore, this process probably appears only in the case of supersaturation for nanoparticles' concentration per volume of dispersion media.

3.6. *In Vitro Antitumor Activity Evaluation of DOX-Loaded Sericin Nanocarriers*

As described above, based on the nanoparticles' size and doxorubicin release profile, the 0.1% sericin formulation was employed in the in vitro biological investigations. To evaluate the viability of MCF-7 breast cancer cells after 6 h and 24 h of exposure to unloaded Ser NPs and DOX-loaded Ser NPs, the quantitative MTT assay was performed. Data were statistically analyzed and graphically represented in Figure 10 using GraphPad Prism 6 software.

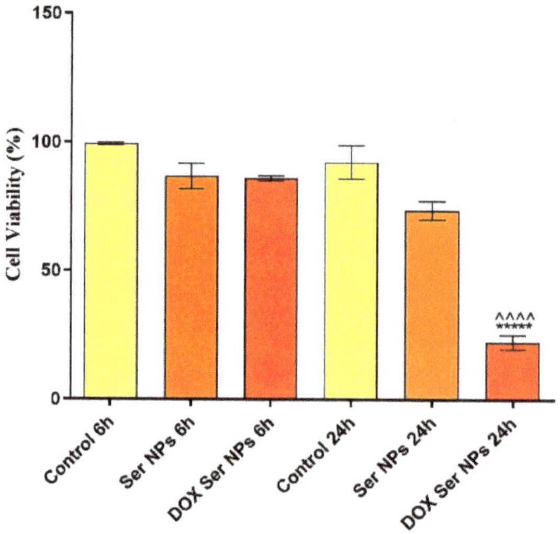

Figure 10. Graphical representation of MCF-7 breast cancer cells viability after 6 h and 24 h of treatment with free and DOX-loaded Ser NPs (Dox Ser NPs vs. untreated control **** $p \leq 0.0001$; Dox Ser NPs vs. Ser NPs ^^^^ $p \leq 0.0001$).

Our data showed that after 6 h of treatment, none of the treatments induced cell viability alterations. Moreover, after 24 h of exposure to free Ser NPs, the viability of the MCF-7 cells remained similar to the control, demonstrating good biocompatibility of the pristine sericin nanocarriers. In contrast, after 24 h of treatment, the DOX-loaded Ser NPs significantly decreased the viability of the MCF-7 cells (**** $p \leq 0.0001$). Moreover, MCF-7 breast cancer cell morphology was investigated by fluorescence microscopy after staining the cytoskeleton fibers with phalloidin–FITC and the cell nuclei with DAPI. The images captured are presented in Figure 11. No alterations were produced by the treatment with unloaded Ser NPs during 24 h, as compared with the untreated cells. In contrast, the treatment with DOX-loaded Ser NPs induced modifications in terms of actin filaments' organization and distribution in the cellular cytoplasm. Additionally, the fluorescence microscopy images captured in the samples treated with DOX-loaded Ser NPs revealed red fluorescence inside the MCF-7 cells. Considering that doxorubicin is well known as a red fluorescent chemical compound, this valuable observation indicates/proves that the DOX-loaded Ser NPs successfully enter the cells.

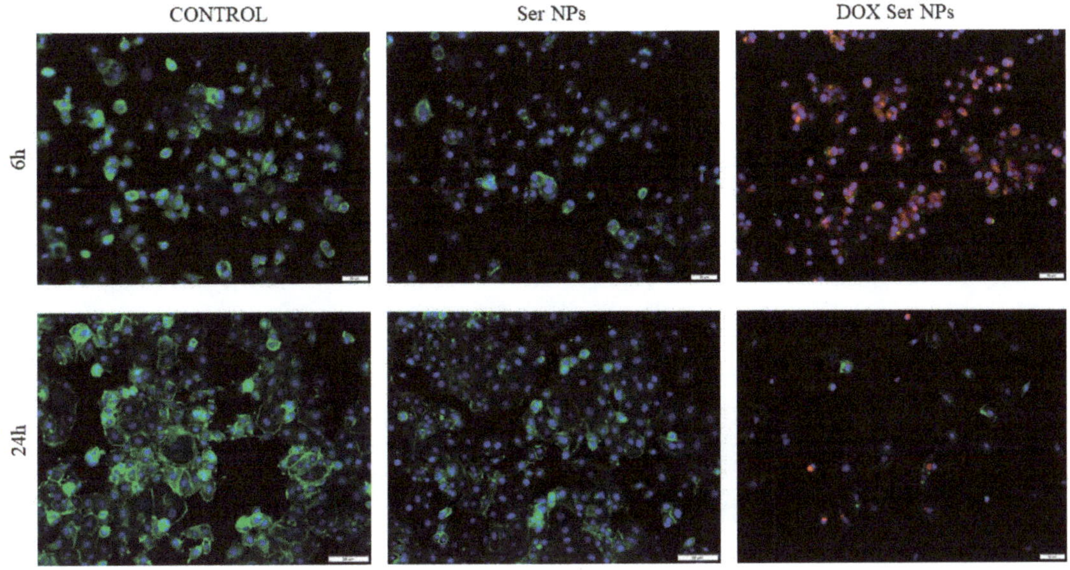

Figure 11. Fluorescence microscopy images of MCF-7 cells treated for 6 h and 24 h with free Ser NPs and DOX-loaded Ser NPs, as compared with an untreated MCF-7 monolayer: green fluorescence–phalloidin–FITC: actin filaments; blue fluorescence–DAPI: cell nuclei; red fluorescence–DOX). Scale bar: 50 μm.

Finally, to assess the genotoxic potential of the DOX-loaded Ser NPs treatment, the comet assay was performed. After fluorescence image processing and data analysis, the DNA damage profile in MCF-7 cancer cells after Ser NPs + DOX treatment was established based on the average length of the comet tails. As presented in Figure 12, the comet-like structures correlated with enhanced DNA migration were identified only in MCF-7 cells exposed for 24 h to the treatment with DOX-loaded Ser NPs. As DNA damage is a hallmark of apoptosis, our data suggest that DOX encapsulation in Ser NPs triggers apoptosis of MCF–7 breast cancer cells.

Sericin nanocarriers have previously been used for breast cancer management [28]. Mandal and Kundu showed that paclitaxel-loaded sericin nanocarriers induced apoptosis in MCF-7 breast cancer cells. Similarly, we demonstrated that DOX-loaded Ser NPs significantly decreased MCF-7 cells viability after 24 h of treatment and altered the morphology of the cells, as revealed by the fluorescent labeling of the cell's cytoskeleton.

Regarding DOX, the literature reports two potential mechanisms of action in the cancer cell: (i) the intercalation into DNA and disruption of topoisomerase-II-mediated DNA repair and (ii) the generation of free radicals producing damages to cellular membranes, DNA, and proteins [55]. Our data showed that the DOX-loaded Ser NPs induced DNA damage in MCF-7 cells, as compared with the pristine Ser NPs, probably due to the toxic effect of the delivered DOX.

a.

b.

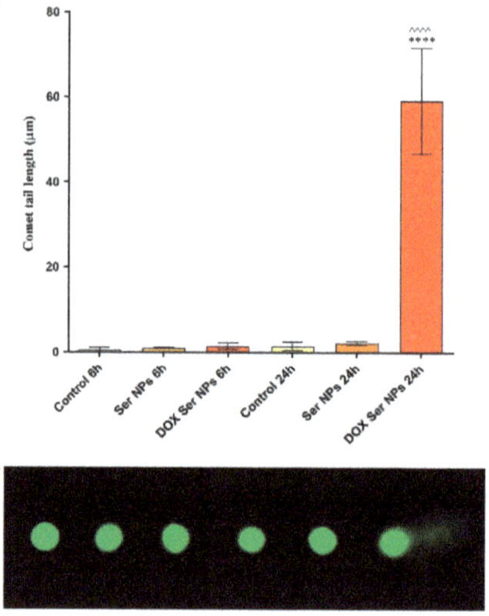

Figure 12. (**a**) Graphical representation showing the comet tail length as an indicator of DNA damage in MCF–7 cell cultures treated for 6 h and 24 h with simple and DOX-loaded Ser NPs (Dox Ser NPs vs. untreated control **** $p \leq 0.0001$; Dox Ser NPs vs. Ser NPs ^^^^ $p \leq 0.0001$); (**b**) representative fluorescence micrographs of the comet-like structures in MCF–7 cell cultures treated for 6 h and 24 h with simple and DOX-loaded Ser NPs.

4. Conclusions

In conclusion, we obtained sericin nanoparticles with a size range between 15 and 40 nm. This dimensional range is below the usual range of the polymeric nanoparticles. The nanoprecipitation proved to be a suitable method for sericin nanoparticles' preparation, loading, and release. The advanced morphological investigation showed the size and size distribution of the nanocarriers with a direct positive influence on the biological investigation. Moreover, we also showed that the DOX-loaded Ser NPs significantly decreased MCF-7 cells viability, altered their morphology, and induced DNA damage, as compared with the unloaded Ser NPs.

Author Contributions: Conceptualization, I.-C.R., B.G., and C.Z.; methodology, I.-C.R., B.G., A.H., and C.Z.; validation, I.-C.R., B.G., A.H., and C.Z.; investigation, I.-C.R., E.T., B.G., A.H., O.G., M.M., and M.C.; writing—original draft preparation, I.-C.R., C.Z., A.H., and B.G.; writing—review and editing, B.G., O.G., and C.Z.; supervision, B.G. and C.Z.; project administration, C.Z. All authors have read and agreed to the published version of the manuscript.

Funding: This work was supported by a grant from the Ministry of Research, Innovation and Digitization, CNCS/CCCDI–UEFISCDI, Project Number PN-III-P4-ID-PCE-2020-1448, within PNCDI III.

Data Availability Statement: The data presented in this study are available on request from the corresponding author.

Acknowledgments: DLS and Circular dichroism were possible due to the European Regional Development Fund through Competitiveness Operational Program 2014–2020, Priority axis 1, Project No. P_36_611, MySMIS code 107066, Innovative Technologies for Materials Quality Assurance in Health, Energy and Environmental-Center for Innovative Manufacturing Solutions of Smart Biomaterials and Biomedical Surfaces–INOVABIOMED.

Conflicts of Interest: The authors declare no conflict of interest. The funders had no role in the design of the study; in the collection, analyses, or interpretation of data; in the writing of the manuscript, or in the decision to publish the results.

References

1. Nagaraju, J.; Goldsmith, M.R. Silkworm Genomics–Progress and Prospects. *Curr. Sci.* **2002**, *83*, 415–425.
2. Kundu, S.C.; Dash, B.C.; Dash, R.; Kaplan, D.L. Natural Protective Glue Protein, Sericin Bioengineered by Silkworms: Potential for Biomedical and Biotechnological Applications. *Prog. Polym. Sci.* **2008**, *33*, 998–1012. [CrossRef]
3. Holland, C.; Numata, K.; Rnjak-Kovacina, J.; Seib, F.P. The Biomedical Use of Silk: Past, Present, Future. *Adv. Healthc. Mater.* **2019**, *8*, 1800465. [CrossRef]
4. Kundu, B.; Rajkhowa, R.; Kundu, S.C.; Wang, X. Silk Fibroin Biomaterials for Tissue Regenerations. *Adv. Drug Deliv. Rev.* **2013**, *65*, 457–470. [CrossRef]
5. Urry, D.W.; Luan, C.-H.; Harris, C.M.; Parker, T.M. Protein-based materials with a profound range of properties and applications: The elastin ΔT t hydrophobic paradigm. In *Protein-Based Materials*; Springer: Berlin/Heidelberg, Germany, 1997; pp. 133–177.
6. Lamboni, L.; Gauthier, M.; Yang, G.; Wang, Q. Silk Sericin: A Versatile Material for Tissue Engineering and Drug Delivery. *Biotechnol. Adv.* **2015**, *33*, 1855–1867. [CrossRef]
7. Radu, I.-C.; Biru, I.-E.; Damian, C.-M.; Ion, A.-C.; Iovu, H.; Tanasa, E.; Zaharia, C.; Galateanu, B. Grafting versus Crosslinking of Silk Fibroin-g-PNIPAM via Tyrosine-NIPAM Bridges. *Molecules* **2019**, *24*, 4096. [CrossRef]
8. Qi, Y.; Wang, H.; Wei, K.; Yang, Y.; Zheng, R.-Y.; Kim, I.S.; Zhang, K.-Q. A Review of Structure Construction of Silk Fibroin Biomaterials from Single Structures to Multi-Level Structures. *Int. J. Mol. Sci.* **2017**, *18*, 237. [CrossRef]
9. McGrath, K.; Kaplan, D. *Protein-Based Materials*; Springer: Berlin/Heidelberg, Germany, 1997; ISBN 3-7643-3848-2.
10. Altman, G.H.; Diaz, F.; Jakuba, C.; Calabro, T.; Horan, R.L.; Chen, J.; Lu, H.; Richmond, J.; Kaplan, D.L. Silk-Based Biomaterials. *Biomaterials* **2003**, *24*, 401–416. [CrossRef]
11. Carissimi, G.; Lozano-Pérez, A.A.; Montalbán, M.G.; Aznar-Cervantes, S.D.; Cenis, J.L.; Víllora, G. Revealing the Influence of the Degumming Process in the Properties of Silk Fibroin Nanoparticles. *Polymers* **2019**, *11*, 2045. [CrossRef]
12. Chirila, T.V.; Suzuki, S.; Bray, L.J.; Barnett, N.L.; Harkin, D.G. Evaluation of Silk Sericin as a Biomaterial: In Vitro Growth of Human Corneal Limbal Epithelial Cells on Bombyx Mori Sericin Membranes. *Prog. Biomater.* **2013**, *2*, 1–10. [CrossRef] [PubMed]
13. Nayak, S.; Kundu, S.C. Silk Protein Sericin: Promising Biopolymer for Biological and Biomedical Applications. *Biomater. Nat. Adv. Devices Ther. Wiley Soc. Biomater.* **2016**, 142–154.
14. Aramwit, P.; Bang, N.; Ratanavaraporn, J.; Ekgasit, S. Green Synthesis of Silk Sericin-Capped Silver Nanoparticles and Their Potent Anti-Bacterial Activity. *Nanoscale Res. Lett.* **2014**, *9*, 1–7.
15. Zhaorigetu, S.; Yanaka, N.; Sasaki, M.; Watanabe, H.; Kato, N. Silk Protein, Sericin, Suppresses DMBA-TPA-Induced Mouse Skin Tumorigenesis by Reducing Oxidative Stress, Inflammatory Responses and Endogenous Tumor Promoter TNF-α. *Oncol. Rep.* **2003**, *10*, 537–543. [PubMed]
16. Zhaorigetu, S.; Sasaki, M.; Kato, N. Consumption of Sericin Suppresses Colon Oxidative Stress and Aberrant Crypt Foci in 1, 2-Dimethylhydrazine-Treated Rats by Colon Undigested Sericin. *J. Nutr. Sci. Vitaminol.* **2007**, *53*, 297–300. [CrossRef]
17. Bari, E.; Perteghella, S.; Faragò, S.; Torre, M.L. Association of Silk Sericin and Platelet Lysate: Premises for the Formulation of Wound Healing Active Medications. *Int. J. Biol. Macromol.* **2018**, *119*, 37–47. [CrossRef]
18. Chouhan, D.; Mandal, B.B. Silk Biomaterials in Wound Healing and Skin Regeneration Therapeutics: From Bench to Bedside. *Acta Biomater.* **2020**, *103*, 24–51. [CrossRef]
19. Baptista-Silva, S.; Borges, S.; Costa-Pinto, A.R.; Costa, R.; Amorim, M.; Dias, J.R.; Ramos, Ó.; Alves, P.; Granja, P.L.; Soares, R. In Situ Forming Silk Sericin-Based Hydrogel: A Novel Wound Healing Biomaterial. *ACS Biomater. Sci. Eng.* **2021**, *7*, 1573–1586. [CrossRef] [PubMed]
20. Arango, M.C.; Montoya, Y.; Peresin, M.S.; Bustamante, J.; Álvarez-López, C. Silk Sericin as a Biomaterial for Tissue Engineering: A Review. *Int. J. Polym. Mater. Polym. Biomater.* **2020**, 1–15. [CrossRef]
21. Aramwit, P.; Kanokpanont, S.; Nakpheng, T.; Srichana, T. The Effect of Sericin from Various Extraction Methods on Cell Viability and Collagen Production. *Int. J. Mol. Sci.* **2010**, *11*, 2200–2211. [CrossRef]
22. Aramwit, P.; Palapinyo, S.; Srichana, T.; Chottanapund, S.; Muangman, P. Silk Sericin Ameliorates Wound Healing and Its Clinical Efficacy in Burn Wounds. *Arch. Dermatol. Res.* **2013**, *305*, 585–594. [CrossRef]
23. Cui, Y.; Xing, Z.; Yan, J.; Lu, Y.; Xiong, X.; Zheng, L. Thermosensitive Behavior and Super-Antibacterial Properties of Cotton Fabrics Modified with a Sercin-NIPAAm-AgNPs Interpenetrating Polymer Network Hydrogel. *Polymers* **2018**, *10*, 818. [CrossRef]
24. Cho, K.Y.; Moon, J.Y.; Lee, Y.W.; Lee, K.G.; Yeo, J.H.; Kweon, H.Y.; Kim, K.H.; Cho, C.S. Preparation of Self-Assembled Silk Sericin Nanoparticles. *Int. J. Biol. Macromol.* **2003**, *32*, 36–42. [CrossRef]
25. Akturk, O.; Gun Gok, Z.; Erdemli, O.; Yigitoglu, M. One-pot Facile Synthesis of Silk Sericin-capped Gold Nanoparticles by UVC Radiation: Investigation of Stability, Biocompatibility, and Antibacterial Activity. *J. Biomed. Mater. Res. A* **2019**, *107*, 2667–2679. [CrossRef]
26. Das, S.K.; Dey, T.; Kundu, S. Fabrication of Sericin Nanoparticles for Controlled Gene Delivery. *RSC Adv.* **2014**, *4*, 2137–2142. [CrossRef]
27. Hazeri, N.; Tavanai, H.; Moradi, A.R. Production and Properties of Electrosprayed Sericin Nanopowder. *Sci. Technol. Adv. Mater.* **2012**, *13*, 035010. [CrossRef]
28. Mandal, B.B.; Kundu, S. Self-Assembled Silk Sericin/Poloxamer Nanoparticles as Nanocarriers of Hydrophobic and Hydrophilic Drugs for Targeted Delivery. *Nanotechnology* **2009**, *20*, 355101. [CrossRef]

29. He, H.; Cai, R.; Wang, Y.; Tao, G.; Guo, P.; Zuo, H.; Chen, L.; Liu, X.; Zhao, P.; Xia, Q. Preparation and Characterization of Silk Sericin/PVA Blend Film with Silver Nanoparticles for Potential Antimicrobial Application. *Int. J. Biol. Macromol.* **2017**, *104*, 457–464. [CrossRef] [PubMed]
30. Hu, D.; Xu, Z.; Hu, Z.; Hu, B.; Yang, M.; Zhu, L. PH-Triggered Charge-Reversal Silk Sericin-Based Nanoparticles for Enhanced Cellular Uptake and Doxorubicin Delivery. *ACS Sustain. Chem. Eng.* **2017**, *5*, 1638–1647. [CrossRef]
31. Li, H.; Tian, J.; Wu, A.; Wang, J.; Ge, C.; Sun, Z. Self-Assembled Silk Fibroin Nanoparticles Loaded with Binary Drugs in the Treatment of Breast Carcinoma. *Int. J. Nanomed.* **2016**, *11*, 4373.
32. Numata, K.; Kaplan, D.L. Silk-Based Delivery Systems of Bioactive Molecules. *Adv. Drug Deliv. Rev.* **2010**, *62*, 1497–1508. [CrossRef]
33. Radu, I.-C.; Hudita, A.; Zaharia, C.; Stanescu, P.O.; Vasile, E.; Iovu, H.; Stan, M.; Ginghina, O.; Galateanu, B.; Costache, M. Poly (Hydroxybutyrate-Co-Hydroxyvalerate)(PHBHV) Nanocarriers for Silymarin Release as Adjuvant Therapy in Colo-Rectal Cancer. *Front. Pharmacol.* **2017**, *8*, 508. [CrossRef]
34. Rivas, C.J.M.; Tarhini, M.; Badri, W.; Miladi, K.; Greige-Gerges, H.; Nazari, Q.A.; Rodríguez, S.A.G.; Román, R.Á.; Fessi, H.; Elaissari, A. Nanoprecipitation Process: From Encapsulation to Drug Delivery. *Int. J. Pharm.* **2017**, *532*, 66–81. [CrossRef]
35. Rabha, B.; Bharadwaj, K.K.; Baishya, D.; Sarkar, T.; Edinur, H.A.; Pati, S. Synthesis and Characterization of Diosgenin Encapsulated Poly-ε-Caprolactone-Pluronic Nanoparticles and Its Effect on Brain Cancer Cells. *Polymers* **2021**, *13*, 1322. [CrossRef] [PubMed]
36. Jara, M.O.; Catalan-Figueroa, J.; Landin, M.; Morales, J.O. Finding Key Nanoprecipitation Variables for Achieving Uniform Polymeric Nanoparticles Using Neurofuzzy Logic Technology. *Drug Deliv. Transl. Res.* **2018**, *8*, 1797–1806. [CrossRef]
37. Salatin, S.; Barar, J.; Barzegar-Jalali, M.; Adibkia, K.; Kiafar, F.; Jelvehgari, M. Development of a Nanoprecipitation Method for the Entrapment of a Very Water Soluble Drug into Eudragit RL Nanoparticles. *Res. Pharm. Sci.* **2017**, *12*, 1. [CrossRef] [PubMed]
38. Zielińska, A.; Carreiró, F.; Oliveira, A.M.; Neves, A.; Pires, B.; Venkatesh, D.N.; Durazzo, A.; Lucarini, M.; Eder, P.; Silva, A.M. Polymeric Nanoparticles: Production, Characterization, Toxicology and Ecotoxicology. *Molecules* **2020**, *25*, 3731. [CrossRef] [PubMed]
39. Sun, H.; Xie, Z.; Ju, C.; Hu, X.; Yuan, D.; Zhao, W.; Shui, L.; Zhou, G. Dye-Doped Electrically Smart Windows Based on Polymer-Stabilized Liquid Crystal. *Polymers* **2019**, *11*, 694. [CrossRef]
40. Deschamps, F.; Isoardo, T.; Denis, S.; Tsapis, N.; Tselikas, L.; Nicolas, V.; Paci, A.; Fattal, E.; de Baere, T.; Huang, N. Biodegradable Pickering Emulsions of Lipiodol for Liver Trans-Arterial Chemo-Embolization. *Acta Biomater.* **2019**, *87*, 177–186. [CrossRef]
41. Zhang, L.; Xie, L.; Xu, S.; Kuchel, R.P.; Dai, Y.; Jung, K.; Boyer, C. Dual Role of Doxorubicin for Photopolymerization and Therapy. *Biomacromolecules* **2020**, *21*, 3887–3897. [CrossRef]
42. Lin, W.; Ma, G.; Yuan, Z.; Qian, H.; Xu, L.; Sidransky, E.; Chen, S. Development of Zwitterionic Polypeptide Nanoformulation with High Doxorubicin Loading Content for Targeted Drug Delivery. *Langmuir* **2018**, *35*, 1273–1283. [CrossRef]
43. Jørgensen, J.R.; Thamdrup, L.H.; Kamguyan, K.; Nielsen, L.H.; Nielsen, H.M.; Boisen, A.; Rades, T.; Müllertz, A. Design of a Self-Unfolding Delivery Concept for Oral Administration of Macromolecules. *J. Control. Release* **2021**, *329*, 948–954. [CrossRef]
44. Lohcharoenkal, W.; Wang, L.; Chen, Y.C.; Rojanasakul, Y. Protein Nanoparticles as Drug Delivery Carriers for Cancer Therapy. *BioMed Res. Int.* **2014**, *2014*. [CrossRef]
45. Naoum, G.E.; Tawadros, F.; Farooqi, A.A.; Qureshi, M.Z.; Tabassum, S.; Buchsbaum, D.J.; Arafat, W. Role of Nanotechnology and Gene Delivery Systems in TRAIL-Based Therapies. *Ecancermedicalscience* **2016**, *10*, 660. [CrossRef]
46. Orlandi, G.; Bari, E.; Catenacci, L.; Sorrenti, M.; Segale, L.; Faragò, S.; Sorlini, M.; Arciola, C.R.; Torre, M.L.; Perteghella, S. Polyphenols-Loaded Sericin Self-Assembling Nanoparticles: A Slow-Release for Regeneration by Tissue-Resident Mesenchymal Stem/Stromal Cells. *Pharmaceutics* **2020**, *12*, 381. [CrossRef]
47. Huang, L.; Tao, K.; Liu, J.; Qi, C.; Xu, L.; Chang, P.; Gao, J.; Shuai, X.; Wang, G.; Wang, Z. Design and Fabrication of Multifunctional Sericin Nanoparticles for Tumor Targeting and PH-Responsive Subcellular Delivery of Cancer Chemotherapy Drugs. *ACS Appl. Mater. Interfaces* **2016**, *8*, 6577–6585. [CrossRef]
48. Liu, J.; Li, Q.; Zhang, J.; Huang, L.; Qi, C.; Xu, L.; Liu, X.; Wang, G.; Wang, L.; Wang, Z. Safe and Effective Reversal of Cancer Multidrug Resistance Using Sericin-coated Mesoporous Silica Nanoparticles for Lysosome-targeting Delivery in Mice. *Small* **2017**, *13*, 1602567. [CrossRef]
49. Das, G.; Shin, H.-S.; Campos, E.V.R.; Fraceto, L.F.; del Pilar Rodriguez-Torres, M.; Mariano, K.C.F.; de Araujo, D.R.; Fernández-Luqueño, F.; Grillo, R.; Patra, J.K. Sericin Based Nanoformulations: A Comprehensive Review on Molecular Mechanisms of Interaction with Organisms to Biological Applications. *J. Nanobiotechnol.* **2021**, *19*, 1–22. [CrossRef]
50. Radu, I.C.; Hudita, A.; Zaharia, C.; Galateanu, B.; Iovu, H.; Tanasa, E.; Georgiana Nitu, S.; Ginghina, O.; Negrei, C.; Tsatsakis, A. Poly (3-Hydroxybutyrate-CO-3-Hydroxyvalerate) PHBHV Biocompatible Nanocarriers for 5-FU Delivery Targeting Colorectal Cancer. *Drug Deliv.* **2019**, *26*, 318–327. [CrossRef]
51. Galindo-Rodriguez, S.; Allemann, E.; Fessi, H.; Doelker, E. Physicochemical Parameters Associated with Nanoparticle Formation in the Salting-out, Emulsification-Diffusion, and Nanoprecipitation Methods. *Pharm. Res.* **2004**, *21*, 1428–1439. [CrossRef]
52. Rao, J.P.; Geckeler, K.E. Polymer Nanoparticles: Preparation Techniques and Size-Control Parameters. *Prog. Polym. Sci.* **2011**, *36*, 887–913. [CrossRef]
53. Blouza, I.L.; Charcosset, C.; Sfar, S.; Fessi, H. Preparation and Characterization of Spironolactone-Loaded Nanocapsules for Paediatric Use. *Int. J. Pharm.* **2006**, *325*, 124–131. [CrossRef] [PubMed]

54. Maaz, A.; Abdelwahed, W.; Tekko, I.A.; Trefi, S. Influence of Nanoprecipitation Method Parameters on Nanoparticles Loaded with Gatifloxacin for Ocular Drug Delivery. *Int. J. Acad. Sci. Res.* **2015**, *3*, 12.
55. Gewirtz, D. A Critical Evaluation of the Mechanisms of Action Proposed for the Antitumor Effects of the Anthracycline Antibiotics Adriamycin and Daunorubicin. *Biochem. Pharmacol.* **1999**, *57*, 727–741. [CrossRef]

MDPI
St. Alban-Anlage 66
4052 Basel
Switzerland
www.mdpi.com

Polymers Editorial Office
E-mail: polymers@mdpi.com
www.mdpi.com/journal/polymers

www.ingramcontent.com/pod-product-compliance
Lightning Source LLC
LaVergne TN
LVHW070144100526
838202LV00015B/1890
9783036589497